A GUN'S CREW OF H. M. SHIP "EXCELLENT."
READY!

To face the Title.

TO THE
LORDS COMMISSIONERS OF THE ADMIRALTY.

A TREATISE

ON

NAVAL GUNNERY.

BY

GENERAL SIR HOWARD DOUGLAS, Bart.;

G.C.B.; G.C.M.G.; D.C.L.; F.R.S.

FOURTH EDITION, REVISED.

LONDON:
JOHN MURRAY, ALBEMARLE STREET.
1855.

The right of Translation is reserved

INTRODUCTION
TO THE FIRST EDITION.

London, 1819.

AWARE of the objections that may attach to the measure of cultivating improvements in warlike practice through the medium of the Press, I must explain the circumstances which have led to the publication of this Work.

Having, during the war, made observations and formed opinions respecting the state of Gunnery in the British Navy, which led me to reflect studiously how this important branch of our martial system might be improved, I occupied myself, for some time after the close of the war, in composing this Work; and in October, 1817, transmitted it to the Lords Commissioners of the Admiralty.

The long absence of the Senior professional Lord (the late Admiral Sir George Hope) from the Board, and the changes which took place upon the lamented death of that distinguished officer, occasioned some delay in taking my papers into consideration; but I was soon afterwards honoured with " the thankful acknowledgment of the Naval Administration for the very able " (they were pleased to say) " and valuable communication I had made ;" but no decision was then formed as to the use that should be made of it.

In November, 1818, I was honoured with a further, and still more flattering, acknowledgment from the

Admiralty, accompanied by a request, that I would permit a Copy of my Work to be retained in the Admiralty Office, with a view to carry into effect the whole, or any part of my plans, hereafter, when those considerations and financial circumstances, which at that time prevented their adoption, might admit. To this communication I returned the following answer :—

Sir, *Farnham, Nov. 29th, 1818.*

    I HAVE had the honour to receive your letter of the 24th instant, in which you acquaint me, by Lord Melville's directions, that his Lordship and the Board think very highly of the manner in which I have treated the important subject of Naval Gunnery, in the several MSS. which I have addressed to the Admiralty—their approbation of many of the suggestions contained in those papers—and their request to be permitted to retain a Copy of the Work at the Admiralty, in order that they may be enabled to carry into effect the whole or any part of my plans, in the event of changes in those circumstances and considerations which at present prevent the adoption of my suggestions.

    I am gratified by this communication; and beg you to present to the Board my consent to their Lordships retaining a Copy of my Work; to which I request copies of all my Letters, including this, may be attached, for the purpose of procuring reference to me, should I be living, at the time *any* adoption of my plans may be contemplated.

    I had formed an intention of making a publication upon this subject; but considering this consent to the request which the Board has made, as inconsistent with the measure of publication, I willingly abandon that intention unless it should have the approbation of their Lordships.

I have the honour to be,
&c. &c.

To Vice-Admiral   (Signed)   HOWARD DOUGLAS.
  Sir Graham Moore, K.C.B.,
      &c. &c.

    In answer to this letter I was soon afterwards acquainted by Vice-Admiral Sir Graham Moore, that he had "communicated my letter of the 29th November to Lord Melville and the Board, and was authorized to acquaint me, that their Lordships did not see any objection to the publication of my Essay on Naval Gunnery," which was accordingly soon afterwards returned to me, for this purpose, with the following letter from the Secretary of the Admiralty :—

Sir, *Admiralty Office, 25th Nov.*, 1819.

My Lords Commissioners of the Admiralty, having had under their consideration your Essay upon the Theory and Practice of Naval Gunnery, submitted to them in your letters of the 15th and 23rd October, 1817, command me to express to you their Lordships' thanks for the communication, and for the attention you have paid to this important subject; and in reference to your intimation of causing it to be published, their Lordships further command me to acquaint you, that they have no objection to your doing so, if you should think proper; and with this view, I am directed to return to you herewith the original Manuscript, and to add, that a Copy of it, together with your letters addressed to the Members of the Board upon the subject, have been retained in this Office.

I am, Sir,
Your most obedient, humble Servant,
J. W. CROKER.

Colonel Sir Howard Douglas.

Upon the receipt of this letter, I applied for and received Lord Melville's permission to dedicate my Work, in print, to his Lordship, as I had already done in MS., and I send it forth, conscious of its defects, as well as of my inability to treat the important subject in the way it merits.

CONTENTS.

	PAGE
Introduction to the First Edition	iii

PART I.

On the Organization and Training of Naval Gunners 1

PART II.

On the Theory and Practice of Gunnery, more particularly applied to
the Service of Naval Ordnance 23
    The Theory of Projectiles in Vacuo 26
    The Theory of Projectiles in Air 56
    The Effect of Splinters 68
    The Effects of Heavy and Compound Shot 72
    Employment of Double and Triple Shot 77
    The Effects of Windage 84
    The Effects of Length of Gun 101
    On Recoil and Preponderance 117
    The Effects of Wads 119
    Penetration of Shot 122
    On Cylindro-conoidal and Excentric Spherical Projectiles . . 148

PART III.

On Ordnance re-bored and newly constructed for the British and
Foreign Navies 181
    1. Bored-up Guns 181
    2. Monster Guns 187
    3. New Guns for Solid Shot 190
    4. New Guns for Shells and Hollow Shot 197
    5. Rifle-guns loaded at the Breech 212
    6. Relative Values of Solid and Hollow Shot 219
    7. Stowage of Shells, &c. 261
    8. On Military Rockets 324
    9. On the Attack of Maritime Fortresses 335

CONTENTS.

PART IV.

	PAGE
On the Service of Guns in Naval Actions	377
§ 1. On the Determination of Distances at Sea	377
§ 2. On Pointing Naval Ordnance	385
§ 3. Locks and Tubes for Naval Ordnance	397
§ 4. On the Practice of Firing at Sea	410
§ 5. On the Manual Exercise of Naval Artillery	440
§ 6. On Gunpowder and Gun Cotton	468

PART V.

On the Tactics of Single Actions at Sea 479

APPENDIX.

(A) On Rifle Muskets 503
(B) The Emperor Louis Napoleon's New System of Field Artillery . 536
(C) Abstract of the French Navy 556
(D) Tables of Gunnery Practice 565
(E) Copy of the Prospectus compiled by Sir S. J. Pechell from the 'Naval Gunnery' for the Establishment on board the "Excellent" 596
(F) Armament of the British Navy, &c. 601
(G) Floating Batteries cased with Iron 608
(H) Observations on the Lancaster Gun 611
(I) Operations in the Black Sea 614

INDEX TO THE WORDS, ALPHABETICALLY ARRANGED 631

ILLUSTRATIONS.

Plates I. and II. to be placed at the end of the Work.

NAVAL GUNNERY.

PART I.

ON THE ORGANIZATION AND TRAINING OF NAVAL GUNNERS.

1. Soon after the termination of the wars arising from the French Revolution (of 1793), the author published the first edition of his work on Naval Gunnery, with the view of drawing public attention to a subject which he considered of vital importance to the interests of the country. This he was induced to do in consequence of the unsatisfactory state of practical gunnery at that time, to which the author attributed some of the disastrous results which ensued. To the remarks which he was then led to offer he would still solicit attention, not for the purpose of conveying censure, which is far indeed from the author's intention, but simply to show in bolder relief the improvements which have since been made, and to hold out encouragement to the exertion of every effort for maintaining and increasing the advantages already gained. The author here takes leave to state, that he has had the most gratifying assurances from officers of every rank, and of the most distinguished character, that the present improved condition of our officers and seamen, with respect to this essential branch of the naval profession, is mainly, if not wholly, to be attributed to the means adopted, conformably to his suggestions, for its cultivation theoretically and practically; and he hopes he shall be able to show, in the present work, the necessity we are under of extending our *Gunnery Establishment*, both as the best means of keeping up and advancing our attainments in the objects

of instruction there, and of placing the country in a state of security in the event of the occurrence of any sudden emergency.

2. When the fleets of Europe, opposed to us in the late war, had been swept from the face of the ocean by the gallant achievements of the British marine, a period of triumphant, undisputed dominion ensued, during which our seamen were not, in general, sufficiently practised in the exercise of those weapons by which that dominion had been gained; but, in the pride of conquest, were suffered, in many instances, to lose much of that proficiency in warlike practice which had been acquired in a long series of arduous service.

3. Reviewing carefully our naval actions with European enemies during the whole of the last war (1793-1815), and comparing them with the battles which were fought in that which immediately preceded, there appears abundant proof that the navies of Europe had, in the later epoch, very much deteriorated in the practice of gunnery. In the later years of Napoleon's reign, though considerable improvements had then been effected in the marine of France, the state of practical gunnery among French seamen was so bad, that we have seen ships, strongly manned, playing batteries of twenty or thirty heavy guns against our vessels, without more effect than might easily have been produced by one or two well-directed pieces; and we have seen some cases in which heavy frigates have used powerful batteries against our vessels for a considerable time without producing any effect at all.

4. The danger of resting satisfied with superiority over a system so defective as that of our former opponents, has been made sufficiently evident. We became too confident by being feebly opposed; then slack in warlike exercise, by not being opposed at all; and lastly, in many cases, inexpert for want even of drill practice; and herein consisted the great disadvantage under which, without suspecting it, we entered, with too great confidence, into a war with a marine much more expert than that of any of our European enemies.

PART I.   ORGANIZATION OF NAVAL GUNNERS.   3

5. Comparative views of warlike skill, as well as of bulk and force, besides other considerations, are necessary to a correct analysis of naval actions. If a vessel meet an enemy of even greatly superior force, it is due to the honour of her flag to try the effect of a few rounds; but, unless in this gallant attempt she leave marks of her skill upon the larger body, whilst she, the smaller, is hit at every discharge, she does but salute her enemy's triumph and discredit her own gunnery.[a] It is not disgraceful that a vessel should be compelled to yield to another of superior force; but it is so, that the enemy should not be made to smart for his conquest; yet the defenders of the devoted vessel could not in justice be blamed, so long as naval gunnery was not a matter of professional cultivation and absolute obligation.[b] We do well to augment our naval strength by the construction of large ships, and by attention to all the details of personal and material; but it is even more important to make it an object of timely consideration how our ships are to be provided, at the beginning of a war, with an adequate proportion of officers, gunners, and seamen trained to warlike practice.

6. The material of our navy is in the finest possible condition. Our ships are greatly improved in every feature of strength and quality. Our ordnance is the best in the world; every species of store and equipment is perfect. We possess excellent seamen, trained by the operations of our commercial navy. Our officers, many of them educated at the public expense, are good navigators, excellent astronomers, and are full of energy, activity, and courage: but all these qualifications do not satisfy the requirements of a good *ship of war*: to them

---

[a] In the action between the President and the Little Belt, in 1811, the latter was heavily damaged, and had eleven men killed and twenty-one wounded; while, on board the President, a boy only was killed, not a shot struck her hull, and only two hit her masts.—*James' Naval History*, vol. vi. p. 10.

[b] On one occasion the gunners of a vessel mistook the elevated sight of their arms (carronades) for the point-blank line; and, when close to the enemy, fired at an elevation of 3½ degrees. The author has witnessed, and he could name distinguished authorities to prove the existence, formerly, of much more serious errors in practice.

B 2

should be added a competent knowledge of the science and practice of gunnery.

7. Till lately this important element was almost wholly wanting. After years of war had afforded us ample opportunities of practice, and yielded us many splendid victories, our gallant officers were often sent forth, with well-earned reputations, and the sacred honour of the British flag, depending upon crews on whom no reliance could be placed, excepting for courage and self-devotion; and under such circumstances it is not surprising that we were, in some instances, severely disappointed. The author trusts he may be allowed some credit for having suggested the remedy to the then existing defects in our naval system, which has since been so successfully adopted; and he takes occasion to assure the officers of the navy who may peruse this work, that any remarks on public events which he may think it necessary to make in the course of this treatise, are not quoted to criticise, but, on the contrary, to justify, or account for, operations which were always most gallantly undertaken, and which could not perhaps have been better executed, with the means at their command.

8. It remains only that we take special care to preserve the high position which we have struggled through years of difficulty to attain; that we not only secure it from decay, but also use the utmost diligence to improve it, by availing ourselves of all the resources of science as they arise, and acquiring those facilities in manual operations which continual practice alone can bestow.

9. It will not, perhaps, be considered improper here to make a few observations on the opinion that nothing which is not coarsely simple can be used in naval gunnery; and, consequently, that no delicate instrumental operations ought to be attempted. Nothing can be more unfounded than such an opinion; instruments which may be considered complicated and unmanageable by persons quartered at a gun for the first time, are used with the utmost facility by well-formed artillerists.

The difficulty which the author's father experienced, even from officers, in procuring the adoption of gunlocks, and many other improvements which he had made in naval ordnance, are proofs how far the want of some general system for the cultivation of the science and art of gunnery causes impediments to the introduction of measures which tend to the advantage of the service, and which, instead of being undervalued, would be eagerly received, were naval officers and seamen taught how to estimate them.

10. In the former editions of this work the author pointed out the important advantages which would result from enlightening by theory and training by practice, during peace, a large proportion of those who are to serve our naval ordnance in war. He showed that, whatever plan might be adopted for the improvement of Naval Gunnery, it should be calculated to instruct officers, master-gunners,[a] gunners' mates, and their crews; and that no measure which provides only for the drill of the men, could effectually improve the service practice : the mere dexterity of a few privates will do little, unless directed by cultivated and well-exercised intelligence on the part of the officers commanding in the ship's batteries, and assisted by a good crew of practical gunners.

11. He observed, that to instruct the rising class of officers in gunnery, a short course of theoretical instruction, showing all the established principles of the science, should be introduced at the Naval College; and that gunnery should also be made an item in the examinations for promotion. To instruct a proportion of officers, midshipmen, master-gunners, gunners' mates, and some seamen, in the practice of gunnery, *Depôts of Instruction* should be formed. So much depends upon the nature and composition of these establishments, that the author must be permitted to show, at some length, the reasons which have guided him in making the following suggestions; and

---

[a] To avoid mistake, the gunners of ships are here designated *master-gunners*.

the objections which appear to him to attach, in great force, to some measures that have been contemplated.

12. It is assuredly important, and even indispensable, in establishing schools for instruction in the practice of Naval Gunnery, that the posts of instructors in them should, in consequence of the opportunities afforded to such persons for the acquirement of great proficiency and for originating improvements, be conferred on those only who are, afterwards, to practise, in the profession, on service, what they cultivate and teach during peace.

13. The fundamental principles, then, that should form the basis of any measure that may be adopted for the improvement of Naval Gunnery are, that no plan which does not provide for instructing officers, master-gunners, gunners' mates, and their crews, as well as drilling seamen in the exercise, can effectually improve the service practice; and, in order to render permanent and effectual the benefits that would result from the formation of Naval Depôts of Instruction, a proportion of intelligent seamen should be engaged for a term of years, and formed into a permanent body, from which the important situations of master-gunners should be filled; and which, in a more extended form, might be made to furnish hereafter a considerable number of expert seamen-gunners to act as captains of guns; or, if not sufficiently numerous to do this, capable, at least, of soon drilling to the established system, the ordinary crews of those vessels into which these trained men may be draughted.

14. When the vessels which have been three years in commission are paid off, and the seamen dispersed, what permanent naval advantages do we reap from the instruction that may have been given? The instruct*ed* are thrown at large upon the world—the instruct*ors*, though improved by the lessons they have given, are still not real-service naval gunners; and it may happen that the system and expertness which they may have acquired, will, in numerous instances, be carried to the aid of foreign nations, at the very time we most require

them ourselves. Ship after ship may be commissioned, and the crews drilled, but, on the expiration of their servitude, these are dispersed. All we can hope for from such a plan is, that, by this mode of introducing successions of trained gunners into the merchant service, we *may* hereafter, in war, recover some for the Royal navy; but in the mean time the rising generation of naval officers, of whom but few can be employed afloat, would be left uninstructed, master-gunners and their crews unimproved, and, most certainly, no permanent advantages gained.

15. Now to remedy these serious objections, we should first engage a certain number of seamen, expressly for the service of the gunner's crew, for periods of five or seven years; renewable at their expiration, attaching a small increase of pay to every such consecutive re-engagement. The advantages held out to volunteers should be, that master-gunners, gunners' mates, and a certain number of seamen-gunners, will, eventually, be incorporated; and that a regular advancement in that department will hereafter take place according to merit; so that seamen-gunners may, if they can read and write, consider themselves in the certain road to the attainment, according to their merit, of the situations of gunners' mates, and master-gunners of ships. Seamen-gunners should receive superior pay, and share prize-money, as gunners' mates do now, or have some other rank superior to that of able seamen.

16. The practicability of forming such an institution resolves itself into this—whether, upon these advantages being made known, a sufficient number of volunteers can be procured to commence such an establishment.[a] The experiment might be easily tried; but the proposal should be accompanied with an explanation that the system provides, eventually, a term of relief, or residence on shore, for men so incorporated. If the experiment

---

[a] It was suggested to the author by a gallant Admiral, that boys who may be educated at the Asylum, or by the Marine Society, might, after practice shall have made them able seamen, become very fit persons for service in this department.

answer the confident and authorized expectation that may be entertained of its success, a selection of naval officers, the best practitioners of the late war, should be named, to conduct the Depôts of Instruction; and the author has every reason to believe that some very distinguished officers would come forward to commence such a system.[a] In this way a number of trained men

---

[a] The following extracts are made from some of the very numerous letters which the author received, containing expressions of entire conviction of the necessity of adopting his plan: the letters are from naval officers of the first distinction, whose names the author is proud to quote. Those officers honoured him by advocating and urging the plan with all the weight of their influence, and were materially instrumental in its introduction and extension.

(Extract.)—SIR P. B. V. BROKE to SIR HOWARD DOUGLAS.

"Broke Hall, July 5, 1818.

" DEAR SIR,—As a brother officer, though charged with the same duties upon a different element, I trust you will excuse this my frank address [the author had not then made personal acquaintance with this eminent officer].

" Your obliging letter and treatise arrived this morning, and I am much flattered by your reference to me.

"I am a most anxious well-wisher of your plan. We are all, indeed, deeply indebted to an officer who takes such pains and bestows such ability upon an object so requisite to the support of our national honour.

(Signed) " P. B. V. BROKE."

(Extract.)—SIR P. B. V. BROKE to SIR HOWARD DOUGLAS.

" Broke Hall, September 23, 1818.

" MY DEAR SIR HOWARD,—A strong proof of the exigency of enforcing some improvement in our naval gunnery is the indifference with which the subject is regarded, after the useful lessons which the late war afforded us; but I hope your exertions will effect much. If you can get one small depôt established, it will soon create an excellence that will excite emulation.

" I am, my dear Sir Howard,
(Signed) " P. B. V. BROKE."

(Extract.)—SIR P. B. V. BROKE to SIR HOWARD DOUGLAS.

" Broke Hall, January 3, 1819.

" MY DEAR SIR HOWARD,—I thank you for your communication, but do most sincerely regret the contents of it; and it is mortifying to see that all our hopes in your spirited and assiduous endeavours are for the present suspended. We must hope that hereafter circumstances may be more propitious to the adoption of your plans.

" But I hope that, however great a sufferer from the non-adoption of your plans, I shall not suffer by their not being published, and request you will have the kindness to order some amanuensis to copy your work for me, reward him liberally, and favour me with an account of the expense. It is a curious plea that the present state of the country does not require the improvement

would always be retained in the service—successions of commanders, and many officers who cannot be employed afloat in a limited peace establishment, would, at the trifling expense of full pay, be improved in this important branch of their military duties; master-gunners and gunners' mates would be trained; and a permanent stock of seamen-gunners brought up, to fill hereafter

proposed: it will be too late to teach our hands to war when the enemy is pillaging our trade, and Lloyd's in an uproar to have ships rigged and sent instantly to sea.

(Signed) " P. B. V. BROKE."

(Extract.)—SIR S. J. PECHELL to SIR HOWARD DOUGLAS.

" Stratford Place, 1818.

" I gave your M.S. to Sir Philip Broke to read, who might do more good than any other man, did he reside in London, with the views which he has expressed.

" Sir Philip Broke has so completely reviewed your work, and been so decidedly of your opinion in every essential part of it, that there remains nothing for me to add but my most earnest wish that something should be done to rescue that part of our duty from the state in which it now is; and notwithstanding you will yet meet with much opposition in attempting to carry your plans into effect, I sincerely hope that you will not lose sight of the object to which you have so ably drawn our attention; and I am inclined yet to hope that perseverance will overcome the prejudice you have had to encounter.

(Signed) " S. J. PECHELL."

(Extract.)—SIR S. J. PECHELL to SIR HOWARD DOUGLAS.

" Albany, June 16, 1830.

" The old Flag Officer has done what even you could not effect, though not to the extent that either of us could wish; but within these few days an order has been given to establish a gunnery school on board the " Excellent " at Portsmouth.

(Signed) " S. J. PECHELL."

[The expression " the old Flag Officer " in the above extract alludes to a celebrated pamphlet which had been recently published, urging the immediate adoption of the author's plan. That pamphlet was attributed at the time to Admiral, the late Sir Charles, Penrose; but it is now known to have been written by an able and gallant officer who, fortunately for the service, and happily for his friends, is living—Rear-Admiral Bowles, C.B. The author's plan was also strongly urged in the leading articles of the 'Times' newspaper of the 17th, 20th, and 24th of May in the same year, 1830.]

(Extract.)—SIR S. J. PECHELL to SIR HOWARD DOUGLAS.

" Admiralty, November 3, 1831.

" We are now likely to do something more. Sir James Graham has approved of the plan, and has appointed the " Excellent " for this service; and I am

these important offices; and should it be extended ultimately in the manner here proposed, it would furnish, besides, a considerable number of very expert captains of guns.

17. In the former editions of this work (see the pages from 17 to 24 and page 28 in those editions) there are given in detail estimates of the composition and strength of the proposed establishment, and also of the expense which would attend its formation, together with indications of the subjects which should constitute a course of practical education for young officers, master-gunners, gunners' mates, and seamen-gunners: these subjects comprehend the details of practical gunnery with the different natures of naval ordnance, both on shore and afloat, and the laboratory operations required for the service; but the proposed measures having been carried into effect, and the system of instruction being now fully organized, it is unnecessary to repeat here what was then pointed out. The prospectus originally drawn up by Sir S. J. Pechell, in conformity to the author's suggestions, and transmitted to him for revision, in that

---

sure you will not refuse me your assistance in drawing up a prospectus for our future sea-gunners.

" I cannot help congratulating you on seeing your labours brought to a satisfactory result, and likely to confer so much benefit on the country.

(Signed) " S. J. Pechell."

(Extract.)—Sir S. J. Pechell to Sir Howard Douglas.

" Admiralty, February 22, 1832.

" I enclose you my compilation [a prospectus] from your book for the establishment on board the " Excellent," and request you will be kind enough to revise, add to it, or correct it, as you may judge fit. The plan is yours, and I may indeed say, the whole of it is yours, as anybody may see in a moment by turning over the leaves of your book. [See Appendix E.]

(Signed) " S. J. Pechell."

(Extract.)—Sir S. J. Pechell to Sir Howard Douglas.

" Admiralty, February 10, 1832.

" After writing to you I almost felt ashamed of having troubled you upon the subject, when I found everything I wanted ready cut and dry in your own work. I had, therefore, nothing to do, taking, as you did, a twenty-eight-gun ship establishment to begin with.

(Signed) " S. J. Pechell."

distinguished officer's letter of the 22nd February, 1832, will be given at length in the Appendix. (App. E.) The following remarks, which contributed in a great degree to the adoption of measures for the systematic cultivation of naval gunnery, are, on that account, restated in this place.

18. As soon as one set of seamen are returned complete in exercise and practice, they should be transferred to commissioned ships, and should there drill the seamen engaged in the ordinary way, according to the general system; so that in this respect these would be as well trained, at least, as by the contemplated plan; and all the permanent advantages of the proposed system would be so far gained.

19. Fresh seamen should be engaged as gunners, and drawn in to the Depôt of Instruction in proportion as trained men are turned over to the guard-ships. These again should, by degrees, transfer to the cruisers a certain proportion of the trained gunners that will have been received from the depôts; which, however, should, together with the guard-ships and home cruisers, always retain a sufficient number of trained men for newly commissioned ships, in the event of a sudden armament. In this manner vast facilities and advantages would be experienced in fitting ships, and in rendering them more immediately efficient. The plan now suggested would provide people not only qualified to assist in fitting the ship, but also to assist in working her;—not only qualified to drill to gunnery the fresh hands, but to examine and arrange all the ordnance equipment, and very soon to make that ship, if properly commanded, a *good man of war*.

20. In all departments of warlike organization, depôts are allowed to be the very hearts of the system, by which improvement is cultivated, circulated, and established. In all services this is recognised and observed; no body can be permanently good, no system uniform, without them. It is to this general measure that the service-efficiency of every branch of our army is mainly to be attributed. It is this which supports the uniform

systematic excellence of the whole machine, however remote some of its parts may be. It is a similar system, connected with the naval profession, which has made the marines what they are, and which has so much improved—perfected, indeed—the Marine Artillery. Detachments of the Marine Artillery are embarked on board the ships of a squadron sent out on a cruise of exercise and practice; and it was no uncommon thing for naval officers fitting out ships to apply for detachments of this corps to drill their seamen to the gun exercise. If such detachments had been drawn from a permanent body composed of *seamen* gunners, trained by *naval* officers, instead of marines, there would have been no comparison between the influence of the two systems on the practice of Naval Gunnery. For the same reason that the Marines have their divisions, the Royal Artillery its schools of practice, and every regiment its depôt, naval gunnery should have its permanent seat of instruction, and store of trained men.

21. The Marine Artillery has been raised to a condition of great excellence by its zeal, talent, and gallantry, and has certainly performed all the service that was contemplated at its formation. The author has witnessed its efficiency on service, and bears willing testimony to all the talent it has put forth, and all the distinction it deserves. It is well constituted, thoroughly instructed, and ably commanded. It is either a corps of good infantry, of scientific bombardiers, or expert field artillerymen; and it should be allowed to retain all these characters, by being kept for mortar service afloat, or for land-artillery service in desultory coast operations.

22. The advantages that would result from such an establishment are beyond calculation. These depôts would become the resorts of zeal and talents—the nurseries of improvements; and numbers of young naval officers of all ranks would resort thither at their own expense. Such is precisely the opportunity which the naval service wants in this branch of the profession. Improvement may thus be cultivated without pursuing it through other departments:—Naval officers would find a field

opened to them, which is now occupied by others. Naval Gunnery would become, as it most certainly should be, an organized department of the naval service, under the direction and control of the Admiralty; and the author feels most certain that this simple measure would lay the foundation of a system which would soon be carried to perfection by the professional genius and zeal which it would call into action.

23. The merits of this plan do not depend upon the limited extent to which we may be obliged to confine it at present, on account of the difficulty of making financial provision for a more general operation.—If it be plainly calculated to do some good, it should not be rejected because, for contingent reasons which attach not to its merits as a system, it cannot, at present, yield its full benefits. The adoption of a good sound system is the question at present under consideration, not its immediate extension. If we found our measure upon a good professional principle, the superstructure may be raised gradually, in proportion as we may require it. The question for consideration is, whether the plan which is suggested does not provide a good professional system for instructing officers, midshipmen, master-gunners, and gunners' mates;—for training a proportion of seamen as captains of guns, as well as for drilling seamen engaged in the ordinary way: whether such a measure would not eminently tend to encourage the professional cultivation of artillery knowledge in a permanent institution, which may hereafter be carried to any extent which circumstances may demand. If it promise such advantages, it will be cheaply purchased by any expense that may attend it.

24. Were this an experimental measure that could not be commenced without first committing the country to vast preparatory expense, we might hesitate about making the trial; but the system may be instituted at a cost that would not amount to the expense of adding a twenty-gun ship to our establishment. The items of the expense of one depôt were severally estimated by the author and given in the former editions, but, as the

question relating to such expenses, though important at the time, has now been decided, it will be unnecessary to re-state either the items or the amount.

25. The author hopes he has succeeded in convincing the reader that the improvement of naval gunnery will mainly depend upon the character of the depôts of instruction. Whether seamen be raised in the ordinary way for three years, and drilled as at present, or be engaged for a longer term, and trained as naval gunners in the manner here proposed, there can be no reason why they should be taught by any but the members of the profession to which the practice really belongs.

26. Such were the observations made by the author, in the first edition of his work, on contemplating the state of gunnery in the British navy at that time. Political circumstances and financial considerations prevented the British Government in 1817 from immediately adopting the proposition which, in that year, was presented by the author to the Admiralty; but, in 1830, the former having materially changed, and the latter no longer existing, a commencement was made by the Naval Administration to put the proposed plan in execution.

27. In that year Lord Melville, to whom the former edition of this work was dedicated, by an Admiralty "Minute," dated June 19, directed that a gunnery-school should be formed on board the "Excellent," on a limited establishment. It was placed as a temporary measure under the superintendence of the late Captain George Smith, Commander of H.M.S. "Victory," an ingenious and zealous officer, well known as the inventor of the moveable lever target.[a] In 1832 the system for

---

[a] This contrivance was intended for the instruction of seamen-gunners in the practice of naval gunnery at sea, without the consumption of ammunition, when, from the floating motions, the ships fired from and at, are more or less unsteady—an ingenious expedient, devised, as Captain Smith communicated to the author, to explain and illustrate the observations contained in pages 226 and 227 of the Naval Gunnery, 2nd edition, by giving the target an artificial motion whilst the gunner is taking his sight. This motion is stopped by a line, leading to the target, connected with the trigger-line; and thus

PART I.   ORGANIZATION OF NAVAL GUNNERS.   15

gunnery instruction was extended and permanently established by Sir James Graham, first under Captain Sir Thomas Hastings, and, since his appointment to the important office he now holds, under Captain, now Rear-Admiral Chads, to both of whom the naval service and the country are deeply indebted for the zeal and ability with which, aided by a staff of meritorious officers, the courses of study relating principally to Naval Gunnery are unremittingly carried on, and the institution brought into so much practical efficiency. The school for instruction in the practice of Naval Gunnery was established on board the "Excellent," which is stationed at Portsmouth for that purpose: it is now under the able superintendence of Captain Sir Thomas Maitland, K.C.H., C.B.

28. In 1839 the building which had been vacated, in consequence of the abolition of the Naval College for the education of young gentlemen destined for the naval service, was appropriated, in combination with the Gunnery-school on board the "Excellent," to the important purpose, as suggested by the author in 1817 (see Art. 16), of accommodating and instructing naval officers of all ranks on half-pay, who might be desirous of availing themselves of the great advantages which

---

there is exhibited, relatively with the line of sight through the centre of the wooden gun, any error which the gunner may have made in pulling the trigger-line at the proper instant to hit the centre of the target.

On Captain Smith's appointment to the superintendence of this commencement of a system of which he, in several interviews with, and letters to the author, had declared himself a zealous and earnest well-wisher, he addressed to the author a congratulatory letter, of which the following is a copy:—

"H.M.S. Excellent, Portsmouth Harbour,
"July 14, 1830.

"SIR HOWARD,—I take the earliest opportunity, after having heard of your return to England, to inform you that the Establishment which you have so long desired to see in our Navy has commenced on the limited plan I had the honour to submit to you; and, if I may be allowed to use a familiar remark, as *ce n'est que le premier pas qui coûte*, I trust you will have eventually the gratification of seeing the system extended as widely as you can desire. You have clearly demonstrated the necessity of it, and in such language as cannot fail in its effect.

"I have the honour to be,
"Sir Howard,
"Your most obedient and much obliged Servant,
(Signed)   "GEORGE SMITH."

those institutions severally and conjointly tender to officers, when not otherwise employed, to improve themselves in the various branches of nautical science and practice which enter so largely into the qualifications indispensable to an accomplished, skilful, and experienced naval officer.

29. The vast importance of these Establishments in promoting the efficiency of the naval service will be immediately appreciated from the following outline of their constitution, and of the professional objects for which instruction is there given. The Naval College is an establishment for twenty-four officers of all ranks, on half-pay, from Captains to Lieutenants; for Marine officers previously to their entering the Marine Artillery, and for twenty-five Mates. The two last classes of officers go through a course of gunnery on board the "Excellent," previously to entering the College; and the Mates compete for a Lieutenant's commission, which is given at Christmas and at Midsummer to him who has attained the greatest proficiency. All the above classes of officers study, under a very able professor, the higher branches of mathematics, including Astronomy. Their course also includes the theory and practice of Steam Engineering, with Mechanical Drawing, Fortification, and Gunnery.

The half-pay officers remain at the College from twelve to eighteen months, and the others, twelve months.

30. The establishment includes twenty Cadets for the Marine forces, who may remain two years unless they attain the required proficiency, and can undergo the examination, in a shorter time. Their course of study comprehends Fortification, Gunnery, History, and the French language.

Field Works are thrown up on a small island in the harbour, for the purpose of training the officers and men, and also for the service of field guns. On the same island there is a small laboratory, where all the students go through a course of instruction in making rockets, tubes, cartridges, fuzes, &c. Field-practice with shot and shells is carried on by them at South Sea.

PART I. ORGANIZATION OF NAVAL GUNNERS. 17

31. Besides the regular duties there are prosecuted, under the direction of the Commander of this Institution, experiments of the most important nature, whenever the prospect of benefit to the service requires that such should be made. Among these may be included experiments for the determination of the relative values of different natures of ordnance for the projection both of solid and hollow shot; the extent of the ranges of projectiles; their powers of penetrating into materials; their crushing and splintering effects when fired against wood and iron; and the effects of ricochets from land and water. It ought to be added, that other nations, particularly France, have, since the termination of the wars of the Revolution, carried on elaborate and highly scientific courses of experiments, with the view of improving their naval-gunnery practice. The details of these experiments, with the investigations founded on them, have been published; and, to the experiments conducted on board the "Excellent," in conjunction with those at Woolwich and at Shoebury Ness, we are indebted for the verification (in some instances the disproval) of the advantages ascribed to the discoveries of foreigners. In these operations a vast amount of science and practical skill has been brought to bear upon every object of national utility within the limits of their scope.

32. Since the institution of the school of naval gunnery on board the "Excellent," the total number of able seamen entered as gunners amounted in 1851 to about 2500. Of these upwards of 2100 passed examination, and obtained certificates of qualification to serve as captains of guns, gunners' mates, and gunners of ships. Of that number 273 were, at different times, appointed as gunners, and 500 as gunners' mates; while no less than 1140 well-trained seamen-gunners were then serving in the British navy as first and second captains of guns, or in other important stations.

33. Seventy or eighty lieutenants, and 150 mates and midshipmen, of those who have at different times been on the establishment of the "Excellent," passed

their examinations, and obtained certificates of proficiency in the theory, practice, and manual exercise of naval gunnery; and of these, about sixty were then serving as gunnery-lieutenants, mates, and midshipmen, on board of ships in commission; and so zealously have officers of all ranks availed themselves of the advantages of resorting to Portsmouth for improvement, that about 460 officers of different ranks, post-captains, commanders, and lieutenants, passed their examinations, and obtained certificates of qualification according to their different attainments in naval gunnery.

34. The present establishment of the "Excellent" consists of a Captain, who is also superintendent of the Royal Naval College, 1 Commander, 12 Lieutenants, and 25 Mates (of these, there are in general 16 or 18 only); there are, besides the master-gunners, 360 seamen-gunners, and 200 boys under instruction, constituting all those who are not in sea-going ships.

35. The recent augmentation, to about 12,000 men, of the Royal Artillery, was made, as stated in Parliament, because none of our foreign garrisons were well provided, and some very inadequately, with this most essential description of troops; and also because the gunners disposable at home were not only insufficient to furnish more than at the rate of one artilleryman for every three guns in our coast defences, forts, and arsenals, as they now exist, but absolutely incapable of furnishing a supply of artillerymen for the additional armament required to man the numerous defensive works which it would be necessary to construct on the occurrence of any serious alarm. In addition to the above reasons for so great an increase in the land-service artillery, it was stated that six years are required to instruct land-service artillerymen, and render them practically efficient. The complements of seamen-gunners for our ships now in commission are too limited; and a long time would elapse before new levies of seamen could be properly trained to the guns, in the event of any emergency which might require an extension of our naval force. Though we have an

PART I.   ORGANIZATION OF NAVAL GUNNERS.   19

establishment for instructing seamen-gunners during peace, in the details of their duty, it is on a scale too contracted to be effectual in war. As our land-service artillery has been augmented to about 12,000 gunners, so the number of seamen under instruction in naval gunnery should be increased to an extent which may suffice to furnish at least 3000. This may be done, not by increasing the establishment of the " Excellent" to that extent, but by retaining the most expert of the men that have been so instructed, and keeping them in active employment in sea-going ships and gunnery frigates,[a] two of which should be added to that important department for this purpose. We may thus have available at any time a number of men well-practised and ready for any emergency; whereas, by casting them loose on the world, the means devoted to the purposes of their instruction are expended in vain, and the ends for which the expense is incurred are defeated, while the men are liable to be picked up and employed for the benefit of rival nations.

36. Soon after the publication of the Third Edition of this work, in which the augmentation of the important establishment of seamen-gunners was advocated (Art. 38 of the former, and Art. 35 of the present edition), the "Edinburgh," of 58 guns, now Admiral Chads' Flagship, was, on the recommendation of that distin-

---

[a] Frigates of instruction for seamen-gunners in the French Navy are appropriated to that important purpose, and occasionally accompany their squadrons and fleets on sea-going service. Thus the Amazone, frégate d'instruction, formed part of Admiral La Lande's fleet in the Mediterranean, in 1840.

The " Minerve" 1st class frigate was specially commissioned for L'Ecole des Matelots Canonniers, in 1848.

The Russians have long since adopted the French practice of forming what are called *Equipages de Ligne* (ships' crews); of these they have 46, each consisting of 1195 men : viz., 27 in the Baltic and at Archangel, 16 in the Black, and 3 in the Caspian Sea. Each "Equipage" is divided into four companies, of which three form the quota assigned to ships of three decks, two-and-a-half to ships of two decks, and one-and-a-half to frigates of 60 guns. Nor are the Russians less active than other maritime powers in teaching the science and improving the practice of naval gunnery. It is scarcely necessary to observe that, on a recent occasion, they gave ample proof that they can effectually handle those destructive weapons which have been introduced into the warlike navies of other nations within the last forty years.

C 2

guished officer, commissioned as an additional gunnery-ship, and placed under the command of Capt. Hewlett, R.N., who, on board the "Excellent," had long acted under the former commander with so much zeal and ability as to prove himself one of the best disciples of the school; and who, when called upon, will be found fully qualified to put in practice his high scientific attainments.

When not at sea, the "Edinburgh" is stationed at Plymouth, where it, in part, accomplishes the object proposed by the author, of having a depôt of trained gunners at each of the great fitting-out ports of the empire. This addition to the gunnery establishment is a step in the right direction; but much yet remains to be done in order to attract seamen to, and retain them in, that important branch of the service. From the subjoined table [a] it will be seen that, according to the present regulation, those seamen who obtain certificates of the first and second classes receive, respectively, only 2d. and 1d. per day above the pay due to their ratings.

[a] A TABLE of the RATE of WAGES in the ROYAL NAVY per MONTH and per YEAR, under the OLD and NEW REGULATIONS.

RATING.	Under the Old Entries.		Continuous Service, or Entry for 10 Years.		Difference in favour of Continuous Service per Year.
	Per Month of 31 Days.	Per Year.	Per Month of 31 Days.	Per Year.	
	£. s. d.	£. s. d.	£. s. d.	£. s. d.	£. s. d.
Chief petty officers . .	3 2 0	36 10 0	3 9 9	41 1 3	4 11 3
First-class working petty officers	2 14 0	31 18 9	3 2 0	36 10 0	4 11 3
Second-class working petty officers . .	2 9 1	28 17 11	2 16 10	33 9 2	4 11 3
Leading seamen . .	2 6 6	27 7 6	2 14 3	31 18 9	4 11 3
Able seamen . . .	2 1 4	24 6 8	2 9 1	28 17 11	4 11 3
Ordinary seamen . .	1 13 7	19 15 5	1 18 9	22 16 3	3 0 10
Second-class ordinary seamen	1 8 5	16 14 7	1 11 0	18 5 0	1 10 5
Boys (first class), and naval apprentices .	0 18 1	10 12 11	0 18 1	10 12 11	No diff.
Boys of the second class	0 15 6	9 2 6	0 15 6	9 2 6	No diff.

Seamen-gunners, who are men trained in the "Excellent," receive 2d. per day in the first class, and 1d. per day in the second class, in addition to all other pay of their ratings.

PART I.   ORGANIZATION OF NAVAL GUNNERS.   21

This is not sufficient to induce seamen to enter the service for the purpose of being trained as gunners rather than as mere seamen on board a ship bound for a port to which they may desire to go, and under officers whom they may prefer; nor is the amount of pension awarded to the class of trained gunners sufficiently liberal. The government and the country should consider what would have been the state of the naval service if the Establishment for training seamen-gunners had not been formed, and the deficiencies even now experienced from its inadequate extension.

While the "Edinburgh" was a gunnery-ship, its establishment contained 250 seamen-gunners and 70 boys. The total establishment on board the "Excellent" and the "Edinburgh" consisted of 610 seamen-gunners with 200 boys: these supplied a limited number of seamen-gunners to every ship of war now in commission; and the whole establishment must eventually be increased to enable this country to keep pace with the other great maritime powers.

Such was the number of seamen-gunners available for immediate service when the country began to prepare for the impending war: to them, however, must be added the men previously trained on board the "Excellent," who were serving in ships on the Peace Establishment. The naval force in commission was then vastly and suddenly increased. The total number of guns carried by the ships in the Black Sea Fleet is 1394, and those in the Baltic Fleet 2200. Supposing the ships' batteries are manned on one side only, half the sum of these, or 1797, is the number of guns on the fighting sides of the ships; and for this vast force we had only 610 men fit to serve as captains of guns in depôt, as none could be spared from the ships previously in commission. This inadequate number furnishes only one trained seaman to each gun, as its captain, whereas there should be three: the captain, the man under him, with the man who serves the vent-tube and lock. Had the Establishment for trained gunners been increased, as urged by the author three years since (see the preceding article), how different would have been the effective condition of the fleets! The present deficiency, which is severely felt, notwithstanding the vast exertions which the naval officers have made to train to the guns men who, even, were not always seamen, cannot be too strongly pressed upon the notice of the naval administration of the country.

37. It should be remembered that, when emergencies shall arise, we shall require both *gunners* well qualified to cope with opponents trained in all the improvements which the practice with solid shot has everywhere attained, and *bombardiers* highly expert in the more difficult and delicate operations of shell-firing—a practice which, if not carried on with the utmost skill and care, may be attended with accidents of a nature capable of causing the loss of a ship, or even of compromising victory in a general action.

38. The extensive establishments at the five principal naval arsenals of France for the organization of *Equipages de ligne* (ships' crews) (Ordonnance of the 11th Oct. 1836), as well as the establishment for the instruction and training of Matelots Canonniers, will hereafter be described in detail. From these reserves, together with the extensive establishments of Marine Artillery and Marines, a very considerable number of ships of the line, frigates, and other vessels may be promptly manned and speedily prepared for service. This would be done by immediately embarking the regulated number of companies, each composed of six petty officers, thirteen marine apprentices, and thirty seamen, and the regulated number of Matelots Canonniers.

PART II.

ON THE THEORY AND PRACTICE OF GUNNERY,

MORE PARTICULARLY APPLIED TO THE SERVICE OF NAVAL ORDNANCE.

39. The discoveries made by Robins, Hutton, and others in the science of gunnery have, within a few years, been so much the subjects of elementary instruction in the public seminaries of Great Britain for the education of men intended to enter the military and naval service of the country; and an artillerist has now in so high a degree the power of attaining the pure science of his profession, that it may seem unnecessary to dwell at length on the mathematical theory of projectiles. This subject will, therefore, be briefly introduced in notes; and it is intended, in the text, merely to exhibit the formulæ which comprehend the rules employed in the practice of gunnery, and to explain the manner in which those rules are deduced from the theory.

40. The theory of gunnery, in its present state, is admitted to be insufficient to serve alone as a basis for the practice of the artillerist; yet, as it may be not unreasonably expected that the theory will ultimately be carried to a degree of perfection corresponding to that which has been attained in other departments of physical science, it should be borne in mind that this object can be gained only by a diligent cultivation of the theory itself; and it may be added that it is only a scientific member of the profession who can direct experiments tending to useful results, or can advantageously introduce in formulæ the data which the experiments furnish.

41. While the processes to be followed in investigating the laws of projectile motion in resisting media were known only to a few men of superior scientific

attainments, it happened unavoidably, the guide to a correct and improved practice being wanting, that the most indispensable precepts relating to construction, armament, and equipment were violated, and the most obvious results of experimental research were unseen or disregarded.

42. A knowledge of the principles of gunnery is more essential to the naval artillerist than to an officer in any other branch of the service;—to him it may indeed be said to be absolutely necessary—and it cannot be doubted that he ought to avail himself of every means in his power to study the theory which should form the ground of his practice. It is true that the construction of a gun and the regulation of its equipment depend not always on an individual officer, but are determined by the authorities in the proper department under the Board of Ordnance; it is also true that tables of the ranges of shot, with different charges of powder and different elevations of the "arm," have been formed from experiments, and published by authority, for the guidance of the artillerist in like circumstances; but these tables are, even now, not sufficiently extensive to meet the vast variety of cases in which, afloat or on shore, the officers of our navy may be called upon to act; and, the charges, the elevations, and even the windages remaining the same, the ranges continually differ in consequence of variations in the quality of the powder, the state of the atmosphere, the figure of the shot, and many other circumstances.

43. A familiarity with all the details of the theory of projectiles can be acquired only by the aid of works professedly treating of dynamics in its highest branches; but a course of study carried on to an extent which the reading of such works implies may not be within the reach of every officer; and, at present, it is perhaps unnecessary. A certain knowledge, however, of the principles is indispensable for all, in order to enable them to comprehend the deductions made from the theory, and apply the formulæ to such novel cases as may arise in the course of their practice.

44. The numerical determination of points in the curve described by a projectile in a resisting medium, even when the analytical formulæ are given, is a work of so much labour that scarcely any practical artillerist would be disposed to execute it for a particular case which may occur, and such officer, confining himself to the subject of ranges only, would probably prefer making approximative interpolations between the ranges which have been experimentally determined; yet even this comparatively simple process cannot be effectively accomplished without a knowledge of the principles on which the ranges are investigated; and an officer of the artillery should no more be unacquainted with these than with the important discoveries which, in gunnery, relate to construction and armament. It is, in fact, of the utmost importance that officers should be familiar with the velocities due to the charges they use, with the principles on which those charges have been regulated, and with the effects which may result from variations in their quantities or qualities. It is particularly essential that those naval officers at whose discretion and on whose application the equipment of vessels with respect to the nature of gun is frequently regulated, should know the laws of the action of powder in guns of different lengths, and with charges of different quantities, also the effects of a plurality of shot as well as of single balls, or shells, of different weights. Lastly, officers should be acquainted with the laws of the penetration of shot of different sizes, when fired at different distances into materials of different kinds, as well as with the effects of the air's resistance on projectiles of different magnitudes and forms when discharged with different velocities.

45. Persons unaware of the force with which a military projectile, when impelled with great velocity, is resisted in its passage through the air, and imperfectly acquainted with the mathematical processes by which the trajectory, or path, in a resisting medium may be approximatively determined, still imagine that the laws of the flight of such projectile may be deduced from the

well-known properties of the parabolic curve, which it would describe in vacuo. This opinion is, however, unfounded, and the results obtained from those properties are seldom of use in the practice of gunnery; yet as the parabolic theory affords useful deductions in some cases, and a knowledge of it constitutes a step in the ordinary courses of military instruction; that theory is, in the subjoined note, briefly explained.[a]

---

[a] Let A, Fig. 1, Plate I., be the centre of gravity of the projectile at the mouth of a piece of ordnance; let A X be a horizontal, and A Y a vertical line, both passing through that point; also let A T, a tangent to the curve at A, be the direction in which the shot is discharged, so that T A X is the angular elevation of the piece. Let this angle be represented by a, and let the initial velocity of the shot be represented by v: let t be any time reckoning from the instant of the discharge, also let x, y be the horizontal and vertical co-ordinates of any point in the curve.

Now, if the force of gravity did not exist, the shot would move with uniform velocity in the line A T, and if b were its place at the end of the time t, we should have

$$x (= A R) = t v \cos. a, \quad b R = t v \sin. a:$$

but, by Dynamics, the deflexion (b P) produced by gravity in the time t is expressed by $\frac{1}{2} g t^2$ ($g = 32.2$ feet); therefore we have

$$y (= P R) = t v \sin. a - \tfrac{1}{2} g t^2.$$

Now, from the first equation, we have $t = \dfrac{x}{v \cos. a}$;

and, substituting for t this value, we have

$$y = x \tan. a - \frac{g x^2}{2 v^2 \cos.^2 a} :$$

again, by Dynamics, $v^2 = 2 g h$ (h representing the height due to the initial velocity), therefore the equation to the trajectory becomes

$$y = x \tan. a - \frac{x^2}{4 h \cos.^2 a} ; \quad (A)$$

and by the theory of curves of the second order, this is the equation to a parabola, the diameters of the curve being parallel to A Y.

When $y = 0$, we have, first $x = 0$, and secondly $x = 4 h \sin. a \cos. a$; or, by trigonometry, $= 2 h \sin. 2 a$. The first value of x corresponds to the point A, and the latter to the point I; this last expresses the value of the range on a horizontal plane.

Differentiating the equation (A) and making $\dfrac{d y}{d x} = 0$, we have $x = 2 h \sin. a \cos. a$, the half-range ($= $ A M), and substituting this value of x in (A), we get

$$y = h \sin.^2 a$$

for the greatest vertical ordinate ($=$ M K), and K is called the vertex of the curve.

PART II. THEORY AND PRACTICE OF GUNNERY. 27

46. But no artillerist requires now to be told that the deductions from the parabolic theory are without value in practice, except when the initial velocities of projectiles are less than two or three hundred feet in a second, when the resistance of the air is very small. In such a case, indeed, the propositions of gunnery may be solved, by the parabolic theory, with tolerable accuracy; but, with greater velocities, the conclusions obtained from it

---

Since $x = t\, v \cos. a$, we have

$$\frac{dx}{dt} = v \cos. a;$$

that is, the velocity of the shot in the horizontal direction at any point in the curve is constant.

Again, since $y = t\, v \sin. a - \tfrac{1}{2} g t^2$, we have $\frac{dy}{dt}$ (the velocity, in a vertical direction, at the end of a given time t) $= v \sin. a - g t$.

From these values of $\frac{dx}{dt}$ and $\frac{dy}{dt}$ we have $\frac{1}{dt} \sqrt{(dx^2 + dy^2)}$ (the velocity at the end of a given time t, in the direction of the curve)

$$= \{v^2 - 2 v g t \sin. a + g^2 t^2\}^{\frac{1}{2}};$$

but, if θ represent the angle which a tangent to the curve at any point makes with a horizontal line, the velocity at that point in the direction of the curve, will be expressed by $\frac{dx}{dt} \cdot \frac{1}{\cos. \theta}$, or $\frac{v \cos. a}{\cos. \theta}$; that is, such velocity varies inversely as the cosine, or directly as the secant of the inclination of the curve, at the given point, to the horizon.

From the value of x we have $t = \frac{x}{v \cos. a}$, and when $x = 4 h \sin. a \cos. a$, that is the whole range, the value of t becomes

$$\frac{4 h \sin. a}{v}, \text{ or } \sin. a \sqrt{\frac{8 h}{g}};$$

this is the expression for the whole time of flight, and it evidently varies with the sine of the given angle of elevation.

Since the product $\sin. a \cos. a$ remains the same for any given value of a and for its complement, it follows that there are two elevations (a and $90° - a$) of a piece of ordnance which, with the same initial velocity, give the same horizontal range.

The elevation of a piece being the same, the horizontal ranges ($2 h \sin. 2 a$) vary with the heights due to the initial velocities; or, since $h = \frac{v^2}{2 g}$, with the squares of the initial velocities; lastly, the initial velocities being equal, the horizontal ranges vary with the sine of twice the given elevation, and such range is evidently a maximum when a, the elevation, is 45 degrees.

are grossly erroneous. It will neither serve to determine the range, when the elevation and initial velocity are given; nor if the time of flight were observed and the range measured, could the initial velocity be computed from those data.

47. The determination of the velocity of shot at the moment of being discharged from a piece of ordnance, the charge of powder being given, would be an easy deduction from the known theory of forces, were it not that much uncertainty exists respecting the value of the expansive force of fired gunpowder; and it was to ascertain that value by comparing the mathematical expression for the initial value of shot with velocities obtained from experiment that Mr. Robins first invented the Ballistic Pendulum, by which such velocities can be measured with great precision.

48. The experiments of Mr. Robins formed an era in the theory of gunnery, but having used only such shot as could be discharged from musket-barrels, it became desirable that experiments of the same nature should be made on a much larger scale than the means of a private individual could afford, and the prosecution of this important subject was entrusted to Dr. Hutton, professor of mathematics in the Royal Military Academy, to whom the country is so much indebted for the high state of theoretical knowledge which the British artillery possesses. These experiments were commenced in the year 1775, and a copy of the report, being presented to the Royal Society, was honoured with the gift of the annual gold medal. Dr. Hutton carried on a second series of the experiments in the years 1783, 4, 5, &c.; and Dr. Gregory, his successor in the chair of mathematics at the Academy, was, in 1817, conjointly with General Millar and Colonel Griffith, charged with the superintendence of a corresponding series, in which a pendulum weighing 7000 lbs. was employed. It should be observed that experiments having the like object in view, but with a different kind of machine, were, in 1764, carried on by the Chevalier d'Antoni, at Turin, and the results at which he arrived differ

but little from those subsequently obtained by Dr. Hutton.

49. The Ballistic Pendulum, Fig. 2, Pl. I., consists of a large block of wood, suspended by iron rods, so as to vibrate on a horizontal axis upon receiving the impact of shot projected into it. In order to measure the arc of vibration Dr. Hutton applied a pointed piece of iron, as at P, to the inferior surface of the block, so that its lower extremity might, during the vibration, trace a curve line on a bed of soft grease which filled a groove in the circular piece A B of wood. The number of degrees in the act of vibration, or the chord of that number of degrees, was measured on a scale of chords which was laid down on the machine; and, from thence, the places of the centre of gravity and the centre of oscillation, together with the momentum of the inertia of the pendulum being previously found, the velocity of the shot at the moment of impact could be determined.

50. In conducting their experiments at L'Orient, from 1842 to 1846, the French engineers made use of a pendulum consisting of a suspended gun in combination with a suspended receiver (*récepteur*), corresponding to the block in the English pendulums; and, in 1843, the United States' government caused a series of experiments to be made, under the direction of Brevet-Major Mordecai, with an apparatus constructed on the same principle as the French machine. It consisted of a gun (a 24-pounder or a 32-pounder) suspended between two piers by a horizontal axis turning on *knife-edges*; and of a hollow frustum of a cone of cast iron suspended in a similar manner; the axes of the gun and cone, when both were at rest, being in the same horizontal line. The hollow frustum was filled with sand, in bags, to receive the impact of the shot; its front being covered with a plate of iron having in the centre a perforation, which was covered with sheet lead, and through this the shot passed after being discharged from the suspended gun. The whole weight of the gun-pendulum was 10500 lbs., and that of the receiver pendulum 9358 lbs. The distance of the centre of gravity in each, from the

axis of suspension, was about 14¾ feet; and the sensibility of the pendulums was such, that when set in motion in an arc of 12 degrees, the receiver-pendulum continued to vibrate about 24 hours, and the gun-pendulum about 30 hours.

51. The distance of the centre of gravity G, Fig. 2, Pl. I., from the axis of suspension at C, in the ballistic pendulum is, in some cases, found experimentally by balancing the whole apparatus, in two different positions, on an edge of a triangular prism; but the most accurate process for its determination is (while the machine is suspended on its horizontal axis at C) by a weight at the end of a string attached near P, and passing over a pulley, to draw the block up till its vertical axis C P is in a horizontal position. Then the known weight of the whole pendulum being supposed to act at G, the required place of the centre of gravity, and being represented by W; also P representing the weight which, by experiment, keeps the machine in *equilibrio* in a horizontal position; we have, by the nature of the lever,

$$P . CP = W . CG \; ; \; \text{therefore} \; CG = \frac{P}{W} CP$$

Thus the required distance of the centre of gravity from the axis of suspension is known.[a]

---

[a] The position of the centre of gravity in each of the pendulums used in the United States' experiments was found by balancing the machine, in different positions, on a steel bar; but the method employed by the French artillerists was nearly as follows:—The axis of the bore being in a horizontal position, a weight w sufficient to cause that axis to take a position, as A B, Fig. 3, Plate I., inclined to the horizon in a certain small angle A D E, represented by θ, was applied, as at H, under one extremity of the piece; the centre of gravity of the whole, which previously was in a vertical line C Z passing through the axis C of suspension, was consequently made to describe an arc subtending an equal angle Z C D. The same weight was then removed to the opposite extremity K, of the piece, by which means the centre of gravity was made to describe an arc subtending an angle θ' on the other side of the vertical line. Then, W being the weight of the pendulum with its gun or mortar; l the length of the pendulum, or the perpendicular distance from the axis of suspension to a line H K, parallel to the axis of the bore, and joining the points H, K, at which the weight w was attached under the piece, if a be the length of that line, also if G represent the required distance of the centre of gravity from the axis of suspension; we shall have

$$G = \frac{w}{W} \cdot \frac{a - l \sin. (\theta + \theta')}{\sin. (\theta + \theta')}$$

In the investigation of this formula, θ and θ' being small angles, cos. θ and

PART II. THEORY AND PRACTICE OF GUNNERY. 31

52. The centre of oscillation, in a vibrating body, is that point in which, if the whole quantity of matter could be concentrated, the time of a vibration about the same axis of suspension would be equal to that of the given body. In order to obtain, experimentally, the distance of the centre of oscillation from that axis, let the body, when suspended freely, be made to perform vibrations of small extent on each side of a vertical plane passing through the axis of suspension, and let the number of such vibrations which are made in a given time, as one minute, be counted. Then, the number being represented by n, since, by Dynamics, the lengths of pendulums are inversely proportional to the squares of the number of vibrations performed in a given time, we have, l the length of a mathematical pendulum vibrating seconds in the latitude of the place of experiment being known,[a]

$$n^2 : (60)^2 :: l : l';$$

l' being the length of the equivalent mathematical pendulum or the required distance (C O, Fig. 2) of the centre of oscillation from the axis of suspension: this distance is supposed to be measured on a line passing through the axis of suspension and the centre of gravity in the body.[b]

53. The centre of oscillation being found, the momentum of inertia with respect to the horizontal axis of suspension may be readily obtained thus:—

Let l be the distance of the centre of oscillation from

---

cos. θ' are, each, considered as unity; and sin. $(\theta + \theta')$ is put for sin. θ + sin. θ'.
—(*Expériences d'Artillerie exécutées à L'Orient*, Paris, 1847.)
The formula for G in the work just quoted (page 8) is more complex than that which is above given, in consequence of the introduction in the former of an additional weight called the *Compensateur*, which is employed in order to insure a perfect horizontality of the axis of the bore when an experiment is to take place.

[a] $l = 39.13929$ inches in the latitude of London.
[b] The observed number of oscillations should, in strictness, be reduced to the number which would be performed by the body in an equal time if the vibrations were infinitely small. For this purpose the observed number n should be multiplied by $\left(1 + \dfrac{a}{8\,l'}\right)$, a being the height of the observed mean arc of vibration, or the vertical distance from its highest to its lowest point.

the axis of suspension; G the distance of the centre of gravity from the same axis; m the mass ($=\frac{w}{g}$) of the body, w representing the body's weight, and g (= 32.2 feet) the force of gravity; then, by Dynamics,

$$l = \frac{\int r^2 dm}{m\,G};$$

and $\int r^2 dm$ (the required momentum of inertia) $= G\,m\,l$ or $G\frac{w}{g}l$.

54. If the block of a ballistic pendulum, when at rest, be struck at any point in its front by a shot projected horizontally against it, the line of projection being also in a vertical plane perpendicular to the axis of suspension, the angle through which the vertical axis of the pendulum is made to recoil by the impact may be measured by the scale on the machine; and this angle, with the other data obtained in the manner just stated, will afford the means, by a comparatively simple process, of obtaining the required velocity of the shot. The formula for this purpose is

$$v = \frac{\sqrt{\{(G\,W+w\,h)(G\,W\,l+w\,h^2)g\}}}{w\,h} 2 \sin \tfrac{1}{2}\theta;$$

in which v is the required velocity of the shot at the instant of striking the pendulum;

G, the distance of the centre of gravity of the pendulum from the axis of suspension;

W, the weight of the pendulum;

w, the weight of the shot;

l, the distance of the centre of oscillation in the pendulum from the axis of suspension;

h, the distance, in a vertical direction, of the point of impact from the axis of suspension;

g, (= 32.2 feet) the force of gravity; and

θ, the angular extent of a vibration on either side of a vertical line drawn through the axis of suspension.[a]

---

[a] Let m represent the mass ($=\frac{w}{g}$) of the ball; then $m\,v$ expresses the quantity of motion in it, and $m\,v\,h$ is the momentum of the ball's motion,

PART II.   THEORY AND PRACTICE OF GUNNERY.

It is evident, other things being equal, that the velocity of the shot varies with sin. $\frac{1}{2} \theta$, or with the chord of the arc of vibration.

---

with respect to that axis, at the time of impact, the pendulum being then at rest.

Now v' representing the angular velocity of the pendulum after impact, $v' r \, dm$ represents the motion of an element dm of the pendulum at the distance r from the axis of suspension, and $\int v' r^2 \, dm$ is the momentum of the motion of the whole pendulum; but v' is constant for all the elementary molecules, and $\int r^2 \, dm$ is (Art. 53) $= \dfrac{GW}{g} l$, or (M being the mass of the pendulum) $= G M l$; therefore $v' G M l$ expresses the momentum of the motion of the whole pendulum. Also $m v' h$ being the motion of the ball in the pendulum, $m v' h^2$ is the momentum of that motion. Then $v' (G M l + m h^2)$ expresses the momentum of the motion of both ball and pendulum after the impact; and, since the momenta before and after impact are equal, we have

$$v' (G M l + m h^2) = v m h:$$

Hence

$$v' = \frac{v m h}{G M l + m h^2} = \frac{v w h}{G W l + w h^2}.$$

By the recoil of the pendulum after impact the centre of gravity in the pendulum is raised vertically to a height expressed by $G (1 - \cos. \theta)$, or $2 G \sin.^2 \frac{1}{2} \theta$, and the centre of gravity in the ball is raised vertically to the height $h (1 - \cos. \theta)$ or $2 h \sin.^2 \frac{1}{2} \theta$; multiplying the first by W and the second by w, we have

$$2 \sin.^2 \tfrac{1}{2} \theta \, (G W + h w)$$

for the force of gravity overcome by the ball and pendulum during the recoil; and, by Dynamics, this is equal to half the *vis viva* of the system. But the *vis viva* being equal to the product of the momentum of inertia by the square of the angular velocity is, for both pendulum and ball, expressed by

$$\frac{v'^2}{g} (G W l + w h^2);$$

therefore $\quad \dfrac{v'^2}{g} (G W l + w h^2) = 4 \sin.^2 \tfrac{1}{2} \theta \, (G W + h w),$

and $\quad v' = 2 \sin. \tfrac{1}{2} \theta \left(\dfrac{(G W + h w) g}{G W l + w h^2} \right)^{\frac{1}{2}}.$

Equating this with the second of the preceding values of v', we obtain

$$v = 2 \sin. \tfrac{1}{2} \theta \, \frac{\sqrt{\{(G W + h w)(G W l + w h^2) g\}}}{w h}.$$

If the distance of the point of impact from a horizontal line passing through the centre of oscillation be very small; then h may be considered as equal to l; and the last equation will become

$$v = \frac{2 (G W + w l)}{w} \sin. \tfrac{1}{2} \theta \sqrt{\frac{g}{l}};$$

or

55. In the United States' pendulums, after l had been determined, as above mentioned, an additional weight (667 lbs.) was applied under the gun and block, by which the distance from the centre of oscillation to the axis of suspension was rendered equal to the distance of the line of fire, and point of impact, from that axis. This weight, representing it by w', is expressed by the formula—

$$w' = \frac{G' W' (l-l')}{b (b-l)};$$

in which W' is the weight of the original pendulum; w' the additional weight to be applied;

G' and G the distances from the axis of suspension to the centres of gravity of the original and of the augmented pendulum;

b the distance from the same axis to the centre of gravity of the additional weight;

l' and l the distances from that axis to the centres of oscillation in the original, and in the augmented pendulum respectively.[a]

---

or (G' being the distance from the axis of suspension to the common centre of gravity of the pendulum and shot)

$$v = \frac{2 (W+w) G'}{w\, l} \sin. \tfrac{1}{2} \theta \sqrt{g\, l}.$$

But $2 \sin. \tfrac{1}{2} \theta \sqrt{g\, l} = v'\, l$, the linear velocity of the centre of oscillation; therefore

$$v = \frac{W+w}{w} G'\, v'.$$

[a] The formula for w' may be investigated as follows :—By the nature of the centre of gravity, $G' W' + b w' = (W' + w') G$; whence $G = \frac{G' W' + b w'}{W' + w'}$. This, being multiplied by $W' + w'$, gives $G' W' + b w'$ for the momentum of gravity on the augmented pendulum. Now, (Art. 53) $\frac{G' l' W'}{g}$ is the momentum of inertia for the original pendulum: also, by Dynamics, the momentum of inertia of w' with respect to a horizontal axis passing through its centre of gravity (supposing w' to be a cylinder or prism) is $\tfrac{1}{2} \frac{w'}{g} c^2$ (c being half the height of the additional weight); and, with respect to the axis of suspension,

PART II.   THEORY AND PRACTICE OF GUNNERY.   35

In the above formula for v (Art. 54) W, G and l may be considered as expressing, respectively, the weight of the augmented pendulum and the distances of its centres of gravity and oscillation from the axis of suspension.

56. In the investigation leading to that formula no notice is taken of the resistance of the air against the face of the pendulum when made to vibrate; but it is shown by Dr. Hutton (Tract XXXIV. Art. 27) that the diminution of the velocity on this account does not exceed $\frac{1}{3377}$ of the whole; and even this is, in part, corrected by the mechanical method employed in determining the centre of oscillation. It is proved also by Dr. Hutton that no sensible error in the computed velocity of shot arises from the time (about $\frac{1}{500}$ of a second in the pendulum which he employed) during which the shot is penetrating into the block.

57. The distance of the ballistic pendulum from the muzzle of the gun being necessarily from 30 to 50 feet, in order that the block may not be affected by the flame arising from the discharge, the velocity of the shot at the moment of leaving the gun is greater than that which is obtained from the formula in Art. 54; and different formulæ have been proposed for the purpose of ascertaining one of these velocities from the other, or the velocity lost by the shot in passing from the gun to the pendulum : before stating these, however, it will be proper to show in what manner may be obtained the coefficient of the square of the velocity in the expression for the retardative force arising from the resistance of the air. (See Art. 59.)

58. The initial velocity of shot has been determined by firing the shot directly against one end of a hollow

---

is $\frac{1}{2}\frac{w'}{g}c^2 + \frac{w'}{g}b^2$. But, c being very small, the first of these terms may be rejected; therefore $\frac{G' W' l'}{g} + \frac{w'}{g}b^2$ expresses, approximatively, the momentum of the inertia of the original pendulum. It follows that

$$\frac{G' W' + w' b}{G' W' l' + w' b^2} = \frac{1}{l}; \text{ whence } v' = \frac{G' W' (l - l')}{b (b - l)}.$$

cylinder closed at the extremities, while the cylinder is made to revolve uniformly on its axis, the shot being fired as nearly as possible parallel to the axis. The rule for determining the velocity by such a machine is very simple.

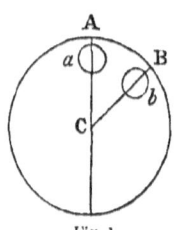
Fig. 1.

Suppose the shot were to enter the nearest face of the cylinder at a, and to emerge from the farthest face at b (b being, by the revolution, brought opposite to a during the time that the shot is passing through the cylinder). Let the length of the cylinder be 24 feet; let the angle A C B be 30 degrees, and let the time of the cylinder's revolution on its axis be $\frac{1}{8}$ of a second: then

$$360° : 30° :: \tfrac{1}{8} : \tfrac{1}{96}\text{th of a second};$$

and the last term is the time in which the shot was passing in a straight line through the cylinder, that is 24 feet; therefore 24 × 96, or 2,304 is the velocity of the shot in feet per second.

59. Since the absolute resistance of any medium against a spherical body moving in it depends upon the surface which is acted on during the motion, upon the density of the medium, and upon some function (suppose the square) of the velocity: let r be the semi-diameter of the body, δ the density of the medium, v the velocity, and let π represent 3.14159 (half the circumference of a circle whose radius is unity). Then, by Dynamics, $r^2 \pi \delta v^2 k'$ will denote the resistance, k' being a constant coefficient, which can be determined only by experiment. But the mass of the shot, supposed to be a solid sphere, being $\frac{4}{3}r^3 \pi \delta'$, in which δ' represents the density of the shot; the retardative force arising from the resistance of the air will become $\frac{3 k' \delta v^2}{4 r \delta'}$, which may be represented by $\frac{k \delta}{r \delta'}v^2$, k being a fraction whose value must be found by experiments. Much doubt exists concerning its true value; but, according to the best experiments which have been made, it should be expressed

Part II.   THEORY AND PRACTICE OF GUNNERY.   37

by 0.225. The specific gravities of the air and the shot may be substituted for δ and δ' respectively in the above formula.

60. The formulæ alluded to in Art. 57 are obtained from the usual differential equation of motion

$$\frac{d^2 x}{d t^2} = -f,$$

in which f may represent the retardative force arising from the resistance of the air. Then, if the resistance be considered as proportional to the square of the velocity, it may be represented by $c v^2$ or $c \frac{dx^2}{dt^2}$ (in which c is put for the coefficient $\frac{k \delta}{r \delta'}$ above); and we have $v = \frac{V}{e^{cx}}$, or $v = V e^{-cx}$.[a]   (a)

In these expressions, V represents the initial velocity, or the velocity of the shot at the muzzle of the gun, x the distance of the gun from the pendulum, v the velocity at the instant of striking the pendulum, and e ($= 2.71828$) is the base of the Napierian logarithms.

Dr. Hutton, making the resistance of the air proportional partly to the square and partly to the first power of the velocity, investigated a formula which is equivalent to

$$x = p \text{ hyp. log.} \frac{V - q}{v - q};$$

whence $v = \dfrac{V - q}{e^{-\frac{x}{p}}} + q.$ (b)

In these expressions p is put for $\dfrac{w}{.000007565 \, g \, d^2}$, ($w =$ the

---

[a] Assume $\dfrac{dx}{dt} = A e^{-cx}$:

then, differentiating and dividing by dt, we have

$$\frac{d^2 x}{d t^2} = A e^{-cx} \times -c \frac{dx}{dt} = -c \frac{dx^2}{dt^2};$$

thus the differential equation of motion is satisfied.

But, when $x = 0$, $\dfrac{dx}{dt}$ becomes V; therefore, making $x = 0$ in the equation for $\dfrac{dx}{dt}$, we have A = V; and that equation becomes $v = V e^{-cx}$, as in the text.

weight of the shot in pounds, and $d =$ its diameter in inches) and q for 231. (Hutton's 'Tracts,' Tract XXXVII., Arts. 78, 79.)

But the officers of the French school at Metz, making the resistance proportional, partly to the square, and partly to the cube of the velocity, have more recently proposed the formula,

$$v = \frac{V}{(1+a\,V)\,e^{\gamma x} - a\,V}; \qquad (c)$$

(See the 'Aide Mémoire d'Artillerie Navale,' 1850) in which (v, V and x being expressed in English feet) $a = 0.0007$ and γ varies with the different natures of shot; for a 30-pounder shot (French), corresponding nearly to an English 32-pounder shot, it is equal to 0.0001034.

For a 32-pounder shot, whose semidiameter is 3.1 inches ($= 0.2583$ feet) when the initial velocity, V, is 1600 feet and x is 50 feet, we obtain, from the first formula above, $v = 1588$ feet. From the formula of Hutton we have, in the same circumstances, $v = 1580$ feet; and from the last formula, $v = 1579$ feet: consequently in a distance equal to 50 feet, such a shot will lose about 20 feet, or $\frac{1}{80}$ of its original velocity. It is evident from the first of these formulæ for v, that, other things being equal, the velocity retained by a shot after having described a given distance in air will be greater as the initial velocity, and also as the diameter or the weight of the shot is greater.

61. Means being thus found of obtaining by experiment nearly the velocity with which a shot is projected from a piece of ordnance, it is proper next to have an expression for the velocity in terms of the charge of powder employed. This subject has been investigated by Dr. Hutton (Tract XXXVII., Art. 160), and his result, when modified so as to introduce in it the weight of the gun and its carriage, is expressed by an equation equivalent to

$$V = \sqrt{\left\{\frac{w}{w+w'} \cdot \frac{2gpr^2\pi}{w'+q}\left(\frac{\lambda'n}{M}\text{common log}\frac{\lambda}{\lambda'} + \lambda' - \lambda\right)\right\}}.$$

In this equation p ($= 2125$ lbs.) represents the pressure

PART II.   THEORY AND PRACTICE OF GUNNERY.   39

of the atmosphere on a square foot, g the force of gravity ($=32.2$ feet), and $\pi = 3.14159$.

r denotes (in decimals of a foot) the semidiameter of the shot, w' its weight, including the wad, and w the weight of the gun with its carriage.

λ represents the length of the bore of the gun, λ' the length of the charge in the bore, and q one-third of the weight of the charge.

M ($=.43429$) is the modulus of the common logarithms, and n is an abstract number expressing the number of times that the pressure arising from the expansion of fired gunpowder exceeds the pressure of the atmosphere. Dr. Hutton found the value of n to be very variable, increasing with the length of the gun; and from a mean of his experiments he assumed it to be 2200. The experiments of Dr. Gregory, in 1815 and 1816, make n equal to 2250; that is, he found the force of fired gunpowder to be equivalent to 2250 times the pressure of the atmosphere.[a]

---

[a] The centre of gravity of the charge moving in the bore with about one-third of the velocity of the ball, the weight of the latter during its movement in the bore may be considered as increased by one-third of the weight of the charge, supposing this to move with the same velocity as the ball; therefore the accelerative force of the gunpowder on the ball, when in contact with the charge, will be represented, in terms of the force of gravity, by

$$\frac{g\,n\,p\,r^2\,\pi}{w' + q}.$$

Now, let x be, at any instant, the distance of the centre of the ball from the extremity of the charge, during its passage along the bore; then, the accelerative force of the fired gunpowder on the ball being inversely proportional to the space it occupies in the bore, $\dfrac{g\,n\,p\,r^2\,\pi\,\lambda'}{(w' + q)\,x}$ expresses the accelerative force on the ball when at a distance x from the charge; but the retardative force of the atmosphere on the ball and charge while in the bore is $\dfrac{g\,p\,r^2\,\pi}{w' + q}$; therefore, finally, the accelerative force on the ball and charge is

$$\frac{g\,p\,r^2\,\pi}{w' + q}\left(\frac{\lambda' n}{x} - 1\right).$$

Making (by Dynamics) $\dfrac{d^2 x}{d t^2}$ equal to this expression for the force; and multiplying both members by $2\,d\,x$, we have

$$\frac{2\,d\,x\,d^2 x}{d t^2} = \frac{2\,g\,p\,r^2\,\pi}{w' + q}\left(\frac{\lambda'\,n\,d\,x}{x} - d\,x\right).$$

Integrating

In this formula the measures of length are all expressed in feet, and the weights in pounds avoirdupois.

62. But, in the employment of the formula this value of n should be diminished on several accounts. First, because of the thickness of the bag containing the powder, it should be diminished in the ratio of the area of a section of the bore to that of a section of the powder: now the thickness of the bag being about $\frac{1}{10}$ inch, the areas of the sections are to one another (R being the radius of the bore in inches), as R^2 to $R^2 - \frac{1}{20} R$, nearly, or as 1 to $1 - \frac{1}{20\,R}$ nearly; therefore, on this account the above value of n should be diminished by a fractional part of it, which is expressed by $\frac{1}{20\,R}$. Secondly, it should be diminished on account of the windage of the gun: now Dr. Hutton estimated that the loss of force occasioned by windage was one-third of the whole force when the lunaric area between a section of the ball and a section of the bore was one-tenth of the latter; therefore, considering the loss as proportional to such lunaric area,

$$\tfrac{1}{10} : \tfrac{1}{3} :: \frac{R^2}{R^2 - r^2} : \frac{10\,R^2}{3\,(R^2 - r^2)};$$

and, on this account, the above value of n must be diminished by a fractional part of it, which is expressed

---

Integrating between $x = \lambda'$ and $x = \lambda$, we have

$$\frac{d\,x^2}{d\,t^2} (= V^2) = \frac{2\,g\,p\,r^2\,\pi}{w' + q} \lambda' n \text{ hyp. log. } \frac{\lambda}{\lambda'} + \lambda' - \lambda :$$

therefore the velocity of the shot at the muzzle of the gun is expressed by

$$\sqrt{\left\{\frac{2\,g\,p\,r^2\,\pi}{w' + q} \lambda' n \text{ hyp. log. } \frac{\lambda}{\lambda'} + \lambda' - \lambda\right\}}.$$

But the accelerative force of the fired powder should be further diminished in the ratio of the sum of the masses of the gun, with its carriage and the shot, to that of the gun and its carriage. If, therefore, this condition be included, the first factor under the radical sign should be multiplied by $\frac{w}{w + w'}$. If, also, $\lambda' n$ be divided by M (the modulus of the common logarithms), the common logarithm of $\frac{\lambda}{\lambda'}$ may be substituted for the hyperbolic logarithm of that fraction; and thus is obtained the equation for V in the text.

by $\frac{3\,(R^2-r^2)}{10\,R^2}$. Lastly, it should be diminished on account of the loss of force by the escape of powder from the vent: now, the semidiameter of the vent being 0.1 inch, if we consider the lost force to bear, to the whole force, the ratio that the area of the vent bears to a section of the bore, the value of n must, on this account, be diminished by a fractional part of it, which is expressed by $\frac{1}{100\,R^2}$.

The sum of the three corrections is
$$\frac{30\,(R^2-r^2)+5\,R+1}{100\,R^2},$$
(R and r being expressed in inches), and this expression denotes the fractional part of n, by which n is to be diminished.

63. Since the weights of the shot and charge are small compared with that of the gun with its carriage, a near approximation to the initial velocity will be obtained by omitting, in the expression for V, the terms $\frac{w}{w+w'}$, and q. The equation then becomes, omitting also the two last terms because of their small effect on the value of the velocity,
$$V = \sqrt{\left\{\frac{2\,g\,p\,n\,r^2\,\pi\,\lambda'}{w'\,M}\,\text{com. log.}\,\frac{\lambda}{\lambda'}\right\}}.$$

64. A gun suspended from a horizontal axis like the block of wood in a ballistic pendulum, has long been occasionally employed to determine by the angular extent of its recoil, the velocity with which a shot is projected by a given charge of powder; and in the experiments alluded to in Art. 50, a gun-pendulum, as it is called, and a ballistic pendulum, were combined for the purpose of comparing together, shot being projected from the former, the initial velocities determined from the recoil of the one and the impact on the other. The formula employed for computing, from the recoil, the velocity of shot at the instant of leaving the gun-pendulum was (V representing that velocity)

$$V = \frac{2 \sin. \frac{1}{2} \theta . \text{W G} \sqrt{gl} - l'c\,\text{N}}{w'l'\frac{\text{R}^2}{r^2} + \frac{1}{2} c'\,l'};$$

in which W is the weight of the gun-pendulum, including that of the shot, and w' that of the ball and wad.

G is the distance from the axis of suspension to the centre of gravity of the pendulum.

l is the distance from the same axis to the centre of oscillation, and l' the distance from that axis to the axis of the gun.

c is the weight of the charge, and c' that of the bag.

θ is the angular extent of the vibration from the vertical position.

R the radius of the bore, r that of the shot, and g the force of gravity.

N represents the velocity communicated to the pendulum by a unit of the charge. (In the United States' experiments it is estimated at 1600 feet.)[a]

---

[a] If v denote the angular velocity of the pendulum, $vl\dfrac{\text{W}}{g}$ will express the motion in the pendulum, and $vl\dfrac{\text{W}}{g}\text{G}$ the momentum of that motion.

$2\,\text{W G} \sin.^2 \frac{1}{2} \theta$ expresses the action of gravity overcome by the ascent of the centre of gravity during the recoil, and $v^2 l\dfrac{\text{W}}{g}\text{G}$ the *vis viva*; therefore, by Dynamics,

$$4\,\text{W G}\sin.^2 \tfrac{1}{2} \theta = v^2 l\frac{\text{W}}{g}\text{G};$$

whence $4g \sin.^2 \frac{1}{2} \theta = v^2 l$, and $v = 2 \sin. \frac{1}{2} \theta \sqrt{\dfrac{g}{l}}$.

This value being substituted in $vl\dfrac{\text{W}}{g}\text{G}$, the latter becomes $2\text{WG}\sin.\frac{1}{2}\theta \sqrt{\dfrac{l}{g}}$, the momentum of the motion of the gun-pendulum.

Now, V being the velocity of the shot at the moment of leaving the gun, we have $\dfrac{\text{V}}{g}w'$ for the motion of the ball and wad; but the elasticity of the fired gunpowder produces in the gun a motion which exceeds this in the ratio of the area of a section of the ball to that of a transverse section of the bore; that is, of r^2 to R^2; therefore $\dfrac{\text{V R}^2}{g\,r^2}w'$ expresses the motion of the gun; and $\dfrac{\text{V R}^2}{g\,r^2}.l'\,w'$ is the momentum of that motion.

Next, assuming the mean velocity of the fired gunpowder at the time the

PART II.   THEORY AND PRACTICE OF GUNNERY.   43

65. The force of fired gunpowder is, however, still very imperfectly known; and the results of experiments with the gun-pendulum are very frequently at variance with those obtained from the ballistic pendulum. By French artillerists the initial velocity of a shot, the axis of the bore being horizontal, has been expressed by a formula, which, when reduced, so that the required velocity shall be given in feet per second, is

$$4200 \left(\log. \frac{\lambda}{\lambda'} \right)^3 \cdot \left[\log. \left(1 + \frac{c}{w'} \right) \right]^{\frac{1}{2}}.$$

The measures of length in this formula are in feet, and the logarithms are those of the common kind. (*Account of Experiments carried on at Gavre*, chap. xix. sec. 7.) It must be observed, however, that the results obtained from this formula are, in general, less than those given by experiments made in England with the ballistic pendulum.

66. The experiments of Sir Benjamin Thompson (Count Rumford) have fully proved that the ignition of gunpowder does not take place instantaneously, and that ball leaves the gun to be $\frac{1}{2}$ V; then $\frac{V c'}{2 g}$ will be the quantity of motion in the cartridge and bag, and $\frac{V c' l'}{2 g}$ is the momentum of that motion.

Again, if it be assumed that the quantity of motion in the fired gunpowder is proportional to the mass of the charge, it may be represented by $\frac{c N}{g}$; then $\frac{c l' N}{g}$ is the momentum of the motion of the fired gunpowder.

Adding together all the momenta relating to the action of the charge, and equating it with the momentum of the gun-pendulum, we have

$$\text{W G sin.} \tfrac{1}{2} \theta \sqrt{\frac{l'}{g}} = \frac{V R^2}{g r^2} l' w' + \frac{V c' l'}{2 g} + \frac{c l' N}{g};$$

whence

$$V = \frac{2 \text{ W G sin.} \tfrac{1}{2} \theta \sqrt{l g} - c l' N}{\frac{R^2}{r^2} l' w' + \tfrac{1}{2} c' l'}.$$

When the centre of oscillation is in the axis of the gun, we have $l' = l$; in which case

$$V = \frac{2 \text{ W G sin.} \tfrac{1}{2} \theta \sqrt{\frac{g}{l}} - c N}{\frac{R^2}{r^2} w' + \tfrac{1}{2} c'}.$$

the whole of the charge in a gun is never in a state of inflammation within the bore, though it is presumed that 8 ounces may be all ignited in $\frac{1}{130}$ of a second. Further experiments have also shown that, for every nature of gun, there is a certain charge which produces a maximum of initial velocity, so that if a greater quantity were employed the velocity would be less: the reason is, that besides the greater quantity of powder which is thrown out of the gun unfired, the powder which is actually ignited takes effect on the shot during a shorter time than it would, if, from a smaller quantity of powder being employed, the motion of the shot in the bore were less rapid. The charge of powder which would render the initial velocity a maximum may be found, mathematically speaking, by differentiating the second member of the equation for V, in Art. 63, considering λ' (the length of the charge) as variable, and making the result equal to zero. There is thus obtained

$$\text{com. log.} \frac{\lambda}{\lambda'} = \text{Modulus of com. log.}; \text{ whence } \lambda' = \frac{\lambda}{2.71828};$$

that is, the length of the charge producing the maximum velocity, with the same gun, is equal to about one-third of the length of the bore.

The charge which produces a maximum velocity is, therefore, greater as the gun is longer, but the increase is found to be not in so high a proportion as that of the length of the gun (*Hutton's Tracts*, vol. iii. p. 78); and Dr. Hutton's experiments have shown that, on account of the rapid diminution of high velocities by the resistance of the air, there is little advantage in point of extent of range by increasing the charge beyond what is necessary to communicate a certain velocity to the ball.[a]

---

[a] Major Mordecai, of the United States' artillery, states that an addition of nine calibres to a gun of sixteen calibres' length adds only one-twelfth to the velocity of a 12-lb. ball, when fired with a charge of 2 lbs., and one-eighteenth when fired with a charge of 4 lbs. It is observed also that, by increasing the charge beyond one-third of the weight of the ball, the recoil is increased in a much higher ratio than the initial velocity of the ball. It may be added, that for every purpose on service, even for that of breaching, the advantage gained by using a charge greater than one-third of the weight of the shot is unimportant, and does not compensate for the inconvenient recoil, or the

67. The force exerted by fired gunpowder on a shot varies with the square of the velocity of the fluid, that is, of the ball which it impels; and that force being also proportional to the quantity of fluid acting at once on the ball, it follows that the square of the ball's velocity, in the gun, varies with the quantity of fluid acting on the ball, that is, with the square of its diameter; and from this law it may be proved that the decrements of the initial velocity of shot in consequence of windage are nearly proportional to the differences of the diameters of the shot in the same gun. From thence may be obtained the equation

$$V = \frac{v-v'}{d-d'}(D-D') + V',$$

expressing the velocity which a shot would have if the windage were zero. Here V denotes such velocity; D the diameter of the bore of the gun, or of a shot which would occupy the bore without windage; d, d', D' the diameters of different balls, and v, v', V' their velocities on being projected from the same gun.[a]

68. The charge being supposed to have a cylindrical

---

destructive strain on the gun and carriage. The initial velocity of shot from 6-pounder and 12-pounder guns, with charges of 1¼ lb. and 2½ lbs. respectively, is about 1450 feet; and this is considered quite sufficient for field service. The initial velocity of canister and spherical case-shot, fired from field-howitzers with the service-charge, is about 1000 feet; and that of shells from howitzers is about 1175 feet.

[a] Since $v^2 \propto d^2$, $v'^2 \propto d'^2$, &c., we have

$$v^2 - v'^2 \propto d^2 - d'^2, \text{ or } (v-v')(v+v') \propto (d-d')(d+d');$$

or again,

$$\frac{v-v'}{d-d'} \propto \frac{d+d'}{v+v'}.$$

But the difference between v and v', also between d and d', being very small, the last fraction is very nearly equal to $\frac{2d}{2v}$; therefore $\frac{v-v'}{d-d'}$ may be considered as constant. Let the value of this fraction be obtained from experiments; then V' being the velocity due to any other shot (projected from the same gun) whose diameter is D', we have

$$\frac{V-V'}{D-D'} = \frac{v-v'}{d-d'},$$

whence $V = \frac{v-v'}{d-d'}(D - D') + V'$, or $V' - V - \frac{v-v'}{d-d'}(D-D')$.

form, and the density of the powder being constant, it follows that λ' (Arts. 61, 63) varies with the weight of the charge; hence the formula for V, Art. 63, indicates that the initial velocities of shot vary, nearly, as the square root of the weight of the charge directly, and as the square root of the weight of the shot inversely. These laws have been confirmed by the results of experiments,[a] and M. Piobert has proposed, for the purpose of determining the initial velocity of shot, when its weight and that of the charge are given, the empirical formula

$$V = v \frac{\sqrt{\log.\left(1 + \frac{c}{w'}\right)}}{\sqrt{\log.\left(1 + \frac{c}{b}\right)}};$$

in which v represents the experimented initial velocity of a shot whose weight is b, and V is the required initial velocity of a shot whose weight is w'; the weight c of the charge being the same for both.

Major Mordecai, of the United States' artillery, has found that the rule agrees nearly with his experiments when the charges do not exceed one-third of the weight of the ball, and the gun has not less than sixteen calibres in length; for higher charges he does not consider it sufficiently correct. If the factor $\dfrac{v}{\sqrt{\log.\left(1+\frac{c}{b}\right)}}$ be represented by M, he finds that, with a 32-pounder gun, the charge being one-third of the weight of the shot, and the windage 0.16 inch, M = 5200; with a 24-pounder, the charge being one-third and the windage 0.14 inch, M = 5400.

---

[a] In the experiments made at Washington with the gun and ballistic pendulums combined for the purpose of ascertaining the initial velocities produced by equal charges of powder in the same piece of ordnance on balls of different weights, it was found that, with a 24-pounder gun and a charge of 4 lbs. of powder, the windage being 0.175 inch, the initial velocity of a shell filled with lead and weighing 27.68 lbs. was 1325 feet; of a marble ball weighing 9.29 lbs., was 2154 feet; and of a lignum vitæ ball weighing 4.48 lbs., was 2759 feet. The two first of these velocities are, in accordance with the formula, nearly in the inverse ratio of the square roots of the weights of the shot; but the two last are nearly as the cube roots of the weights inversely.

PART II.   THEORY AND PRACTICE OF GUNNERY.   47

In the British artillery service, use is frequently made of the empirical formula

$$V = 1600\sqrt{\frac{ac}{w'}},$$

in which c is the weight of the charge, and w' that of the shot; a is a coefficient to be determined by experiments. Those of Dr. Hutton indicate that a varies between 2.1 and 2.5 as the length of the gun, and also as the ratio of the weight of the charge to that of the shot increases. The experiments of General Millar, in 1817, show that, with a windage equal to 0.202 inch, the value of a should be 2.8, and that on reducing the windage to 0.075 inch it should be 3.55. On computing the initial velocities of the shot from the ranges given in the Table of the experiments made at Deal in 1839, and also from those in the Tables of the practice on board the "Excellent" from 1837 to 1847 (Tables I., II., III., IV., V., VI., Appendix D), the mean values of a for different windages were as follow :—

Windages.	Values of a.
0.233 in.	3.2
0.2	3.4
0.175	3.6
0.125	4.4
0.09	5.

From these, by means of the formula for V, Art. 67, it is found that when the windage is zero, $a = 6.66$; and it may be observed that the numbers in this table are nearly conformable to the law of the decrements of velocity, as stated in that article.

For carronades, in which the windage varied from 0.061 to 0.078, the mean value of a was 4.5.

69. Gunpowder when ignited expands with equal forces in every direction, and consequently it acts equally upon the bottom of the bore and upon the ball during the passage of the latter along the cylinder, supposing it to fit tightly. Hence, neglecting the allowance which should be made for the frictions of the ball and of the gun-carriage, the velocity of the recoil will be, to that of the shot, inversely as the weight of the gun to that of the shot : that is, if V represent the initial velocity,

in feet, of the shot, w the weight of the shot, and W the weight of the gun with its carriage, the velocity of the recoil, in feet per second, will be expressed by $\frac{Vw}{W}$.

Thus, supposing the initial velocity of a 24 lb. shot to be 1600 feet per second, and the weight of the gun with its carriage to be 57.7 cwt. or 6462 lbs., we shall have 5.9 feet per second for the velocity of recoil.

70. In contemplating the nature of the resistance to the flight of shot or shells it must be remarked that, in the motion of a body through the air, no particle of that fluid can be disturbed without moving others to a considerable distance about it, while the displaced particles take time to fall back into the space which they before occupied. As the moving body passes on, there is left behind it a kind of vacuum more or less complete according to the degree of the body's velocity; and when the ball moves quicker than the air can rush into the space left behind, the vacuum becomes perfect. Now there is a certain limit to the velocity with which air can rush into a vacuum, viz., about 1300 or 1400 feet in a second,[a] and, consequently, when the velocity of a ball is greater than this, it is manifest that the resistance will be very great; for there being then no pressure of the fluid behind the ball, while that which is in its front is in a state of condensation from the particles there not being able immediately to escape, the ball will be resisted by the whole pressure of the condensed air on its fore part.

71. The resistance, besides depending on the velocity, diameter, and the weight of the projectile, is affected by so many circumstances which cannot be duly estimated, that experiment alone can determine it, and this only to a certain extent. If shot could be discharged so accurately as to hit a ballistic pendulum at considerable distances, the loss of velocity occasioned by resistance might be easily found; but such a degree of accuracy

---

[a] 1366 feet, when the barometer stands at 30 inches.—See Hutton's Tracts, vol. iii. p. 195.

cannot be obtained, and the ballistic experiments have hitherto only furnished us with these results at different distances as far as 300 feet, beyond which shot cannot be directed with sufficient accuracy to hit the block.

72. The method of determining the resistance of the air by the ballistic pendulum did not, it was found, answer with velocities under 300 feet in a second, on account of the balls rebounding from the block instead of entering into it. To ascertain the resistances to smaller velocities, Mr. Robins had recourse to experiments with his *whirling machine* (Fig. 4, Pl. I.). This ingenious contrivance consists of a brass barrel, B C, moveable on its axis, and furnished with friction wheels so as to reduce the friction to an inconsiderable quantity. A light, hollow cone, A F G, is placed upon the barrel, with the vertex A in the termination of the axis; a fine wire, A H, supports the arm, G H, upon which the body, whose resistance is to be tried, is fixed. A silk line is wound upon the barrel, and thence leads, in a horizontal direction, to the pulley L, over which it is passed, and a proper weight M hung to its extremity. If the weight M be left at liberty, it will descend with accelerated motion, causing the body P to revolve with increasing velocity, until the resistance on the arm G H, and on the body P, become nearly equal to the weight M, when the motion of both will be nearly equable. Thus, when the machine has acquired an equable motion, which it usually does in five or six turns, ascertain first, by counting a number of turns, in what time one revolution is performed. Then remove the body P and the weight M, and find, by trials, what smaller weight will cause the arm G H to revolve in the same time as when P was fixed to it; and the difference of the two weights is, obviously, equal, in effort, to the resistance of the air on the revolving body. Reducing this weight in the ratio of the length of the arm to the semidiameter of the barrel, we shall have the absolute quantity of resistance.

73. With this machine Mr. Robins ascertained that the resistance of the air to a 12 lb. iron ball, moving

with a velocity of 25 feet in a second, is not less than half an ounce avoirdupois; and that the resistance of the air, within certain limits, is nearly in the duplicate proportion of the velocity of the resisted body, that is, as the square of the velocity.

A light, hollow globe, the size of a 12 lb. shot, was fixed at the end of the arm, and a weight of $3\frac{1}{4}$ lb. hung at M. Ten revolutions being first made, the succeeding twenty were performed in $21\frac{1}{2}''$. The globe was then removed, and a thin plate of lead, equal in weight to the globe, placed in its room; when it was found that a weight of 1 lb. caused the arm to move quicker than before, making twenty revolutions in $19''$, after ten turns had been suffered to elapse. Now twenty revolutions in $21\frac{1}{2}$ seconds, the radius of revolution being 51.75 inches, give a velocity of $25\frac{1}{4}$ inches in a second; whence it is evident that the resistance on the globe is not less than the effect of $2\frac{1}{4}$ lb. placed at M; and, the radius of the barrel being nearly ($\frac{1}{30}$) part of the radius of the circle described by the centre of the globe, it follows that the resistance of the globe is not less than $\frac{1}{30}$ part of $2\frac{1}{4}$ lb. or $\frac{1}{30}$ of 36 oz., which is nearly $\frac{3}{4}$ of an ounce.

In a second experiment, weights in the proportion of 1, 4, 9, 16, were hung on at M; and, after ten revolutions, the following observations were made:—

With $\frac{1}{2}$ lb. at M the globe turned 20 times in $54\frac{1}{2}''$.
That is 10 times in $27\frac{1}{4}''$.
With 2 lb. it turned . . . . 20 times in $27\frac{1}{4}''$.
With $4\frac{1}{2}$ lb. . . . . 30 times in $27\frac{1}{4}''$.
With 8 lb. . . . . 40 times in $27\frac{1}{4}''$.

Thus it appears that the revolutions in the proportions 1, 2, 3, 4, correspond to resistances in the proportions 1, 4, 9, 16, which shows that the resistances are as the squares of the velocities.

74. From this machine and from the ballistic pendulum the following useful results have been obtained:—

1st. When the motions are slow, the resistance to balls of equal magnitude and weight is nearly proportional to the square of the velocity, conformably to the elementary theory of resistances in fluids; for, in this, it is assumed (the balls having equal diameters and densities or weights) that the resistance depends on the number of fluid particles displaced by the moving body in a given time, and on the reaction of each particle against the body, both of which effects vary with the velocity of the body. But the exponent of the power of the velocity expressing the resistance gradually increases as the velocity increases; and, when the shot moves at the rate of 1400 or 1500 feet per second, that exponent attains a maximum, being then 2.125; beyond such velocity its exponent decreases.

2nd. If balls have equal weights, but different diameters, and move with equal velocities, the resistance varies nearly with the surfaces or with the squares of the diameters, increasing a little above that proportion when the diameters are considerable. Hence, if the velocities also are different, the resistance is proportional to the surface and to the square of the velocity; or, r representing the semidiameter of the shot, and v the velocity, the resistance varies with $r^2 v^2$.

3rd. If balls have equal diameters and different weights or densities, the resistances vary directly as the squares of the velocities, and inversely as the weights; or, w representing the weight, the resistance varies with $\frac{v^2}{w}$.

4th. Projectiles terminating with conical or conoidal heads in front experience less resistance than bodies with flat or with hemispherical ends, the diameters being equal.

5th. The resistances experienced by bodies, in moving through fluids, are considerably affected by the forms of the posterior surfaces of bodies; and much useful information on this subject is contained in the "*Nautical and Hydraulic Experiments*" of Colonel Beaufoy, London, 1834.

75. When a body descends in air from a state of rest, its velocity increases for a time by the action of gravity on it; but since the resistance of the air increases also while the velocity increases, it must at length become equal to the accelerative power of gravity, which is con-

stant; after which the body will move uniformly with the velocity acquired at that time. This is called the *terminal velocity* of the body.[a] Making therefore, for a solid shot, the expression $\frac{k\delta}{r\delta'}v^2$ (Art. 59) equal to g, the force of gravity ($= 32.2$ feet), we have

$$v^2 = \frac{r\delta'g}{k\delta}, \text{ or}$$

v (the terminal velocity) $= \sqrt{\frac{r\delta'g}{k\delta}}$;   (*a*)

and from either of these equations we have

$$k = \frac{r\delta'g}{v^2\delta}.$$

It being supposed that the value of k is known, we have a formula for computing the terminal velocity, and conversely, if the terminal velocity were known, the value of k might be found.[b]

76. It has been observed (Art. 59) that the formula for the absolute resistance of the air against a spherical body moving in it is represented by $r^2\pi\delta\,v^2k'$; the resistance is, therefore, directly proportional to the square of the diameter of the shot and to the square of the velocity. Hence the resistance which would be experienced by a shot of any magnitude, and moving with any given velocity, may be found by proportion, if the resistance actually experienced by a ball of given dimen-

---

[a] In strictness a terminal velocity is never exactly attained; but, in a short time from the commencement of the descent, the body acquires a velocity which is extremely near being uniform.

[b] Since, by Dynamics, $\frac{v^2}{2g}$ expresses the height due to the velocity v, it follows that $\frac{r\delta'}{2k\delta}$ is the height due to the terminal velocity of a shot.

The weight of a solid shot in the British service being, to that of a shell, as 1.42 to 1; the terminal velocity of a shell will be expressed by the second member of the equation,

$$v = \sqrt{\frac{r\delta'g}{1.42\,k\,\delta}};\qquad .\qquad .\qquad (b)$$

and the height due to the terminal velocity of a shell by $\frac{r\delta'}{2.84\,k\,\delta}$.

sions and moving with a given velocity be considered as known. Now, from experiments both with the ballistic pendulum and the whirling machine, Dr. Hutton formed a table (Tract xxxvii. Art. 17) of the resistances experienced by a ball 2 inches diameter, moving with different velocities from 5 feet to 2000 feet per second. Therefore, the actual resistance experienced by such ball, when moving at the rate of 1000 feet per second, for example, being 22.63 lbs., if it were required to find the resistance experienced by a 24 lb. ball, whose diameter is 5.6 inches, when moving with a velocity of 1600 feet per second, the proportion to be used would be

$$(2 \times 1000)^2 : (5.6 \times 1600)^2 :: 22.63 : x;$$

the last term x expressing the required resistance in pounds.

If the given velocity is one which may be found in Dr. Hutton's table, the resistances will be proportional to the squares of the diameters merely. No notice is here taken of the continual diminution of density which takes place in the atmosphere from the surface of the earth upwards; but when bodies fall from great heights, also when they are projected vertically or at considerable angles of elevation, it becomes necessary to attend to that circumstance.

77. It is proved by writers on pneumatics that the densities of atmospherical strata, of equal and indefinitely small thicknesses, reckoning from any point down towards the earth, form an increasing geometrical series of terms, while the depths of the strata from the same point downwards constitute an increasing arithmetical series; and Dr. Hutton has shown by an approximative process (Tract xxxvii. Art. 60) that the density of the air at any distance x, in feet, from the surface of the earth upwards, may be represented by $\frac{c-x}{c+x}$, in which $c = 55000$ nearly.

78. From the formulæ which have been given above (Art. 75), it is evident that the terminal velocity of a body descending in a resisting medium may be com-

puted when the value of a certain constant k is known: but as the terminal velocity of a shot takes place when the absolute resistance which it experiences in the medium is equal to the weight of the shot, it follows that the terminal velocity of a shot may be found from the table of resistances given by Dr. Hutton (Tract XXXVII. Art. 17). For, assuming that the resistances are as the squares of the velocities, if we select from the table a resistance, 0.71 lb., which differs but little from the weight 17.477 oz. ($= 1.092$ lb.) of the shot on which the table is formed, and take out the corresponding velocity; the terminal velocity of the same shot may be found by proportion, thus:—

$$.71 : 1.092 :: 200^2 : 61504,$$

and the square root of the last term ($= 248$) is the terminal velocity of the shot.

But, from the formula (a) Art. 75, other things being equal, the terminal velocities vary with the square roots of the diameters of the shot; therefore, if it were required to find the terminal velocity of any other shot, another proportion must be made: thus if, for example, it were required to find the terminal velocity of a 24 lb. shot whose diameter is 5.612 inches, that of the above shot being 2 inches, we have

$$\sqrt{2} : \sqrt{5.612} :: 248 : 415.43;$$

and 415.43 is the terminal velocity of the 24 lb. ball.

The height due to the terminal velocity may be found by the usual formula $h = \dfrac{v^2}{2g}$: and, in this manner, Dr. Hutton's table (Tract XXXVII., Art. 69) was formed.

The terminal velocity of a shell is easily derived from that of a solid shot of equal diameter; and for this purpose, agreeably to the formula (b) Art. 75, Note b, it is only necessary to divide the terminal velocity of the shot by $\sqrt{1.42}$; or, which is equivalent, multiply it by 0.8392.[a]

---

[a] The celebrated Carnot, in his proposition for defending places by means of vertical fire, has entirely overlooked the effect of the resistance of the air in

79. The subject of the penetration of shot into materials is one of considerable importance in the theory of gunnery; and

$$p = \frac{2\,v^2\,r\,\delta}{3\,\mathrm{R}\,g}$$

is a formula expressing the depth to which shot, on striking with a given velocity, will penetrate into an object whose resisting force is supposed to be known. In this formula

- p represents the depth penetrated;
- v the velocity of the shot at the instant of striking;
- r the semidiameter, and
- δ the density of the shot; also
- g ($= 32.2$ feet) denotes the force of gravity.

If the above formula be put in the form $p = \dfrac{2\,v^2\,r\,\delta\,g}{3\,\mathrm{R}\,g^2}$, the specific gravity of the shot may be substituted for δg. The value of R must be determined by experiment. When the resisting material is the same,

$$p \propto r\,v^2\,\delta ; \quad . \quad . \quad . \quad (a)$$

also, when shot of the same density is used, and the resisting material is the same,

$$p \propto r\,v^2 ; \quad . \quad . \quad . \quad (b)$$

or, in the last case, the depth penetrated varies with the

---

producing the terminal velocities of falling bodies; and the author, in his observations on Carnot's work, has endeavoured to expose the fallacy of a system erroneous in principle; any application of which must, for other reasons also, be either enormously expensive or extremely insecure.

The wall which that engineer proposed to be built in the main ditch of a fortress, at the foot of the rampart, being covered by a counterguard, would no doubt form a serious obstacle to the besiegers, supposing it to be entire at the time of an assault: but the author early foresaw that this would not be the case: and it was in consequence of his representations that His Grace the Duke of Wellington, when Master-General of the Ordnance, caused a trial to be made in order to determine whether or not it was possible to breach the wall by a fire of heavy ordnance directed over the covering-work. For this purpose, in 1823, a construction similar to that proposed by Carnot was executed at Woolwich; and in the following year, after six hours' firing of solid shot from eight 68-pounder carronades, and live shells from three 8-inch and three 10-inch howitzers, at the distances of 400 and 500 yards, a practicable breach 14 feet wide was effected in the wall; the tops of the counterguard and of the work in rear were also entirely degraded by the shot which struck them.

diameter of the shot and with the square of the velocity at the instant of striking.

From the hypothesis of M. Poncelet, that the resistance of a material struck by a shot is proportional to the square of the diameter of the projectile, and from a comparison with the results of experiments, it has been found ('*Expériences d'Artillerie exécutées à Gavre,*' chap. 21, sect. 3) that the depth of penetration into oak may be expressed by a formula which, when transformed so that the penetration shall be obtained in English feet, is

$$4.612 \, r \, \delta \, \log. \left\{1 + \frac{v^2}{1076696}\right\}, \quad . \quad . \quad (c)$$

v being, in feet per second, the velocity of the shot at the time of impact, r the semidiameter of the shot in decimals of a foot, and ε the specific gravity of the shot, that of water being unity.[a]

80. The differential equations for the determination of the trajectory of a shot in a medium, which produces a resistance proportional to the square of the velocity, may be seen in the '*Journal de l'Ecole Polytechnique,*' tom. iv., art. *Ballistique*; in Poisson's '*Traité de Mécanique,*' tom. i., no. 211, and in many other treatises on Dynamics; they are not capable of being integrated in finite terms, but the values of the co-ordinates, for a number of points in the curve, may be determined by the method of quadratures, as it is called, in which the differences in the density of the atmosphere at different heights may be introduced, and thus the curve may be approximatively traced. An example is given in Le Gendre's '*Exercices de Calcul Intégral,*' tom. i. p. 330.

---

[a] The volume of the space penetrated by a shot in any material is frequently represented by the *vis viva*, or active force of the shot in motion; in which case it is proportional to the product of the mass of the shot multiplied by the square of its velocity. That is, w' representing the weight of the shot, v its velocity, and g ($= 32.2$) the force of gravity, the volume of penetration varies with $\frac{w'}{g}v^2$; or, r being the semidiameter, and δ the density (for which may be substituted the specific gravity) of the shot, it varies with $r^3 \, v^2 \, \delta$.

The co-efficient in the formula (c) in the text is conformable to the recent experiments indicated in the '*Aide Mémoire d'Artillerie Naval*,' Paris, 1850.

When, however, the elevation of the piece of ordnance is small (not exceeding 10 degrees) the equation to the trajectory is comparatively simple; and in treatises on dynamics it has the form

$$y = x \tan. a - \frac{1}{8 c^2 h \cos.^2 a}(e^{2cx} - 2cx - 1),$$

in which x is the horizontal and y the vertical ordinate of any point in the curve, the origin being in the axis of the bore at the muzzle of the gun; a is the angular elevation of the gun, h the height due to the initial velocity of the shot, e ($= 2.71828$) the base of the Napierian logarithms, and c is put for $\frac{h \, \delta}{r \, \delta'}$, the co-efficient of the square of the velocity in the expression (Art. 59) for the retardative force arising from the resistance of the air, or (Art. 75, Note b) half the reciprocal of the height due to the terminal velocity of the shot.

The following more simple expression for y, which is sufficiently accurate for small elevations, is obtained by substituting for e^{2cx}, its development by Maclaurin's theorem; extending the series to the fourth term only, since c is a very small fraction, and h is very great compared with x:—

$$y = x \tan. a - \frac{x^2}{4 h \cos.^2 a}\left(\frac{2}{3} cx + 1\right),$$

or

$$y = x \tan. a - \frac{g x^2}{2 V^2 \cos.^2 a}\left(\frac{2}{3} cx + 1\right);$$

V representing the initial velocity.

81. If the range on a horizontal plane were alone required, it might be obtained from either of the above formulæ on putting for y, the height of the axis of the gun, at the muzzle, above that plane, prefixing to it a negative sign; the first equation for y in the preceding article may then have the form

$$e^{2cx} - 1 = (8 c^2 h \sin. a \cos. a + 2c) x + 8 c^2 h y \cos.^2 a \quad . \quad . \quad (a)$$

When the horizontal plane passes through the axis of

the gun at the muzzle, y being then zero, the same equation becomes

$$e^{2cx} - 1 = (8\, c^2 h \sin. a \cos. a + 2c)\, x \quad . \quad . \quad (b)$$

The second equation for y (Art. 80) gives, in the same case,

$$x = \frac{3}{4c}\left\{\left(\frac{32}{3}ch \sin. a \cos. a + 1\right)^{\frac{1}{2}} - 1\right\} \quad . \quad (c)$$

For the *point blank* range, when $a = 0$, or the axis of the gun is horizontal, y being the height of the gun above the horizontal plane on which the shot falls,

$$e^{2cx} - 1 = 2cx + 8c^2 hy \quad . \quad . \quad . \quad (d)$$

The equations (*a*), (*b*), (*d*) being transcendental, the value of x, or the required range, can be obtained from either of them only by tentative processes.

82. If x represent the whole range, when consequently $y = 0$, we have from the second equation for y (Art. 80),

$$\sin. 2a = \frac{1}{2h}\left(\frac{2}{3}c\, x^2 + x\right);$$

whence h (the height due to the initial velocity)

$$= \frac{2cx^2 + 3x}{6 \sin. 2a}:$$

or V (the initial velocity)

$$= \sqrt{\left\{\frac{gx}{3 \sin. 2a}(2cx + 3)\right\}} \quad . \quad . \quad (e)$$

Differentiating the same equation for y, we have, for the tangent of the inclination of the trajectory to the horizon at any point in the curve,

$$\frac{dy}{dx} = \tan. a - \frac{cx^2 + x}{2h \cos.^2 a};$$

which, on putting tan. a in the form $\frac{\sin. 2a}{2 \cos.^2 a}$, and substituting for sin. 2 a the above value (when $y = 0$), becomes at the furthest extremity of the range, θ representing the angle of descent,

$$\tan. \theta = -\frac{4cx^2 + 3x}{12 h \cos.^2 a} \quad (f)$$

PART II.   THEORY AND PRACTICE OF GUNNERY.   59

From the expression for the initial velocity above, we have

$$c = \frac{3}{2} \frac{V^2 \sin. 2a - gx}{gx^2};$$

and from this formula, by means of experimental velocities and ranges, the value of c (the co-efficient of the air's resistance) may be found.

In the 'Aide Mémoire Navale' (p. 531) there is given an expression for y, which is equivalent to the second and third values of y in Art. 80, and differs from either of them only by the substitution of a constant K, for $\frac{c}{V^2 \cos.^2 a}$; this consequently restricts the application of the formula to the same piece of ordnance, with equal shot and equal charges. Here a is supposed to represent the true elevation of the trajectory at the mouth of the gun, which is always found to exceed by a few minutes the elevation of the axis of the piece.[a]

83. The time t, of describing any portion of the trajectory, reckoned from the instant of the discharge, is expressed by the equation

$$t = \frac{1}{c\sqrt{2gh}\cos.a}(e^{cx}-1):$$

and when for x is substituted the whole range, the value of t becomes the whole time of flight.

84. The longitudinal and lateral deviations of projectiles from the object aimed at are presumed to depend on the following causes, independently, it must be understood, of the irregularities which may arise from the friction and rebounding of the shot against the surface of the bore:—First, the inequality of friction exercised on opposite sides of the shot's path by the air

---

[a] The following empirical formula is given in the 'Aide Mémoire Navale' (p. 535) for the ranges of shot when the elevation is between 10° and 30°:—

$$P(\sin. 3a)^{\frac{2}{3}} = X.$$

Where P is the range, when the elevation is 30°, considered as that which produces a maximum range, and supposed to have been determined by experiment; a is the given angle of elevation between those limits, and X is the required range.

which it passes through; next, the want of exact sphericity in the shot; and again, the want of perfect homogeneity, on which account the centres of gravity and of the figure are not coincident. To these may be added the diurnal rotation of the earth. The phenomena of rotatory motion in bodies have during more than a century been made the subject of mathematical investigation by many distinguished men of science, particularly in their relation to Physical Astronomy. The rotation of projectiles has been, in a manner purely analytical, treated by Euler and La Grange; but M. Poisson has made the rotations and deviations of military projectiles in particular, the subjects of three elaborate *Mémoires* in the '*Journal de l'Ecole Polytechnique*,' tom. xvi.

85. With respect to the deviation caused by the earth's rotation, it is shown that when the line of aim is directed from north to south, and from south to north, the lateral deviations are westward and eastward respectively; and when directed from west to east, and from east to west, the lateral deviations are southward and northward respectively; that is, in all cases, towards the right hand of the soldier. M. Poisson shows that, in latitudes corresponding to the central parts of Europe, a 10-inch shell weighing 112 lbs., when fired in any vertical plane at an elevation of 45 degrees, with an initial velocity equal to 400 feet per second, will, on this account, at the distance of 1200 yards, deviate from the mark between 2 feet 10 in. and 3 feet 10 in. A 12-inch shell, weighing 200 lbs., fired at an equal elevation with an initial velocity equal to 900 feet per second, will, at the distance of 4400 yards, deviate from the mark between 15 and 30 feet. Hence, in order to strike an object, the projectile should, if this cause of deviation were alone considered, be fired in a vertical plane, making with a vertical plane passing through the piece and the object an angle equal to about three minutes towards the left hand of the soldier.

86. The equations for the motions of a projectile of a given figure, involving the consideration of a rotation on an axis and a non-coincidence of the centres of

gravity and of figure, are too complicated to allow even an approximate value of the unknown quantity to be deduced, with the simplicity requisite for practical utility; and on this account M. Poisson, in his investigations, confines himself to the case in which the projectile differs but little from a sphere and from perfect homogeneity. With respect to a spherical and homogeneous projectile revolving on an axis during its motion of translation, his analysis leads to the conclusion that, if the rotation take place about a vertical axis, the horizontal deviation induced by it is to the left or right of the vertical plane of projection, according as the anterior hemisphere of the shot turns from left to right, or from right to left (the spectator looking from the fire-arm towards the object); this deviation vanishes when the projectile turns on a horizontal axis. When the axis of rotation is horizontal, and perpendicular to the plane of projection, there is induced a vertical deviation of the shot, the effect of which is to elevate or depress the projectile, and consequently to augment or diminish the range according as the anterior part of the projectile turns, from above, downwards, or from below, upwards: this deviation vanishes when the axis of rotation is vertical, and also when a vertical plane passing through the axis of rotation is at right angles to the vertical plane of projection. In other positions of the axis of rotation the vertical deviation bears a constant proportion to the horizontal deviation.

In a Paper on the Deviation of Projectiles, by Professor Magnus, of Berlin, which is published in the Memoirs of the Royal Academy, Berlin (1852), and translated in 'Taylor's *Scientific Memoirs,*' May, 1853, the writer conceives that the observed deviations of shot having a movement of rotation on an axis, as well as of translation in space, may be explained by the direct resistance of the atmosphere combined with a rotation induced in the air immediately about the shot. M. Magnus observes, from experiments ingeniously devised, that, on one side of the projectile, the force arising from the motion of the air which is generated by the rotation, has the same direction as the force of resistance arising from the reaction of the air in the motion of translation; and that, on the opposite side, these forces have contrary directions. When the reacting force is great, compared with the other, the effect on the shot is nearly the same as if the latter did not rotate on its axis; but, when these are nearly equal, a pressure of the air takes place on the side of the shot at which the direct motion of the shot is coincident with the movement of rotation, while on the opposite side the air, being expanded, is strongly driven off. Hence, supposing a spherical shot during its progressive motion to rotate so that the axis of rotation is always in a normal to the trajectory, and in a vertical plane; when the front of the shot turns from left to right, the observer being behind the gun, a decrease of pressure takes place on the right hand, and an increase on the left, and the shot deviates to the right; on the contrary, when the front of the shot turns from right to left, there is a decrease of pressure on the left, and an increase on the right, and the deviation is to the left.

A like explanation is given of the deviations produced when the axis of rotation is horizontal and perpendicular to the plane of the trajectory. If the upper hemisphere turns in the direction of the motion of translation, the projectile descends, and the range is less than if there were no rotation. If the upper hemisphere turns in a direction contrary to the motion of translation, the shot rises and the range is increased.

87. According to Poisson, if a spherical and non-homogeneous shot be discharged from a piece of ordnance without any rotation, the straight line joining the centres of gravity and of figure will, in consequence of the resistance of the air, have its position continually changed by a movement on the centre of gravity, till it coincides with a tangent to the trajectory.[a] If the ball were projected

---

[a] The author trusts he shall be excused for mentioning in this place that, in consequence of the above observations, together with what has been elsewhere in this work stated concerning the importance of shot being fabricated with the utmost attention to perfect sphericity, and uniformity of density

with a small initial velocity of rotation, that line might make entire revolutions about the centre of gravity, or it might make continual oscillations on all sides of a tangent to the path. In this case the excentricity, considered as small, would have an insensible effect on the movement of translation, and consequently would produce only insensible deviations; in order that these deviations might become sensible, it is necessary that the shot should be projected with a great initial velocity of rotation.

---

throughout the mass, the attention of Mr. D. Napier, an eminent mechanician and manufacturer of steam-engines, was attracted to that subject as early as the beginning of the year 1831, as appears by a letter to the author bearing that date; and the circumstance led him to the invention of the ingenious method of forming leaden bullets for rifles or common muskets, by strong compression, from masses of solid lead which have previously been cast in the form of cylinders. By this process the air-bubble, and the irregular inequalities which always occur in bullets when cast, do not exist in balls so formed; one uniform density is produced in them, and the centre of gravity of the ball is coincident with its geometrical centre. It is easy to perceive how much the efficiency of infantry-fire, and indeed of any in which leaden balls are used, would be increased by the employment of missiles formed on this principle.

The following is a brief description of the process:—The lead is first formed by casting it in cylindrical ingots, each three feet long and three quarters of an inch diameter, and each of these is then passed between the circumferences of two pulley-shaped rollers, those circumferences being notched for the purpose of holding and drawing the ingot or cylinder forward. By this operation the cylinder is compressed in the direction of a diameter as much as a quarter of an inch, the diameter at right angles to this being correspondently increased. The cylinder is then passed between two rollers similar to the former, but having on each of their circumferences a number of hemispheroidal indentations: the compression of the cylinder now takes place in the direction of the longest diameter, and the indentations just mentioned cause the cylinder to assume the appearance of a chain of bullets, each having the form of an irregular spheroid, and each being connected with the next by a flat portion of lead about a quarter of an inch thick, which is in part cut through by the machine. Each separate bullet is, lastly, received in a fixed die, and, a moveable die being forced down upon it, the third compression gives to it a perfectly spherical form, the bullets being now connected together only by a portion of lead about the thickness of a wafer, so that they may be easily separated from one another.

Simple musket-bullets, belted rifle-balls and the Minié shot are executed by the machine in the like manner; and since steam-power has recently been applied to this important purpose, it is hoped that the supply of balls so formed will be sufficiently ample to meet all the wants of the service.

It is worthy of remark that, in cast-iron solid shot there is a great want of that homogeneity which is essential to concentricity of gravity and figure: very few such shot, when floated in mercury, rest, as all of them ought to rest, indifferently in any position; but it is highly desirable that some mode should be discovered of obtaining perfect homogeneity, and the importance of this quality will be particularly shown hereafter in treating of the subject of excentric spherical projectiles.

If the projectile be nearly spherical, and if it have that rotation on an axis which is produced by being fired from a rifled barrel, the deviations arise from two causes: the first is due to the spiral movement, by which the axis of the shot and the axis of rotation are not coincident; and the other to the rotation of the shot. The effects of this rotation will be different according as the shot is compressed or elongated in the direction of the axis of the bore. The result of the investigations alluded to in Art. 84 is, that an elongated ball deviates to the *right* of the soldier when the rotation of the upper part is from left to right, and *vice versâ*. A flattened ball (one whose shorter axis is in the direction of the bore) deviates to the *left* when the rotation of the upper part is from left to right, and *vice versâ*.

88. When pointed, or cylindro-conoidal shot are projected from rifled cannon or muskets, it is observed that they have always a lateral deviation towards the right hand of the soldier, and never towards the left. The deviation is, however, always less than that of a spherical projectile fired from a smooth bore; and it is manifest that this is caused by the rifle-grooves having a direction from the left to the right of the soldier, in the upper part of the barrel. It would, however, be proper that experiments were made with barrels rifled in the contrary direction, or from right to left, in order to ascertain if the deviations would, or would not, in that case, be always towards the left hand.

M. Magnus, in a paper which has been quoted before (Art. 86), describes some ingenious experiments from which he ascertained that the axis of an elongated, cylindro-conical shot is not exactly in the direction of a tangent to the trajectory, but makes with it a small angle. He shows that, if the elongated shot had no rotation, the resultant of the air's resistance, passing through the axis between the centre of gravity and the apex, that resistance would tend to raise the apex above a tangent to the trajectory; and that, when a rotation towards the right is combined with the projectile motion, the apex is not elevated, but the longitudinal axis of the shot makes an angle with a vertical plane passing through the axis of the fire-arm, such that the apex is towards the right; the resistance of the air thus presses the centre of gravity towards the same side, or towards the right hand of the soldier, and causes the lateral deviation which is observed, at the same time it causes the apex of the shot to tend below the direction of a tangent to the curve. Professor Baden Powell, in a Paper read at the Royal Institution in London, March, 1854, observes that the Minié projectile being hollowed at the base, its centre of gravity will be nearer the apex than in a solid shot; consequently that the resistance of the air may tend to depress the apex and cause a lateral deviation of the shot towards the left of the soldier.

89. Whatever be the causes which produce the lateral deviations of projectiles, the effects may be considered as being directly proportional to the velocity v of the projectile in its trajectory, and inversely proportional to its diameter $2r$, and its density δ; the deviating force is therefore represented by $k\dfrac{v}{r\delta}$, in which k is a constant depending on the state of the atmosphere and on the nature of the projectile; and the formula for the mean deviations resolved in vertical and horizontal directions is found to be, for each of those ordinates, at a distance x from the gun,

$$\frac{kv}{r\delta c^2 V^2}(e^{cx}-cx-1),$$

in which V denotes the initial velocity, and c has the same value as above stated, Art. 80.

By deductions from experiments with various natures of French ordnance, it is found that, for solid spherical shot, k may be expressed by 0.00125, and, for hollow spherical shot, by 0.00135 ('*Suite des Expériences d'Artillerie à Gavre*,' Paris, 1844).

90. It has been stated that in the parabolic theory the angular elevation of the piece which gives the greatest

F

range is 45 degrees; this is not the case when the shot moves in a resisting medium; for then the maximum range depends upon the resistance as well as upon the initial velocity. The best method of finding the elevation which would give the maximum range is to calculate, by quadratures (see the works quoted in Art. 80), the ranges due to several different angles of elevation; and when, thus, two limits of the elevation have been found between which the greatest range is produced, the exact elevation required may be determined by the usual process of interpolation.[a]

In this manner it has been found, by the writer of the *Mémoire* on Ballistics in the '*Journal of the Ecole Polytechnique*,' tom. iv., that when the initial velocity is 1595 feet per second, the elevation which, by the formula, gives the greatest range is 31° 59′ 6″; but, in proportion as the initial velocity is diminished, the angle which gives the maximum range is greater. In fact, when the initial velocity was 500 feet per second, the elevation producing the maximum range was 42° 14′. This augmentation of the elevation, with the diminution of the velocity, is exactly what might be anticipated, since the resistance of the air would diminish with a diminution of velocity, and the trajectory would approach nearer to a parabolic form. It may be observed that, in the Deal experiments, the greatest ranges were obtained (conformably to the deductions from theory) when the elevations were 32°. See Table I. Appendix D.

91. Writers on Dynamics have shown, that the

---

[a] By differentiating an equation of the trajectory, considering x and a (the horizontal co-ordinate and the angle of elevation respectively) as variable quantities, and making $\dfrac{dx}{da}$ equal to zero, Dr. Robison obtained (*Encycl. Brit.*, 5th edit., Art. PROJECTILES) an equation equivalent to

$$\frac{4mh}{\sin. a\,(4\,m\,h\sin. a + 1)} = \text{hyp. log.} \frac{4\,mh + \sin. a}{\sin. a};$$

and the value of a obtained from this equation gives, approximatively, the elevation producing the maximum range. In this manner was formed the table for maximum ranges given in Hutton's Tracts (Tract XXXVII., Art. 118).

ascending and descending branches of the trajectory are not similar to one another, as they are in the parabolic theory; the tangents at points in the ascending branch making smaller angles with a horizontal line than those at corresponding points in the descending branch. The former branch, if continued downwards, has an asymptote which is inclined to the horizon, and the latter has an asymptote which is in a vertical position. It follows that, when the axis of the gun is inclined to the horizon, the vertex, or highest point in the trajectory, is nearer to the remote extremity of the range than to the place of the gun.

92. The differences between ranges in a resisting medium and those which should result conformably to the parabolic theory are exhibited in the following Table; and from these it appears that the ranges, instead of varying as the squares of the velocities (Art. 45, Note), are in a lower proportion than the velocities; while, from 800 feet to 1600 feet (the most ordinary velocities with guns), the ranges are nearly as the square roots of the velocities.

The Table, which has been computed from theory, exhibits the ranges corresponding to several given velocities with which a 24 lb. shot may be projected in a resisting medium, and also the ranges in vacuo, the elevation of the piece being in both cases 45 degrees.

Initial Velocities in Feet per Second.	Ranges in Air, in Yards.	Ranges in Vacuo, in Yards.
200	350	415
600	1,760	3,731
800	2,230	6,632
1,000	2,650	10,362
1,600	3,754	26,528
2,000	4,340	41,450
2,600	5,180	70,050

From the experiments carried on at Deal in 1839 the range of a 56-pound shot, with a charge of 16 lbs. (windage = 0.175 inch), consequently having an initial velocity of 1622 feet per second, and fired at an elevation of 32°, was 5720 yards. By the parabolic theory it would have been 24,475 yards, or nearly 14 miles.

93. Having given the principal formulæ for projectile motion, as well in air as in vacuo, it is intended now to offer a series of observations on the most important circumstances in the practice of naval gunnery, and to state the means by which the highest degree of efficiency in it may be attained.

These subjects are, for the sake of convenience, disposed under the following heads:—

    I. The effect of Splinters.
   II. The effects of Heavy and Compound Shot in increasing the force of Impact.
  III. The employment of Double and Triple Shot in action.
  IV. The effects of Windage on correct Firing, and on the Velocity of Shot.
   V. Effects of the Length of Gun on the Velocity of Shot.
  VI. On Recoil and Preponderance.
 VII. The effects of Wads.
VIII. Penetration of Shot into Materials.
  IX. On Cylindro-conoidal, and Excentric Spherical Projectiles.

94. I.—*The effect of Splinters.*—It appears to have been a general rule to use the full charges of powder at the commencement of an action, and to reduce them as the guns became heated. This is a very proper precaution to guard against accidents in continued, quick firing; but a general rule to commence action with the full charges may, in some cases, interfere with the important effect that might be produced from splinters. The prodigious ravages occasioned by splinters, in naval actions, are such, that we should study as much as possible, consistently with other views, to reap the fullest effect from so destructive an agent; and this depends very much upon the degree of velocity with which the balls penetrate.

95. In close action, shot discharged from large guns with the full quantity of powder tear off fewer splinters than balls fired from the same nature of guns with reduced charges. This may be familiarly exemplified by firing a musket or a pistol charged with a bullet through panes of glass, at different distances, or with different charges. Superior velocity will make a clean round hole, without breaking or even cracking the plate; but a certain reduced celerity will dash the glass to pieces.

In firing into masses of timber, or any solid substance, *that* velocity which can but just penetrate will occasion the greatest shake, and tear off the greatest number of, and largest, splinters; for, as in the brittle glass, the parts struck by a solid shot, moving with great velocity, are driven out before they communicate motion to the circumjacent parts of the substance. This is particularly the case with respect to the impact of shot on plates of iron, as will be shown hereafter. Whereas by shot of inferior weight, or moving with less velocity, or by both of these circumstances duly combined, the rents, shakes, and ravages made in perforating timber will be more extensive; a greater number of splinters, and those of larger size, will be torn off, and apertures more difficult to plug will be made: while, in the perforation of iron plates, large and ragged fractures will be formed, seams and rivets will be broken to a greater extent, and fragments will be driven on board, which are far more destructive of life than if the shot had been discharged with velocity sufficient to cause it to perforate both sides of the ship. In close actions with heavy guns charges may therefore be, with advantage, greatly reduced. In fact, the velocities communicated to solid shot by charges exceeding one-third of their weight are too great to produce the maximum effect from splinters; and charges of one-sixth, from heavy guns, are found quite sufficient, at 600 yards, to drive single and even double shot through the thickest part of the side of any ship. Many other examples might be given here, but the subject will be fully treated in the section on the Penetration of Shot.

96. The above observations require, however, to be modified according to circumstances; their justness depending on the position of the enemy and on the degree of steadiness with which the firing is carried on during an action. If no greater force were wanted than that which is capable of penetrating the side of an enemy's ship with the greatest effect from splinters, those observations will hold good. Vessels armed with the heavier natures of guns, in close, steady, broadside action may certainly use reduced charges, if, as in attacking an enemy at anchor, the relative positions do not rapidly change: also, since the velocity of a shot from a 32-pounder, at close action, with the service charge, is more than sufficient to penetrate any mast, we are at liberty to sacrifice some of the superabundant power, by diminishing the charge, for the purpose of obtaining an increased effect from the splinters which may be driven off from masts, spars, or any other timber, as well as from the ship's sides. But, while we keep this object in view, we should particularly provide for the destruction of masts, dismounting of guns, and ruining their carriages, breaking beams, and penetrating into masses of timber in firing obliquely, or at great distances (see Art. 104), in which cases full charges are required. Nothing, moreover, can be spared in force with any nature of gun up to the 12-pounder inclusive: with the higher natures, charges of one-quarter of the weight of the shot may be occasionally used with advantage, but in no case for raking or diagonal fire.

97. To increase still further the effect from splinters, *hollow shot* have been proposed, and these may be used with great advantage from guns of large calibre. For whilst, by their magnitude, fractures or apertures of great extent may be made in the side of an enemy's ship, their relatively lower momenta, arising from the diminished weight, render them far more conducive to the formation of destructive splinters than solid balls of equal diameters. But it may be doubted how far hollow shot are sufficient substitutes for solid shot at great distances; the requisite accuracy and certainty

of practice, that is, the greatest power of range with the least elevation, being attainable only by the heavier projectiles.

It may also be doubted whether or not we are right in the general adoption of shell-guns for the bow and stern armament of steamers, in preference to long and powerful solid shot guns, possessing, in a higher degree, power of range, accuracy and penetrating force, in distant firing; and whether, in the armament of our navy in general, we have not so reduced the number of guns, in proportion to the *displacement*, in consequence of the introduction of shell-guns into the broadside batteries of some classes of ships (the " Thetis " and " Castor " class in particular), as to place those vessels in disadvantageous circumstances when opposed to such as, though equal to them in size, carry a greater number of guns, and which, consequently, possess a degree of superiority, from the power of making a plurality of discharges in a given time, and from the facility with which the less unwieldy guns may be worked relatively with the number of men forming the crew according to the present establishment. The " Thetis " and " Castor," the displacements of which are, respectively, 1528 and 1283 tons, carry only 36 guns (see Art. 249). The " Indefatigable," with a displacement of 2043 tons, has only 50 guns. Our largest frigate, the " Vernon," with a displacement of 2082 tons, carries only 50. Where are our 60-gun frigates? The French had (in 1850) twelve 60-gun frigates afloat, and five were being built:[a] they had, moreover, twenty-five 50's. The number, names, and ratings of the ships and vessels afloat, building, and in commission, have not been given in L'Etat Général de

---

[a] *Frigates Afloat.*—L'Andromaque; La Belle Poule; La Didon; La Forte; L'Indépendante; L'Iphigénie; La Minerve; La Persévérante; La Renommée; La Sémillante; L'Uranie; La Vengeance. (See Notes by the author on ' L'Enquête Parlementaire sur la Situation et l'Organisation des Services de la Marine Militaire.' Paris, Jan. 30, 1852.)

Frigates Building, January, 1850.—La Victoire, at *L'Orient*; L'Entreprenante, at *L'Orient*; La Sémiramis, at *Rochelle*; La Pallas, at *L'Orient*; La Guerrière, at *Brest*. The Immortalité and the Melpomène, both of them first-class frigates of 60 guns, have, since, been laid down.

la Marine since 1850; but, according to a statement in L'Enquête Parlementaire, the French had then (1852) fifty-six frigates and thirty-four corvettes. (See Appendix C.) We had then only twenty-six frigates above the class of 44 guns, and of this class only twelve! The war armament of the United States' frigate "Congress," and eleven others of that class, is upwards of 50 guns; and of the "Independence," 60; the peace armament of this ship is 54 guns. The first-class frigates in the Russian navy carry 60 guns; the second class, upwards of 50. The Dutch have their 60-gun frigates, though in peace they carry only 50. In the Turkish navy there are ten of from 54 to 64 guns. In the French service, the razées carry 58.[a] Our "Warspite," reduced from a 76, originally with a displacement of 2490 tons, but lighter now, at the load-line, by 2 feet 4½ inches, equivalent to 436 tons, carries only 50 guns. The "Arrogant," with a displacement of 1872 tons, far from retaining the full gunnery power due to that tonnage, is reduced to 46 guns.[b]

98. II.—*The effects of Heavy and Compound Shot in increasing the Force of Impact.*—To gain an increase of penetrating power, or to augment the force of the blows with which shots strike, balls made of heavy material may be used with advantage. A shell, filled with lead, will produce a greater blow than an iron shot of the same diameter, discharged with an equal quantity of powder; for the penetrating power, which (Art. 79) varies with the product of the density and the square of the velocity of the body, will, in this case, be proportional to the weight of shot. The former may also be made to range farther than the iron ball, from being better able to overcome the resistance of the air, and, consequently (see Art. 60), retaining longer the superior velocity communicated by a greater charge

---

[a] An accurate account of the armament of foreign navies will be given on a future occasion.

[b] The whole subject of the armament of the British Navy will be fully considered on a future occasion.

of powder.[a] From this an increase of accuracy also results, because the same range may be produced with less elevation; and accuracy is gained as elevation is reduced.

From experiments made in France it was found that a shell filled with sand, and weighing about 46 lbs., being projected with a charge of 9 lbs. from an 8-inch Paixhans gun, with an initial velocity of 1200 feet per second, and at an elevation equal to 1°, ranged twice as far as a shell filled with lead weighing 75 lbs. when projected with a charge of 7 lbs., the initial velocity being 840. But, at an elevation of 3°, the ranges were nearly equal; and at 5° (the weights and charges being as before) the range of the shell filled with lead surpassed that of the other, amounting to 1800 yards, while the range of the shell filled with sand was only 1400 yards. This proves that great initial velocity is not alone essential to the power of range, but that much depends upon the weight, or inertia, of the shot, by which the velocity of projection is longer retained. We shall revert to this subject in treating of hollow shot.

99. Heavy, or, as they may be called, compound shot, may be used with great advantage against vessels, such as we are told have been constructed for floating batteries, with sides so thick as to be proof against the known penetration of ordinary shot; for balls made of heavier matter (strong shells run full of lead) will penetrate farther, and be more formidable in every way. If the sides of these vessels be made of solid masses of timber, then red-hot solid shot should be used; and if they lodge, the effect will be rendered the more certain; but as the furnaces for heating shot on board of ships cannot supply as many balls as may be discharged, cold shot must also be used. The effect of these will be more or less formidable, in proportion to their powers of penetration and their momenta, both of which will be

---

[a] The like is evidently true in comparing the effects of solid and hollow shot.

greater with the compound heavy shot than with solid iron balls.

The weight of a 56-pounder shell, having its cavity filled with lead, is 65.443 lbs.; for the whole diameter being 7.475 inches, and the thickness of the metal one-sixth of the diameter, the weight of the shell itself is 38.827 lbs., while the weight of the solid ball of lead which occupies the cavity is 26.616 lbs. Now (Art. 68) the velocities of balls of equal diameter, but different weights, and discharged with equal quantities of powder, being inversely as the square roots of the weights, the velocity of a homogeneous 56-pounder shot will be to that of the shell when filled with lead, as $\sqrt{65.443}$ to $\sqrt{56}$; so that, the velocity of the solid shot with a charge of 14 lbs. being by computation 1596 feet per second, the velocity of the shell filled with lead, with the same charge, would be 1476 feet per second. But the velocities of balls of equal weights being directly as the square roots of the charges, or the charges being as the squares of the velocities; if we increase the charge from 14 lbs. in the ratio of the squares of those numbers, we get 16.36 lbs. for the charge, which will render the velocity of the shell filled with lead equal to that of the homogeneous 56-pounder shot. This is a simple way of increasing the weight of metal, as it is popularly called; and on particular occasions it may be used with considerable advantage.

In Table IV. (Appendix D.) there will be seen the results of experiments made at Deal in 1839 to determine the ranges of homogeneous shot and of shells filled with lead from a 56-pounder gun whose dimensions are given. From these it appears that with charges of 17 lbs., at an elevation of 15°, the first graze of a solid shot was 4001 yards, and at an elevation of 35°, 5200 yards; while, at the former elevation, the first graze of a shell filled with lead was 4029 yards, and at the latter elevation 5600 yards.[a]

---

[a] At the siege of Cadiz the French used shells filled with lead, which, discharged with a velocity of about 2000 feet per second from howitzers (one of which is now placed as a military trophy in St. James's Park), ranged to the

In chasing, or in flight, heavy compound shot, fired from long guns with charges increased in the proportion just shown, will increase the power of range, and consequently the chance of carrying away a mast or a spar, at a very considerable distance.[a] For such special purposes as these, it may, therefore, be expedient to provide a few compound shot, with directions how to use, and injunctions to economize them.

100. Oblong shot, of a cylindrical form and terminating at each end in a hemisphere, may be used occasionally either against ships or in breaching batteries by land. The great uncertainty of such shot, however, from irregularities in their flight, would not, at great distances, be compensated by the advantages contemplated; but they might sometimes be used at moderate distances with great effect, particularly by vessels carrying only 9 or 12-pounder guns; for the weight of metal may thus be increased so much as to be capable of carrying away a mast, which a round shot, discharged from the same gun with equal velocity, could not bring down; and the oblong shot would make large irregular fractures in the side of an enemy's ship which it would be very difficult to plug up.

The windage of oblong shot should not be greater than .14 of an inch (the windage of a carronade), so that, in fact, a 12-pounder gun discharging a solid of iron of 24 lb. weight (which its oblong shot might be), with a charge of 4 lb., would be capable of producing as great an effect as a 24-pounder carronade; and if the oblong shot were to strike with its larger dimensions, the effect would be much greater. Oblong shot should be cylindrical, with hemispherical ends. For a 12-pounder, a shot of the shape above indicated, and exactly

---

distance of three miles. Greater ranges than this have been obtained with shells in experiments recently (1852) carried on at Shoebury Ness in this country. See Art. 188.

[a] A very distinguished naval commander mentioned to the author that he knew a person who had served in an American privateer which, having been out of shot, and being unable to procure a supply of iron balls, used leaden shot as substitutes. This person always mentioned with great surprise the superior effect of leaden balls.

equal to the weight of two 12-pounder balls, should be 2.935 inches long in the cylindrical part, and the total length 7.338 inches. In close action the chance of hitting with separate balls is so great, that two round shot should then be preferred to a single oblong projectile; but when the distance is such, compared with the magnitude of the object fired at, that from the divergence of the balls, both cannot hit, it may frequently prove advantageous to increase the weight of metal with the smaller nature of guns in the manner proposed.

The proposition for using oblong shot is not new. Some very extensive experiments were made at Landguard Fort, in 1776, to ascertain the comparative ranges and accuracy of long and round shot with 42, 24, 18, and 12 pounders. It is not intended to notice the trials made with the three higher natures of ordnance, because it is not necessary to increase the momenta of their balls for any naval service, but some of the experiments made with the 12-pounder are given in the following Table :—

Nature of Shot.	Weight of		Diameter of Shot.	Elevation.	Recoil.		First Graze.
	Powder.	Shot.			Ft.	In.	
	lbs.	lbs. oz.	Inch.				Yds.
Round . . .	5	11 8	4.3	P.B.	4	5	498
Long . . .	5	24 3	4.4	P.B.	7	4	335
Round . . .	5	11 9	4.3	1°	4	4	818
Long . . .	5	24 3	4.4	1°	6	2	774
Round . . .	5	11 10	4.24	5°	4	5	1,789
Long . . .	5	23 3	4 4	5°	6	0	1,879

The weight of the oblong shot was more than double that of the ball, but the windage of the former was about one-tenth of an inch less. The ranges with the oblong shot were, of course, considerably less than with the balls; but it does not appear from these trials that the variations, or errors, were greater than they are with round shot; and it follows that, within the limit of elevation in the Table, we may, on suitable occasions, safely adopt a practice which would increase the effect in a considerable degree. Oblong shot would occasion

PART II.   THEORY AND PRACTICE OF GUNNERY.   77

great ravages in the revetment-wall of a fortress, and be more likely than round shot, by the great shake they would occasion, to bring it down after the part intended to be breached is disunited from the adjoining parts by round shot fired with full charges. There can be little doubt, also, that oblong shot are far preferable, in naval warfare, to bar-shot.

If an oblong shot, twice the weight of a round shot of equal diameter, be fired with the same charge, the velocity of the former will be less than that of the latter in proportion as the square root of the weight is greater. That is, the weights being as 2 to 1, the velocities will be as 1 to $\sqrt{2}$. But the effect of impact, measured by the volume of penetration, being as the weight of the shot and the square of its velocity (Art. 79, Note), it follows that, with equal charges, the effect of the oblong shot will be just equal to that of a round shot of half its weight.

101. III.—*Employment of Double and Triple Shot in Firing.*—The reason which has been given for the occasional use of oblong shot leads to a consideration of the practice of firing two or more balls at once from the same gun, and it is here intended briefly to notice its defects as well as the case in which it may be advantageous.

It must be observed, first, that when two equal, or unequal, shot are projected at one time, they issue from the gun with different velocities which may be determined by the theory of the collision of bodies. Thus, the velocity communicated by the expansive force of powder to a single shot equal to either of the two if these are equal, or to that which is nearest to the charge if not so, being determined, suppose by the second formula in Art. 68 (the denominator under the radical sign being the weight of that shot), let that velocity be represented by V; then, e representing the force of elasticity, which from the results of some experiments may be expressed by the fraction $\frac{1}{2}$, if v represent the required initial velocity of the shot nearest the charge and v' that of the other; also, if w and w' represent the weights of those shot respectively, we have, by dynamics,

$$v = \frac{w-ew'}{w+w'}V \quad \text{and} \quad v' = -\frac{w+ew'}{w+w'}V.$$

When the two shot are equal to one another, these formulæ become

$$v = \frac{1}{2}(1-e)V \quad \text{and} \quad v' = \frac{1}{2}(1+e)V,$$

or

$$v = \frac{4}{9}V \quad \text{and} \quad v' = \frac{5}{9}V.$$

Now the velocity of a shot from a 32-pounder medium gun, whose length is 9 feet and weight 50 cwt., with a charge of 8 lbs., is, by computation, equal to 1693 feet per second (windage = 0.198); while with two balls and an equal charge the velocities would be 753 feet and 941 feet. The greatest of these velocities would, therefore (since the velocities are as the square roots of the charges), be the same as that of a shot propelled singly by only 2½ lbs. of powder. But the point-blank range of a 32-pounder gun, with a charge of 2½ lbs., would be less than 300 yards, while, with a charge of 8 lbs., it is 610 yards; thus, with the former charge and a single shot, or with the latter charge and two shot, there would be required an elevation exceeding one degree to obtain ranges equal to the point-blank range produced in a single shot by a charge of 8lbs. In using two shot, therefore, there is a loss of accuracy, not only arising from the divergence of the balls, but also from the greater elevation which is required. This should deter us from adopting such practice, excepting at close quarters, when it is scarcely possible to miss; and never with carronades after one breeching (for they should always have two) is broken; nor, as in fighting the weather side, when the inclination is much in the direction of the recoil.

102. In practice with two shot there are some other causes of irregularity, which it is proper here to point out, that will further show how little it can be depended

upon, excepting in close action. Besides a difference of range, frequently amounting to above a hundred yards, there is also a lateral divergence, which, in firing at ships, at long ranges, or at any objects whose magnitudes are not sufficient to subtend the divergence of the balls, incurs a considerable risk of missing entirely, and a certainty of not hitting with both. If the balls touch each other on quitting the gun, which very frequently happens, some deviation must arise, according to the nature of the collision. If it be direct, one ball will be accelerated, and the other retarded by the blow. But this will rarely happen—it is more likely that the blow will be oblique; from which both balls will diverge, very considerably, in directions compounded of the projectile velocity, and the force and direction of the collision.

In firing two shot, another cause of deviation, still more destructive of accuracy, is operating during the passage of the balls along the bore; particularly when the windage is great. The shot nearest to the charge impelling the other, will evidently be forced to one side of the cylinder, as at A, pressing the outward ball to the other side, B, of the bore. From this it is manifest that the outward ball B must receive an oblique impetus on its departure from the muzzle, which will disturb, from reaction, the direction of the other. This deviating cause may either operate in a lateral or in a vertical direction: in the former case it will affect the line, in the latter case the elevation of the shot's departure, and consequently the length of range; or the error may exist in both directions.[a]

Fig 2.

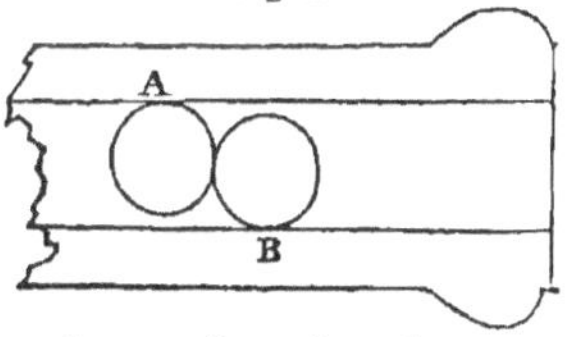

103. It has been found, from experiments on board the "Excellent" (1847), that, in firing triple shot with three times the elevation required for a single shot, the

---

[a] It is said to be a rule in the naval service of the United States not to fire double shot, even at 400 yards, from guns of less calibre than 32-pounders (Ward, p. 82); and never with 32-pounders under 46 cwt. (Bureau of Ordnance, 1845).

balls took very good directions, but one fell rather short of the mark, and the others over or beyond, and that the carriage did not appear to be distressed. The gun used in firing the three shot was a 32-pounder, 9 feet 6 inches long, and weighing 56 cwt.; the charge was 6 lbs. It would probably be unsafe to fire three shot from a 32-pounder, weighing but 50 cwt., even though the charge were reduced to 5 lbs.; and if the charge were less than 5 lbs. the penetrating effect of the three shot would be uncertain, at least beyond 50 yards. It has been observed that, on running alongside of an enemy, the first broadside might be given with triple shot; and it is supposed that triple shot might answer the purpose against flotillas of gun-boats at distances not exceeding 400 or 500 yards.

104. After a few firings from the same gun a certain diminution of the charge becomes necessary; for, the gun being already heated, the powder on being introduced ignites more completely than at the commencement of the firing, and thus its expansive force is increased: hence considerable provision of reduced charges among the quantity of *filled* ammunition should always be made; it would even be convenient if some regulated proportions between the numbers of the full and of the reduced charges, to be used with single and with double shot, were made and adhered to. In the United States' service there are used two reductions of the full charge, or three charges in all, viz., $\frac{1}{3}$rd, $\frac{1}{4}$th, and $\frac{1}{5}$th of the weight of the shot, and the cartridges of all three are painted with different colours. In the French service, of the total of filled ammunition, $\frac{5}{10}$ths are full charges, $\frac{3}{10}$ths are charges first reduced, and the remainder are still further reduced. Care should, however, be taken that the diminished charges are not rendered too low, since the penetrating force may, at the same time, be so far reduced as to render the shot incapable of perforating a side of an enemy's ship. This circumstance occurred in the action between the Shannon and the Chesapeake, on the termination of which some shot from each of the ships were found sticking in a side of the other. In

this case either the charges were too much reduced, or the weight of metal projected was too great. Had these shot perforated the side which they struck, the most serious consequences might have ensued.

105. In all naval actions the essential point is to ensure penetrating force sufficient to drive the projectiles through, or, at least, into the opponent's ship, whatever may be her position. This point appears to have been somewhat lost sight of on several occasions during the late war; and it may be feared that, in carrying so far as is proposed the use of hollow shot and a plurality of projectiles, the momentum will be too much diminished, particularly in the employment of shell-guns. Naval actions are subject to such sudden and unforeseen mutations in the positions of the contending ships, and are liable to such great alterations in the distances of the ships, that until the affair becomes close and the struggle is near its termination, no two rounds can well be fired exactly under the same circumstances. An affair which had commenced in parallel order and in direct broadside battle, at well-ascertained distance, may speedily change, altogether, its circumstances and aspects. The enemy may be suddenly brought into a position in which he is exposed to oblique, diagonal, or raking fire; and for this an experienced commander should always be prepared. In every case he should ensure the penetration of his projectiles into and through the whole dimension of the enemy's ship, however she may be struck. Accordingly, he must not carry too far the theory that the greatest ravages which can be produced by a projectile on the object which it strikes is when the striking velocity is just sufficient to penetrate a stationary body, because if this be too strictly followed in direct action, there would be no perforation at all in the event of the enemy being struck obliquely. It is not possible to foresee, it would not be prudent to prescribe, what particular descriptions of projectiles or what particular charges may be required to take proper advantage of different positions; but it has often happened that penetrating force has been improperly sacrificed to

the practice of double or triple shotting; and thus when, by superior skill, the antagonist's ship has been suddenly brought to a position in which it is exposed to a diagonal or a raking fire, the penetrating force has not been sufficient to perforate it, or otherwise do upon it effectual execution. Seamen are prone to overload their guns with shot in close action; and many instances of this may be found in descriptions of naval engagements which might have terminated differently had the projectiles, which only stuck in the side, gone through the ship. This practice accounts, in some degree, as Captain Simmons observes in his '*Ideas as to the Effect of Heavy Ordnance, &c.*' (page 69), for the comparatively little damage done to the hulls of line-of-battle ships in general actions.

106. In the action between the Chesapeake and the Shannon, iron bolts, as well as chain and bar shot, were found sticking in without penetrating the side of the latter, and it is remarkable that there was found sticking in the Chesapeake a pump-bolt which had been fired from the Shannon, no doubt without authority. Five 18-pounder round shot struck without penetrating into the Chesapeake; but all the 32 lb. shot (carronades) that struck perforated the sides; none of these last struck the masts high up, and only one the bowsprit; but some 18-pounder round shot struck the masts 20 and 25 feet above the decks. So far it appears that the carronades were, probably in part owing to their small windage, truer than the 18-pounder guns.

The report from which these facts were taken was made by the carpenter of the Shannon, by order of Lieutenant, now Rear-Admiral Wallis, an officer of the highest distinction and merit, who served on board the Shannon during the action. The effect of the shot on these ships, with respect to penetration, affords many useful practical deductions, and will accordingly be given in full on a future occasion.

107. Double shotting may be used with all 32-pounder guns above those of 32 cwt., at distances not exceeding 400 or 500 yards, but the most efficient prac-

tice with two shot is at 300 yards; the 32-pounders of 32 cwt. and 25 cwt. should not, however, be so used beyond 200 or 250 yards. Shell-guns of 65 and 60 cwt. cannot use double shot beyond 200 yards; and at that distance they can only use hollow shot, with a charge of 5 lbs.

With charges of 6 lbs. for the 32-pounders of 56 cwt. and 50 cwt.; of 5 lbs. for those of 45 cwt. and 42 cwt.; 4 lbs. for those of 32 cwt., and 3 lbs. for those of 25 cwt., the guns recoiled with a considerable degree of violence; the charges, therefore, have been reduced to 5, 4, 3, and $2\frac{1}{2}$ lbs. respectively. With these charges the recoil was found to be easy; the guns may be fired throughout an action without injury to their tackling, or to the ring-bolts, and at 400 yards the penetration is sufficient.

With double loadings of round-shot and grape, when the shot is put in first, the projectiles range more together than when the reverse process is used; such loading requires, however, more elevation to be given to the gun than when single shot are used, on account of the grape-shot impeding the flight of the other.

A double load of grape from the same gun ranges tolerably well together for 300 yards. With a double load of case-shot, even with half a degree more elevation than when a single load is used, a great many balls will not range above 100 yards to the first graze; within this extent they lose much of their velocity, and few reach an object at 200 yards.

A 32-pounder gun of 56, or of 50 cwt., double shotted with charges of 6 lbs., requires, at 400 yards, $1\frac{1}{2}°$ of elevation; at 300 yards 1°, and at 200 yards half a degree; and, in general, half a degree must be added with any double loading, to the elevation required with single shot.

For round and grape, at 400 yards, there is required $1\frac{1}{2}°$ of elevation, and at 200 yards half a degree. These projectiles range well together at a target, but they should not be used at a greater distance than 150 yards, on account of their dispersion, and the differences of their striking velocities and penetrating forces.

With single grape (a plurality of projectiles) at 400 yards, the elevation required is 1°, a full charge of powder being used. With double grape, at 400 yards, and the reduced charge, the elevation required is 3½°: at that distance double grape scatters so much as to make very bad practice.

108. There is some difference, particularly with heavy guns, in the times required for loading with single and with double shot; and when vessels are engaged so closely as to be within double shot action; when, consequently, the most rapid firing should take place, it is evidently of the utmost importance to reduce as much as possible the time of loading. Other things being equal, a ship which can, in the shortest time, give a second broadside, will have a great advantage over her opponent; and, in order to effect this, it is recommended to have ready a few rounds of double shot sewed up in thin canvas, with a little oakum between the shot, and the whole secured by a line passed round between them, no wadding being used.

109. IV.—*The effects of Windage on Correct Firing and on the Velocity of Shot.*—Windage is the difference between the diameter of the shot and the calibre of the piece of ordnance; and its amount for the different natures of ordnance is given in Table XVII., Appendix D.

From experiments which have been made on this important subject it appears that very great differences in the initial velocities of shot arise from very small differences in the windage: that, with the degree of windage formerly established in the British service, no less than between one-third and one-fourth of the force of the powder was lost; and that, as balls are often less than the regulated size, it frequently happened that half the force of the powder was lost by unnecessary windage.

110. In the account of the Artillery Practice carried on at Gavre, between the years 1830 and 1840, there is given a method of determining by experiment, approx-

imatively, and from an empirical formula, the effect of windage on the initial velocity of shot. The substance of the method is given in the subjoined note, in which the weights and dimensions are expressed in English denominations.[a] From this it may be shown that a windage equal to 0.0512 inch reduces the initial velocity in the ratio of 1.14157 to 1, and a windage equal to 0.3189 inch reduces it in the ratio of 1.6377 to 1; comparison being made, in both cases, with a velocity resulting from the windage being zero.

111. From subsequent experiments at L'Orient with the 30-pounder long gun, it appeared that the loss of velocity on account of windage was independent of the weight of the projectile, and was proportional to the lunaric area between a great circle of the shot and a transverse section of the bore. Now R being the semi-diameter of the bore, and r that of the shot, $\dfrac{R^2 - r^2}{R^2}$ will express the ratio between the lunaric area and the area

---

[a] In the experiments, spherical iron balls of two different diameters, but of weights all accurately equal, were projected from the same piece of ordnance (30-pounder, French, long gun), each ball, of both natures, with two different charges of powder. The diameter of the smaller ball was 6.1734 inches, and of the larger 6.4415 inches; the diameter of the bore was 6.4923 inches; consequently the windage of the smaller ball was 0.3189 inch, and of the larger 0.0512 inch. The axis of the gun was horizontal, and the initial velocities of the balls, computed from the measured ranges, were as follow:— With a charge equal to 10 lb. 13.4 oz., the mean velocity of the smaller, in feet per second, was 1060, and of the larger 1588. With a charge equal to 5 lb. 6.5 oz., the mean velocity of the smaller was 932 feet, and of the larger 1263 feet.

Now, with equal charges and equal weights of shot, the velocity depends on the windage, which may be represented by $\dfrac{a}{A}$; in which $a =$ the diameter of the shot, and A the diameter of the bore. Then, if V denote the initial velocity on the supposition that the windage is zero, and V' any one of the above computed initial velocities, we should have the values of V on substituting, successively, the above values of V' in the empirical equation

$$V' = V\left[1 - \left(1 - \dfrac{a^2}{A^2}\right)^{\frac{3}{2}}\right].$$

When the charge is 10 lb. 13.4 oz. we obtain for the smaller ball 1732 feet, and for the larger 1729 feet. Also, when the charge is 5 lb. 6.5 oz., for the smaller ball 1529 feet, and for the larger 1519 feet.

of a section of the bore; and the loss from windage may be represented by

$$K \cdot \frac{R^2 - r^2}{R^2},$$

K being a constant which depends on the charge.

The following is a table of the values of K for the 30-pounder gun, with different charges:—

Charge.	K for Solid Shot, weighing 32 lbs.	K for Hollow Shot, weighing 23 lbs. 13 oz.
lbs. oz.		
2 3	813.4	891.5
5 8	789.7	881.9
11 0	655.3	710.6

Major Mordecai found, however, that the above formula does not give results in accordance with his experiments at Washington; and he agrees with Dr. Hutton in considering that the loss of velocity by windage is proportional to the linear value of the windage, or to the difference between the diameters of the bore and shot. (See Art. 67, and the Note.) He also found that, with a given windage and corresponding charges, the loss of velocity decreased as the calibre of the piece was increased.

In some experiments which were made in Sweden in 1846, with cylindro-conoidal shot projected from a rifled gun, the loss of velocity occasioned by windage was, in part, avoided by pasting paper round the shot. The initial velocities of balls strapped to sabots are rather greater than those of balls without sabots, probably because the force of the charge is increased in consequence of the escape of the gaseous fluid, by the windage, being in a great measure prevented.

112. The prejudicial effect which great windage has upon accuracy arises from the reflections of the ball in passing along the bore. These will manifestly be greater in proportion as the difference between the diameter of the shot and the calibre of the gun increases. From

these reflections a shot acquires a sort of zig-zag motion, and does not generally quit the cylinder in the direction of its axis. If the last bound be upon the lower part of the bore, the angle of the shot's departure will be increased; if above, diminished: these affect the length of range. If the last reflection be on either side, right or left, of the bore, the direction will be altered; and in every case the friction arising from these rubs will give the ball an irregular whirling motion, productive of inequalities in the resistance of the air, unless the rotation be at right angles to the direction of the projectile's flight. A high degree of windage may have been necessary when guns were imperfectly bored, and shot incorrectly cast; but proofs of the accuracy of both are now so rigid, that no allowance need be made for any errors of construction.[a]

113. It is well known that the diameter of the shot is the datum from which the quantity of windage and the calibre of the gun were originally determined. Thus the diameter of the shot was supposed to be divided into 20 parts—one of these was allowed for windage, and the calibre of the gun made equal to the sum of that diameter and windage. Thus the diameter of a 24 lb. shot being 5.547 in., $\frac{1}{20}$ of it, or .277, was allowed for windage, and $5.547 + .277 = 5.824$ in., was the calibre of the gun.

This degree of windage was exclusively observed, till the proprietors of the Carron Company, availing themselves of the discoveries made in the ballistic experi-

---

[a] The statements contained in the five articles succeeding that to which this note belongs, concerning the defects and inconveniences arising from the great and anomalous windages formerly allowed in the British artillery, led in 1817 to the appointment of a select committee at Woolwich for the purpose of taking the subject of windage into consideration (see Art. 117), and the result was the adoption of a lower and more uniform windage for both services; but, as something in this respect remains to be done, the author has retained the whole of the articles on windage which were given in the former editions of this work, partly because those articles serve in some degree to show, historically, the progress made in correcting the evil, and partly because they may be useful in pointing out the importance of giving the most scrupulous attention to accuracy in boring guns as well as in the formation and preservation of the shot.

ments in favour of reduced windage, determined to apply the principle in the construction of their carronades; but as no alteration was made in the windage for guns (which should have been done by increasing the diameter of the shot), the Carron Company lessened the calibres of their new ordnance, to admit ordinary shot with reduced windage, by which great confusion and complications arose in the windage, not only of shot but of shells also. The shells for guns were of the same diameter as the shot, but carronade shells were smaller by about .1 of an inch. Now if carronades admit ordinary shot, there can be no reason why their shells should be smaller;—that is, there can be no reason why the windage should be greater for a shell than for a shot. This indeed is practically admitted by the windage allowed for mortars and howitzers, which is only .15 of an inch for all natures, excepting the $4\frac{2}{5}$ inch, which is .2 inch. The considerations which should be provided for in determining the degree of windage are, 1stly, the expansion of shot by a white heat; 2ndly, the incrustation of a little rust; 3rdly, the foulness of the cylinder in continued firing; and 4thly, the thickness of the tin straps for wooden bottoms.

114. The degree of expansion of shot by white heat is found to be about $\frac{1}{70}$ of the diameter of the shot for a 24-pounder; $\frac{1}{76}$ for a 16-pounder; and $\frac{1}{82}$ for a 6-pounder.

The French allow 1 line 6 parts for the windage of heavy or siege ordnance, and 1 line for field ordnance. The former, reduced to English measure, is .133 of an inch, and the latter about .088, which are not much more than one-third of ours.

The calibre of a French 24-pounder is 5 inches, 7.625 lines, and the windage (= 1.5 lines) or about $\frac{1}{45}$ of the calibre of the gun.

The calibre of a French 8-pounder is 3 inches 11 lines, or 47 lines; and the windage is one line, or $\frac{1}{47}$ of the calibre of the gun.

If then the windage must be proportioned to the shot or gun, it may safely be decreased to $\frac{1}{40}$, or at most to $\frac{1}{35}$ of the calibre. These degrees of windage, and the

PART II.   THEORY AND PRACTICE OF GUNNERY.   89

corresponding diameters of balls, are shown in the following table; in the seventh column of which the present degree of windage is inserted to show the differences.

Nature of Gun.	Calibre of Gun.	Windage $\frac{1}{35}$ of Calibre.	Diameter of Shot $\frac{34}{35}$ of Calibre.	Windage $\frac{1}{70}$ of Calibre.	Diameter of Shot $\frac{39}{40}$ of Calibre.	Present Windage $\frac{1}{20}$ of Shot's Diameter.
Prs.	in.	in.	in.	in.	in.	in.
42	7.018	.2	6.818	.175	6.843	.33
32	6.41	.183	6.227	.16	6.250	.30
24	5.823	.166	5.657	.145	5.678	.27
18	5.292	.151	5.141	.132	5.16	.25
12	4.623	.132	4.491	.115	4.508	.22
9	4.200	.12	4.08	.105	4.095	.20
6	3.668	.105	3.563	.092	3.576	.17
4	3.204	.092	3.112	.08	3.124	.15
3	3.013	.086	2.927	.075	2.938	.14
1	2.019	.058	1.961	.05	1.969	.09

115. But it may be repeated, why should the difference between the diameters of the ball and cylinder, which may be necessary to permit the shot to enter freely, vary with different cylinders? The impediments to its entrance bear no relation to the magnitude of the cylinder or ball, excepting the expansion of the latter by heat; and that, in the largest balls, is only, on an average, $\frac{1}{70}$ of the calibre, or $\frac{1}{2}$ of the proposed windage $\frac{1}{35}$. The windage of carronades recognises this reasoning; for it is fixed at .15 of an inch for the three higher natures, viz. for 68, 42, and 32-pounders; .14 is allowed for 24-pounders; .12 for 18 and 12-pounders; and .15 is the allowance for the windage of all the higher natures of mortars and howitzers.

Suppose the windage of a 24-pounder carronade (.14 of an inch) were adopted for all the higher natures of ordnance; this in the 24-pounders would be $\frac{1}{43}$ of the calibre; but the French allowance (which, as we have seen, does not vary with the nature of the gun) is only $\frac{1}{35}$ of the calibre in the 24-pounders. Now no degree of foulness from quick firing, and no coating of rust that should be suffered to remain, or can well remain on a shot, is equal to .07 (about $\frac{1}{15}$) of an inch, which a windage of .14 would allow. If, therefore, we adopt .14 for the windage of heavy guns, we proceed upon a certainty in the improvement; but, perhaps, .13 of an inch is windage sufficient for all natures of guns and carronades, from the 68-pounder carronade to the 12-pounder inclusive; and downwards from the 9-pounder gun inclusive, .1 or .11 may be recommended. The proportions which these degrees of windage bear to the calibre of the gun are shown in the Table below.

Nature of Ordnance.	Calibre of Ordnance.	Proposed Windage.	Proportion of proposed Windage to Calibre of Gun.
Prs.	in.	in.	About
42	7.018	.13	$\frac{1}{53}$
32	6.41	.13	$\frac{1}{49}$
24	5.823	.13	$\frac{1}{44}$
18	5.292	.13	$\frac{1}{40}$
12	4.623	.13	$\frac{1}{35}$
9	4.2	.1	$\frac{1}{42}$
6	3.668	.1	$\frac{1}{36}$
4	3.204	.1	$\frac{1}{32}$
3	3.013	.1	$\frac{1}{30}$
1	2.019	.1	$\frac{1}{20}$

This important reform would admit of a considerable saving of powder; for those velocities which at present require charges of $\frac{1}{4}$ the weight of the shot, might be produced with smaller quantities of powder. We should also have greater accuracy in practice, and consequently greater effect; but a series of experiments should be made, before this scale is altered, to ascertain whether shot of the proposed magnitude will roll freely home, repeating these trials with a great many new shot, and in a great many guns. Also trials should be made with the new shot heated to white heat, and with cold shot under the various circumstances of foulness from quick firing, and the incrustations of rust.

116. But whether a new scale of windage be adopted or not, it is important that shot be protected as much as possible from those vicissitudes, and from that treatment, which are continually operating to diminish their size,

and consequently to increase the windage. Every possible precaution, therefore, should be used to keep shot from rust by painting or greasing them, and to keep them as dry as possible—perhaps some better way of storing them might be found than throwing them promiscuously into the shot-lockers. When shot are cleaned, the rust should be carefully rubbed off, and not *beaten* off with hammers. This mode of cleaning shot, as practised in the arsenals, should be strictly prohibited; and also the more destructive method of putting a number of rusty balls into a rotatory machine, in which they are literally ground to very reduced magnitudes, and the windage prodigiously increased. It is useless to consider the effects produced by a minute regulation of windage so long as such practices continue. These precautions should be strictly observed, for if the degree of windage demand rigid and nice restriction, it is no less important to secure it from increasing, and this can only be done by carefully protecting the shot from all destructive agents and operations.

117. The preceding remarks (Arts. 111–116) on windage, having been brought under the consideration of the Master-General of the Ordnance in 1817, his Lordship referred the paper to the consideration of a select committee of artillery officers, who stated in their Report that "they were very desirous that experiments should be made with a view to ascertain to what extent the benefits which the author of this work had anticipated could be realised." The committee, therefore, proposed to the Master-General to be permitted to make a course of experiments on this subject, commencing with field artillery, and for that purpose recommended that a proportion of shot of various increased magnitudes should be provided. These measures having been approved, a course of experiments, founded upon the suggestions communicated by the author, was instituted accordingly.

After repeated trials with a 6-pounder, a 9-pounder, and a 12-pounder, at 300, 600, and 1200 yards, it was proved, "that, with charges of powder one-sixth less than usual, the larger shot, and smaller windage, produced rather the longest range."

Recourse was also had to the ballistic pendulum, to discover the proportional excess of momentum of the larger balls over the smaller; and the result, after a very satisfactory course of experiments, assisted by the scientific research and well-known mathematical abilities of Dr. Gregory of the Royal Military Academy, corroborated the trials by ranges, leaving no doubt of their accuracy.

In consequence of these trials the Committee fixed the quantity of windage for field guns at one-tenth of an inch, the same as had been suggested by the author. (See the Table, Art. 115.)

Now it is clear that this improvement may either be applied to save one-sixth part of the quantity of powder provided for field-service, without diminishing the power of range, and consequently to economise, without detriment, the means of transport of ammunition; or the alteration may be applied to produce longer ranges, if this be preferred to the economical consideration. This preference has, very properly, been given, and the established charges adhered to accordingly.

118. A great collateral advantage has followed from this correction of windage. It was at first apprehended that the increased effects arising from the additional weight of shot and diminished windage would injure brass guns; but it is quite the reverse. With the reduced quantum of windage guns are much less injured, and will last much longer than formerly; and this has been so well ascertained that, in consequence of this correction, it is now proposed to abandon the wooden bottoms to which shot were fixed for the purpose of saving the cylinder, substituting for them the paper cap taken off the end of the cartridge. This being put over the ball is quite sufficient to keep it from rolling or shifting; whilst, by supporting or fixing it thus, the centre of the ball coincides with the axis of the cylinder, and the space for windage is reduced to a complete annulus, which admits of the percussion from the charge

being equally received, and which prevents, or very much reduces, that injury, or indentation, which the cylinder receives when the ball touches it on the lower part only; for when the space for windage is all above, the action of the powder is exerted in a manner to produce that effect on the bore, near the seat of the ball, which is soon discovered in brass guns. With iron guns the expedient of putting a paper cap over the ball is of less importance as far as the preservation of the bore is concerned, iron guns being not so susceptible of injury as those of brass; but with respect to accuracy of fire, the expedient will not fail to be productive of good effect, and it is important that it should be adopted in Naval Gunnery.

119. Experiments were also made for the purpose of ascertaining what advantage would arise from diminishing the windage of heavy iron ordnance in the manner proposed.

For this purpose larger shot, having various degrees of diminished windage, were fired from an iron 24-pounder, ship or garrison gun, with the service charge and with reduced quantities of powder; when it appeared so advantageous that the benefits arising from a reduced rate of windage should also be extended to the higher natures of guns, that the Committee recommended a diminution of windage to .15 of an inch for all natures of siege and garrison guns from the 12-pounder upwards.

The importance of this measure is perhaps sufficiently established by this decision; but as it is moreover necessary, in cultivating improvements in practical science, to consult the experience of actual service as much as possible, and to state any strong facts that may have occurred to guide and support us in such pursuits, it is important to exhibit here the opinion given by a late very gallant and highly distinguished artillery officer

---

^a Lieutenant-Colonel Sir Alexander Dickson, K.C.B., who commanded the battering-train at all the sieges in the Peninsula.

"Most fully," says this officer, "do I subscribe to your proposal for diminishing the windage of our ordnance, by casting shot of a larger size than the

in a letter on this subject, and to show upon what solid grounds that opinion was formed.

120. There could be no doubt of the advantages, in accuracy of practice, and either in economy of powder or in increase of effect, that would result from extending this improvement to naval guns; but here we were met by a very formidable obstacle, which, so long as it was permitted to exist, prevented the adoption of the improvement both in the naval and in the land-service artillery. This obstacle arose from our having adopted a nature of ordnance (carronades) of smaller calibre than guns of the same nature. From this extraordinary anomaly it resulted that shot for carronades, considerably above gauge, were received in the arsenals and distributed in the service. Ships were actually sent to sea stored with new and perfectly clean shot, which, when required for use, were incapable of being introduced into the carronades, the· incrustation of rust, which in those days was permitted to form itself on the shot in the racks, being frequently thicker than the amount of the windage.

---

present. Experiments are certainly necessary to ascertain how small this windage may be made with safety; but the regulation on this head, as practised by the French, affords an excellent scale to follow, and the rates you propose, of .13 inch for heavy ordnance and .1 inch for light ordnance, would, I am convinced, answer perfectly well, and be conducive to future accuracy.

"I remember a very convincing proof of the advantage of high shot, which I think is worth mentioning.

"The battering-train was assembled at Almeida previous to the siege of Ciudad Rodrigo, and there being a great deficiency of transport to bring forward the shot from the rear, it became a very important object to collect as many as possible from the shot belonging to the fortress, of which there was a considerable quantity, and of an infinity of calibres. In order, therefore, to obtain every shot that would answer, gauges of the full diameter of our 24-pounder bore were used, and every shot was selected as serviceable that passed through these gauges; so that many of the highest balls chosen would not, when heated, go into the gun; and this I ascertained by trial. When this operation had been completed, the selected shot were again tried with a gauge rather smaller than the ordinary 24-pounder shot gauge, and all that went through it were rejected as too small. The number of very high shot, however, amounted to two or three thousand; and as I know they were brought forward in the latter part of the siege, it has always been to the use of these shot that I have attributed the singular correctness of the fire in making the smaller breach; for although the battery was from 500 to 600 yards distant from its object, every shot seemed to tell on the same part of the wall as the preceding one. Now, this was not the case in firing with common shot at such a distance; for some struck the wall high and others low, although the pointing, as the best gunners have assured me, was carefully the same."

The matter could not rest thus. Land-service shot of magnitudes corresponding to the windage, .15 of an inch (the quantity fixed by the Committee for siege and garrison ordnance), cannot be issued to the navy, because they cannot be used with carronades; and as this species of ordnance is frequently used in land-batteries also, two sorts of shot were rendered necessary for land-service. The only perfect remedy for these evils was to re-bore, or *ream out*,[a] all existing carronades to equality with the calibres of guns; to introduce one uniform system of windage for all natures of ordnance; and, consequently, to have but one scale for the gradations of shot.[b]

Great objection was at first raised to the proposition, but it was soon afterwards carried into effect. The advantage thus obtained seems to have led to the practice of boring up the old guns to the next higher calibre; in order that, with a diminution of windage, the weight of metal projected might be increased. To the first of these objects the bored-up guns owe their success, and the favour with which the expedient has been received.

We shall, hereafter, inquire whether bored-up guns are efficient substitutes for new ordnance of the calibre to which they have been enlarged, and whether they are not still to be considered as imperfect guns which, however useful they may be as garrison artillery, should, except on particular occasions, be abandoned as naval ordnance.

121. The differences between the calibres of guns

---

[a] A term for scraping out.

[b] The author proposed the measure to the late General Millar some time before the publication of the first edition of this work; and a letter, of which the following is an extract, forms part of a considerable correspondence which took place on the subject.

Extract of a letter from the late Major-General Millar to Sir Howard Douglas, dated Woolwich, March 2nd, 1817.

"I have well considered the subject, and agree with you that the only radical way to overcome the difficulty about windage is to bore up the carronades as you originally suggested. In forwarding your idea I have proposed to General Sir Thomas Blomefield to have one prepared in this way, to show the committee when they meet, as a matter of experiment, to ascertain the practicability of the proposition, and what may be the cost should the measure be generally adopted."

and carronades of corresponding natures are shown in the following Table:—

Nature of Ordnance.	Gun.	Carronade.	Difference of their Calibres.
	Calibre, in.	Calibre, in.	in.
42-pr.	7.018	6.84	.178
32-pr.	6.41	6.25	.16
24-pr.	5.823	5.68	.143
18-pr.	5.292	5.16	.132
12-pr.	4.623	4.52	.103

To equalize the calibres of guns and carronades, therefore, the thickness of metal of the latter would only be reduced by half the differences in the last column of the preceding table, or about .07, or .08 of an inch. This trifling reduction of metal in the chase only might certainly be made without incurring any greater risk of bursting than at present: for it is a fact pretty well known, that although numerous accidents have occurred in action from the bursting of guns, yet carronades have never been known to fail; they break their breechings, draw their bolts, and split their sides, but do not burst; and this power of resisting their charges may be established by referring to the original proof returns.

The weight of metal required to be taken out in order to enlarge the calibres of carronades to an equality with guns of the corresponding natures is very small, as may be seen in the following table; and though such diminution of weight is not in itself desirable, it cannot be objected to, considering the advantages to be derived from the equalization of windage.

Nature of Ordnance.	Weight of Carronade.			Weight of the Metal required to be taken out from Carronades.	
	cwt.	qrs.	lbs.	lbs.	
42-pr.	22	1	0	26.2	
32-pr.	17	0	14	17.1	or about 1 lb. per cwt.
24-pr.	13	0	0	12.8	
18-pr.	9	0	0	9.7	
12-pr.	5	3	10	5.2	

122. When the former edition of this work was published, our arsenals were stored with vast numbers of carronades which it would have been wasteful to condemn and expensive to replace, while many circumstances were favourable to the measure of enlarging the calibres. First, it was found that the calibres of great numbers of carronades had become so enlarged by rust and wear that they could be rendered serviceable only by being reamed out; and, secondly, to enlarge them by boring, it was found unnecessary to transport them to the boring-houses, for machines were devised to perform the operation as they lay on the skids. It may be added that a change, which had been suggested by the late General Millar, had then been made in the chambers of carronades by forming there a hollow frustum of a cone, connecting the cylindrical bore with the hemispherical bottom of the chamber. The existing carronades were, therefore, bored up and chambered in this manner; but being still, in some measure, an imperfect nature of arm, their retention in the service was always viewed as a temporary measure, to be abandoned when more perfect ordnance could be substituted for them.

123. The chamber is a recess formed to contain the charge of powder, at the lower extremity of the bore of a piece of ordnance, in the direction of the axis and of less diameter than the bore. The conical, or gomer, chamber was proposed expressly for mortars, with a view of placing the centre of gravity of the projectile in the common axis of the chamber and bore; a position which, on account of a mortar having considerable elevation, the shell necessarily takes in descending on the conical surface. This puts the projectile, till it begins to move by the force of the powder, in contact with the conical contraction of the chamber, and allows the windage, in a plane passing through the centre of the shell perpendicularly to the axis of the bore, to assume a form correctly annular, consequently it enables the whole charge to act directly upon the projectile. But this is not the case when a piece of ordnance is in,

or nearly in, a horizontal position, since then the projectile rests on the lower part of the bore, and the space between its surface and that of the bore takes a lunaric form, so that the axis of the charge does not pass through the centre of the projectile.[a] On this account, and because, in horizontal firing, the cartridge, on being rammed home, is apt, when reduced charges are used, to shift from its place in the conical chamber, the French reject the gomer form for howitzers, and make the chamber perfectly cylindrical. This is the case with the Paixhans gun, which is employed in the naval services both of France and of the United States, and the diameter of the chamber is equal to that of the bore of a gun of the next lower calibre.

Even the cylindrical chamber is attended with some disadvantages; it must be large enough to contain the full charge, and, therefore, when reduced charges are used, there will be a void space between the powder and the ball, which will affect unfavourably both the extent of the range and the accuracy of the firing.[b] The same space is occupied by a quantity of the ignited fluid before the projectile is moved, and this increases the velocity of the recoil. It may be added that the ordinary sponge cannot, at the same time, clean the bore and the chamber of a piece of ordnance; and thus accidents from explosions are very likely to occur in using it. The evil may, in a great measure, be obviated by the employment of a sponge of a conical form, nicely fitting the chamber and carefully used; the extremity of this entering the contracted space at the bottom of

---

[a] In order to remedy this defect, it is the practice on board the "Excellent" to set the shot well home with a wad, thus forcing it up the lower part of the conical surface, and, as far as possible, into the chamber.

[b] This is a fact well known to all who have had much experience in mortar and howitzer practice; and the author has found the defect so great, that in order to remedy it, he was long since led to insert, between the charge and the shot, a tompion, or piece of wood having the form of the unoccupied part of the chamber. This expedient appears to have been adopted in the French artillery service; and it has been ascertained that a tompion so placed has caused an increase in the initial velocity of the shot.—*Expériences d'Artillerie exécutés à Havre*, 1830-1840. *Première partie*, ch. xii. sect. v., p. 61. Paris, 1841.

the bore, and being carefully turned on its axis, more completely removes the residuum of the charge. Yet the most efficient form for a broadside gun is evidently that in which the diameter of the bore is equal throughout its whole length, it being understood that there is a sufficient thickness of metal at the lower extremity of the gun to resist the charge.

124. With respect to the quantity of windage which may be considered proper for all natures of iron ordnance from the 12-pounder upwards, there cannot be any doubt that .15 of an inch, as already voted for siege and garrison ordnance, would be quite sufficient. Consulting experience on this head, we know, that .14 of an inch is the quantum regulated for the 24-pounder carronades. We are equally certain that this allowance has been very rigidly observed; for any departure from it would be an abandonment of the principle which renders carronades so efficient, with such small charges. We know also that the windage of all French iron ordnance, reduced to English measure, is only .133 of an inch; and the American windage is rather a copy of the French practice than of ours.

125. It has been apprehended that any diminution in the windage of naval guns would occasion more frequent accidents from bursting than at present. This opinion is founded upon the fact, that the action of the charge is increased by its taking place in a more confined space than when the windage is greater. To this it is only necessary to reply, that the danger cannot be greater than at present, if the economical tendency of the measure, as fully admitted in this objection, be taken advantage of as it ought to be; for satisfied, as appears from what has been said, that the velocities of shot, according to our present practice, are quite sufficient (rather, indeed, too great than too little), and knowing that the increased action of the same charge when inflamed in a more confined space, would communicate greater velocity than with the present windage, we should, in reducing it, diminish also the charges to such quantities as should be proved by experiment to be capable of maintaining

the practice given in our present tables, and put the savings of powder in the pockets of the public. In this way, guns would not be more liable to burst than with the present system, if shot be carefully proved.

126. The preceding observation introduces an important remark on the modes of *gauging shot*. With ring-gauges, shot not perfectly spherical may pass in one direction, or section, but not in all; and may therefore jam in rolling into the cylinder of the gun. Shot-gauges should therefore be cylindrical, as in the French service; for balls which roll freely through gauge-tubes, cannot possibly jam in the cylinder of the piece, if it be perfect.

127. What has been said on the subject of windage shows that very minute differences in the magnitude of the shot produce, by the variations in the quantity of windage, very great differences in the practice. It follows, therefore, that to preserve shot as much as possible in their primitive magnitude and condition, is no less important than the nice and very minute investigations from which the proper dimensions of balls have now been determined; and consequently that it is of vast importance to protect shot on board of ships from the destructive effects of rust. The ease, and trim, and sailing of vessels, do not admit of their shot being stowed, in large quantities, any where above the water-line, or near the extremes. The great store of shot, therefore, must be in the lockers, where they are at present kept; and *there* it is impossible to prevent their becoming, in some degree, corroded by rust: but a sufficient number of balls for at least fifteen double-shotted rounds, should always be kept on deck. Now these balls being the first for use, should be preserved from rust as effectually as possible, and cleaned with every degree of care. Shot from the lockers should not be used till the deck shot shall have been expended, and the uppermost balls in the lockers should be frequently examined, cleaned from rust, and greased or painted: by such means, together with frequent inspections, an enlargement by rust would be effectually prevented.

This would contribute much to accuracy of practice; for, considering the prodigious resistance of the air upon the surface of a cannon ball, moving with the usual service celerity, it is evident that the most minute protrusion or inequality on that surface must occasion a very irregular whirling motion, and consequently a great deviation from the intended direction. The rifle principle is a contrivance to reduce the errors arising from the unequal action of resistance on any superficial inequalities in the ball, by giving it a rapid rotation perpendicular to the axis of projection; but it is found that a ball of uniform consistency or homogeneity, smooth and perfectly spherical, may be projected from a plain barrel with as much accuracy as a shot from a rifle. (See the Note to Art. 87.) Numerous very delicate experiments with musket-barrels show that accuracy depends very essentially upon the ball being smooth and perfectly spherical; and this being the case with respect to the minute inequalities on small bodies, it may easily be imagined what must be the effect of large and rough protrusions of thick coats of rust, upon the surface of a cannon ball whose velocity is resisted by such a vast force. Independently, therefore, of any consideration relating to windage, it is of great importance to keep shot as clean and as smooth as possible.

128. V.—*The effects of the length of Gun on the velocity of Shot.*—It is known, both from theory and practice (Arts. 61 and 66, Note), that, with equal charges, and guns of equal weight, but of different lengths, the velocity of the shot increases with the length of the bore, but in a small proportion only—the velocity varying, as appears from experiments, in a ratio between the square and cube roots of that length; and again, experience shows that the velocity of the ball increases with the charge up to a certain amount, which depends on the nature of the gun; and that, by further increasing the charge, the velocity gradually diminishes till the bore is quite full of powder, the recoil always increasing with an in-

crease of charge. These circumstances lead us to consider the relative merits of long and short guns.

The partiality in favour of short guns for naval service had its origin in a misapplication of some of Mr. Robins's maxims and observations. That distinguished civilian, by whom the theory of gunnery has certainly been much improved, having had no practical knowledge of artillery service (as he observes in the preface to his 'Proposal to Lord Anson,' printed in 1747), it is not surprising that the maxims which he deduced from his own investigations should frequently tend to a diminution of that accuracy and practical efficacy which persons better acquainted with the profession would have studied to increase.

129. Mr. Robins says, that neither the distance to which a bullet flies, nor its force at the end of its flight, are much increased by very great augmentations of the velocity with which it is impelled; and therefore that, in distant cannonade, the advantages arising from long guns and great charges are but of little moment. Another of his maxims is, that whatever operations are to be performed by artillery, the least charges with which the object in view can be effected should always be preferred.

In conformity with these maxims, Mr. Robins became a great advocate for reducing the length and weight of guns, for the purpose of enabling ships to carry ordnance of larger calibre.

Mr. Robins also asserts that the objections which attach to short guns in batteries on shore, viz. ruining the embrasures, hold not at sea, where there is no other exception to the shortness of the piece than loss of force, which, in the instances here proposed, is altogether inconsiderable; but this is a very erroneous conclusion, for with carronades it is frequently impossible to place them so that, when fired, the flash should not burn the rigging, or fire the hammocks in the barricade, without bringing the vent under the rail over the port, so as to expose the hammocks to the vent-fire, and the lock to injury when the piece is fired under depression. This is a very serious and a very well-supported contradiction of Mr. Robins's conclusion; and the danger and inconvenience arising from the shortness of carronades have been so often noticed, that it would be very expedient to make hereafter the 24, and particularly the 32-pounder carronade, a little longer in bore, and to add something to the flash-rim.

130. The advocates for the short-gun system support their theory by quoting the deductions from the ballistic experiments, from which it appears that the superior velocity of shot discharged from long guns is reduced to an equality with the velocity of balls from short guns, after passing over certain spaces, and that the extreme ranges do not much differ; but the main principle which should govern our choice of naval guns is, to prefer those which, with equal calibre, possess the greatest point-blank range; and the practical maxim for using them should be, to close to, or within that range, and depend upon precision and rapidity of fire. This is the most simple and the most efficacious use of artillery: it avoids all the difficulty of determining, and the uncertainty in regulating elevation, and it is therefore of the greatest importance in naval gunnery.[a]

131. In the maxims above alluded to, the advantages arising from the superior accuracy of long guns at moderate distances appear to have been overlooked. From experiments carried on at Sutton Heath in 1810, it was found that the greatest point-blank range of a 24-pounder, whose length was 9 feet 6 inches, was, to the first graze, 275 yards, and the mean of three such ranges from the same gun was 248 yards; while the least point-blank range of a 24-pounder, whose length was only 6 feet 6 inches, was, to the first graze, 200 yards, and the mean of three such ranges with the same

---

[a] Dr. Hutton found that, with a gun of 20 calibres in length, the velocities continued to increase with the increase of charge till the latter amounted to ⅞ths of the weight of the shot, and occupied ¼th of the length of the bore; after this the velocities decreased. With a gun of 15 calibres in length, a charge equal to one-half the weight of the shot, and which occupied .275 of the length of the bore, gave the greatest velocity.

gun was 222 yards. The charge was the same (6 lbs.) in all the cases. Again, from the Deal practice in 1839, with 32-pounder guns of the same length respectively as the 24-pounders just mentioned, with equal charges (6 lbs.) and equal windage (.175), the elevation being 1 degree, the range of the longer gun was 853 yards, while that of the shorter one was only 734 yards. A superiority of range with the longer guns does not always take place, but the above circumstances will serve to show that, at moderate distances, a long gun might be laid point-blank, while a short one might require an elevation of half a degree.

132. A comparison of the point-blank range of Sir William Congreve's 24-pounder with a gun of equal calibre and of the old pattern, appears at first sight to militate against the above conclusion; but on examination it will be found that this is not the case. The first of these guns, which weighs 40½ cwt., is 7 feet long, and, from its great thickness about the chamber, the *dispart*[a] subtends an angle of not less than 5 degrees. The other, which weighed 50 cwt., was 9 feet 6 inches long. The windages were equal, and the charges one-third of the weight of the shot; and from experiments made at Sutton Heath in 1813, the point-blank ranges of the Congreve gun varied between 505 and 640 yards, while those of the heavier gun were about 370 yards. The Congreve gun was also tried with two 24-pounders of Sir Thomas Blomefield's construction, one 8 feet and the other 7 feet 6 inches long, and weighing respectively 43 cwt. and 40 cwt.; and at point-blank ranges, the first graze of each of these two last was between 370 and 400 yards. To Sir William Congreve certainly belongs the merit of having first proposed a different distribution of the metal in his guns, adding much to the part about the place of the charge, and diminishing the quantity in the chase. He also placed the trunnions

---

[a] The difference between the semidiameter of the gun at the base-ring and at the swell of the muzzle. This, the length of the gun being the radius, is the tangent of an angle varying generally from 1° to 2¼°, which the line of metal makes with the axis of the bore.

farther back than they were in the old ordnance of a corresponding nature; but his assumption that "the propelling or reacting power of a piece of ordnance may be increased by augmenting the quantity of metal about the charge, though a greater quantity be taken from the other parts, and consequently that a lighter gun may have a greater propelling power than a heavier one, by a judicious distribution of the metal,"[a] is not supported by facts. In the extreme ranges of the shot, the advantage was invariably in favour of the 50-cwt. gun; and with elevations of $2\frac{1}{2}$ and 5 degrees, at which trials were made, the excess of graze was sometimes on the side of the Congreve gun, and sometimes on that of the others. The presumed superiority of the former at the first grazes, when fired point-blank, is ascribed to the *liveliness* of its recoil and to the preponderance of metal at the breech, by which, in being fired, the muzzle of the gun is thrown up. (Captain Simmons, 'A Discussion on the present Armament of the Navy,' p. 10.)

133. With respect to Mr. Robins's maxim (Art. 129), that the force of a ball, at the end of its flight, is not much increased by great augmentations of the velocity with which it is impelled, it may be observed that, if the force of impact alone were the object, in naval actions, the full charges of powder need seldom be used. A charge equal to one-sixth of the shot's weight would be sufficient to drive a ball from any large gun through the side of a ship at 1100 yards; but with this small charge the gun would require twice as much elevation to produce such a range as would be necessary with the full charge, and thus the accuracy of the fire would be sacrificed. To ensure this great object, the charge should be such as will allow the elevation to be the least possible; nor should reduced charges be employed unless the action take place within point-blank range or at close quarters.

The superiority of long guns over short ones, in respect of extent of range, will be readily seen on referring

---

[a] See 'A concise Account of the new Class of 24-pounders proposed by Sir William Congreve,' 1831; *Introd.*, p. i.

to Table III. Appendix D; and it may be depended on that a well-commanded vessel fitted with long guns, will, by keeping at a distance just within the range of such guns, have great advantages over a vessel of superior rate, if fitted only with short guns of equal and even larger calibre.

134. The preference which, during the last war (1793-1815), was given to guns of reduced lengths, less weight relatively to that of the shot, and to comparatively low charges of powder, is, even now, not wholly abandoned. The cause of the preference at the time lay in the practice which then prevailed of pointing guns by the *line-of-metal*,[a] as it is called, which, on account of the dispart, makes a greater angle with the axis of the gun as the latter is shorter; thus causing the axis to be more elevated, and consequently the range to be greater than when a longer gun is used. In the former editions of this work the author objected to this cause of preference, on the ground that the greater elevation induces less accuracy in the practice; and in fact the method of pointing by the line-of-metal is now abolished in the British artillery, the tangent scale being generally introduced, though the *ligne de mire* is still, or was till very recently, retained in the French naval service, the artillerymen allowing for the angle of the dispart according to the elevation or depression of the piece. The preference, however, which is still given to short and, as the author considers them, inferior pieces, leads him in this place to make some observations on the relative merits of guns which, with equal charges, require different elevations to produce equal ranges.

135. In determining the preference which should be given to guns of different lengths, weights, and calibres, and in estimating the charges proper for them, the most important point to be kept in view is the degree of probability of striking an object at a distance which allows the projectile to retain sufficient force for producing the most destructive effects.

---

[a] A line drawn from the top of the base-ring to the top of the muzzle.

136. Exclusive of errors in pointing and laying guns, and those caused by occasional variations in the state of the atmosphere, the irregularities which are produced while the shot is in the bore, by friction, rotation, windage, and the imperfect influence of the charge, with many other circumstances, combine in unknown and variable degrees which theory cannot determine, to affect the probability of striking the object aimed at; and it is only by extensive experience, or by reference to tables exhibiting the mean results of numerous trials under many different circumstances, with all the natures of ordnance in use in the service, that correct judgments can be formed with respect to the probable deviation of the projectile from the vertical plane in which it should move, or the errors, in excess or defect, in the extent of the range, so that the proper means may be used to obtain the requisite accuracy.

M. Piobert, in his valuable work entitled 'Traité d'Artillerie' (tom. ii. p. 270), has made many useful observations on the lateral, as well as the longitudinal, deviations of projectiles; and he has formed, from numerous experiments carried on during several years with various natures of ordnance, and both with solid and hollow shot, some interesting tables exhibiting such deviations, together with the probabilities of striking targets or other objects, of different superficial extent and at various distances. Thus, with 24, 16, and 12-pounder guns (French), at the distance of 656 yards, the charges being one-sixth of the weight of the shot, the average number of hits in 100 rounds were respectively 6.2, 4.9, and 4.7; while, with the same guns and charges equal to one-third of the weight of the shot, the number of hits per cent. in the same object and at the same distance were 8.6, 5.6, and 7.2. On these tables, in part, the author has founded his conviction that the best means of obviating both the longitudinal and lateral deviations is the employment of heavy guns of their respective natures, with full charges, at low elevations, the shot being solid and accurately formed, and the windage as small as possible.

137. Of the two kinds of error, the lateral deviation is by far the most pernicious in naval gunnery; for, in this respect, beyond certain limits the shot would be wholly wasted; while, with respect to the error in extent of range, there is a considerable probability that it may be redeemed by the production of some secondary effect.

The probability of striking an object fired at depends on its superficial magnitude (in a vertical plane perpendicular to the plane of fire). With like guns the probability of striking, at different distances, diminishes rapidly as the distance increases; but against bodies of various magnitudes, at the same distance, the probability increases with the surface when the latter is small (not exceeding 3 or 4 feet square), while it increases in a much higher ratio than the surface when this is great. There is, moreover, a magnitude (the length of a side of a ship, for example) which will exceed by two or three times the extent of any error arising from the lateral deviation at the given distance, and then such an object cannot be missed, provided the error in the length of range should not cause the trajectory to intersect too high or too low, a vertical line passing through the object.

138. When an object moves with great velocity in a trajectory having small elevation above the horizontal plane (the level of the sea, for example), it is evident that, whether the shot fall short of or go over the point aimed at, the errors in the extent of range will correspond to very small errors in a vertical direction at the place of the object; and that the latter errors are so much less as the elevation of the gun is less, or as the trajectory is less incurvated. Hence there may be considerable differences in the length of the range, and yet a great probability that the shot will be intercepted by some part of the body or rigging of the ship, or other object fired at.

If the first graze of the shot falls short of the object, the angle between the horizontal plane and the descending branch of a trajectory having small elevation, is more favourable to the ricochet than the greater angle formed by a more elevated trajectory would be; and it is manifest that a shot, moving in a trajectory of the first kind, might take effect on the hull or rigging of a ship which in the other it might entirely miss. Again, if the first graze should fall beyond the object, the shot whose path is least elevated, if it pass over the hull of a vessel, will scarcely fail to strike some part of the rigging; while one whose trajectory is higher would pass above the whole. The like observation will hold good with the equal ranges obtained from a long and a short gun; the latter, which has greater elevation than the former, describing a higher curve. These two cases suppose the actual elevation not to be disturbed by the floating motion; but it is evident that an error of this nature must, in general, be to the disadvantage of the shorter gun.

139. Ricochet firing, at elevations under 3 degrees or $3\frac{1}{2}$ degrees, with shot from guns near the level of the water, may take place at distances not exceeding 600 yards with considerable effect. But in this practice it is essential that the angles of graze should be as small as possible; therefore great charges should be used, and the elevation of the gun should not exceed 2 degrees. The penetrating power of shot fired à ricochet is, however, much inferior to that of shot fired directly; and therefore in this respect, as well as for correctness of fire, the direct practice is, to a certain extent, the most advantageous.

In firing to windward, the steep sides of the waves are unfavourable to practice; for shot striking those sides do not rise, or they soon lose their velocities; also ricochet firing against boats will be found most effective when the water is smooth.

Table X., Appendix D, exhibits the results of the ricochet practice with solid and hollow shot from on board the "Excellent," in 1838. The extents of the ranges were such as were made before much deflection took place.

140. It has long been imagined by many practical artillerists that the ranges of shot, when fired over a

broad valley or over water, are less in extent than when fired over a level or a gently undulating ground, the charge and nature of the ordnance being alike in both cases; and M. Piobert, in his *Traité d'Artillerie*, has given a table of the comparative ranges of shot in such circumstances, which appear to confirm the opinion. This officer, in endeavouring to explain the cause of the difference in the ranges, suggests that when a shot is fired point-blank, or at a low elevation, over ground which is either level or which rises in the direction of the line of fire, the air driven off in every direction about the ball in its flight must be condensed between its path and the ground; and this condensed air reacting against the ball may prevent it from descending so rapidly as it descends when the repelled air is enabled to escape freely: thus it may be that the ranges are longer when the fire is made over a level plain or gentle eminences than when it is directed across a deep valley, which affords room for a greater diffusion of the air. When shot is fired over the sea, the convex surface of the water, which is depressed in every direction about the gun, and the absence of such inequalities of surface as exist on land, readily permit the escape of the air under the trajectory. Hence there is little or no reaction from below against the shot; and to this circumstance M. Piobert ascribes the supposed inferiority of the ranges over sea compared with those over land.

But from experiments made at Gavre in France, in 1843, it was found that, *cæteris paribus*, the ranges at sea are very nearly equal to those obtained on land. The general opinion of the inferiority of the former appears therefore to be unfounded; and it is observed by the writers of the "Report" on those experiments, that the ranges stated in the work of M. Piobert, which are in favour of that opinion, have been taken from an old work on gunnery, and are not the results of any known experiments.

141. The ordnance employed at Gavre to determine the extent of ranges over sea and land, and also to ascertain the amount of the longitudinal and lateral de-

viations of projectiles in naval gunnery, were a long 30-pounder gun, weighing 61 cwt. 1 qr. 4 lbs., the diameter of the bore being 6.484 inches, and the windage 0.1772 inch; and a canon obusier of 80 (*liv.*), weighing 72 cwt. 3 qrs. 16 lbs., the diameter of its bore being 8.799 inches, and the windage 0.0945 inch. Solid shot only were projected from the former, and hollow shot with wooden bottoms from the latter. The ordnance was placed in a barge (*ponton*) above 60 feet long and 24 feet wide, which was moored at convenient stations. The lines of fire were directed from west to east, and allowance was made for the oscillations of the vessel. In each day's practice twenty rounds were fired, and the table below (p. 112) exhibits the mean results.

In the Table the lateral deviations have been referred both to the line of fire, or rather to a vertical plane passing through the produced axis of the gun, and to another line which is called the line of mean direction. This last was determined so that the sum of the deviations to the right might be equal to the sum of those to the left; and if there were no cause of the shot deviating constantly one way, these two lines, in a mean of many firings, would coincide. On comparing the seventh column with the second, it appears that the deviations of all the lines of mean direction, except one, took place towards the side opposite to that from which the wind blew, which proves that the deviations were affected by the action of the wind. The excepted case is in the last horizontal line but one of the Table, and at the time of this experiment the wind was light. The extreme lateral deviations appear to be, in some cases, more than twice as great as the mean deviations; and this may be accounted for partly by the windage of the ordnance, and partly by the wind acting at times across the line of fire.

142. The near agreement of the ranges over land and sea, in the following Table, appears to be at variance with the results of experiments made on land and at sea with the artillery employed in the British service. On comparing the ranges over the land, from the experi-

Nature of Ordnance.	Wind.	Elevation.	Ranges over Sea.	Longitudinal Deviation.		Deviation of the Line of Mean Direction from the Line of Fire.	Mean Lateral Deviation of the Projectile referred to the		Extreme Lateral Deviation referred to the Mean Direction.	Ranges on Land, from Experiments at Gavre.
				Mean.	Extreme.		Line of Mean Direction.	Line of Fire.		
		° ′ ″	yards.	yards.	yards.	feet.	feet.	feet.	feet.	yards.
30-pr. solid shot, charge, 11 lb. 0.5 oz. {	S., high.	3 5 48	1453	179	432	32.82 left hand, or northwards.	187	33.4	42.6	1355
	S.W., high	4 17 50	1687	201	476	41.01 ditto	31.8	44.3	62.3	1724
	E.N.E., light	5 48 40	1980	76	214	7.816 right hand, or southwards.	20.3	22.6	14.9	2032
	N.	6 50 2	2229	70	229	43.98 ditto	28.9	49.8	75.4	2247
Canon obusier, No. 1; hollow shot; charge, 7 lb. 11.6 oz. {	N.N.E.	4 46 17	1369	39	179	26.6 ditto	16.4	29.2	39.4	1361
	N.E.	6 41 28	1635	111	177	4.0 left hand, or northwards.	27.2	29.2	75.4	1675
	N.N.W.	8 47 50	2177	111	210	4.1 right hand, or southwards.	61.3	62.3	141	2017
	W.S.W.	10 4 50	2479	124	323	11.1 left hand, or northwards.	66.2	73.8	223	2208

ments at Deal in 1839, with those over water, from the experiments on board the "Excellent" in 1838, the nature of the ordnance, the charges, &c., being alike, there were found to be only three in which the ranges over the sea exceeded those over land. In all the others, the ranges over land were the greatest. In many of them the difference was as much as one-tenth, and in some it amounted to one-sixth of the whole range. This is in some measure due to the height of the guns above the ground, in the Deal experiments, being greater than that of the guns of the "Excellent" above the water; but the cause of the discrepancy, whatever it may be, requires to be investigated by experiments purposely conducted, with precisely the like natures of ordnance and with equal charges, in order, should the variations be found to be only such as may be ascribed to accidental circumstances, that the same tables of artillery practice may with confidence be used in both services.

143. In estimating the probabilities of correct firing, all the sources of error should be anticipated and allowed for; and it may be observed here, that it is near the further extremity of the range that the velocity of the lighter and smaller kind of projectile is most diminished, consequently, that, in this part of the trajectory, the causes of the two kinds of deviation are most effective in producing them. Hence it follows that the two kinds of deviation may be much diminished by the employment of those guns which, with the least elevation, have the power of throwing solid shot, with their momenta nearly unimpaired, to the greatest distance; and this fact is what it was intended in the present article to establish. Force may not always be deficient when missiles are projected from the shorter and lighter nature of gun, yet the probabilities of striking objects at considerable distances are much higher when ordnance of a contrary description is used.

This subject will be further considered in treating of hollow shot, and of the lighter natures of ordnance employed in the British service.

144. The disadvantages we suffered from the numerous frigates of the inferior classes which we had in our navy during the late war were severely felt. From not having had a sufficient number of frigates capable of bearing the nature of armament which was likely to be opposed to them, we were obliged, in common prudence, to employ two vessels where one of superior force might have answered better. Frigates capable of bearing the 8 feet 24-pounders, or 32-pounders of 50 cwt., are certainly vast and expensive ships; but they perform their service at a cheaper rate, and in a much more certain and creditable manner to the nation, than two vessels of very inferior rate. The full service of two ships acting against traders and privateers cannot be effected without frequently parting company; and then the honour and interest of the flag they bear are at risk, from exposure to unequal combat; whilst the enemy's force, against which these two inferior vessels may have been sent, remains always entire. Two ships, once divided, may be beaten in detail by a vessel by no means a match for both together; but the honour of the large ship is always secure, and the chance of capture remote. She *may* meet a superior; but if so, to take her would be creditable—to yield, no disgrace. A fleet of line-of-battle ships must keep together, resisting all temptations to chase single ships; but two vessels, sent to cruise on predatory service, will only exert half their powers by keeping company; for this restrains separate exertion, halves the chance of meeting an enemy, and when one is discovered, there always ensues a chance of the two vessels parting company in the chase.

145. The American frigates of the large class all carried long guns. The "President" had, on the main-deck, 24-pounders (length $8\frac{1}{2}$ feet, weight $48\frac{1}{2}$ cwt.); on the quarter-deck and forecastle, 42 lbs. carronades (length 4 feet 4 inches); both English calibre. The "Chesapeake" mounted on the main-deck, twenty-eight 18-pounders (length 7 feet 8 inches, weighing 39 cwt. 1 qr.), and twenty 32 lbs. carronades, nearly of British pattern. The "Essex," on the contrary, was armed

almost exclusively with carronades; and the effect of these we shall examine hereafter.

146. Much of the reasoning contained in the preceding articles attaches to carronades. At close quarters they are a formidable species of ordnance; but at long ranges, on account of the inferiority of their charges, they are no match for long guns, even of smaller calibre (compare the ranges of shot from carronades with those obtained from other ordnance in Tables V., VI., VII., and VIII., Appendix D); and any vessel fitted exclusively with carronades might, undoubtedly, be destroyed or captured by a vessel of very inferior rate mounting long guns, if her commander knew how to avail himself of the great superiority of his weapons.

147. The very mortifying situation in which the gallant Sir James Yeo found himself, in September, 1813, on Lake Ontario, shows the danger of the carronade system of armament. Sir James states, in his letter of the 12th September, "the enemy's fleet of eleven sail, having a partial wind, succeeded in getting within range of their long 24 and 32-pounders; and, having obtained the wind of us, I found it impossible to bring them to close action. *We remained in this mortifying situation five hours*, having only *six guns in the fleet that would reach the enemy*. Not a *carronade was fired*. At sun-set a breeze sprung up from the westward, when I manœuvred to oblige the enemy to meet us on equal terms. This, however, he carefully avoided."

In Sir James Yeo's dispatch of the 15th November, he complains in strong terms of the want of long guns in the Lake Erie squadron.

Captain Barclay states, in his letter of the 12th September, 1813:—"The other brig of the enemy, supported in like manner by two schooners, and apparently destined to engage the 'Queen Charlotte,' kept so far to windward as to render the 'Queen Charlotte's' 24-pounder carronades useless, whilst she and the 'Lady Prevost' were exposed to a heavy and destructive fire from the 'Caledonian' and four other schooners, armed with *long* and *heavy guns*."

The action of the "Phœbe" with the American frigate "Essex" illustrates and confirms, though happily in a reverse sense, all that has been said on the serious dangers of the carronade system of armament, opposed to a vessel, fitted with long guns, commanded and managed as the British frigate was. The "Phœbe" mounted long 18-pounders on the main-deck, and 32-pounder carronades on the quarter-deck and forecastle. The "Essex" had forty 32-pounder carronades, and only six 12-pounder guns (length 6 feet $8\frac{7}{16}$ inches). Captain Porter, of the "Essex," says—" The 'Phœbe,' by edging off, was enabled to choose the distance which best suited her long guns, and kept up a tremendous fire, which mowed down my brave companions by the dozen." Again—" The enemy, from the smoothness of the water, and the impossibility of reaching him with our carronades, was enabled to take aim at us as at a target; his shot never missed our hull, and my ship was cut up in a manner which was perhaps never before witnessed."

This action displayed all that can reflect honour on the science and admirable conduct of Captain Hillier and his crew, which, without the assistance of the "Cherub," would have insured the same termination. Captain Porter's sneers at the "respectful distance" the "Phœbe" kept, are, in fact, acknowledgments of the ability with which Captain Hillier availed himself of the superiority of his arms; and this brilliant affair, together with the preceding facts, cannot fail to dictate the necessity of abandoning a principle of armament exposed to such perils; and to teach the importance of adapting the tactics of an operation to the comparative natures and powers of arms.

148. The defects of carronades, and the danger of employing this imperfect ordnance, are now generally felt and admitted; that ordnance, however, rendered important service in its time, for it taught us practically the great value of a reduced windage, the advantages of quick firing, and the powerful effects produced at close quarters by shot of considerable diameter striking

a ship's side with moderate velocity. Let it be observed, also, that the carronade-armed "Glatton," under Captain Trollope,[a] proved victorious over a superior force by dint of the heavy blows inflicted at close quarters; while the defects of the same nature of ordnance, in long ranges, on the occasions mentioned in the preceding article, exposed the gallant commanders of the British ships to the severest mortifications, and the naval service of the country temporarily to reproach. The like reproach, but in a still more serious degree, will undoubtedly be incurred if that imperfect arm be still retained in the British service, and its place be not supplied, at least in large vessels, by a superior nature of ordnance. In the French navy the carronade may be said to exist only in name; for the *carronade à tourillons* is in reality a powerful gun.

149. VI.—*On Recoil and Preponderance.*—It has been found, by suspending a gun in the same manner as the block of a ballistic pendulum is suspended, that, on attaching weights to the piece in order to diminish its recoil, the initial velocity of the shot discharged was the same as when the gun was allowed to recoil to its full extent; and even when the recoil was entirely prevented, the velocity of the shot was unaltered. It is evident, however, that some uncertainty in artillery practice must exist in consequence of the irregularity of the recoil; and it was at one time supposed that, by mounting a gun on a non-recoil principle, the shot might be projected to greater distances, but the experiments above alluded to serve to show that this is not the fact. A formula has been given (Art. 69), by which the velocity of the recoil may be computed, and it is there shown that the motion of the shot must be very slightly affected by the recoil if the latter were steady; but it is evident that an imperfect equilibrium on the trunnions, or an accident happening to the gun or its carriage at the moment of firing, such as a bed breaking, a quoin flying out, or a truck coming off, will sen-

---

[a] James's Naval History (1847), vol. i. p. 335.

sibly affect the length as well as the direction of the range.

The following Table contains the results of some experiments made to determine the extent of the recoil of sea-service iron guns on ship carriages, upon a horizontal platform.

Shot and Charges.	Elevation.	32-pr.		24-pr.		18-pr.	
		ft.	in.	ft.	in.	ft.	in.
1 shot, charge ⅓ . . .	2°	11	0	11	0	10	6
2 shot, charge ⅓ of a single shot.	4°	19	6	18	6	18	0
2 shot, charge ¼ ditto . .	7°	11	6	12	0	12	0

In an experiment lately (1848) made at Woolwich, on a platform inclining 2½ degrees, two hollow shot, weighing together 112 lbs., were projected with a charge of 5 lbs. of powder from an 8-inch gun, weighing, with its sea-service carriage, 78 cwt.; the initial velocity must consequently have been about 709 feet per second, and the velocity of the recoil, by the formula Art. 69, 9 feet per second. Now a 32-pounder gun, weighing with its carriage 65 cwt., projects a solid shot, the charge being 10 lbs., with an initial velocity of 1600 feet per second, and the velocity of recoil is 7 feet per second. By observation the recoil of the 8-inch gun, on the inclining platform, was 14 feet, and that of the 32-pounder 9 feet; and these extents are, as by theory they ought to be, nearly proportional to the velocities of recoil.

150. On the discharge of a gun, the unsteadiness of the recoil and the *jump* of the piece by the striking of the breech on the quoin or on the elevating screw, are greatly affected by what is called the *preponderance*, an element which depends much on the position of the trunnions. If these are placed too far forward, the projection of the gun in a ship's port may be too small, while the breech is rendered too heavy; if too far back, the muzzle droops on firing, and violent shocks are produced on the quoin or screw; and these effects are

much aggravated when the common axis of the trunnions is below that of the bore. In order that such inconveniences may be avoided as much as possible, the pressure of the breech, when the axis of the bore is horizontal, is usually made equal to one-twentieth of the whole weight of the gun. The preponderance of a piece of ordnance is a very important feature in gunnery, on account of its effects in producing shocks, and in rendering the practice inaccurate ; and the amount of it is now made a subject of a particular proof.

151. It is remarkable that, in the experiments which were carried on at Shoebury Ness, in June, 1852 (Art. 188), the higher elevation of the ordnance always gave the greater recoils, though the charges and weight of the shot were equal and the guns the same. At an elevation of 32°, for example, the recoil was 4 feet 1 inch, which is nearly double the recoil (2 feet 2 inches) at an elevation of 28°. The circumstance is so contrary to what should take place on mechanical principles, if the platform were, in all the experiments, horizontal, that it must be ascribed to some particular effect produced in the platform itself. May not a gun fired at a considerable elevation acquire a jumping motion up and down, thus shaking the carriage irregularly, and making it recoil on the slides more than it would at a less elevation?

With respect to the strain produced in the carriage of a gun by the discharge of shot, it may be observed that experiments made in France show that the initial velocity of the shot, and, consequently, the velocity of the recoil, increases with the elevation of the gun ; and this is sufficient to account for the increase of the strain with the increase of the elevation. At 10° of elevation the strain is greater than, with double the charge, it is when the axis of the gun is horizontal. It is evident, therefore, that, in order to have the strain as little as possible, the lowest elevations should be given which will afford the required range. The strain on a gun is supposed to vary with the sine of the elevation.

152. VII.—*The Effects of Wads.*—Experience has proved that different degrees of ramming, or different dimensions of wads, make no sensible alteration in the velocities of the ball as determined by the vibrations of the suspended gun. Stout, firm, junk-wads,[a] so tight as with difficulty to be rammed into the gun, have been used: sometimes they were placed between the powder and ball, sometimes over both, but no difference was discovered in the velocity of the ball. Different degrees of ramming were also tried without wads. The charge was sometimes set home without being compressed, sometimes rammed with different numbers of strokes, or pushed up with various degrees of force: but the velocity of the ball remained the same. With great windage, the vibrations of the pendulum were much reduced, although tight wads under the shot were used; so that wads do not prevent the escape of the inflamed powder by the windage, nor under any circumstances occasion any sensible difference in the velocity of the ball.

153. It follows that tight wads should not be used in action, because they increase the difficulty of loading, without doing any good: they may be used in loading guns before an affair; but, with a view to quick firing, the wads used in action should be no tighter than is necessary to prevent the charge from shifting while the gun is run out. Occasionally the operation of ramming home a tight wad has taken up two or three minutes. This is a very serious and unnecessary waste of time. When a wad is so 'high,' or tight, as to require the force of a *blow*, instead of a *push*, it spreads from the stroke, and consequently becomes harder, and more difficult to be moved. Care should therefore be taken to provide easy wads, correctly gauged, and made so as to spread under a pressure sufficient to retain the charge from shifting, upon being set up with a smart blow.

154. From experiments made in France (1842-1844)

---

[a] Wads are generally made of rope-yarn rolled up in a cylindrical form; they are usually placed immediately on the shot, and they fit the gun rather tightly, so that they may prevent the shot from shifting its place or rolling out. A grummet wad consists of a piece of rope formed like a ring, the external diameter being equal to the calibre of the gun.

with a 30-pounder and a 12-pounder, long guns, projecting solid and hollow shot, with and without wooden bottoms, it appears that, generally, the charges being equal, the cartridge which has the greater diameter causes the shot to have a greater initial velocity. This, however, was not always the case; and it was found that, with the 30-pounder gun, the initial velocity was the greatest when the diameter of the powder within the cartridge was 5.95 inches (Eng.), and, with the 12-pounder gun, 4.33 inches (Eng.). In 1845 the diameters of the *mandrins* (moulds) used in making cartridges for the French canons-obusiers were regulated conformably to the above results. (See Arts. 217, 218.)

It may be observed that the velocities were greater when the hollow shot had wooden bottoms than when they had not, in the ratio of 1 to 0·98 nearly.

155. From experiments made on board the "Excellent" in 1847, it was found that a grummet wad is more efficient than one of junk, in preventing the cartridge from shifting its place in the bore when the guns (a 32-pounder and a 10-inch and 8-inch shell gun) were run out with a strong jerk. From the experiments alluded to, it must be inferred that, both with junk and grummet wads, the cartridge is more liable to change its place in the guns with conical chambers than in those which have none; and the fact affords strong evidence of the disadvantages of using chambered ordnance (see Art. 123) for broadside shot-guns in quick firing, the chambered ordnance requiring the introduction of a wad over the cartridge before the shot is put in, whereas in a cylindrical bore the shot may always be set home in immediate contact with the cartridge.

Several valuable experiments have recently been made by Major Mordecai, of the United States' artillery, on the deviations of shot from the object when wads of different kinds were employed: the gun being a 14-pounder, the windage 0.143 inch, and the charge 6 lbs.

These deviations were found to be the greatest when junk wads were used, whether one wad was placed on the ball only, or one on the powder and another on the ball. The deviations were also considerable when hay wads were used; and they were the least when grummet wads were employed, whether one only was placed on the ball, or there was placed one on the powder and another on the ball. The deviation was rather greater when a sabot was next to the powder, and a grummet wad over the ball.

156. With respect to small arms, it is found that wads of different kinds have different effects upon the projectile by modifying the action of the charge; and from experiments which have been made in the United States with a musket-pendulum, the following results have been obtained.

With a charge equal to 77 grains, a musket-ball wrapped in cartridge-paper, and the paper crumpled into a wad, the velocity of the ball was 1342 feet; and when two felt wads, cut from a hat, were placed on the powder with one on the ball, the velocity was 1482 feet. With a charge equal to 140 grains, two felt wads being placed on the powder and one on the ball, the velocity was 1525 feet; when cartridge paper was used, crumpled into a wad, the velocity was 1575 feet; and when one wad of pasteboard was placed over the powder with another on the ball, it was 1599 feet. These results seem to indicate that wads made of the stiffest materials are the most advantageous.

157. VIII.—*Penetrations of shot into materials.*—The experiments which have been made to determine the penetration of shot into different substances, have led to very various and often discordant results, arising, no doubt, from inequalities in the consistency of the materials or in the degrees of elasticity, but all of them tend to confirm the laws which have been stated in Art. 79. In that article there is given a formula (*c*) for computing the depth of the penetration of shot into oak timber; and Table XVI., Appendix D., contains the computed amounts of penetration, the shot being fired from different natures of French ordnance and at different

distances of the gun from the object. The Table is abridged from the *Aide Mémoire Navale*, Paris, 1850; the weights and measures being expressed in English denominations. The velocities at the time of impact are computed from the formula (*c*), Art. 60; and the depths penetrated, from the formula (*c*), Art. 79.[a] M. Piobert, in his *Traité d'Artillerie*, has given Tables professing to contain the results of experiments made in France on the depths of the penetration of shot into masonry, wood, earth, and water, from guns of different calibres, with various charges of powder, and at different distances from the object struck.

158. In order to determine the circumstances approaching as nearly as possible to those which may occur in the naval service, many experiments were made in 1838 on board of H. M. S. "Excellent," under Captain Sir Thomas Hastings, by firing both solid and hollow shot against the Prince George hulk, which was moored at the distance of 1200 yards. The guns were laid at small angles of elevation, generally between 2 and 3 degrees, except when the shot was intended to strike the hull after ricocheting on the surface of the water, when the elevations were 1 degree; and the following is a brief statement of some of the most remarkable effects which were produced. It should be observed that the shots often entered the side of the hulk diagonally, and pierced or splintered parts of the structure at different places in the interior; the depth penetrated is expressed by the sum of the distances in solid wood, which the shot passed through or deeply furrowed.

Several 18-pounder shot, with charges of 6 lbs. of powder, penetrated to depths varying from 21 to 33

---

[a] In order to have the penetrations of shot into other media, the numbers in the Table must be multiplied by 1.64 for firm earth, consisting of sand and clay in equal portions; by 1.3 for sand and gravel intermixed; by 3.21 for earth newly raised, and 0.41 for sound masonry. Experiments in France have shown that wood, in consequence of its elasticity, recovers nearly its original volume after being penetrated, and thus closes up the cavity formed by shot on striking it. Experiments have also shown that a projectile remains embedded in oak only when it has penetrated to a depth nearly equal to its diameter.

inches, according to the state of the wood, and there stuck. The effects produced by 32-pounder shot were much greater; with a charge of 6 lbs. one passed through the ship's side diagonally in indifferent wood (30 inches), then, continuing its course, it struck a knee on the opposite side of the ship, making one large splinter, and, after rebounding, it struck the combings on the side which it first entered. Another, with an equal charge, passed through the bow portsill, and struck upon the iron knees of the cable-bitts, which it broke in two; the standard of the bitts was also shaken to pieces. The shot itself was broken in pieces, and the fragments cut the chain-cable in two places. With charges of 8 lbs. the 32-pounder shot penetrated to depths varying from 22 to 48 inches.

Two 32-pounder shot fired singly with charges of 10 lbs. 11 oz., entered at the same place so as to render it impossible to distinguish their separate effects; together, after penetrating through the ship's side in firm wood, they shattered a sound wooden knee: they then passed across the deck, cutting down a wooden stanchion 6 feet long and 8 inches square under the beam; this they shattered in pieces, causing many splinters, six of which were very large, and one of them swept the deck as far as the pumps. One of the two shot penetrated its own depth in sound wood on the opposite side of the ship, and there stuck; the other struck and splintered a port on the opposite side, after which it rebounded against the side which it first entered.

A 68-pounder shot (solid), with a charge of 10 lbs., penetrated diagonally through an oak chock of sound wood, which it shattered to pieces, flattening a large iron bolt; from thence it passed into the ship's side in good wood, making a total penetration of 46 inches.

159. Many hollow shot were fired with remarkable effects from 68-pounder guns, making penetrations which varied from 25 to 56 inches. One of these, with a charge of 8 lbs., penetrated the side of the hulk, passing through 28 inches of good wood, tore out the iron hook which holds the port-hinge, and fractured the

after side of the port, driving the splinters about the deck. It rent away the end of a beam, grazed the deck, passing through two planks, and cutting down a stanchion 8 inches square, making several large splinters; it then struck against the opposite side of the ship, from whence it rebounded against that which it entered.[a]

Hollow shot from a 68-pounder carronade, with a charge of 5 lbs. 8 oz., penetrated to depths varying from 28 to 31 inches. One of these pierced through the ship's side (24 inches) just above the water-line, in fair wood, and then passed through the ceiling in the orlop (7 inches thick), consisting of very strong wood, shattering it considerably.

160. In 1848 some experiments were made to try the penetration of shot into water, when fired with small angles of depression towards its surface. In these experiments three targets were placed vertically in the water, 8 feet asunder, the nearest being about 37 yards from the mouth of a gun, which was a 32-pounder, and charged with 10 lbs. of powder.

When the gun was depressed 7 degrees, the shot struck the first target at the water's edge, and, passing through it, rose from thence and pierced the other targets at the heights of 12 and 18 inches above the water. Again, the water having risen 2 feet, the gun, charged as before, was fired, when the shot striking the water at 8 feet short of the nearest target, rose and passed through the three targets successively: the first at 6 inches from the water, and the others successively higher. The same gun being fired with a depression of 5 degrees, the shot passed through the first target under water, grazed 4 feet, and rebounded from the second target.

The gun was fired once with a depression of 9 degrees, when the shot did not come up again. It passed

---

[a] At 800 yards, with heavy guns, a charge of one quarter of the weight of shot may always be used; at 500 yards the charge may be reduced to one-sixth; and, within 400 yards, two shot at once may be used with advantage.

through the first target at 2 feet under water, and, grazing along the mud, rebounded from the second target, having entirely lost its force. With a depression of 7 degrees, the shot being fired into the water where it was 1 foot deep, rose, after grazing about 8 feet along the mud, and at length fell at the distance of 400 yards. Being fired with the same depression into water 2 feet deep, the shot did not reach the mud, but immediately rose, and finally fell about 600 yards off.

161. In order to ascertain if shot reflected from water would damage a ship, shots from a 32-pounder gun, with a charge of 10 lbs. and a depression equal to 7 degrees, were fired from a *lump* alongside of the Leviathan, and the following are some of the effects produced:—At the distance of 16 yards, the shot struck the water at 4 feet from the ship's side, and in one experiment it lodged in the cutwater; in another it indented the ship's side, and in both cases it struck at 18 inches below the water-line. At the distance of 36 yards, with a depression of 5 degrees, the shot struck the water at distances from the ship's side varying from 2 to 15 feet; and ricocheting, entered the ship at distances above the water line varying from 2 inches to 3 feet.

In consequence of the loss of force which the balls, in all these experiments, sustained by striking the water, it has been inferred that, if a shot be fired with such a depression as a ship's gun will bear, it will not penetrate into water more than 2 feet; and consequently that it will be impossible to injure a ship by firing at her under water. The correctness of this inference we must however be permitted to doubt till further experiments have been made. It is highly probable that conoidal shot would penetrate to a certain depth into the water, and strike the ship below the water line.

162. From the tables of ricochet practice made on board the "Excellent" in 1838, against the "Prince George" hulk, at the distance of 1200 yards, the following particulars are extracted:—A shot fired from a

32-pounder at an elevation of half a degree, with a charge of 10 lbs. 11 oz., penetrated, after one graze from the water, the after port-timber of one of the ports, to the depth of 28 inches in very good wood, shattered the head of the rider, started the plank between the ports, and passed over to the opposite side, where it penetrated to the depth of 10 inches. This shot also broke the beam-clamp, a piece of good wood 6 inches thick.

A 68-pounder hollow shot, with a charge of 12 lbs. and an elevation of half a degree, after five grazes struck a chock of solid wood 4 feet 8 inches thick under the fender, and shattered it in pieces. It struck also a large iron bolt, which it flattened. Another, with a charge of 10 lbs. and an elevation of 1 degree, after two grazes penetrated the ship's side diagonally (34 inches), in tolerably good wood, below the chocks, and lodged behind a cluster of iron knees on the orlop-deck, which were shaken considerably. The planking on the outside of the ship was also started. A third, after two grazes, struck a chock used for the sheers, tearing off a piece 6 feet long, 1 foot deep, and 2½ feet broad. It then penetrated 11 inches deep in the ship's side, in bad wood.

A 68-pounder shot, with a charge of 8 lbs. and an elevation of 1 degree, after two bounds, penetrated to the depth of 24 inches, close to the side of a port just above the lower port-sill, in bad wood, started the inside planking, and tore off a piece which splintered. One of the splinters, a very large one, was thrown beyond the main-hatchway to the opposite side of the deck. The shot having crossed the deck, struck a corner of the main-hatchway combings, and tore out a large piece on each side, destroying the use of the combings. It struck a winch-handle which was lying on the deck, and drove one end of it through a port-scuttle. After striking the combings, the shot grazed a beam and fell on the deck. Another shot, with a charge of 7 lbs. and an elevation of 1 degree, after three bounds, penetrated through the ship's side diagonally (29 inches), shattered the ceiling, and made several splinters.

A hollow shot from a 68-pounder carronade, with a charge of 5 lbs. 8 oz. and an elevation of 4 degrees, after one bound, struck the upper surface of a bulwark, and went overboard.

Ricochet practice is now much better understood than formerly, and is daily becoming more important in the service. When made with large ordnance it is susceptible of great accuracy on level ground, or on the surface of smooth water.

163. In 1838 experiments were made at Gavre with two solid balls fired at once against a butt of oak timber, in order to determine the different penetrations of the shot, and the distances between their centres at different distances from the piece. The diameter of each shot was 6·283 inches (English), and its weight 33 lbs. 4·6 oz. Three different natures of ordnance were used: a long 30-pounder (French) gun, a canon-obusier of 30, and a 30-pounder (French) carronade. One ball was in contact with the charge, and the other in contact with the former. In the carronade, a grummet wad was placed over the shot which was farthest from the charge, but in the other pieces a junk wad. The following tables exhibit the results:—

TABLE I.

Nature of Ordnance.	Charge.		Initial Velocity.		Velocity at Striking.		Penetration.	
			Ball nearest the Charge.	Ball farthest from the Charge.	Ball nearest the Charge.	Ball farthest from the Charge.	Ball nearest the Charge.	Ball farthest from the Charge.
	lb.	oz.	ft. per sec.	ft. per sec.	ft. per sec.	ft. per sec.	inches.	inches.
Long 30 pr. .	8	4.3	938	1194	889	1129	29.61	42·12
Canon-obusier of 30 . .	4	6·4	761	958	718	905	21.06	30.51
30-pr. Carronade . .	3	8.4	727	856	689	810	19.68	25.59

THEORY AND PRACTICE OF GUNNERY.

TABLE II.

Distances of the Butt.	Long 30-pr. Charge 8 lbs. 4.3 oz.		Canon Obusier of 30. Charge 4 lbs. 6.4 oz.		30-pr. Carronade. Charge 3 lbs. 8.4 oz.	
	Distances between the Centres.		Distances between the Centres.		Distances between the Centres.	
	Horizontal.	Vertical.	Horizontal.	Vertical.	Horizontal.	Vertical.
Yards.	ft. in.	ft. in.	ft. in.	ft. in.	ft. in.	ft. in.
55	1 0.2	0 8.3	0 8.6	0 7.9	0 5.9	1 3.7
109	1 11.2	1 7.3	2 3.5	2 0.	0 11.8	1 5.7
164	2 0.8	4 0.4	1 7.7	2 9.8	1 5.7	2 5.9
219	3 1.4	4 6.3	4 7.9	6 3.6	3 6.9	2 1.6
274	2 4.3	9 3.	4 0.	6 6.3	5 0.6	2 11.4
328	4 8.6	9 4.6	4 3.6	5 6.5	7 6.5	4 4.7
383	5 3.4	11 0.2	4 11.4	10 10.7		
437	6 8.	13 8.9	9 8.5	11 6.4		

From all these experiments, it is evident that the ball which was in contact with the charge had the least velocity and the least penetrating power (see Art. 101). It is further remarkable that, at distances beyond 200 yards, the vertical dispersion greatly exceeds the horizontal dispersion.

164. In the summer of 1835 some experiments were carried on at Metz, in order to ascertain the effects of cannon-shot on plates of forged iron, carefully manufactured by rolling; and the following account of them is extracted from the *Mémorial d'Artillerie*, No. 5, page 361. The dimensions and weights are given in English denominations :—

No. 1 plate was 48.8 inches long, 18.4 inches wide, 1.44 inches thick, and weighed 343 lbs. No. 2 plate was 40 inches long, 16 inches wide, 1.72 inches thick, and weighed 220 lbs. No. 3 plate was 62.4 inches long, 18 inches wide, 3.08 inches thick, and weighed 581 lbs. The shot were fired from 12-pounder and 24-pounder guns (French), which were placed at the distance of 66½ feet from the plates.

From the 12-pounder gun, with a charge equal to 2 oz. and a velocity equal to 340 feet per second, impressions from 0.16 inch to 0.2 inch were made in No. 1 and No. 2 plates. The shot was unaltered in form, and rebounded back to a distance of 117 feet. From the

K

24-pounder gun, charge 8.8 oz., and a velocity of 463 feet, the shot produced a crack in the rear face of No. 2 plate; with a charge of 1.1 lb., and a velocity of 633 feet, the shot completely pierced No. 1 plate, and detached a fragment.

From the 12-pounder gun, charge 1.1 lb., velocity 866 feet, the fire being directed against No. 1 plate, also with a charge equal to 2.2 lbs., and velocity 1216 feet, the fire being directed against No. 3 plate, the shot did not pass through, but produced deep impressions and cracks towards the bottom of the indentation. The shot were broken, and the plates were brought to a heat which produced a deep blue colour at the places of fracture. From the 24-pounder gun, charge 4.4 lbs., and velocity 1266 feet, the shot did not pierce No. 3 plate, but produced cracks diverging from the centre of the impression; but with a charge equal to 6.6 lbs., and velocity 1500 feet, the shot perforated No. 3 plate, carrying out a portion corresponding nearly to its own diameter. With charges above 4.4 lbs., the shot passed completely through No. 1 and No. 2 plates, and the diameters of the holes were found to approach that of the shot in proportion as the velocity of the latter was greater. In one case the hole appeared as if punched in a form very nearly circular.[a]

---

[a] This warrants the condemnation pronounced by high authority on iron steamers in France :—" De tous les navires à vapeur, les plus impropres à la guerre sont ceux en fer."—(*Les Principes et l'Organisation de la Marine de Guerre*, par le Général Du Bourg, ancien officier de la Marine, p. 313, Paris, 1849.) The French were the first to propose the introduction of the use of iron in the formation of steam-vessels armed with "*la nouvelle arme*," the canon-obusier. Colonel Paixhans entertained an opinion that such vessels might thus be made shell-proof, and accordingly propounded this in his work (*Nouvelle Force Maritime*, page 7), in which it is announced that the steam-vessels so armed and fitted were to be made proof against artillery by being *cuirassés en fer!* This proposition was taken into consideration by the Comité Consultatif de la Marine, to which the whole of the project was referred, for a Report; and seriously entertained and discussed by them (see section No. 19, pp. 92 *et seq*. in the work just quoted); and this it was that led to the construction of iron steamers in our naval service. The proposition of rendering ships impenetrable to shot, in this or any other manner, was found to be impracticable by any means available in ship-building. The French soon afterwards came to the conclusion that, of all vessels that can be constructed, those made of iron are the most improper for war; and they have long since desisted from any attempts to introduce iron vessels into their naval service.

PART II.   THEORY AND PRACTICE OF GUNNERY.   131

165. From the experiments made at Metz in 1834, it appears that masses of cast-iron above 1 yard square and 13 inches thick, do not resist the shock of balls fired against them with even moderate velocities, having been fractured not only at the point of contact, but also at points considerably distant from thence. It was found also that the sides of a traversing gun-carriage of iron, whose dimensions exceeded those of any carriage which can be employed on service, were broken by an 8-pounder ball, having a velocity of 492 feet, which proves that carriages of this nature would, if struck, be rendered unserviceable; and that a collision which, with a wooden carriage, would have damaged only an accessory part, without requiring its being replaced, would, with a cast-iron carriage, have a more fatal effect. Not only is the object struck destroyed in a short time, but the fragments scattered in different directions are highly dangerous.

166. In August, 1840, some experiments were made at Woolwich, in order to determine the effects of shot upon an iron target lined with a composition of caoutchouc and cork, the invention of Lieut. Walker of the Royal Marines. The thickness of the iron was $\frac{5}{8}$ inch, and of the composition 9 inches. The gun was a 32-pounder, and was placed at the distance of 40 yards from the target.

With a charge equal to $\frac{1}{4}$ lb. of powder the shot rebounded, making only a small indentation; but with a charge equal to $\frac{1}{2}$ lb. the shot penetrated the iron and lodged in the composition: had the shot been heated, the composition would have been set on fire, when by no immediate possibility could the shot have been taken out, so that the consequences might have been fatal. With a charge of 1 lb. the shot passed through both the iron and the composition, producing a clean, round hole, the disk which was struck out being nearly a circle; and with a charge of 10 lbs. there was formed a clean hole, but the fragments struck out were very numerous, small, and jagged.

The target was then turned so as to present to the

K 2

gun the face armed with the composition; when, with a charge of 1 lb., there was formed an irregular fracture 13 inches long; and, with a charge of 10 lbs., the target was pierced and greatly torn, the mean diameter of the perforation being 18 or 19 inches. It does not appear, therefore, that any advantage would be gained by thus lining or covering the sides of an iron vessel.

167. In 1840 a series of experiments were also carried on at Gavre, in France, with grape-shot, consisting of 10 balls of 4 lbs. each, disposed in two layers between two plates of wrought iron, through which passed a central stem; the balls were confined laterally by three iron hoops, one above another, and the upper plate was kept close over the balls by means of a screw. The grape, which was fired from a canon-obusier of 80, with and without a wooden bottom, weighed in the former case 72 lbs. 12 oz., and in the latter 67 lbs. 12 oz. And the following is a table exhibiting the dispersions of the balls at different distances from the gun:—

Distance in Yards.	Grape with Wooden Bottoms. Dispersion in Feet and Inches.			Grape without Wooden Bottoms. Dispersion in Feet and Inches.		
	Horizontal.	Vertical.	Mean.	Horizontal.	Vertical.	Mean.
	ft. in.	ft. in.	ft. in.	ft. in.	ft. in.	ft. in.
109	3 2	9 7	6 4.5	6 4	7 7	6 11.5
164	11 9	13 6	12 7.5	12 7	11 7	12 1.
219	15 1	15 4	15 2.5	16 0	18 1	17 0.5
274	18 0	18 6	18 3.	19 7	21 4	20 5.5
328	22 11	28 6	25 8.5	26 7	29 3	27 11.

From the experiments alluded to, it may be deduced that the mean deviation, at the distance of 1 yard from the gun, is—

For grape-shot with wooden bottoms . . . 0.96 inches.
For   ditto   without ditto . . . . . 1.04 ,,

From experiments made on grape-shot, the balls being contained in a canvass bag as usual, when fired from the same nature of ordnance, the mean dispersion, at 1 yard from the gun, was found to be 0.94 inches. An iron plate in front of the ball causes, therefore, the dis-

PART II.   THEORY AND PRACTICE OF GUNNERY.   133

persion to be greater than it would be without such plate. A wooden bottom causes a diminution of the dispersion: but such bottoms are very liable to be broken to pieces in the gun, by which the dispersion of the balls is rendered irregular.

On the whole, however, grape-shot confined between iron plates appears to offer no advantages over that which is in canvass, and at sea the hoops would soon be corroded by rust.

The mean dispersions of grape shot from a canon-obusier of 80, at a distance of 109 yards, are as follow :—
A grape-shot, consisting of 48 balls, each 1.85 inch diameter, 13 feet 4.5 inches. A grape-shot consisting of 10 balls, each 3.199 inch diameter, 7 feet 9.6 inches; and a grape-shot, consisting of 6 balls, each 4.798 inches diameter, 7 feet 9.2 inches. To the extent of 437 yards the dispersions were proportional to the distances from the piece of ordnance.

Grape, consisting of 48 balls, each 1.85 inches in diameter, were fired from a canon-obusier of 80, with a charge of 8 lbs. 13 oz., the balls being disposed in three layers, and mounted on a wooden bottom of a conical form, and the whole weighing 47 lbs. 9 oz., when the mean dispersion at 1 yard from the gun was 1.9 inch ; the greatness of the dispersion would not permit such shot to be fired at a greater distance from the butt than 219 yards. By applying a wrought-iron plate, instead of the wooden bottom, the mean dispersion at 1 yard was 1.6 inches.

Grape, consisting of 64 balls of the above size, with a wrought-iron plate, were fired from the same nature of ordnance, and with the above charge, when the mean dispersion at 1 yard was 1.75 inch. It is evident, therefore, that the dispersion increases when the number of balls is increased.

In 1837 a number of experiments were made at Gavre, with different natures of ordnance, on the penetration into oak of balls constituting grape or case-shot, the guns being at 54½ yards from the butt; and the following is a Table of the results :—

Nature of Ordnance.	Nature of Shot.	Charge.		Initial Velocity.	Velocity at the instant of Impact.	Mean Penetration.
		lbs.	oz.	Feet.	Feet.	Inches.
Canon-Obusier of 30 {	15 balls, 2.1 inch diam. .	4	6.4	764	715	7.56
	120 balls, 1.05 ditto . .	4	6.4	758	666	3.35
Canon-Obusier of 80	48 balls, 1.85 ditto . .	8	13.	924	854	8.5
Long 30-pr. (Fr.) .	15 balls, 2.1 ditto . .	8	1.5	886	830	9.61
Canon-Obusier of 80	10 balls (4 liv.) 3.199 do.	8	13.	924	882	14.8
Long 30-pr. (Fr.) .	120 balls, 1·05 ditto . .	8	1.5	945	830	4.76

168. When a cluster of grape and one round-shot are fired at the same time from one piece of ordnance, it is found that if the grape is placed in contact with the charge, the ball being beyond it, the plate of the grape is often fractured, and when not broken in pieces, it takes a figure decidedly curved with the convexity towards the round shot; the balls of grape are frequently broken by their mutual percussions, and the stem is either broken or forced out of its place. The like effects are produced, but in a much less degree, when the shot is placed in contact with the charge, the grape being beyond. It was ascertained from experiments made at Gavre in 1838 that, when the grape was in contact with the charge, the dispersion of the balls was twice as great as when the round-shot was in such contact. It follows that the dispersion of grape-shot is diminished by augmenting the thickness of the plate; and it is found that with the same piece of ordnance the dispersion of grape-shot is nearly proportional to the number of balls which compose the cluster. Also, the dispersion is smaller in proportion as the piece of ordnance is longer.

169. The effect produced by shot when fired against iron steamers was remarkably exemplified on the "Lizard" during the operations which took place in the Paraná in 1846, when it was found that, on being struck, the plates of the ship bulged, and the perforations were so irregular and jagged that, for the purpose of stopping them, the common plugs were quite useless. This circumstance suggested the expedient of employing

what has been called a *parasol plug*, which consists of an iron bolt furnished with arms of the same metal and covered with thick canvass well tarred. On being thrust through the shot-hole, and then forcibly drawn back, the head expanded, and thus, the aperture being covered, the leak was closed. In consequence also of the ship being struck, the splinters and rivets detached by the shot flew about like grape, and nearly all the men killed and wounded suffered from this cause. Grape-shot fired at a distance of 200 yards pierced the side; and persons present, who were highly capable of judging, concurred in opinion that a 32-pounder shot would have gone through the sides of three or four iron steamers, doing damage which would be successively greater in those more remote from the ship first struck, till the force were spent. A remarkable circumstance is said to have happened to the "Alecto" at the same time. An infantry soldier fired his ramrod at her, when, like a dart, it went point foremost quite through the nearest side of the funnel, but being prevented by the button from passing through the other side, it fell down in the interior.

170. In the years 1849-1851 some experiments, of which notice is given below, were carried on at Portsmouth under the direction of Captain, now Rear-Admiral, Chads, C.B., in order to try the effects of shot in and upon iron ships; and a distinct knowledge of all the facts established by these trials is a matter of such immense importance, that the author thinks it necessary to present to the reader a summary of all the experiments that have been made, to ascertain exactly the effects of shot of various descriptions on plates or ribs of iron.

The first experiment, made on the 6th of November, 1849, was for the purpose of testing the resistance of iron plates against musketry, canister, and grape shot. Oak planking was also fired at, to make a comparison between the two materials. A marine's percussion musket was used: charge, $4\frac{1}{2}$ drs.; distance, 40 yards.

Iron plates, $\frac{1}{2}$ Oak plank, 1 inch	All passed through.
Iron plates, $\frac{5}{8}$ Oak plank, 2 inch	4 in 6 passed through.
Iron plates, $\frac{3}{4}$ Oak plank, 3 inch	Both musket proof.

Canister, 100 yards : 6 lbs. charge.

Iron plates, $\frac{3}{4}$ Oak plank, 3 inch	Passed through.
Iron plates, $\frac{1}{2}$ Oak plank, 4 inch	Canister proof.

Grape, 200 yards : 6 lbs. charge.

Iron plates, $\frac{4}{8}$ „ $\frac{5}{8}$ „ $\frac{6}{8}$	All passed through.
Oak plank, 4 inch „ 5 inch	All passed through.
„ 6 inch	Generally passed through.

Experiments were made in June, 1850, against two sections of the "Simoom,"[a] $\frac{5}{8}$ inch thick, placed 35 feet apart : the guns and charges were those used in all steam-vessels.

The result was, that two or three shot, or sometimes even a single one striking near the water-line of an iron vessel, must endanger the ship.

Another most serious evil is, that the shot breaks, on striking, into innumerable pieces, which pass into the ship with such force as to range afterwards to a distance of 400 or 500 yards; hence the effect on men at their quarters would be more destructive than canister shot, supposing them to pass through a ship's side, as when the plates are only $\frac{3}{8}$ inch thick.

Rear-Admiral Chads stated, in a letter to the author, that out of seventeen 32-pounder shot which struck the iron butts at the distance of 450 yards, with charges varying from 2½ to 10 lbs., sixteen were shivered to pieces on passing through the first side, and became a cloud of langrage too numerous to be counted.

Experiments were further made on the 11th of July, 1850, against an iron section similar to the "Simoom;" it was filled in and made solid, with 5½-inch oak timber

---

[a] This ship is clinker-built; and its sides are thicker than those of any other iron vessel in the navy.

between the iron ribs, and 4½-inch oak planking above the waterways, which were 1 foot thick, and with 3-inch fir above the port sills; these were strongly secured to the iron plates by bolts.

The results were as follow :—

The holes made by the shot were not so irregular as on the former occasion, but as clear and open. All parts of the shot passed right through the iron and timber, and then split and spread abroad with considerable velocity; parts of the iron plates, and a few very small pieces of shot, were sometimes retained in the timber.

With low charges the shot did not split into so many pieces as before.

With high charges the splinters from the shot were as numerous and as severe as before, with the addition, in this and the former case, of the evil to which other vessels are subject—that of the splinters torn from the timbers.

On the 13th of August, 1850, an iron section similar to the "Simoom" was prepared with a covering of fir plank on the outside, of the thickness of 2, 3, and 4 inches in different parts.

The result of this experiment was similar to the last, when the wood was on the inside, with the exception of the splinters from the wood.

The holes made by the shot were regular, of the full size of the shot, and open.

Every shot split on passing through; those between the ribs into a few pieces only; those that struck on the ribs into a great number; in both cases, when combined with the splinters of the iron side, the effect must prove highly destructive.

A comparison as to the effect of shot on iron and timber was made by firing 8-inch hollow shot, and 32-lb. solid shot, at a butt built for experimental shell-firing, with timber having 6-inch plank on the outside, and 4-inch within; the result was, that the splinters from the wood were trifling when compared with those from the iron.

Again, experiments were made on the 10th October,

1850, against an iron section similar to the "Simoom," lined on the inside with a composition called by Mr. Walters, the inventor, kamptulicon; the result was the same as the former trials when lined with wood. It neither prevented the shot from breaking into numerous small pieces, nor did the hole close up, as was anticipated, after the shot had passed through.

171. Experiments were made on the 5th July, 1851, at a butt constructed with oak and fir uprights, to represent the timbers of a ship; one half covered with $\frac{3}{8}$ths, and the other half with $\frac{4}{8}$ths sheet-iron.

The first practice was with the iron outside (Fig. 3): 32-lb. solid shot, and 8-inch hollow shot, with distant charges, passed through the $\frac{3}{8}$ths iron without splitting.

Fig. 3.

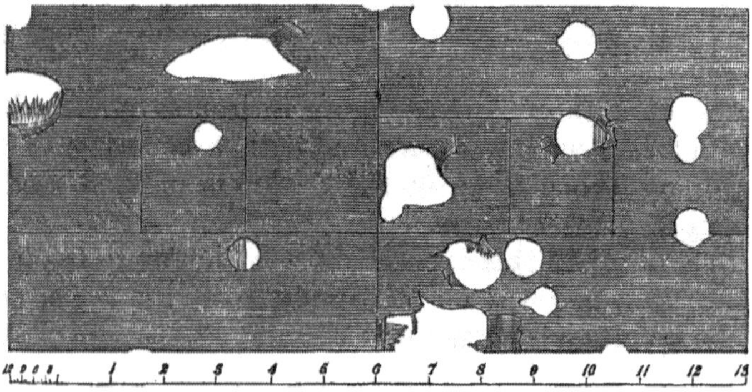

It has not been thought necessary to describe, in detail, the fractures and ravages made by the shot. The delineations, faithfully reported, *sautent aux yeux*, and carry conviction to the mind.

In firing against the $\frac{4}{8}$ths iron plates, two of the shot, on striking a double part where there was a joint, split into four or five pieces, but not in so destructive a manner as on the former occasion against the $\frac{5}{8}$ths plates.

Some shot that passed through this part were picked up and found to be starred or cracked, showing that $\frac{4}{8}$ths is the extreme thickness of iron which should be employed in constructing iron vessels, in order to prevent the breaking of the shot.

The shot on passing through the $\frac{4}{8}$ plate iron made clear, clean holes, of their own diameters, without rending the iron further, but the disc struck out was invariably broken into numerous pieces. Fig. 4 represents the reverse of the target, and exhibits the effects on the timbers.

Fig. 4.

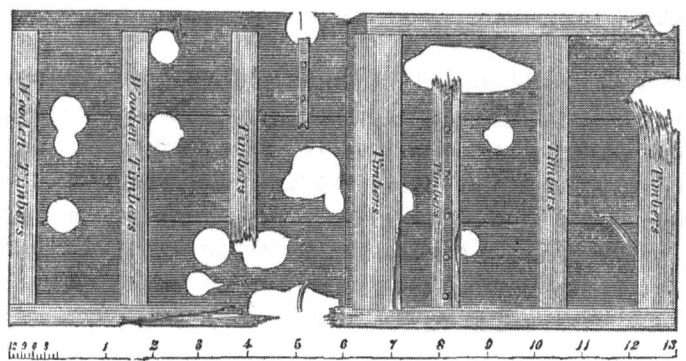

The butt was subsequently turned, so as to represent the off-side (Fig. 4), where the shot would strike the timbers before passing through the iron: reduced charges were used to diminish the velocity of the shot to what it would be after having passed through the front side.

A 32-pounder shot passed through an oak timber, carrying away the iron at the back in a fearful manner, laying open a space of 19 inches by 12, and curling the iron up so as to preclude the possibility of stopping it if under water.

An 8-inch shot also struck a fir timber, with almost a similar result; the hole was 2 feet 6 inches by 10 inches.

On the 11th and 12th of August, 1851, experiments were made the particular objects of which were to endeavour to diminish or obviate the destructive effects produced as above on iron plates by the impacts of shot, and to prevent the shot from splitting into pieces. For these purposes a butt was constructed to represent a section of the "Simoom," formed of iron ribs of $\frac{5}{8}$-inch iron, 4½ inches wide, and 11½ inches apart, instead of iron plates: these were covered with 5-inch teak planking on the outside and 2-inch on the inside (Figs. 5 and 6). The breadth of the 5-inch planking was $10\frac{1}{2}$ inches, and of the 2-inch, $9\frac{2}{3}$ inches.

This butt, 10 feet long by 8 in depth, was fixed between piles firmly driven at the distance of 450 yards from the guns, and with the outside face towards them.

No. 1. Solid shot, 32-pounder, 56 cwt. gun, 10 pound charge.—Hit direct about 3 feet from the centre of the butt, making an open hole (1, Fig. 5) the size of the shot with a jagged edge through the 5-inch teak; it then cut directly through an iron rib, carrying away 7 inches of it; the shot then, with the pieces of the rib, passed through the 2-inch teak close to the edge of a plank, carrying away an extent of 3 feet in length by $1\frac{1}{2}$ in height.

Fig. 5.

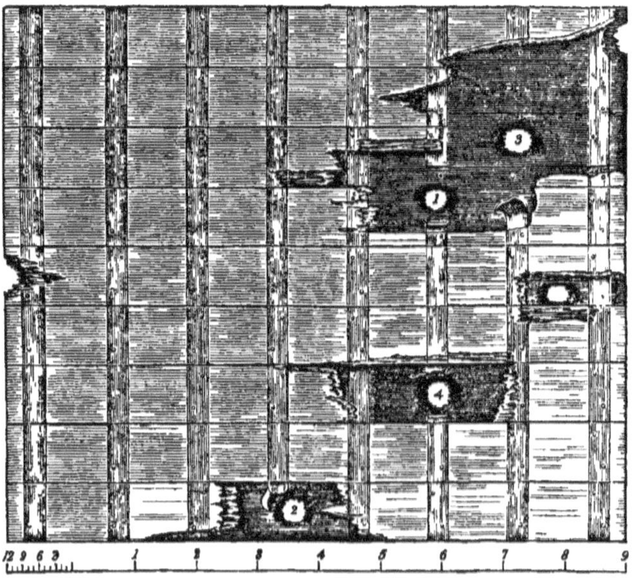

A large number of splinters both of wood and iron were found; the iron ones ranging from 200 to 400 yards: the shot passed on to 1300 or 1400 yards.

No. 2. Solid shot, 32-pounder, 56 cwt. gun, 10 pounds

PART II.   THEORY AND PRACTICE OF GUNNERY.   141

charge.—Hit direct 9 inches from the bottom (Fig. 5), and about the middle part of the butt, making an open hole the size of the shot with a jagged edge through the 5-inch teak; it then cut through the back of an iron rib, carrying away 10 inches of the lower part of it; the shot then, with the pieces of the rib, passed through the 2-inch teak in the centre of a plank, carrying away an extent of 3 feet in length by 1 in height, which probably would have been greater but that the shot struck so near the bottom of the butt and centre of a plank.

A large number of splinters both of wood and iron were produced; the iron ones ranging from 200 to 400 yards: the shot passed on to 1500 or 1600 yards.

No. 3. Hollow shot, 56 lbs., 8-inch, 65 cwt. gun, 10 pounds charge.—Hit from ricochet 25 yards short, about 2 feet from the side and 3 from the top of the butt, making an open hole the size of the shot with a jagged edge through the 5-inch teak; it then cut directly through an iron rib, carrying away 26 inches of it; the shot then with the pieces of the rib passed through the 2-inch teak at the edge of a plank, carrying away an extent of 3 feet in length by about 2 feet in height.

A large number of splinters both of wood and iron; the iron ones ranging from 200 to 400 yards; the shot passed on to 1300 or 1400 yards.

No. 4. Hollow shot, 56 lbs., 8 inch 65 cwt. gun, 10 pounds charge.—Hit direct, about 3 feet from the centre of the butt, making an open hole the size of the shot with a jagged edge through the 5-inch teak; it then cut directly through an iron rib, carrying away 8 inches of it; the shot then with the pieces of the rib passed through the 2-inch teak, in the centre of a plank, carrying away an extent of 3 feet in length by 1 in height.

A large number of splinters both of wood and iron; the iron ones ranging from 200 to 400 yards; the shot passed on to 1400 or 1500 yards.

A butt of the same description reversed was then

used, to try the effect of shot passing off the opposite side of a ship fitted in this manner.

For this purpose 5 lbs. charges were used with the 8-inch, and 4 lbs. with the 32-pounder gun, to allow for the decreased velocity of the shot after having passed through the first side with a 10-pounds charge.

With solid shot, 32-pounder, 56 cwt., 4-pounds charge.—Hit direct 2 feet from the bottom and about the middle part of the butt (B, Fig. 6), making an open hole the size of the shot with a jagged edge through the 2-inch teak; it then cut directly through an iron rib, carrying away about 7 inches of it; the shot then with the pieces of the rib passed through the 5-inch teak near the edge of a plank, making an open hole 12 inches in length by 6 in height, but the wood much splintered to 3 feet in length by 14 inches in height.

Fig. 6.

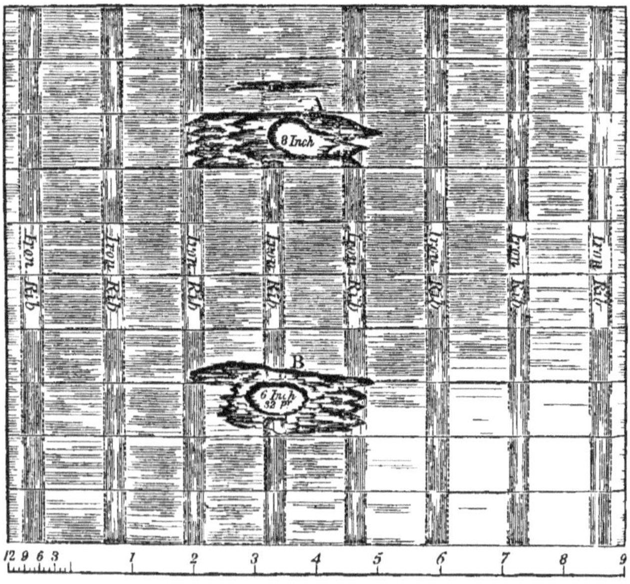

A great number of wooden splinters, but not many iron ones observed; the shot passed on to 700 yards.

Hollow shot, 56 lbs., 8 inch 65 cwt. gun, 5-pounds charge.—Hit direct, at A, Fig. 6, about 2 feet from

PART II.   THEORY AND PRACTICE OF GUNNERY.   143

the top, in the centre of the butt, making an open hole the size of the shot with a jagged edge through the 2-inch teak; it then cut through the back of an iron rib, carrying away about 18 inches of the upper part of it; the shot then, with the pieces of the rib, passed through the 5-inch teak in the centre of a plank, making an open hole 12 inches long by 9 inches wide, but the wood much splintered to 3 feet in length by 1 foot in height, and the next plank above half broken through in the centre.

A great number of wooden splinters, and some of iron, ranged from 100 to 200 yards; the shot passed on to 900 yards.

The holes these two shots made on passing out, it would have been almost impossible to have stopped.

Thus, it appears that the destructive effects of the impacts of shot on iron cannot be prevented. If the iron sides are of the thickness required to give adequate strength to the ship (⅞, or at least ¾ of an inch), the shot will be broken by the impact; if the iron plates be thin enough to let the shot pass into the ship without breaking, the vessel will be deficient in strength; the shot will do its work, particularly in oblique or raking fire, more effectively than its splinters, and in passing out, make apertures more difficult to plug or stop than in passing in. When a clean hole is made by a shot penetrating an iron plate, the whole of the disc struck out by the shot is broken into numerous small pieces, which are driven into the ship with very destructive effects; and if the plate be so thick (viz. upwards of ¾ of an inch) as to cause the shot to break on striking, the fragments will nevertheless pass into the ship as in the case of a concussion or percussion horizontal shell (Arts. 257, Note, and 261), and so produce a terrific compound effect by the fragments of both. The author has now before him a heap of the specimens of both, consisting of small, ragged, deadly splinters, varying in weight from two or three ounces to as many pounds, which he picked up in experiments he witnessed. Better to have a neat, clean hole made in one's body, than to be lacerated by the cruel and, in most cases, incurable wounds produced by splinters detached from iron sides. The expedient of combining wood and iron, either by substituting timber for the iron ribs, or the reverse—outside planking for the iron plates—makes the matter worse. The pieces of ribs struck off, sometimes of great length, pass on with the shot, to produce more extensive ravages elsewhere.[a]

172. From what has been stated, it cannot be denied that iron vessels, however convenient and advantageous in other respects, are utterly unfit for purposes of war. This opinion has been confirmed by the decision of a mixed committee of officers of the naval artillery and engineers, who were assembled under the authority of the Admiralty and Board of Ordnance to consider how far it might be possible to carry into effect a plan for arming the Contract Mail Packet Steamers, and to report whether or not the terms of their contract have been observed by the several companies as regards the adaptation of their vessels for war-purposes. From this category the committee entirely and unanimously

---

[a] Descriptions of some of the splinters of Iron Plates, Rivets, and Shot, picked up in the space between the Targets set up to represent a section of an iron vessel. The splinters were taken promiscuously from innumerable fragments found there when the firing ceased.

Description.	Greatest Dimension.	Least Dimension.	Weight.
	Inches.	Inches.	Ounces.
Splinter of a plate, extremely jagged and pointed	$1\frac{1}{2}$	$\frac{3}{8}$	$\frac{1}{4}$
Splinter of a plate near a rivet, bent, jagged, and pointed.	2	$\frac{3}{8}$	1
Splinter of a plate near a rivet	$1\frac{6}{8}$	$\frac{3}{8}$	$1\frac{1}{4}$
Piece of a rivet, a good deal flattened, bent, and jagged.	$1\frac{2}{8}$	$\frac{3}{8}$	$1\frac{3}{4}$
Splinter of a shot much jagged	$1\frac{1}{2}$	$\frac{1}{2}$	$2\frac{1}{4}$
Splinter of a plate extremely jagged and pointed	2	$\frac{5}{8}$	$2\frac{3}{4}$
Splinter of a plate, with part of a rivet attached, of very irregular shape, pointed and jagged in all parts.	$2\frac{1}{4}$	Too irregular to be defined.	$3\frac{1}{4}$
Splinter of a shot, very jagged, and two of the edges sharp.	$1\frac{3}{4}$	Ditto	$3\frac{1}{2}$
The larger specimens varied from $3\frac{1}{2}$, $4\frac{1}{2}$, and 6 ounces, to nearly as many pounds; all of them were extremely jagged, pointed, and, in some cases, sharp at the edges.			

reject iron vessels. Exclusive of these they found that of the fifty-three vessels belonging to the Peninsular and Oriental and the West India Mail Companies, eight only (of wood) were capable of being effectually fitted to receive an armament especially directed to the object of defence; though some of the others might be fitted to serve as armed packets or armed troop-ships. In almost every case it was considered impracticable to have a pivot-gun either forward or abaft on account of the sharp form of the bow and the great rake of the stern. To put the vessels in a condition to carry and fire the guns assigned to them, it was estimated that the expense of alterations, and the fittings of ports with a magazine and shell room, would vary from 600*l.* to 800*l.*

173. It becomes now a question of serious import to decide whether iron steamers which have been condemned as ships of war, on account of the extensive ravages produced by the impacts and penetrations of shot upon and in vessels formed of that material, are fit to be employed as transports for the conveyance of troops and stores in the event of war.

In vessels made of iron the weight of the whole material is considerably less than that of vessels of the same dimensions but made of timber. With equal displacement (the weight of water displaced) therefore, the former can carry a greater weight of cargo, their capacity for stowage being greater on account of the thinness of their shell. But when an iron vessel is bilged and becomes filled with water, the superior weight of the material of which she is formed is a momentous consideration. Timber, when immersed in water, loses as much of its weight as is equal to that of the water displaced by it, and it floats. A cubic foot of oak timber has a buoyancy, when immersed in salt water, of 76 oz., and a cubic foot of fir, a buoyancy of 450 oz.; but the excess of the weight of a cubic foot of iron over an equal volume of salt water, is 6180 oz., and, with this force, the iron sinks. When, therefore, an iron ship is bilged, having lost its power of floating, the weight of the iron tends to break and destroy it, unless it be stranded on a smooth and shelving bank or beach. This was the case with the "Great Britain," which rested, throughout her length, upon the beach on which she grounded; but when an iron vessel strikes upon, and is perforated by a rock, as was the case with the "Birkenhead," she becomes bilged, and there she remains with deep water at her extremities: she then becomes filled, either wholly or partially, and the iron, deprived of buoyancy, exerts a prodigious force to break the vessel's back and sink the portions which are not in contact with the rock. Even if the portion which is not in such contact should be furnished with compartments which may not permit the water to enter, the difference between the power of floatation in that waterborne portion, which acts in a contrary direction to the weight of the iron, will constitute a strain which no iron vessel can resist, tending to fracture it at the section which divides the filled from the unfilled portion. Should there be no water-tight compartments, or should these not effectually act, then an iron vessel, perforated near the midship's section, with deep water under her extremities, will infallibly be destroyed by the weight of the iron at both ends, acting upon the fulcrum on which the vessel rests.

When an iron vessel parts, all the fragments of the material of which she is composed sink, and nothing floats to save life but a few loose spars: whereas the fragments of a timber ship float, and many, perhaps most of the crew, may thus be saved. In the fearful wreck of a transport, in which the author once was, the whole of the quarter-deck, from the mainmast to the stern, with the top-sides to which it was attached, came on shore like a raft, when the vessel broke up. What had been our fate if that transport had been formed of iron?

Applying these observations to the melancholy case of the "Birkenhead," and on a full consideration of the facts as taken from the report of the proceedings of the court martial on the surviving officers and crew of that ship, there cannot be the least doubt that, had she been a wooden vessel, she would have held together long enough in a sea so smooth as to permit all her hands to be saved. To this conclusion it may be objected, by persons who have not duly considered the difference of circumstances, that the "Avenger," though constructed of timber, went to pieces even more quickly than the "Birkenhead," when, going at full speed, she struck upon a reef of rocks; and that, with the exception of a small boat's crew, all hands perished; but, on that occasion, the sea broke over the reef with such violence as to render it impossible that any vessel could resist it, or that any of the crew could be saved by clinging to the fragments into which it was broken.

174. Having several of these vessels on hand, the Admiralty have acted wisely in employing them as troop-ships at present. Four have accordingly been appointed to, and fitted for, that important service. The armaments have been removed from the main-decks, and a few light guns placed on the upper-decks, where timber is substituted for iron in the formation of their bulwarks; this removes some of the objections to iron sides, inasmuch as the men fighting the guns on those decks would not be liable to suffer more from an enemy's shot than in a wooden ship.

There can be no doubt of the advantages and expediency of employing those vessels as troop-ships in time of peace. Their safety from conflagration; their speed with steam power; the advantages of the screw, which admits of such vessels having great sailing power, and, on the whole, their great capacities and capabilities of voyaging, which have been so well exemplified in their performances, prove that the best possible use is being made of those vessels by so employing them during peace. But, in war, it is not enough that the people on the upper decks will not be exposed to the destructive effects produced by shot on iron, by the substitution of timber for that material, in the upper works; for the main-decks appropriated to the troops will still be liable to the terrific effects described and delineated as above:

and although, by the happy expedient of water-tight compartments, the vessel, however desperately damaged, may be prevented from sinking and the troops from being drowned, the latter cannot be protected from being killed and wounded in great numbers, or captured, in the event of falling in with, and being attacked by, any enemy's vessel armed with even a very few heavy guns.

How far iron ships are fit and safe as packets, in times of war, is another question, which will have to be well considered. The Government packets are, by their contracts, to be so built and fitted, that they may, in the event of hostilities, be converted into vessels of war: if so, it is clearly demonstrated by the facts established in these most important proofs, that vessels subject to these stipulations must not be made of iron: for of this we are sure, that iron vessels are, and will be found, unfit for all purposes of war. Whilst, therefore, the best and the most is being made of our existing iron ships, employed as they now are, we should be prepared with other means to carry on that most important service in war. We have noble ships, which have whole decks available for the reception of troops, great space for the stowage of stores—vast steam ships, possessing full steam, full sailing, and full gunnery powers, admirably adapted to these services, highly capable of executing them quickly, safely, and easily; prepared for any contingency, whether to fight or run, and therefore best suited to carry out that grand maxim—that promptitude, celerity, strength, and certainty in the modes and means of communication, either to succour and support our remote possessions, or to defend the centre of the empire, is, in fact, a practical multiplication of the forces of the state.

175. IX.—*On Cylindro-conoidal and Excentric-spherical Projectiles.*—Sir Isaac Newton has given in the *Principia* (lib. ii., schol. to prop. 34) an indication of the form of a solid body which, in passing through a fluid, would experience less resistance than a body of equal

magnitude and of any other form. He imagined that this might be of use in ship-building, and it is evident that the principle is equally applicable in the theory of projectiles. Investigations of the differential equations of the curve may be seen in Vince's *Fluxions*, in Airy's *Tracts* (p. 237), and in the writings of other mathematicians. The body is a solid of revolution, and the differential equation is—

$$y = C \frac{dz^4}{dy^3\,dx}$$

in which C is a constant.

The form of a section through the axis of the solid is given in the annexed diagram. A B is the axis, and in the direction of that line the solid is to move; y is any ordinate, as D C; and dx, dy, dz, are elementary portions, E F, E D, D F respectively. The end B, as well as A, of the solid is a plane surface; for the numerator of the fraction in the above equation will, evidently, be always greater than the denominator, and therefore y, the ordinate to the curve, can never be zero.

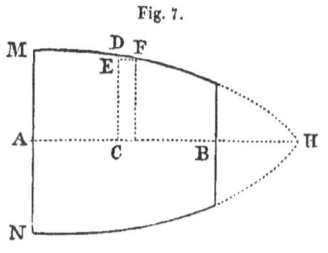

Fig. 7.

176. It is plain, however, that the minimum of resistance would not be obtained with a shot of an elongated form, when discharged from a musket or piece of ordnance, unless the axis A B can be kept in the direction of the trajectory. This may be accomplished if the shot be caused to have a rotatory motion on that axis by being discharged from a rifled bore; and without such rotation, not only will the axis perpetually deviate from the direction of the path, but the projectile will even turn over.

177. The advantages of this form of shot are, that when rotating on their longitudinal axes, and moving with their smaller extremities in front, they experience less resistance from the air than spherical projectiles of the same diameter. To this form alone are to be referred

the long range with the great momentum and penetrating power of the projectiles for rifle-muskets (Appendix A.), which have recently been introduced in the British and foreign military services.

178. Hollow cylindro-spherical projectiles, filled with powder and furnished with percussion caps at their points, were proposed some years ago for rifled muskets, by Captain Norton, for the purpose of penetrating into and exploding the ammunition packed in tumbrils and limber-boxes, and the experiments made with them met with some success; but, on account of the great difficulty of insuring the ignition of the charge on the shot striking, these missiles were found so uncertain that they have not been adopted.

179. The desire of obtaining the advantages which are stated in Art. 100 to be possessed by oblong shot, in combination with those alluded to in Art. 177, probably led to the formation, for heavy ordnance, of shot uniting the cylindrical with the conical or conoidal figures. Cylindro-conical and cylindro-ogivale (cylindro-conoidal) shot and shells, fired from rifled cannon, have recently attracted so much of the attention of professional and other persons, that perhaps some account of these projectiles may be looked for in a treatise which, though dedicated more particularly to the service of naval artillery, is intended to explain generally the theory and practice of military projectiles.

Iron rifled guns for the purpose of discharging such projectiles were, in 1846, invented by Major Cavalli, of the Sardinian artillery, and by Baron Wahrendorff, a Swedish noble. These are loaded at the breech, and experiments for the purpose of testing the merits of their shot have been carried on both at Åker in Sweden and at Shoebury Ness in this country. (See Arts. 233–235.)

The projectiles alluded to are represented in the subjacent figures 8 and 9; the first is designated cylindro-conical, and the other cylindro-conoidal; their entire lengths are about $16\frac{3}{4}$ and $14\frac{3}{4}$ inches, respectively, and their greatest diameter $6\frac{1}{2}$ inches: each has two pro-

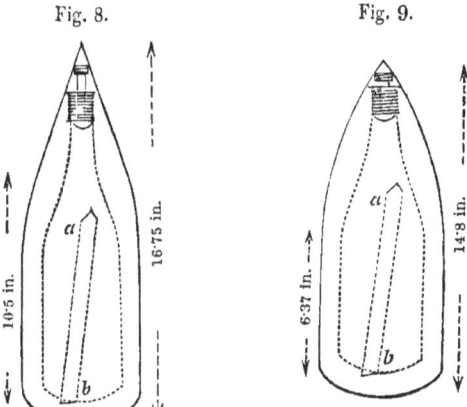

Fig. 8. Fig. 9.

jections, $a\,b$, directly opposite to one another, and ¼ inch deep, which enter the grooves in the rifled bore. (These projections make an angle of 7° 8' with the axis of the shot.) Each shot, if hollow, weighs about 69 lbs.; and if solid, about 101½ lbs. The hollow projectiles of M. Cavalli are each furnished with a double percussion agent; one part is the ordinary percussion cap, and the other consists of four small capsules of glass containing sulphuric acid, placed about the neck of the shot, the intervals being filled with chlorate of potash, while the upper part of the neck is filled with common gunpowder, and stopped with wax.[a] It is said that the shock produced by one of these projectiles, when let fall on a pavement from a height of several metres, is not sufficient to cause explosion, and that this will not take place when the shot fired with a small charge strikes earth; it being only when fired with a service charge sufficient to break the glass globes by the shock that explosion takes place. M. Cavalli admits, however, that inconvenience and uncertainty exist in using such means, from the difficulty of combining the chemical ingredients so that the shell may resist the shock produced by the explosion of a service charge of powder, and yet be acted on by the shock of impact, which, in general, is

---

[a] The projectiles are fully described in a pamphlet by M. Cavalli, entitled 'Mémoire sur les Canons se chargeant par la Culasse,' &c., Paris, 1847, p. 28. The glass globes are not shown in the above figures.

much less than that of the explosion. This difficulty attaches to all such modes of procuring the explosion of shells on striking a resisting body, and, in point of fact, has hitherto stood in the way of the production of a good concussion fuze or an efficient percussion shell.

180. Dr. Hutton recognised, from his experiments in 1796, that there was a certain position of a projectile in which its deviations were less than in other positions, but no development of the discovery appears to have taken place till M. Clement, in 1808, remarked, not only that the irregularities in the lengths of the ranges depended on the same cause as the lateral deviations, but also that the *obus* (a shell whose exterior and interior surfaces are not concentric) occasionally afforded ranges longer and more correct in proportion as the distance of its centre of gravity from the centre of the figure was greater. At length, between 1835 and 1840, a series of experiments were carried on in Belgium with great care, and with ordnance of different calibres, from which it was ascertained that when the centre of gravity was above that of the figure, the range was greater than when it was below, in the ratio of 2467 to 1315; and that when the centre of gravity was to the right or left, the deviation of the shot was in like manner towards the right or left.

181. These results were confirmed by experiments which M. Paixhans carried on at Metz in 1841.[a] This officer caused to be projected from a 12-pounder (Fr.) field-gun a homogeneous spherical shot with the usual *sabot* at an elevation of 4°; and afterwards, with an equal charge and elevation and a like sabot, a ball having in it a cavity which caused the centre of gravity to be at a certain distance from the centre of the figure, and diminished the weight of the shot from 13.24 lbs. to 11 lbs. The following were the effects observed :—

When the centre of gravity was above the centre of the figure the ranges were the longest, and, when below, the shortest; when to the right or left hand, the devi-

---

[a] See 'Constitution Militaire de la France,' note p, p. 243. Paris, 1849.

ations were also to the right or left. The mean range which, with the usual shot, was 1640 yards, was, with the shot whose centres of gravity and of figure were not coincident, the centre of gravity being upwards, equal to 2140 yards, being an increase of 500 yards. The mean of the lateral deviations, which with common shot was $\frac{13}{100}$ths of the range, with the last mentioned shot was only $\frac{8}{100}$ths. Similar results were obtained at the same time with shells 0.15 metres (5.9 inches) diameter, both with high and low charges, the elevation being 4°; when, with a charge of 1 kilogr. (2.206 lbs.), the mean range of the ordinary shell was 1275 yards; while with shells whose centres of gravity and of figure were not coincident, the centre of gravity being upwards, it was 1756 yards. The mean lateral deviation, which with common shells was $\frac{16}{1000}$ths of the range, was, with the last-mentioned shell, reduced to $\frac{8}{1000}$ths only. M. Paixhans remarks, however, as an anomaly, that with this howitzer the lengths of the ranges were more nearly equal when the ordinary, than when the other kind of shell was used.

In 1842 M. Paixhans made experiments with his canon-obusier of 80 livres, from which he found that, with a charge of 13.24 lbs. of powder and an elevation of 10°, the usual shell had a range equal to 2560 yards, while the non-concentric shell, its centre of gravity being placed upwards, had a range equal to 3198 yards: with a charge of 17.6 lbs. the range of the latter shell was 3540 yards.[a]

182. M. Paixhans infers from these experiments that if this method of placing shot, in which the centres of gravity and of figure are not coincident, should be employed with projectiles which are not spherical (as the cylindro-conoidal shot), the advantages above mentioned will be obtained in a much higher degree; and he adds that experiments with such shot would cast light upon a part of the theory of projectiles which is at present involved in considerable obscurity. At the same time

---

[a] 'Constitution Militaire de la France,' note D, p. 246. Paris, 1849.

there are artillerists who go so far as to say that the influence of excentricity in the projectile, on the form of the trajectory it describes, is such as to permit what has hitherto been considered an imperfection, and a cause of error, to be made the means of regulating and compensating the deviation to which all shot are subject, and thus to render the practice with excentric shot more precise than that of concentric spherical balls, whether solid or hollow.

This assertion, if true, would entirely overthrow what has ever been stated as an essential condition of accuracy in the flight of projectiles, and would impugn what has been inculcated in this work (Art. 87, with the Note, and Art. 116), on the importance of making spherical projectiles as perfect in form and homogeneity as possible, and of preserving them, with the most scrupulous care, free from rust, perfectly clean, and well lacquered.

No doubt the form of the trajectory of an excentric spherical projectile will be very much influenced by the position which its centre of gravity occupies when placed in the gun; and to M. Paixhans is due the credit of showing that the deviations may thus be made to take place either in vertical or lateral directions with tolerable certainty; and if the amount of the deviations were in every case constant, the practical value of such projectiles would be most important.

It has been said by a foreign artillerist that we "kept secret the remark made by Hutton, in 1796, that when a projectile is placed in a gun, there is a certain position of its centre of gravity, with respect to the axis of the bore, which diminishes the lateral deviations and increases the range" (which it does when the centre of gravity is placed upwards), "and that in 1813 the secret was communicated to the Prussians, who accordingly made some important experiments with projectiles of different degrees of excentricity, and discovered a method by which the lateral deviations might be diminished." It is added that "in 1826 and 1827 experiments were made at Mayence to ascertain the effects of excentricity, and it is stated that, from these, the theory of the flight of spherical projectiles was proved to be erroneous; but the results were kept secret, and little was done to improve the practice."

But there is no "obscurity" as to the cause of the rotations which produce these deviations, nor any difficulty in understanding the manner in which they may be made to act in particular directions; and it will be shown hereafter that the experiments which have been made in this country with excentric spherical projectiles, far from overthrowing, confirm, in a remarkable degree, the soundness of the theory, and the truth of the maxim, that of all spherical projectiles, those are the most accurate in their flight, which are most perfect in sphericity and homogeneity; and it will be seen that the only practical advantage resulting from the employment of excentric shot and shells, is the increase of range which takes place when the centre of gravity is placed upwards; it must be remembered, however, that this increase is variable in amount, and that it can only be of use in firing at great distances, as against arsenals, roadsteads, or fortresses, so large and crowded that the objects can scarcely be missed.

When the centre of gravity is not coincident with that of figure, the projectile is made to revolve *ab initio* on the former centre; thus occasioning a compound motion in the flight of the projectile. Now, the shot being placed in, and fired from the gun, with the line joining the centres of gravity and figure in a vertical plane, and perpendicular to the axis of the gun; if the centre of gravity be above that of figure, the shot on leaving the gun must continue, as when passing along the bore, to turn over on a horizontal axis perpendicular to the plane of projection, the resultant of the projectile forces in the powder causing the front to turn in a direction from below upwards: hence the resistance engendered at the anterior surface of the projectile, by the rotation, gives rise to a force of reaction which produces a motion of the shot upwards. The resistance engendered at the posterior surface of the projectile, by the same rotation, gives rise to a force which produces a motion of the shot downwards; but, as the density of the air is greater in front of, than behind the ball, on the whole the motion is upwards, and this increases the extent of the range.

If the ball moves in a direct line, and its axis remains always in the direction of that line, the air will be compressed in front of, and rarefied behind the projectile, but will remain symmetrical with respect to density on the sides around the axis of rotation; and no deviation, or error of *derivation,* as M. Tamisier names it (because it is derived from the rotation of the projectile), will be engendered.

When the centre of gravity is below the axis of the bore, the resultant of the projectile forces will cause the front of the shot to turn from above downwards, and, a rotation in this direction continuing, the range will be diminished. In like manner, when the centre of gravity is placed on the right or left hand of the axis of the bore, the shot will turn on a vertical axis, and produce deviations to the right or left hand respectively. The cut represents a front view of the muzzle of a gun, with the projectile in it. The points *a, b, c, d,* show the different positions of the centre of gravity in a vertical section through the geometrical centre of the shot.

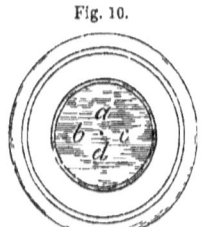

Fig. 10.

183. With respect to the ricochet of excentric spherical projectiles, there can be no doubt that the rotation which causes deflection in the flight must act in a similar manner to impede a straightforward graze. When an ordinary well-formed homogeneous spherical projectile, upon which probably very little rotation is impressed, makes a graze, the bottom of the vertical diameter first touches the plane and, immediately, the projectile acquires, by the reaction, a rotation upon its horizontal axis, by which the shot rolls onwards throughout the graze, favourably for a straight-forward second flight.

PART II.   THEORY AND PRACTICE OF GUNNERY.   157

But in the case of an excentric spherical projectile, placed with its centre of gravity to the right or to the left, its rotation upon its vertical axis during the graze must occasion a fresh deflection in its second flight; and it is only when the centre of gravity is placed in a vertical plane passing through the axis of the gun, that the rotation occasioned by touching the ground will not disturb the direction of the graze, though the extent of range to the first graze will be affected more or less, accordingly as the centre of gravity of the projectile may have been placed upwards or downwards. In the former case the rotation of the projectile conforms with that produced by the graze, and will not therefore

Fig. 11.

Fired with the Centre of Gravity Upwards.

retard the motion so much as when the centre of gravity is placed below. In the latter case, the projectile turning in the reverse direction tends to increase the resistance of the plane to the progress of the shot during the graze. In neither of these cases is there any tendency to produce deflection if the medium struck be uniform.

184. Although convinced of the soundness of the theory respecting the homogeneity of spherical projectiles and the truth of the practical maxims founded thereon, as well as aware of the deviations occasioned by excentricity, the author deemed the subject one of so much importance as to require the fullest experimental investigation; and he addressed accordingly to the Lords Commissioners of the Admiralty, and to the Master-General of the Ordnance,[a] an earnest recommendation that courses of experiments should be insti-

---

[a] July 9, 1850.

tuted on board Her Majesty's ship "Excellent," and at Shoebury Ness, to ascertain the amount of the deviations, vertical and horizontal, in the flight of excentric spherical projectiles when placed in the gun with the centres of gravity and figure in different positions with respect to the axis of the bore, and to compare the flight of such projectiles with that of projectiles formed with the utmost possible attention to homogeneity and sphericity. The object in having such experiments made was to ascertain whether the deviations of the former were so regular as to admit of being allowed for in pointing the gun; and whether any result might appear to disprove the maxim that spherical and homogeneous projectiles are the truest in their flight. The recommendation was, with the utmost promptitude and liberality, carried into effect; and courses of experiments were instituted both at Portsmouth and Shoebury Ness.

The following are the results of the experiments made on board the "Excellent" at Portsmouth.

The positions of the centre of gravity are stated with respect to the geometrical centre of the shot; and the ranges with the deflections, which are expressed in yards, are the means of those obtained from the different rounds fired in the like circumstances.

July 18, 1850. With a 32-pr. gun of 56 cwt., the quantity of metal removed from one side of the shot being 1 lb. :—

Position of the Centre of Gravity with respect to the Centre of the Shot.	Charge 8 lbs., Elevation 2° 7½'.		Charge 10 lbs., Elevation 3° 36'.	
	Range.	Deflection.	Range.	Deflection.
On the Right . .	1032	6 right.	1474	20 right.
On the Left . .	1163	7 left.	1479	24¾ left.
Upwards . . .	1433	7 right, 3¼ left.	1991	20 right, 6 left.
Downwards . .	980	5 right, 3 left.	1499	2 right, 6¼ left.
Inwards . . .	1150	4¼ right.	1608	5¼ right.
Outwards . . .	1097	4⅜ right.	1428	9 right.
Concentric . . .	1160	4 right, 3 left.	1624	1 right.

PART II. THEORY AND PRACTICE OF GUNNERY. 159

July 26, 1850. With an 8-inch gun, 9 feet long, of 65 cwt., the quantity of metal removed from one side of the shot = 5 lbs. 5 oz. :—

Charge 10 lbs., Elevation 5°.			Charge 10 lbs., Elevation 2° 30'.	
Position of the Centre of Gravity.	Range.	Deflection.	Range.	Deflection.
Right	1617	14⅔ right.	1040	5 right, 1 left.
Left	1684	17 right.	1028	2 right, 4¼ left.
Upwards . . .	1940		1146	4½ left.
Downwards . .	1562		957	3 right, 3¼ left.
Inwards . . .	1740	8⅔ left.	1126¾	1½ right, 1 left.
Outwards . . .	..		1080	8½ left.
Concentric . . .	1702	3⅓ right.	1033	2¾ right, 2 left.

August 7th, 1850. A 32-pr. gun of 56 cwt., 9 ft. 6 in. long. Quantity of metal removed from one side of the shot = 1 lb. :—

Charge 8 lbs., Elevation 12°.		
Position of the Centre of Gravity.	Range.	Deflection.
Upwards . .	3614	Variable.
Downwards . .	2676	Variable.
Coincident . .	3058	16½ right, 29 left.

An 8-inch gun of 65 cwt., 9 ft. 6 in. long. Quantity of metal removed from one side of the shot = 3 lbs. :—

Charge 10 lbs., Elevation 10°.		
Position of the Centre of Gravity.	Range.	Deflection.
Upwards . .	3612	Variable.
Downwards . .	2480	Variable.
Coincident . .	2765	13 right, 13¼ left.

185. The experiments were carried on under the direction of Captain (Rear-Admiral) Chads, C.B. The shot were rendered excentric by a hole drilled in each of them on one side, and afterwards filled up with a wooden plug, or covered with a brass plate.

From the preceding abstract it is evident that the vertical deviations arising from the centre of gravity being placed upwards produced considerable augmentations of range, and, though these were variable in amount, yet since they might become of great practical utility for particular purposes, it was desirable to ascertain whether a like increase of range might not be obtained when excentric shells are projected at great elevations from the most powerful guns.

186. The author, in consequence, addressed to the Master-General of the Ordnance (Memorandum of Sept. 5, 1850) a recommendation that experiments should be made at Shoebury Ness with a 32-pounder gun, with solid shot carefully selected, and with shot made excentric by removing some of the metal, as in the previous experiments, and likewise with excentric 68-pounder shot and excentric 8-inch shells at moderate and at great ranges. He also recommended that experiments should be made in order to ascertain how much an excentric projectile revolving (as when the centre of gravity is placed sideways in the gun) upon its vertical axis would be deflected from its original course on grazing the ground.

Experiments with these particular objects in view being immediately ordered to be made, they accordingly took place at Shoebury Ness, under the direction of Captain Walker, R.A., and the following Tables contain an abstract of the results.[a]

Sept. 16, 1850. With a 32-pounder gun of 60 cwt., charge 10 lbs., elevation 2° 30′, when the shot was strapped, and its axis placed accidentally in the bore, the mean of 10 rounds gave for the range 1392 yards,

---

[a] In the preceding and following tables the means of the deflections towards the right and left hand are given separately. This method has been preferred because it exhibits the mean ratio of the deflections in opposite directions. The usual process is to add together (algebraically) all the deflections, considering those on the right and left to have contrary signs, positive and negative, and to divide the sum by the whole number of rounds; but this process exhibits an appearance of accuracy which does not always exist. Some artillerists have preferred taking the arithmetical mean of the deflections to the right and left without distinction of direction.

PART II.   THEORY AND PRACTICE OF GUNNERY.   161

and for the means of the deflections $6\frac{11}{14}$ yards to the right, $4\frac{1}{3}$ yards to the left. When the axis of the shot was placed in coincidence with the axis of the bore, the mean range was 1416 yards, and the mean deflection $5\frac{2}{3}$ yards to the right.

Sept. 17th. With the same gun, charge 8 lbs., elevation 2° 30', when the axis of the shot was placed accidentally the mean range was 1299 yards and the mean deflections $12\frac{1}{3}$ yards to the right, 7 yards to the left; and when the axes of the shot and bore were in coincidence, the mean range was 1296 yards; the mean deflections were 10 yards to the right, $2\frac{2}{3}$ yards to the left.

Sept. 19th and 20th. With the same gun the following results were obtained. (The first column in the three tables which immediately follow shows the position of the centre of gravity with respect to the geometrical centre of the shot. The distances are in yards.)

Position of the Centre of Gravity.	Mean Ranges.	Mean Deflections.	Mean Ranges.	Mean Deflections.
Charge 8 lbs., Elevation 2° 30'.			Charge 10 lbs., Elevation 2° 30'.	
Below	1137	$6\frac{1}{4}$ right, 4 left.	1287	10 left.
Above	1570	$6\frac{3}{4}$ right, $20\frac{3}{4}$ left.	1633	$9\frac{3}{4}$ right, 21 left.
Right	1395	$9\frac{3}{4}$ right.	1346	$13\frac{1}{4}$ right.
Left	1252	$17\frac{1}{4}$ left.	1364	$19\frac{3}{4}$ left.
Inwards . . .	1355	$6\frac{3}{4}$ left.	1393	4 left.
Outwards . . .	1322	$13\frac{1}{4}$ left.	1386	$2\frac{1}{4}$ right, $3\frac{1}{3}$ left.

October 3rd and 9th. With a 32-pounder gun of 56 cwt.

Position of the Centre of Gravity.	Mean Ranges.	Mean Deflections.	Mean Ranges.	Mean Deflections.
Charge 10 lbs., Elevation 12°.			Charge 10 lbs., Elevation 2° 30'.	
Accidental . . .	3006	59 right, 80 left.	1401	$7\frac{1}{3}$ right, 2 left.
In the axis of the bore	3114	72 right, 54 left.	1397	$4\frac{1}{6}$ right, 6 left.
Right	. .	. .	1257	16 right, $6\frac{3}{4}$ left.
Left	. .	. .	1244	$8\frac{3}{4}$ left.
Upwards . . .	3498	185 right.	1594	$22\frac{3}{4}$ right.
Downwards . .	2598	$24\frac{1}{4}$ right, 178 left.	1180	$4\frac{1}{4}$ right, 2 left.

M

October 10th and 12th. With an 8-inch gun, 65 cwt., and the weight of the solid shot = 68 lbs., from which 4 lbs. of metal were removed in order to render the shot excentric.

Position of the Centre of Gravity.	Charge 10 lbs., Elevation 5°.		Charge 10 lbs., Elevation 10°.	
	Mean Ranges.	Mean Deflections.	Mean Ranges.	Mean Deflections.
Accidental . . .	1820	16¼ right, 18⅓ left.	2616	4⅔ right, 23½ left.
In the axis of the bore	1762	15⅔ right.	2702	80¼ right, 32 left.
Right	1726	53¼ right.	2558	181 right.
Left	1608	15 right.	2332	57 left.
Upwards . . .	2207	26⅔ right, 4⅔ left.	3339	141 left.
Downwards . .	1462	20 right.	2051	66 right.

In October, 1850, Capt. (Rear-Admiral) Chads made, on board the "Excellent," some important experiments with ten-inch shells, which he rendered excentric by boring two holes in each, diametrically opposite to one another, stopping up one with 5 lbs. of lead, and the other with wood. The interior was occupied by a quantity of sand and ashes equal to the weight of the powder required to fill the shell. The lead was introduced in a melted state, and some of it was allowed to flow about the interior of the aperture, which was of a conical form with the smaller diameter inwards. Shells thus prepared, and others prepared in a similar manner with 6 lbs. and with 7 lbs. of lead, also common shells, were fired at elevations of 15° with 12 lbs. of powder; and, while the mean range of the latter was 3073 yards, the others, when the centre of gravity was upwards, ranged from 3200 to 3550 yards. These important results were produced by an excentricity which amounted only to 0.15 inch—a circumstance which will no doubt be noticed, and lead to the trial of a greater excentricity, in future experiments.

187. The method adopted in making experiments with excentric shot was, first to ascertain the position of the centre of gravity, by floating the projectile in mercury, and marking its culminating point or vertex; then a mark upon the shot diametrically opposite to

that point, gave the direction of the axis in which the two centres lay; thus the shot, when fixed to a wooden bottom, could be placed in the gun with the centre of gravity in the several positions, upwards or downwards, right or left, inwards or outwards, in which the flight was to be tried.

On making these experiments, it appeared that not above one iron shot in a hundred, when floated in mercury, was indifferent as to the position in which it was so floated, but turned immediately until the centre of gravity arrived at the lowest point, and consequently that not one shot in a hundred was perfect in sphericity and homogeneity. This defect demands the most serious consideration. It has been remedied, as we have seen (Art. 87, Note), in the fabrication of leaden bullets, and there can be no doubt that, if duly notified to the contractors for the supply of iron shot, mechanical skill and ingenuity would discover some mode, by the employment of wrought iron,[a] of providing more perfect

---

[a] A proposition, emanating from an officer of the Royal Engineers (Captain Symmons), of great acquirements and intelligence, and much experience in the manufacture of wrought-iron girders and other heavy articles, has recently been made for fabricating guns of wrought iron, which, with equal powers of resistance to guns of cast iron, may be much lighter, and more capable of resisting the explosion of the charge. The reader will perceive by the Note, Art. 211, that Mr. Daniel Treadwell, of the United States, constructed a 32-pounder sea-service gun, in great part of pieces of wrought iron welded together, and compressed by a hydrostatic machine of great power, the whole mass formed into a gun by turning and boring—therefore this is not a new proposition, so far as material is concerned; but it is anticipated that the means recently discovered for welding and consolidating bars of wrought iron into uniform masses will allow of a gun being produced of that material, sufficiently sound and tenacious to withstand the action of repeated discharges of shot by the agency of gunpowder.

With respect to shot, it may be stated that a manufacturer of great experience is now prepared to make shot of wrought iron for experiment. These will not be so apt to split or break as shot of cast iron; they may also be made more spherical, and perfectly homogeneous: it is, therefore, extremely desirable that experiments should be instituted to try that important proposition. Wrought-iron shot may be more liable to rust and corrode than cast-iron; but by proper precautions this may be prevented. If heated to a certain temperature, and then plunged into a vessel containing pitch, a very enduring lacker will be formed.

In a letter from Messrs. Fox, Henderson, and Co., addressed to Captain Symmons, it is stated that a wrought-iron shot 6.2 inches diameter would weigh 34.24 lbs., and one 6.3 inches diameter would weigh 35.92 lbs. Cast-iron shot of those diameters weigh, respectively, 31.89 lbs. and 33.46 lbs.

shot, as well in figure as in density. The process by which this may be effected will no doubt render the shot much more costly, and, if it were necessary for general service, the expense would prove a fatal objection to the suggestion, but this is not required. Ordinary shot, however imperfect in these respects, are sufficient at distances and under circumstances which do not require a high degree of perfectibility. But when a successful result may depend upon the effect of one or two rounds—suppose from the pivot-gun of a large steamer, which may have just attained a favourable position, not likely soon to be regained, if the effect of its shot should fail—economy in expense becomes the worst of prodigality. It is now proved, beyond all doubt, that perfect concentricity is indispensable to accuracy in the flight of spherical projectiles; and no expense, therefore, should be spared in providing at least a few correct shot for such occasions.[a]

188. On analyzing the experiments both at Portsmouth and Shoebury Ness, it appears that the flight of the ordinary solid shot was the most true, the lateral deflections being frequently but one half, sometimes one third or one fourth only, of the deflections of the excentric shot; that these last deflections were always in the direction in which the centres of gravity of the shot were placed in the gun;[b] and that the increases or diminutions of range caused by the vertical deviations

---

Also that the price of 200 wrought-iron shot weighing 35 lbs. each (in all 3.12 tons) would be (5s. 5½d. each) 54l. 12s. If the shot were galvanized, the additional cost would be (1s. 3½d. each) 13l. 2s. Hence the total cost of 200 wrought-iron shot 6.2 in. diameter, galvanized, would be 67l. 14s.

[a] It is rather singular that the French, while professing to attach great importance to the perfect sphericity and concentricity of projectiles, as being essential to their steadiness of flight and accuracy of practice, should, in their recent formations, have departed from the principle. It being found that the windage of the canons-obusiers was not sufficient to admit such shells as were strapped by slips of metal to the wooden bottom; in order to remedy this inconvenience they diminished the equatorial diameters of the shells by 0.5 millim. (.019 inch), extending the reduction over a zone about 32 millim. (1.26 inch) broad. They have thus brought the shell to the form of a spheroid, and have, thereby, entirely destroyed the good qualities above mentioned.

[b] There is one exceptional case only in the whole of the trials.

were produced, respectively, as the centres of gravity of the shot were placed upwards or downwards. It appears also that the lateral deviations, though, in general, constant in direction, were very variable in amount. In the experiments carried on at Shoebury Ness, on the 9th of October, 1850, the vertical deviation of a 32-pounder shot, with a charge of 10 lbs. and an elevation of $2\frac{1}{2}°$, produced an increase of range equal to above 400 yards. And, on the following day, the vertical deviation of a 68-pounder shot from an 8-inch gun of 65 cwt., with a charge of 10 lbs. and an elevation of 5°, produced an increase of range nearly equal to 750 yards. There can be no doubt, therefore, that from a 68-pounder gun of 95 cwt., with a full service charge, and an elevation of 15° or 20°, a range may be obtained, which, under certain circumstances, may be of very great importance to the naval service of this country.

In compliance with the recommendation of the author, as above stated (Art. 186), four guns of 112 cwt., mounted on carriages which admit of elevating the gun up to 32° inclusive, were prepared as proposed by Captain (Rear-Admiral) Chads, and the experiments were resumed at Shoebury Ness in July, 1851.

The shells were made excentric by being cast with a solid segment, a, weighing 4 lbs., of the interior sphere, left in the shell (fig. 12), by which the weight of the shell was increased to 91 lbs. The figure shows the position of the excentric shell as placed in the gun with the centre of gravity upwards, for the purpose of determining the longitudinal deflection, or increase of range, thence arising.

Fig. 12.

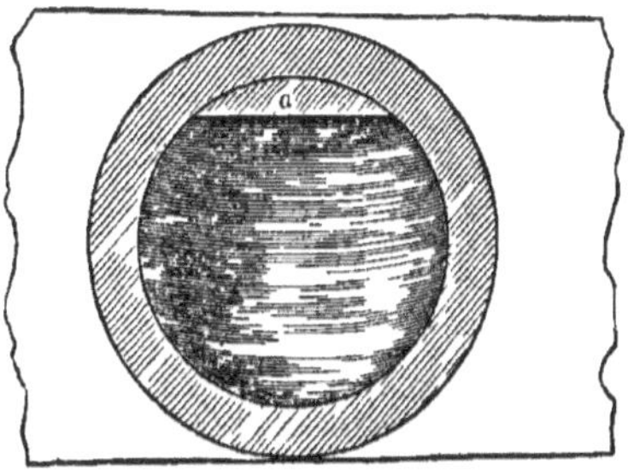

RESULT of COMPARATIVE TRIALS from 10-inch Guns of 112 cwt., 10½ feet in length, with Excentric and Concentric Hollow Shot of 91 lbs. and 87 lbs. respectively, and 15 lbs. Charges of Powder, taken at Shoebury Ness, in the year 1851.

Elevation.	Nature of Projectile.	Ranges.	Increase of Range obtained with Concentric Shell.	Deflections to Right.	Deflections to Left.	Times of Flight.
°		Yards.	Yards.	Yards.	Yards.	
2	Excentric . .	1192	..	$1\frac{2}{5}$	$3\frac{2}{3}$	$3\frac{24}{40}$
	Concentric . .			Not taken.		
4	Excentric . .	1830	145	$9\frac{3}{5}$	$9\frac{1}{3}$	$6\frac{10}{40}$
	Concentric . .	1695		9	$10\frac{2}{3}$	$5\frac{27}{40}$
6	Excentric . .	2485	..	6	$18\frac{1}{8}$	$9\frac{15}{40}$
	Concentric . .			Not taken.		
8	Excentric . .	3024	559	$48\frac{2}{3}$	61	$12\frac{10}{40}$
	Concentric . .	2465		$25\frac{1}{3}$	13	$9\frac{17}{40}$
10	Excentric . .	3472	..	$23\frac{1}{3}$	50	$14\frac{32}{40}$
	Concentric . .			Not taken.		
12	Excentric . .	3805	621	37	116	$16\frac{39}{40}$
	Concentric . .	3184		$72\frac{1}{3}$	54	$13\frac{19}{40}$
16	Excentric . .	4558	749	210	$145\frac{2}{3}$	$21\frac{29}{40}$
	Concentric . .	3709		30	89	$15\frac{32}{40}$
20	Excentric . .	5076	939	$87\frac{1}{2}$	$105\frac{1}{2}$	$25\frac{34}{40}$
	Concentric . .	4137		157	109	$19\frac{18}{40}$
24	Excentric . .	5311	706	162	255	$28\frac{13}{40}$
	Concentric . .	4605		45	280	$24\frac{3}{40}$
28	Excentric . .	5566	916	185	221	$32\frac{26}{40}$
	Concentric . .	4650		89	198	$24\frac{27}{40}$
32	Excentric . .	5536	670	361	139	$34\frac{33}{40}$
	Concentric . .	4866		321	298	$28\frac{12}{40}$

It appears from the preceding Table, that the excentric shell reached its greatest power of range at 28°, whilst that of the concentric shell continued to gain up to 32°; it remains to be seen whether at higher elevations the concentric shell would gain upon or overtake the other.

The deflections with both projectiles at these high elevations were very great, but those of the excentric cannot be said to have been in every case greater than those with the concentric shell; seeing that at 12° and 20° it was the reverse, and at 32°, whilst the deflection to the right was nearly equal, that to the left was more than twice as much with the concentric as with the excentric projectile.

In these experiments both the excentric shells and hollow shot were occasionally found to be cracked when recovered after firing. In prosecuting these experi-

PART II.   THEORY AND PRACTICE OF GUNNERY.   167

ments it may therefore be necessary to increase the thickness of metal in both these projectiles. The increase of weight thus rendered necessary, the charge remaining the same, will reduce somewhat the initial velocity : in some cases it may increase, in others diminish the range (Arts. 238, 243) ; but the projectile will be enabled the better to overcome the resistance of the air, and though the striking velocity may be less, the momentum, and consequently the penetrating power, which varies with the density of the projectile (Art. 79), will be greater, than that of the lighter shell.

It is very important that these experiments should be carried out to their utmost practical extent with shells made still more excentric, by increasing the weight of the solid segment a, fig. 12, left in the shell, not only with the 10-inch guns, but, as recommended by the author in his letter of the 5th September, 1850, addressed to the Master-General of the Ordnance, with the 68-pounder guns of 95 cwt., from which it may be expected that useful results may likewise be obtained.

It must, however, be remarked, that firing at these high elevations partakes so much of the conditions of vertical shell firing as to be liable to all the uncertainties of that practice with ordinary projectiles (Art. 260), and is such that there would be little probability of hitting a ship or battery; but in shelling fortresses, arsenals, crowded roadsteads, or any extensive space, these extreme ranges might be very serviceable.

The season being then too far advanced to prosecute the experiments with excentric shot and shells; it being, moreover, necessary that excentric shells should be cast on purpose, and that there should be provided a carriage which might admit of high elevations as well as resist the shock of a discharge with the maximum service-quantity of powder, those experiments were unavoidably deferred.

In June, 1852, experiments were resumed at Shoebury Ness with a 10-inch gun, of 116 cwt., carrying an excentric shell of 100 lbs., and the following is an abstract of the results.

	Yards.		Yards.
Charge 16 lbs., Elevation 28°.		Charge 16 lbs., Elevation 32°.	
Mean range in 10 rounds	5342	Mean range in 23 rounds	5675
Greatest range	5473	Greatest range	5860
Mean deflection, right	334	Mean deflection, right	107
,, left	380	,, left	284

It is deserving of remark that, after the greatest range (5860 yards) had been obtained, at the next discharge the gun burst. This was at the 54th round.

In December of the same year further experiments were carried on at the same place with a 68-pounder gun of 95 cwt. and a charge of 12 lbs.

Nature of Projectile.		Mean Deflection.
	Yards.	Yards.
Elevation 24°		
(Hollow shot) Mean Range =	5744	
Greatest =	6500	
(Shell fitted with Moorsom's plugs) Mean Range =	5297	} 105 right, 9 left.
Greatest =	5470	
(Ditto with Moorsom's fuzes) Mean Range =	5587	40 right, 125 left.
(Ditto with wooden plugs) Mean Range =	4637	250 right, 350 left.
Elevation 26°		
(Hollow shot) Mean Range =	5368	95 right.
(Shell fitted with Moorsom's plugs) Mean Range =	5415	73 right, 33 left.
(Shell with plug cut off) Mean Range =	5767	} 122 left.
Greatest =	6000	
(Shell with wooden plug) Mean Range =	4950	200 left.
(Shell with wooden fuze) Mean Range =	5300	

All the results above stated prove decisively the correctness of the deductions from theory, and of the practical maxim that errors in sphericity and homogeneity in a shot are causes of its deviation from a correct path; and it follows that spherical and homogeneous projectiles, being the most simple, and quite indifferent to the position in which they are placed in the gun and rolled home, as well as to that in which they pass through the atmosphere, are decidedly to be preferred to the others.

In ricochet firing, whether the rebounds take place from water, as in the experiments made on board the "Excellent," or on land, as in those carried on at

Shoebury Ness, the shot when revolving on a vertical axis, instead of making a straightforward graze, suffered deflections which were invariably towards the same side of the line of fire as the centre of gravity; and at every graze up to the fourth a new deflection took place. (See Art. 184.)

189. Excentric as well as irregular shot can only be made to describe undeflected curves by having rotations on their axes, on the rifle principle, by which the tendencies to deviate may be compensated; and from all that has been above adduced it may be unhesitatingly asserted that the employment of shot, purposely rendered excentric, can only exaggerate the errors occasioned by the imperfections in the sphericity and homogeneity of ordinary spherical shot. Could these conditions be practically fulfilled, no deviations could take place, except from causes quite independent of figure and centreing; but the expectation of producing corrections by increasing the causes of error must for ever be vain and delusive. When, however, mere extent of range is contemplated, that important object may be gained by excentric projectiles, since the recent experiments, concurrently with theory, show that if the centre of gravity is placed directly above the centre of figure the range is greatly increased.

The results of these very curious and instructive experiments fully explain the extraordinary anomalies, as they have heretofore been considered, in length of range and in the lateral deviations: these have been attributed to changes in the state of the air, or the direction of the wind, to differences in the strength of the gunpowder, and to inequalities in the degrees of windage. All these causes are, no doubt, productive of errors in practice, but it is now clear that those errors are chiefly occasioned by the excentricity and non-homogeneity of the shot and the accidental positions of the centre of gravity of the projectile with respect to the axis of the bore. The whole of these experiments furnish decisive proof of the necessity of paying the most scrupulous attention to the figure and homogeneity of solid shot, and the concentricity of shells; and they exhibit the remarkable fact that a very considerable increase of range may be obtained without an increase in the charge, or elevation of the gun.

The abstract given in Art. 186 of the Shoebury Ness experiments is formed from the Report made to the Master-General of the Ordnance by the Select Committee to whom the consideration of those experiments had been referred; and, in concluding the subject, the author has great satisfaction in stating that the Report alluded to is perfectly in accordance with the views which induced him to recommend that those experiments should be instituted. The Report states that, though no useful application of the excentric principle can be made in general service, and that its use is limited to cases in which a more extended range may be required than is practicable under ordinary circumstances, yet the Committee expresses a decided opinion that the very interesting experiments which have been made, agreeably to the author's suggestions, must be considered as highly instructive, since they prove that the most influential among the various causes of the deflection of shot, in their flight, is the want of perfect homogeneity in their material.[a]

190. A very ingenious invention has lately been patented by Mr. Lancaster, for causing a shot to rotate on its axis throughout the range, by firing it from a carbine or a piece of heavy ordnance, having an elliptical bore of small excentricity. The helix, which has a turn of about one-fourth of the periphery in its length, is not uniform, being nearly rectilinear to a certain distance from the bottom of the bore, and increasing gradually but rapidly in curvature from thence to the muzzle. In August, 1851, some experiments were carried on at Shoebury Ness, with an 8-inch cast-iron gun of this nature, 10 feet long and weighing 96 cwt. The ellipticity of the bore, or the difference between

---

[a] The Report was communicated to the author by order of the Master-General of the Ordnance, Jan. 11th, 1851.

the axes, was 0.27 inch : the shot, of which a section perpendicular to the axis was elliptical, was hollow, its form cylindro-conoidal, and it weighed 75 lbs. An elliptical piece of felt, rather larger than the base of the projectile, was fixed to that base, with a plate of iron of the like form, but smaller, below it. This felt performed the office of the patch in the ordinary rifle, and it had the effect of taking away the windage without creating much friction.[a]

The following Table contains the results of the experiments referred to :—

Rounds.	Charge in lbs.	Remarks.
1st 2nd 3rd	10	The shell broke on leaving the gun.
4th 5th 6th 7th	5	The shell ranged 800 or 900 yards. The shell broke on leaving the gun. The shell, which was of wrought iron, broke. The shell, which was of cast iron, broke.

No loaded shells were tried.

It will thus be perceived that all the shells broke in, or on leaving the gun, except that used in the fourth round. The others were all blown to atoms, as if they had been large iron canisters for holding powder, instead of being designed for projectiles. That round alone afforded the means of forming an opinion concerning the efficiency of the inventor's method of producing rotation : the shell came out of the gun entire, but it wabbled on to the distance of 800 or 900 yards, in a manner which seemed to preclude the expectation that any useful effect would be obtained from this nature of ordnance with *shells* or *hollow shot*. If solid shot, or shells so strong as not to break, be used with the maximum charge of 16 lbs., it would probably be a severe trial for the gun.

In December, 1852, some further experiments were carried on at the same place, with a 68-pounder gun of

---

[a] It will be seen in Art. 206 that, in the United States, shot have been enveloped in felt in order to prevent the abrasion of the bores of the guns.

92 cwt., formed as above described, and carrying a spheroidal shell. The ranges and deflections were as in the following Table :—

Charge in lbs.	Elevation.	Range to the first Graze in Yards.	Deflection in Yards.	
			Right.	Left.
10	2°	1340	..	7
10	5	2290	5	..
10	10	3540	2	..
10	15	4400	..	50
12	15	4400	..	50
12	15	4800	10	..
10	17	5600	5	..

Two 8-inch round shot fired from the same gun, with charges of 12 lbs., and at elevations of 15°, ranged, one of them to the distance of 3200 yards, with a deflection to the right of 50 yards, and the other 3350 yards, with a deflection of only 2 yards, to the right.

At the eighth round, the oval or spheroidal shell stuck in the gun.

Some further experiments were, in the present year (1854), carried on with Lancaster's projectiles; and it appears that the results are highly favourable to the principle on which ordnance of the nature above described is constructed.

191. In December, 1848, some experiments were made at Woolwich, in the presence of Colonels Dundas and Chalmer, with a shell weighing 57½ lbs., of a conoidal figure, having curvilinear grooves on its convex surface, the invention of a Captain Thistle, of the United States' army. Four rounds with this shell being fired from a 32-pounder gun, with a charge of 4 lbs. of powder, the ranges were found to be 550 yards, except in one instance, when the shell ranged 950 yards. Four rounds were next fired from an 8-inch gun (65 cwt.), with the usual shells, each 56 lbs., and charges of 4 lbs., when the mean of the ranges was about 870 yards. Two rounds of solid 32 lbs. shot, of the usual spherical form, were then fired from the 32-pounder gun, when each shot ranged about 1000

PART II. THEORY AND PRACTICE OF GUNNERY. 173

yards. All the rounds were fired at the same elevation, 3½ degrees. The superiority of the spherical projectiles is in this case manifest; though it is said that the conoidal shot went forward like an arrow, with its point in front.

The experiments were resumed January 15th, 1849, when Captain Thistle's conoidal shells, projected from a 32-pounder gun, were compared with spherical shells weighing 48 lbs., projected from an 8-inch gun as before. The charges were one-twelfth of the weights of the respective shells, and the elevations were 3 degrees. At the first round Captain Thistle's shell ranged 1155 yards, and entered the butt after ricocheting on the ground; and, at the next, the spherical shell ranged 1010 yards. At the third round Captain Thistle's shell ranged 1000 yards, but, on this occasion, instead of going straight forward, *it turned over several times in its flight.*

192. In March, 1849, Dr. Minesinger, an American, exhibited at Woolwich a spherical ball (23 to the lb.), having attached to it a four-grooved tail resembling the first screw-propellers with four leaves. The ball was fired from a gun 5 feet 7 inches long, with a percussion-lock. The gun contained at the breech a space for a chamber (an old construction), and five of these chambers, each 3 inches long, being previously loaded, might be successively introduced in the barrel, no ramrod being necessary. Each chamber has a projecting nipple, on which is a percussion-cap. The grooved tail of the ball is placed next to the charge, and 20 rounds may be fired in a minute.

Fig. 13.

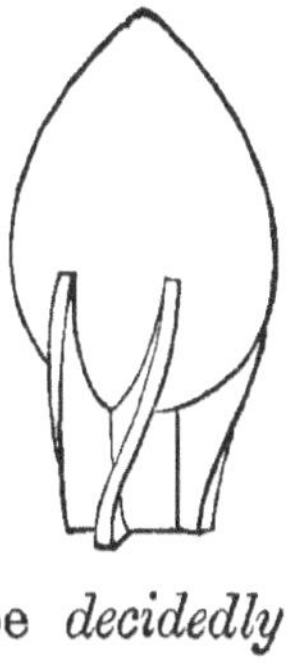

Dr. Minesinger exhibited also an oblong shot or shell (fig. 13) for a 32-pounder gun, with a four-grooved tail similar to that of the musket bullet; it weighed 46 lbs.; when used as a shell it is cast hollow, and has a copper cap placed on the end. Neither of these projectiles was found to answer the proposed purpose. Dr. M.'s shot were fired from a British 32-pounder of 56 cwt., with a charge of 10 lbs., and were found to be *decidedly*

inferior to the service balls used with that nature of ordnance.

193. The bombardment of a fortified place is more destructive to the non-combating inhabitants, generally very numerous, than to the defenders of the works, who, when not on duty, are lodged in shell-proof barracks, and produces less effect upon the military works than upon the dwellings of the citizens. In fact, the object of the bombardment is to compel the garrison to surrender, not by the injury which it may sustain, but by the slaughter and misery inflicted on the unoffending inhabitants; and no retrospect can be more afflicting than that which intrudes on the mind of one who, like the author, has witnessed the horrors of a bombardment.

When the very existence of a nation is menaced, when, consequently, self-defence, active as well as passive, is a paramount consideration, the infliction of the greatest possible injury on the aggressor, by any means whatever, becomes justifiable; but, in a war of policy, which should be directed rather against the chief of a state and his forces than against his subjects, a measure, the dreadful consequences of which fall upon these alone, must inevitably engender among them feelings of bitter animosity against the nation by which they have been so cruelly outraged.

194. But, if it be determined that a place, whether it be an inland fortress or a naval arsenal, shall be bombarded, the ordnance used should be of large calibre, and mortars should be employed rather than howitzers. When the object is to crush buildings or destroy shipping, bomb-shells, fired at considerable angles of elevation, should be used, in order that the momentum acquired in their descent may be sufficient to penetrate magazines or casemates down to the foundations, and there exploding, set fire to the buildings and create havoc and disorder among the troops; or should they fall on ships, either pierce through and sink them, or, by exploding, blow them up. Neither percussion nor concussion shells, fired at comparatively small elevations,

PART II. THEORY AND PRACTICE OF GUNNERY. 175

will serve this purpose ; and it may be doubted whether any time fuzes can withstand the shock of the charge which must be used in firing excentric shells (Art. 188) from a 68-pounder gun, or a 10-inch shell-gun, to obtain a range of 5000 or 6000 yards. But it is known that the fuze of a 13-inch shell will resist the shock produced by the explosion of 20 lbs. of powder in a sea-service mortar, and the fuze of a 10-inch shell will resist the shock produced by 10 lbs. of powder. The effect of the recoil of a 10-inch shell-gun downwards on the deck of a steamer, severely damages the vessel by straining it in every joint ; but no distress is caused by the discharge of a 13-inch shell in a mortar-ship, in which the ordnance is placed on a solid mass of material resting on the bottom of the vessel.[a]

195. For the bombardment of a naval arsenal, the 13-inch mortar is a most valuable piece of ordnance ; and, as its transport by sea is neither so difficult nor so costly as by land, the considerations which prevent, or impede to a great extent the employment of these, and of 10-inch mortars, in the land-service, are not sufficiently cogent to warrant the disuse of such ordnance in naval expeditions. It appears to the author that a large fleet fitted out specially for the purpose of bombardment, should be provided with a certain number of bomb-ships. These are very inexpensively furnished, draw little water, and may be towed to their fighting

---

[a] The ranges of the 13-inch and 10-inch sea-service mortars differ very little from each other, the former being 4200 yards, and the latter 4000 yards, the elevations in both cases being 45 degrees. (Table XXVII., Appendix D.) The weight of a 13-inch mortar is 101 cwt., that of a 10-inch gun 86 cwt., and of a 68-pr. gun 95 cwt. The weight of a 13-inch shell, empty, is 190 lbs., and that of its bursting powder 6 lbs. 12 oz.; the weight of a 10-inch shell, empty, is 85 lbs., and that of its bursting powder 2 lbs. 12 oz. Lastly, the weight of an 8-inch shell, empty, is 41 lbs., and that of its bursting powder 1 lb. 14 oz. (*Artillerist's Manual.*) The practice of these large shells, from their comparatively small initial velocities, with respect to range, time of flight, and velocity in the curve, may be determined with tolerable precision from the parabolic theory. (See Art. 45, Note.) A range of 4000 yards, or about $2\frac{1}{4}$ miles, is quite adequate to the bombardment of any arsenal, with its magazines, barracks, dock-yards, and basins : these could scarcely be missed at every discharge, and a few 13-inch shells would suffice to crush the buildings and destroy the shipping.

positions by small steam-tugs: an officer and a few marine artillerymen suffice for the service of the mortars: they are also small objects for the enemy to fire at; and, if destroyed, there is comparatively little loss, in value, and small risk of life. Seven bomb-ships were attached to the Baltic fleet in 1801, and, from these, at the time of forcing the passage of the Sound, shells were thrown into Cronenberg with considerable effect, while the fleet sustained no injury whatever from the enemy's guns. Admiral Nelson made use of those bomb-ships at the battle of Copenhagen; and to this bombarding power was greatly indebted in bringing those very critical operations to a successful termination. In the British naval service, though bomb-ships no longer exist, 13-inch and 10-inch sea-service mortars are retained (see Table XVII., Appendix D): from this it appears that horizontal or howitzer shells, fired from the pivot guns of steam frigates, are supposed to be efficient substitutes for mortars in bombardments of fortresses and naval arsenals: but this is not so; and therefore to employ large and costly steam ships in this way, is to incur a risk of very great loss in property and life, without the power of accomplishing the peculiar conditions required in bombardments, and which mortar shells can so much more effectually fulfil. (See the preceding article.)

The bombardments by the French of St. Juan d'Ulloa, in 1838, and of Vera Cruz, in 1839, prove sufficiently that our neighbours have not abandoned the use of sea-service mortars; and, from the *Aide Mémoire Navale* we learn that such mortars are considered indispensable for bombardments. The French possess several mortar-ships (*bombardes*), as "La Dore," "L'Acheron," "Le Vesuve," "Le Cyclope," &c., each carrying two mortars and complements of 500 shells.

The following Table gives the ranges, deflections, &c. of a 12-inch French sea-service mortar; and to these are added the number of shells in every hundred fired, which, in experiment with different mortars, fell within rectangular areas of various magnitudes.

Part II.   THEORY AND PRACTICE OF GUNNERY.   177

TABLE of PRACTICE with a French Mortar of 32 Cent. (12½ inch), 1840. Elev. 42°.

Charges.	Ranges.	Time of Flight.	Mean Deviation.				Charges.	Ranges.	Mean Deviation.			
			Longitudinal.		Lateral.				Longitudinal.		Lateral.	
lbs. oz.	Yards.	"	Yds.	ft.	Yds.	ft.	lbs. oz.	Yards.	Yds.	ft.	Yds.	ft.
1 1·6	279	7.33	5	1	8	2	1 15.7	545	9	2	15	1
2 3.2	601	10.6	10	3	16	1	2 5.8	654	12	0	17	1
3 4.8	984	13.5	16	1	24	0	2 10.7	763	13	0	19	2
4 6.4	1333	15.8	21	2	32	2	2 15.4	872	15	1	22	0
5 8.	1651	17.85	27	1	40	1	3 5.	981	16	1	24	0
6 9.6	1924	19.53	32	2	50	0	3 9.6	1090	18	2	27	1
7 11.2	2191	20.65	38	0	59	0	4 0.	1199	19	2	29	1
8 12.8	2458	22.55	43	2	67	2	4 5.1	1308	20	2	31	2
9 14.4	2694	23.6	49	0	75	1	4 11.2	1417	21	2	35	0
11 0.	2909	24.4	54	1	84	0	5 1.1	1526	24	0	38	0
12 1.8	3113	25.1	58	2	91	2	5 7.2	1635	26	0	41	1
13 4.	3271	25.6	62	0	99	0	5 14.2	1744	28	1	44	2
11 5.3	3405	26.15	65	1	105	2	6 5.3	1853	30	2	48	0
15 7.	3497	26.6	68	2	111	0	6 12.3	1962	32	2	51	1
16 8.6	3589	27.03	73	0	116	2	7 3·4	2071	35	0	54	2
17 10.4	3671	27.47	74	0	120	0	7 10.4	2180	37	0	57	2
18 12.	3744	27.85	75	1	124	1	8 1.9	2289	40	1	62	0
19 13.6	3817	28.22	76	1	127	1	8 9.	2398	42	1	65	1
20 15.4	3889	28.6	77	1	129	2	9 1.	2507	44	2	68	2
22 1.	3961	28.92	78	1	132	0	9 8.8	2616	47	0	73	0
23 2.6	4026	29.23	79	2	134	0	10 1.2	2725	50	0	76	1
24 4.2	4091	29.53	80	2	136	1	10 10.	2834	52	1	79	2
25 5.9	4151	29.8	81	2	137	1	11 3.4	2943	54	1	85	0
26 7.6	4211	30.05	83	0	139	1	11 12.	3052	56	2	89	1
27 9.2	4267	30.3	84	0	140	2	12 5.1	3161	59	0	93	2
28 10.7	4316	30.5	85	0	141	2	13 3.8	3270	62	0	99	1
29 12.4	4349	30.65	86	0	142	2	14 2.9	3379	64	1	103	2
30 14.1	4360	30.75	87	1	144	0	15 3.9	3488	67	2	111	0
							16 10.5	3597	71	0	116	2
							18 3.2	3706	74	0	122	0
							19 13.6	3815	76	1	127	2
							21 7.2	3924	78	1	131	0
							23 4.3	4033	79	2	134	0
							25 13.	4142	81	2	137	1
							27 3.9	4251	84	0	140	2
							30 14.	4360	87	1	144	0

The arrangement of this Table is the same as in the *Aide Mémoire d'Artillerie Navale*, from which it was taken; the intention being that the Charges, Ranges, &c., for which the time of Flight is given, may be easily distinguished from the others.

Table of Practice with a Mortar, &c.—*continued*.

Number per cent. of Shells which struck the Ground within—

	1. A Rectangle of indefinite length in the direction of the Line of Fire, and whose Breadth was, in Yards,								
	2	5½	11	22	33	44	54	65	87
Mortar of 32 c. (12½ in.) pointed at an object distant 654 yards	24.	35.5	53.5	74.	86.5	93.	96.	98.	99.5
Mortar of 32 c. and 27 c. at 654 yards (12¼ in. and 10½ in.)	17.5	26.	40.	60.	72.5	84.5	91.5	96.	99.
Mortar of 27 c. (8½ in.) at 654 yards	13.	20.	30.5	48.	62.	75.	84.5	91.4	96.5
Mortar of 15 c. (6 in.) at 327 yards	11.	27.	50.	80.5	95.	99.	100.	100.	100.
Ditto ditto at 654 yards	..	..	..	..	..	..	..	..	..

	2. A Square whose Side was, in Yards,								
	2	3¼	11	22	33	44	54	65	87	109
Mortar of 32 c. (12½ in.) pointed at an object distant 654 yards	0.72	1.95	5.78	15.8	27.1	39.5	50.4	60.3	72.6	80.
Mortar of 32 c. and 27 c. at 654 yards (12¼ in. and 10½ in.)	0.53	1.43	4.32	12.8	22.7	35.9	48.1	59.0	72.3	80.
Mortar of 27 c. (8½ in.) at 654 yards.	0.33	0.9	2.53	7.7	15.1	24.4	34.2	44.3	59.8	71.5
Mortar of 15 c. (6 in.) at 327 yards	..	..	..	..	..	..	..	..	..	..
Ditto ditto at 654 yards	0.11	0.68	2.56	10.3	21.4	34.9	48.	62.8	85.3	100.

196. It might have been foreseen that numerous gun-boats would be required to act with the naval force sent to the Baltic sea, as no expedition intended to act in shallow waters was ever undertaken without being accompanied by such craft. Numerous gun-boats, each armed with a powerful gun, formed part of the equipment for the naval force sent to Walcheren in 1809, and were found extremely useful, on all occasions, in shallow inland waters. The day after the landing of the troops, a gallant and efficient attack was made by those boats on the fort of Ter Veer, then invested on the land side, which ended in the place being taken.

After the action at Copenhagen, in 1807, the Danes lost nearly all their line-of-battle ships and frigates, but they still possessed some stout brigs of war, and a great number of well-armed gun-boats; and during the calms which frequently prevail in the Danish waters, the latter were particularly destructive to the British cruisers and convoys.

On the 4th of June, 1808, during a calm in the Great Belt, the "Tickler" gun-brig, commanded by Lieutenant Skinner, was attacked by 4 Danish gun-boats, and after a conflict of four hours, in which the commander was killed, she was obliged to surrender. A few days afterwards, the bomb-vessel "Thunderer," Captain Caulfield, and the gun-brig "Turbulent," Lieutenant Wood, with a convoy of 70 vessels, were attacked by 25 Danish gun-boats, when the "Turbulent" was captured, with 10 or 12 of the merchant-ships: and on the 2nd of August, the gun-brig "Tigress," Lieutenant Greenswood, was taken by 16 Danish gun-vessels. On the 25th of October, in the same year, the British 64-gun ship "Africa," with a homeward-bound convoy of 137 sail, was attacked by a Danish flotilla, consisting of 25 large gun and mortar-boats, and 7 armed launches; nearly all the convoy escaped, but the "Africa" was severely damaged; and, had the daylight continued two hours longer, she must have surrendered or sunk. —*James's Naval History*, vol. v. pp. 74-76.

On the 7th of July, 1809, a British squadron of 4 ships, under Captain Thomas Byam Martin, while cruising off the coast of Finland, discovered a Russian flotilla of 8 gun-boats, and a fleet of merchant vessels, at anchor under Porcoln Point. The position which these had taken was one of extraordinary strength, being between two rocks, which served as covers to their wings, and from whence a destructive fire of grape could be poured upon any boats that should approach them. Notwithstanding this, the boats of the ships, 17 in number, advanced close up to the gun-vessels without firing a musket, and, boarding them, carried all before them: 6 of the vessels were taken, 1 sunk, and 1 escaped; 12 merchant-vessels also, laden with powder and provisions for the Russian army, were taken. On the 25th of the same month, 17 boats from a British squadron, under Captain Dudley Pater, attacked 4 Russian gun-boats, and, after a sanguinary contest, succeeded in capturing 3 of them.—*Ibid*, pp 180-182.

The Russians have now in the Baltic numerous gun-boats, which, keeping in comparatively shallow water, among the islands, where no large ship can approach them, are ready at every moment to issue forth and seriously annoy the most formidable fleet of line-of-battle ships on the coast.

PART III.

ON ORDNANCE RE-BORED, AND NEWLY CONSTRUCTED, FOR THE BRITISH AND FOREIGN NAVIES.

I. BORED-UP GUNS.

197. THE practice of *reaming-out* guns, or boring them up, first took place in the British service in the year 1830, when about 800 guns, 24-pounders, 7 feet 6 inches long, which had been made according to the construction recommended by Sir William Congreve, and about as many more guns, also 24-pounders, of Sir Thomas Blomefield's construction, were bored up to 32-pounders for the navy. (See Table XVII., Appendix D.) The practice was afterwards extended to iron guns of all natures, from the 9-pounder to the 32-pounder inclusive, by enlarging the bore of each to the next, and, in some cases, to the second higher calibre, and leaving reduced windages.[a] This may be considered as a temporary expedient to increase the weight of metal projected from such guns as were then on hand in this country, at a time when the advantages of large calibred ordnance were not absolutely decided on, and when the Government was not prepared to sanction the expense of casting new guns for projecting the heavier natures of shot or shells. But now that the vast advantages of heavy shot are well understood, and that France, the United States, and other countries have determined on arming their ships with new, long and powerful ord-

---

[a] The French, on the recommendation of General Paixhans, had previously decided to bore up their long guns to the next greatest calibre: but all guns so treated have been superseded in their naval service by new guns and canons-obusiers.

nance, the time is come in which the half measure, as it may be called, should be abandoned; that the bored-up guns in the service, with such exceptions as will be mentioned presently, should be superseded, and that good and efficient guns should be provided of calibres equal to those to which the bored-up guns were enlarged. By such a measure only can it be expected that in a future war our ships should be able to contend successfully with those of nations which have adopted in their navies the most approved natures of ordnance.

198. The great advantage arising from the reduction in the windage, together with the increase in the weight of metal projected, procured immediately for the bored-up guns a certain degree of consideration, which, it is to be apprehended, has caused the imperfections arising from the diminution of the quantity of metal in the gun itself to be overlooked. The reduced windage indeed permits, in some respects, equal effects to be produced with lower charges of powder than were used with the original gun. For instance: the old 24-pounder gun, whose windage was .211, when bored up to a 32-pounder, with a windage equal to .123, and charged with $7\frac{1}{3}$ lbs. of powder, would, with rather higher elevations, propel its shot as far as the gun in its original state with its usual charge of 8 lbs. The initial velocities would be the same (about 1600 feet per second), and the force with which the shot would penetrate into any material would be, at all practicable distances, in favour of the bored-up gun. Again, if the latter were charged with $6\frac{1}{2}$ lbs. of powder, by which an initial velocity equal to 1490 feet per second would be produced, it would be found that the force of penetration is at first greatest in the original 24-pounder gun, with the charge of 8 lbs.; at the distance of about 400 yards from the object the penetrating forces become equal; and at all greater distances, the penetrating force of the larger shot of the bored-up gun is the greatest.[a]

---

[a] It is to be regretted that, when the old guns were bored up to a higher calibre, measures were not taken to introduce uniformity in the quantity of windage, and that great anomalies in this respect were allowed to subsist.

199. It is not, however, from a comparison of the effects produced by the bored-up guns with those produced by the gun in its original state, that the relative values of the former, and of a gun equal in calibre but of more perfect construction, are to be determined. The advantages with respect to the power of penetration, which appear to be so much in favour of the bored-up gun, are perhaps more than counterbalanced by the defect arising from the diminution of the weight of metal in the gun, by which its recoil is rendered considerably greater than that of a gun of equal calibre and of the original formation. This circumstance, besides producing greater strain on the carriage, renders the gun more unsteady, and the practice, in consequence, more uncertain. If, in order to diminish the recoil, smaller charges of powder be employed, the penetrating power of the shot will be diminished in a corresponding degree.

200. Captain Simmons, in his "*Discussion on the Present Armament of the Navy,*" has calculated a Table (page 18), in which are shown the relative penetrating powers of a 24-pounder gun, and the same gun reamed out to a 32-pounder, the charge of the latter being diminished so far that the initial velocities are proportional to the weights of the guns with their carriages; when, consequently, the recoils or strains are rendered equal to one another. In this state the penetrating force of the bored-up gun falls even below that of the original 24-pounder, till the distance of the shot from the gun is about 3000 yards; but at such a distance as this, the elevation of the gun being necessarily high, the practice must be extremely uncertain. It is thus of the utmost importance, as well with respect to the extent and

---

Thus the diameter of the bore of the old 32-pounder weighing 56 cwt., and of those weighing from 48 to 50 cwt., is 6.41 inches, while Congreve's and Blomefield's 24-pounders of 39, 40, and 41 cwt. were bored up (as 32-pounders) to 6.35 inches; and again, Blomefield's 24-pounder of 32 cwt. and his 18-pounder of 25 cwt. were bored up, as 32-pounders, to 6.3 inches. The calibre of Mr. Monk's gun (A) is 6.375 inches. Thus there have been introduced in the service four different diameters of bores for 32-pounder guns, consequently four different windages for shot of equal diameters.

accuracy of the range as to the penetrating power of the shot, that a gun should contain a certain mass of metal with relation to the weight of the projectile; a deficiency in that mass causing the advantages of a diminished windage to be in a great measure lost; the shaking of the piece before the projectile quits it, particularly when high charges of powder are used, unavoidably producing irregularity in the initial direction of the shot. The perfection of a gun consists in an union of both qualities, steadiness and the least possible windage; for only by these combined can it be rendered capable of throwing shot to the greatest distance within battle range, with the least elevation, and, consequently, with the greatest amount of useful effect.

201. Since the country now possesses a considerable number of new guns of large calibre and improved constructions, no reason exists for retaining bored-up guns as part of the armament of the British navy, except, perhaps, for brigs and other small vessels, or for private ships hired into the service. The French have no bored-up guns on the broadsides of their ships of the line or frigates;[a] nor are there any bored-up

---

[a] The French ships are armed, or are appointed to be armed, with the following natures of gun, exclusive of canons-obusiers (the weights and dimensions are expressed in English denominations):—36-pounders (Fr.), length 9 ft. 7 in., and weight 69 cwt., the weight of the shot being 43.21 lbs.; 30-pounders (Fr.), No. 1, length 9 ft. 3 in., and weight 59 cwt.; 30-pounders (Fr.), No. 2, length 8 ft. 6 in., and weight 49 cwt., the weight of the shot for both being 34 lbs.; and 24-pounders (Fr.), length 9 ft. 4 in., and weight 41½ cwt., the weight of the shot being 26 lbs. 10 oz. The decree of the 27th of July, 1849, by which the 50-pounder gun, 10 feet 2 in. long, and weighing 84 cwt., had been ordered to be introduced into the French naval service, was rescinded by the commission which drew up the "*Enquête Parlementaire*" (see tom. ii. p. 405 of that work), and 30-pounder guns were ordered to be substituted for them: this was done on the ground that, as the weight of the 50-pounder is far greater than that of a 30-pounder, by the substitution of the lighter but very efficient gun, vessels may be armed with a greater number of pieces; and, from the circumstance, it is evident that the commission attached more importance to the fire of the smaller ordnance than to that of the heavier nature, which would, necessarily, be fewer in number. It appears by the proceedings of the commission of Gavre, 1848, that three other new pieces of ordnance have been adopted in the French naval service. 1st. A canon-obusier (No. 3) of 80 (*livres*); its charge for hollow shot being 2.5 kil. =5 lbs. 8 oz., and for solid shot 2.6 kil.=5 lbs. 12 oz.: these are somewhat shorter than the pieces of the same nature already established. 2nd. A new

guns in the United States navy, except the 8-inch guns, which were bored-up from 42-pounders. All the guns retained in that service are either the most efficient of the old ordnance, or certain others whose use has not yet been discontinued.[a] A great prejudice exists in the United States against the practice of boring-up, since a 42-pounder bored-up to a 64-pounder, the diameter being enlarged from 7 to 8 inches, with a charge of 14 lbs., burst on board the "Fulton;" the service charge was 8 lbs.

The long 42-pounder so reamed out, and carrying a ball of 64 lbs., is said to afford, with a charge of 12 lbs., a better range at the second graze than the gun in its original state (Ward, *United States' Navy*, p. 105); but at the first graze, and at elevations not exceeding 3 degrees, the long 42-pounder with a charge of 12 lbs., which the reamed-out gun could not sustain, is greatly superior to the other.

Since the publication of Lieutenant Ward's Treatise there have been introduced, in 1851, a new shell-gun 8 feet 10 inches long, and weighing 63 cwt., and another 8 feet 4 inches long, weighing 55 cwt. Besides these, in 1850, a new 64-pounder unchambered gun 10 feet 8 inches long, and weighing 105 cwt., was adopted as a pivot-gun for vessels of all descriptions. Thus has been realized Lieutenant Ward's announcement that a gun possessing, in the highest degree, extent of range, accuracy of fire and power of penetration, was being prepared for purposes which essentially require those important faculties.

30-pounder (Fr.) gun (No. 3), its charge, with solid shot, being 3 kil.= 6 lbs. 10 oz. (reduced charge 2.5 kil.=5 lbs. 8 oz.), and for hollow shot, 2 kil. =4 lbs. 6 oz. The commission ascertained the ranges of a new 60-pounder gun, with solid shot and charges of one-third of the shot's weight: they also compared together the effects produced by this gun, the new 50-pounder and the new 30-pounder guns, with respect to range, and to the ravages made by their shot in targets representing the sides of ships.

[a] These are long 24-pounders, length 9 ft. 4½ in., and weight 50 cwt., and 18-pounders, length 8 ft. and weight 38 cwt., on board of the "Constellation," "Macedonian," and other ships.— *Ward*, p. 36. It is perhaps to be regretted that the long 24-pounder (length 9 ft. 6 in., and weight 50 cwt.) should have been discarded from the British navy.

202. The defects of bored-up guns were sensibly experienced on board the "Sesostris" steam-ship, Captain Ormsby, commander. This vessel was armed with 32-pounder guns, which had been bored-up from 24-pounders; and it was found that, after a few hours firing, the breechings were destroyed, and the bolts drawn so that her ordnance became unserviceable, even with greatly reduced charges.

203. Bored-up guns, besides being subject to the defect just mentioned, are found to be not altogether safe when fired with two shot; there are even some of them for which double-shotting was always prohibited; and, where the practice was allowed, very reduced charges were considered indispensable. The following Table shows the charges with single shot for bored-up guns. No charges for double shot are now officially given.[a]

Nature of Gun.		Proof Charge in Pounds, single Shot.	Service Charge in Pounds, single Shot.	Shot.	Wad.
32-pounder	cwt. cwt. 39 and 40	12	6	1	1
	32	10	5	1	1
	25	9	4	1	1
18-pounder	22	7	3	1	1
	20	7	3	1	1
	15	5	2	1	1

All the bored-up guns, with single shot, have been proved with charges equal to about one-third of the shot's weight; but it has been found that, after being bored-up, many of the old 24-pounders with two shot, when fired with a charge of 11 lbs. of powder, burst, after having stood, with a single shot, a charge of 18 lbs. In order that such guns may be considered safe when double-shotted, it is evident that they should be tried, so shotted, with quantities of powder considerably greater than the reduced charges stated in the Table. This uncertainty respecting the safety of the bored-up

---

[a] It is no longer contemplated to use double shot with any nature of bored-up guns.

guns is surely a strong argument against them; and though it would be much better if all were discarded, it is satisfactory to find that the only bored-up guns which are now to be retained in the Royal Navy are the 32-pounders of 41 cwt., 40 cwt., and 39 cwt., which have been so converted from Blomefield's and Congreve's 24-pounders; and 18-pounders, which have been bored up from 9 and 12-pounders for the armament of small vessels. The 32-pounders just mentioned are ultimately to be replaced by the 32-pounder new gun, 8 feet long, and weighing 42 cwt.

II. MONSTER GUNS.

204. In 1810, the French having caused some large pieces of ordnance to be cast at Seville, formed with them batteries at the Trocadero Point, from whence they threw shells into Cadiz, a distance of more than 5000 yards. (Art. 100, Note.) In order to obtain this range, it was necessary to use enormous charges, and to fill the shells nearly with lead. These pieces of ordnance were the invention of Colonel Villantroy, of the French artillery. They were a sort of long howitzer, with trunnions, and were capable of being fired at great elevations with high charges; they also possessed great powers of range, even at low angles, with hollow shot; and it has been said that the construction of these howitzers suggested to Colonel Paixhans the principle on which he afterwards formed his canons-obusiers. It is, however, a remarkable circumstance that the Emperor Napoleon was, in fact, the originator of both natures of ordnance; since, in addressing the Ministre de la Marine, Décrès (1807), he writes to the following effect :—" I desire that you will cause to be cast for trial, in the foundry at Douay, a gun which may project shells of 8 inches diameter. Provide also some solid shot of 78 lbs. to be used with those pieces, and try the ranges and effects which may be obtained. Cause also to be cast shells and hollow shot of 48 lbs., and let them be tried. Such projectiles, fired from batteries

of twenty pieces of the above description, should produce great effects."[a]

At the siege of the citadel of Antwerp, in 1832, the French used a mortar whose calibre was 24 inches, and weight nearly 7 tons, and it projected a shell which, with its bursting charge (= 99 lbs.), weighed 1015 lbs. The effect, however, was not so great as was anticipated, and the mortar afterwards burst while being fired.

205. In the years 1842 and 1843 there were cast in England, for the Pacha of Egypt, three large pieces of ordnance, of which one was a simple gun, without chamber, 12 feet long, having a calibre of 10 inches (windage .16 inch), and weighing 11 tons; this was intended to project solid shot, each weighing 128 lbs., or shells, each weighing 82 lbs. The second was a howitzer gun 13 feet long, having a calibre of 15.3 inches (windage .2 inch), with a chamber on Paixhans' principle, and weighing 18 tons; it threw solid shot, each weighing 460 lbs., or shells, each weighing 326 lbs. The third was a mortar, having a calibre of 20 inches (windage .2 inch), and weighing 13 tons; its chamber was conical, and similar to those of the British sea-service mortars; it threw a solid shot weighing 1030 lbs., or a shell weighing 658 lbs.

206. In the year 1845 there was cast at Liverpool, for the United States' steam frigate Princeton, a gun, having a bore 13 feet long, with 12 inches calibre (windage .25 inch); the gun weighed about $7\frac{1}{2}$ tons, and it was intended to project from it solid shot weighing 213 lbs., or shells weighing 152 lbs. The gun is without a chamber; and a remarkable circumstance is, that the shot is to be enveloped in felt, in order to prevent the abrasion of the bore; this expedient appears to have been advantageously adopted with other guns of considerable magnitude in the service of the United States, windage being allowed accordingly. It is said

---

[a] Thiers' 'History of the Consulate and the Empire.'

that this gun was intended to replace that which some time before had burst on board the same vessel, by which accident several persons, among others the Secretary of State, lost their lives.

207. The Pacha's 130-pounder gun was fired at Deal, in July, 1842, from a carriage similar to those which are used on board the British steam-vessels, with charges equal to 26 lbs. (considered as the established charge), 29 lbs., and 32 lbs. of powder; but as very little additional extent of range was obtained when the higher charges were employed, those only which resulted from the charge of 26 lbs. are given in the subjoined Table. The other ranges in the Table are results of computation, with the charges there stated, which are such as are supposed to be employed on service. In all the ranges the height of the gun above the plane on which the shot is supposed to fall is 16 feet :—

Description of Ordnance.	Weight of Shot or Shell in Pounds.	Charge in Pounds.	Elevations.				
			Point blank.	5°	10°	15°	20°
			Ranges in Yards.				
Pacha's 130-pounder {	Shot, 128	26	558	2151	3253	4040	4669
	Shell, 82	16	420	1767	2532	3130	3440
Pacha's 15-inch gun {	Shot, 460	45	333	1235	2130	2810	3370
	Shell, 326	45	367	1553	2534	3110	3500
Pacha's 20-inch mortar {	Shot, 1030	45	..	700	1035	1693	1970
	Shell, 658	45	..	907	1716	2078	2479
United States' gun {	Shot, 213	30	360	1617	2600	3267	3730
	Shell, 152	30	400	1853	2759	3370	3924
French mortar.	Shell, 1015	30	..	400	834	1033	1317

On comparing the above ranges with those which were obtained in the practice on board the "Excellent," (Tables V., VI., Appendix D.), it will readily be seen that little or no advantage would be gained, in this respect, by the employment of such unwieldy pieces over the more manageable ordnance in use in the British service.

III. NEW GUNS FOR SOLID SHOT.

208. It has been observed (Art. 132), that Sir William Congreve was the first who proposed to diminish the quantity of metal in the chase of a gun and increase the quantity about the seat of the charge, in order to gain a greater degree of strength without increasing the weight of the gun. The pieces of ordnance constructed by that officer were 24-pounders, $7\frac{1}{2}$ feet long, and weighing from 40 to 42 cwt.; and a certain number of these constituted the armament of the "Eurotas" frigate at the time she engaged the French frigate "Clorinde," which was equipped with long 18-pounders.[a] These experimental guns certainly did not do so much execution in proportion to the duration of the action (nearly two hours) as had been done on many other occasions by an equal number of long 18-pounders, or in proportion to what she, the "Eurotas," suffered from the fire of the "Clorinde." This, perhaps, was owing in part to deficiency in the gunnery of the British frigate, but the main defect was in the short 24-pounder guns, which, however they may have succeeded in the experiments at Sheerness (when they *bounded* but a little more than the long 24-pounder against which they were tried), acted most violently on their carriages when heated with continued firing in that protracted action. This is ascribed partly to the greatness of the windage, partly to the charge (one-third of the weight of the shot) being too high, and again, to the diminution of the preponderance of the breech by the trunnions being placed so far back.

General Sir Thomas Blomefield, about the same time, procured to be executed a considerable number of 24-pounder guns on nearly the same principle; but as neither of these kinds of ordnance were considered as having succeeded, others were cast upon a principle which was brought forward by Mr. Monk in 1838. That principle consists in maintaining the proportion

---

[a] James's Naval History, vol. vi. pp. 272, 273.

which the weight of metal in a gun should bear to that of the shot[a] (about 1¾ cwt. in the gun to each pound in the weight of the shot); at the same time increasing the thickness of metal round the cylinder of the charge, and diminishing it in the chase.

Mr. Monk's proposition having been approved, he first applied his method in the construction of a 56-pounder, 11 feet long, and weighing 98 cwt. (Table XVII., Appendix D.); and in this he so far reduced the thickness of metal in the chase, that he was able to appropriate an additional quantity, amounting to about 10 cwt., in the part immediately surrounding the cylinder of charge; thus he considerably increased the thickness of the gun to a certain distance in front of the load, and formed a piece of ordnance much stronger in the part where strength is most required, and yet lighter than any other iron gun of the same calibre before made. The windage in this gun was reduced from .235 inch, or about $\frac{1}{33}$rd of the diameter, to .175 inch, or about $\frac{1}{44}$th of the diameter.

209. At that time, 1838, the heaviest solid shot gun in the naval service was the 32-pounder, 9 feet 6 inches long, and weighing 56 cwt., the 42-pounder having been discarded, and not yet replaced by one of improved construction (see Art. 214); but shell-guns of large calibre for hollow shot and shells, had then recently been introduced for the armament of steamers.[b] Whether Mr. Monk designed his 56-pounder gun as a rival to the 8-inch shell-gun or not, does not appear; but that his guns, and some others of large calibre (the 68-pounder of 95 cwt.), do enter into very favourable competition with the shell guns, and prove themselves to be more efficient in distant firing, and therefore preferable for the bow guns of steamers, is a truth which forces itself upon our serious consideration, and to this subject we shall accordingly apply ourselves hereafter. His immediate object in bringing forward his 56-pounder was to

---

[a] As in the 32-pounder gun for general service.
[b] In 1824 the 10-inch gun; in 1825 the 8-inch gun, weighing 50 cwt.; and in 1838 the 8-inch gun of 65 cwt.

obtain by it more efficient and accurate practice at great ranges for general service, but more particularly for coast defences, in which artillery having the greatest powers of range seaward is of the utmost importance. But the 56-pounder of 87 cwt. has now become a naval gun, and on this account it will hereafter be particularly compared with shell guns and others which are employed in the naval service.

210. In Table IV., Appendix D., will be found the ranges obtained from the 56-pounder on its trial at Deal, in 1839, with charges of 16 lbs. and 17 lbs. of powder, at different elevations; and it may be remarked, that at an elevation of 32 degrees, with a charge of 16 lbs., its shot ranged to the distance of 5720 yards; exceeding the range of a 32-pounder, with a charge of 12 lbs., at nearly the same elevation, by 860 yards (Table I., Appendix D.). Experience showed that the 16 lbs. charges gave ranges, with solid shot, as long as, if not longer than charges of 17 lbs.; and 16 lbs. of powder were consequently adopted as the maximum charge for such guns.

A heavy 42-pounder, whose weight ($80\frac{3}{4}$ cwt.) was 216 times the weight of its shot, was compared with the 56-pounder, whose weight ($90\frac{1}{2}$ cwt.) was 236 times that of its shot, at an elevation of 15 degrees, the greatest which could be given to the 42-pounder; and likewise with the 10-inch gun of 85 cwt., with hollow shot, when the following ranges were obtained:—

<pre>
The 56-pounder (Table IV.) . . . . 4087 yards.
The 42-pounder   ,, ... 3732   ,,
The 10-inch gun (Table I.) 3546   ,,
</pre>

The trials between the 56 and 32-pounders were continued up to an elevation of 33 degrees, when, at the sixtieth round, the 32-pounder burst.

A 68-pounder of 110 cwt. was then prepared, and afterwards one of 112 cwt. The former was tried with a charge of 18 lbs. ($\frac{1}{4\frac{4}{7}}$ of the weight of the shot), and the latter with one of 20 lbs. ($\frac{1}{4}$ of the weight of the shot), both of them with solid shot: but their ranges were not superior, if equal, to those of the 56-pounder, with a charge of 16 lbs., or $\frac{1}{37}$ of the weight of the shot. One of the most valuable guns in the service is the 68-pounder which has been constructed by Colonel Dundas; its length is 10 feet, and it weighs 95 cwt. Its greatest service charge is 16 lbs., and its proof charge 25 lbs. (See fig. 12, plate II.)

211. It had been intended to bore-up all the 6-feet and 9-feet 24-pounders to 32-pounders, but, in the trials made with several, they failed, partly through a diminution of the windage from .21 inch to .15 inch, and partly from the increase in the weight of the shot; on which accounts, though the diminution of the weight of metal was inconsiderable in itself, the strength of the guns was so much reduced that they could not resist the charges; it therefore became necessary to provide new 32-pounders, medium guns, as they were called, in order to complete the gradation to the old standard, the 32-pounder, 56 cwt., and 9 feet 6 inches long.

Mr. Monk accordingly applied his method in the construction of the new 32-pounder of 50 cwt., 9 feet long; and he produced the excellent gun marked A (Table XVII., Appendix D., see fig. 13, plate II.); though no heavier than the old 24-pounder, 9 feet 6 inches long, it is thicker in the cylinder of charge than the 32-pounder, weighing 56 cwt., and its shot ranges very nearly as far with 8 lbs. of powder as the shot from the old long 32-pounder ranges with its charge of 10 lbs., though the gun is shorter by 6 inches. This very efficient gun has now generally superseded the old 24-pounders of 50 and 48 cwt. in the naval service.

Mr. Monk next applied his method, with some modifications, to the construction of the 32-pounders, 8 feet 6 inches and 8 feet long, marked B and C (Table XVII., Appendix D.). Of these, with the guns marked A, no less than 4279 have been proved without a single failure, although the proofs to which they were subjected were more severe than those applied to the old guns, from the windage being less, and the charges more by 2 lbs. than double the full service charge. All these guns now enter largely into the armament of the

British navy, and, though not superior in range either to the 18 or 24-pounders (old guns) of equal weight and with equal charges, they have great advantages over these from the superior magnitude and momentum of their shot.

In Table I., Appendix D., will be found the ranges obtained at Deal, in 1839, from the new 32-pounder guns A, B, and C. The windage of these guns is 0.175 inch; but on introducing them in the Navy, that of the 9 feet gun was increased to .2 inch, and of the 8 feet gun, to 0.198 inch.[a]

212. Table XVIII., Appendix D., exhibits a comparative view of the ranges of French guns of 30 (livres), long and short, with those of English 32-pounder guns. The ranges of the French guns are obtained by interpolations from the general Table founded on the experiments made at Gavre, between 1830 and 1840; while those of the English guns are obtained from the Tables of Experiments, carried on in the year 1838, on board the "Excellent;" and, from the comparison, the following important circumstances are deduced.

With charges of 10 lbs., and elevations not exceeding 8 degrees, the ranges of the English gun are superior to those of the French gun, though the windage of the former is rather greater than that of the latter; but the differences are the greatest between the elevations of 1°

---

[a] Mr. Daniel Treadwell, of the United States, executed, for experiment merely, in 1844, four 32-pounder sea-service guns on a new construction: they consist of a number of rings or hollow cylinders one within another; the interior part of each ring, equal to about one-third of the thickness of the ring, was of steel, and the exterior part of iron; these parts, as well as the different rings, were welded together, and compressed by a hydrostatic machine, which is said to exert a power of 1000 tons, so that the pores of the metal are closed, and the metal itself condensed, to a degree not to be attained by hammering; the whole was then formed into a gun by turning and boring.

The bores are 5 ft. 10 in. long, and the weight of each gun is less than 1900 lbs. One of the guns bore a succession of charges commencing with 8 lbs. of powder and one shot, and ending with 12 lbs. of powder, 5 shot, and 3 wads.

The evil attending the employment of these guns would be the greatness of the recoil. The inventor contrived a means by which he conceived it might be checked; but it is doubtful whether it would effectually serve the purpose.

and 6°. With charges of 7 lbs. and 6 lbs. the English gun maintained its superiority, in respect of range, at all the elevations up to 9°; the differences being the greatest when the elevation was less than 7°. But it is a remarkable anomaly, that when the charges were 8 lbs. and the elevations more than 4°, the ranges of the English gun were considerably less than those of the French gun. It may be remarked, in addition, that, conformably to theory, the ranges of a long gun almost always exceed those of one which is shorter.

Table XIX., Appendix D., contains an abstract of some experiments made with different canons-obusiers of 80 (livres), as well as with French 36-pounder and 50-pounder guns; and, on comparing the ranges of these last guns with those of the English 32-pounders in Table XVIII., Appendix D., it appears that the ranges of the latter, with a charge of 10 lbs., and even of 8 lbs., always exceed those of the French 36-pounder with a charge of 13 lbs.; and up to an elevation of 4° the English 32-pounder, of 56 cwt., and a charge of 10 lbs., gives ranges exceeding those of the French 50-pounder gun, with its charge of 17 lbs. 10 oz.

213. Throughout the late and previous wars the 42-pounder gun had formed the armament for the lower decks of some of our line-of-battle ships, and the regulation to that effect continued in force till 1839, when the 32-pounder of 50 cwt. (one of Monk's guns) was substituted for it. Now, neither in the French nor the Russian navy has the 42-pounder nor the 36-pounder been discarded; and, by a decree of the French Government in May, 1838, no less than one thousand eight hundred and sixty-eight 36-pounder guns have been cast for the lower decks of the ships of the line built previously to 1834;[a] while, in the United States'

---

[a] The Navy of France consisted in 1852 of :—ships of the line, 27 afloat, of which 11 are now fitted with screw-propellers, and 23 building; frigates, 38 afloat and 20 building; corvettes, 34 afloat and 5 building; and a vast number of smaller vessels, exclusive of an extensive force in steam-ships.— *L'Enquête Parlementaire*, vol. i. p. 415. See also Appendix (C). An accurate account of the French Navy, with the armament of the ships, will be given on a future occasion.

Navy, not only have the 24-pounder guns been retained, but the 42-pounders, 9 feet long, and weighing 70 cwt., are mounted on the lower decks of ships of the line.

214. Three 42-pounders, each of 67 cwt., were constructed in 1846 for the naval service, and tried, first at Portsmouth to ascertain their ranges, and then at Woolwich to prove *à outrance* their strength. One of these guns was constructed by Colonel Dundas, another by Mr. Monk, and the third is intermediate between the two others.

In March, 1846, the three guns were received on board the "Excellent" for the purpose of being compared with each other, and with a 32-pounder of 56 cwt., in respect of stability, steadiness, and extent of range. After trials continued during eight days, it was found that all the 42-pounders were admirable guns, alike steady in rapid firing with the established charges ($10\frac{1}{2}$ lbs. with single shot, and 6 lbs. with double shot), and that they have advantages in these respects over the 32-pounder guns, whose charge is 10 lbs.; their recoil also was not so great as that of the latter gun. It was found, however, that they work heavier with 15 men than the 32-pounders with 13 men; and that, in training and elevating, the latter were rather easier. The ranges of all the guns were nearly the same, the 32-pounder alone appearing to have very slightly the advantage with double shot.

With respect to the strength of the construction of the 42-pounder gun, that which had been most often fired, projected 305 shot. No. 1 gun stood 40 rounds, with two shot and two junk-wads, the charge being $10\frac{1}{2}$ lbs.; it bore 10 rounds, with three shot and three junk-wads, the charge being 12 lbs.; but it burst at the fifth round, with three shot and three junk-wads, and a charge of 14 lbs. No. 3 gun stood 40 rounds with two shot and two wads, the charge being $10\frac{1}{2}$ lbs.; it stood 10 rounds, with three shot and three wads, and with an equal charge; but it burst at the first round with three shot and three wads, the charge being 12 lbs. No. 2 gun stood 18 rounds, with two

PART III.   GUNS FOR SHELLS AND HOLLOW SHOT.   197

shot and two wads, the charge being 10½ lbs., when it burst.

Mr. Monk, the constructor of No. 2 gun, could only account for the less enduring form of his gun by ascribing it to some inferiority in the metal. The specific gravity of a small cube of metal taken from the same position in each gun was tried. The result was :—

>   No. 1 gun . . 7.2375
>   No. 2 gun . . 7.3112 } The sp. gr. of water being 1.
>   No. 3 gun . . 7.1954

Mr. Walker, the founder, stated that the guns were cast from an exactly similar mixture, a certain portion of old gun-metal, some remelted iron pigs, and some pigs as he purchased them. Mr. Walker is satisfied there is not a shade of difference between the metal of which the three guns were cast. The carriages stood well till the guns were burst, when they were, by the explosion, destroyed. The conclusion was, that the early bursting of No. 2 gun is attributable to an inferiority of construction. The charges of powder were taken fresh from the laboratory daily; the mean diameter of the shot was 6.775.

Gun No 1 (see Table XVII., Appendix D.), Colonel Dundas's construction, was recommended for adoption in the naval service on account of its greater endurance, and such were introduced accordingly as lower-deck guns in the "Blenheim" and "Ajax," steam guard-ships; but having been found too heavy, requiring crews so large as to overcrowd the fighting-decks of the small 74-gun ships of which the steam guard-ships were composed, those guns were removed and the 56 cwt. 32-pounders substituted.

IV. NEW GUNS FOR SHELLS AND HOLLOW SHOT.

215. Large calibred guns for the projection of shells and hollow shot were introduced into the Naval Service of Great Britain subsequently to the adoption of the Paixhans guns in France in 1824 (see the section on

Foreign Armaments and the French Shell System, in a forthcoming work) as analogous or corresponding pieces of ordnance.ª First, in that year the 10-inch gun, 9 feet 4 inches long, and weighing 85 cwt., was formed; but, being found too heavy for ordinary ships, the 8-inch gun, 6 feet 8½ inches long and weighing 50 cwt., was introduced. This gun was, however, deemed too light and short for the armament of great ships of war; and at length, in 1838, the 8-inch gun, 9 feet long and weighing 65 cwt. (see Fig. 14, Plate II.), was brought forward. This has now become the favourite arm in the British Naval Service, and it enters largely into the armament of ships of all rates and classes, for broadside batteries as well as for the pivot-guns of steamers.

Table XVII., Appendix D., contains the weights and dimensions of British shell-guns, with their service charges; and in Table V., Appendix D., are the ranges obtained from them by experiments carried on, on board the "Excellent," in 1838. The supposed inability of shell-guns to bear the great charges necessary for propelling solid shot, which, being 10 inches and 8 inches in diameter, weigh 130 lbs. and 68 lbs. respectively, is the cause that, till recently, those natures of ordnance were restricted to the firing of hollow shot, or shells, with charges not exceeding 10 lbs.,[b] and that

---

ª The original Paixhans *canon à bombes* is about 8 cwt. heavier than the British 8-inch shell gun (65 cwt.), while it is 4 inches longer; but by a judicious distribution of the metal, the French gun was so much reinforced about the chamber, or place of the charge, that it could bear being fired with solid shot weighing from 86 to 88 lbs. avoirdupois and charges amounting to 10 lbs. 12 oz. It was found also, by experiments made in the Roadstead of Brest, that the same gun could bear without detriment a double loading, either two shells weighing together 133 lbs. nearly, with a charge of 10 lbs. 12 oz., or two solid shot weighing together from 164 to 176 lbs., with a charge of 21 lbs. 8 oz.

ᵇ See 'Experiments on board of H.M.S. "Excellent," 1832 to 1849,' page 11. With 8-inch chambered guns of 65 cwt. and a charge of 10 lbs., are used hollow shot plugged with iron weighing 56 lbs., and shells weighing 48 lbs., which, with the bursting powder (2 lbs. 11 oz.), make up nearly 51 lbs. The common shells used with the 8-inch guns for land service weigh 46 lbs., and, when charged with the bursting powder, 48 lbs. Hollow shot, used with the like guns, is made up to the same weight as the filled shell, viz. 48 lbs., including its iron plug. Shells are cast thinner than the projectiles which are designated hollow shot on account of the large quantity of powder which

double-shotting was forbidden. This restriction no longer exists with respect to the 8-inch guns of 65 and 60 cwt. (Art. 246.)

216. There remains, apparently, an anomaly in the windages respecting the projectiles used in the naval and land services, with the 8-inch shell gun weighing 65 cwt., whose diameter is 8.05 inches. The average diameter of sea-service hollow shot and shells is 7.925 inches; the windage is, consequently, 0.125 inch; whereas, in the land-service, the 8-inch shell gun is used with the mortar and howitzer shells, whose average diameter is 7.86 inches, and windage 0.19 inch. The difference in the windages is small, but the extent and the accuracy of the ranges are considerably affected by it; and there exists no good reason why the windages should not be equalized by the adoption of the same magnitude and weight of shot and shell for both services. Until this is done, mistakes may be made in the issues, according to the present practice, of shot and shells of different diameters and weights for the same calibre of ordnance in the land and sea services.

217. Shell-guns in the British Service are chambered in the Gomer form (see Fig. 14, Plate II.), the bore, at the extremity nearest the breech, being contracted in the form of a frustum of a cone as in the carronade mentioned in Arts. 122, 123. This form of chamber, which was originally intended for mortars, is not well adapted for howitzers or shell-guns, which are to be fired horizontally, on account of the difficulty of retaining in their proper places the reduced charges, in cartridges which do not fill the chambers, and which are liable to slide down the surface of the conic frustum on which they lie, thus causing the gun to miss fire. Such failures are reported to have occurred in the expe-

---

would otherwise be required to burst them; and it may be observed here that a smaller quantity of powder sufficed to burst an 8-inch shell when the common metal fuze was screwed in, than when Freeburn's wooden fuze was applied. The diameter of the hole for the metal fuze was 0.86 inch, and that for the wooden fuze was 1.15 inch. It appears that shells fitted with metal fuzes burst with greater violence, and are more destructive, than those which are provided with wooden ones (see Art. 273.)

riments made on board the "Excellent" (October 12, 1838) and other ships; they have been frequently witnessed by the author, and on one occasion this took place at Woolwich (July, 1849), when, though the tubes acted, an 8-inch gun missed fire several times, and could not be made to go off but by pouring loose powder through the vent into the chamber. The chambers of the French canons-obusiers are cylindrical, connected with the bore of the gun by an enlargement in the form of a frustum of a cone (Figs. 14, 15, and 16, pp. 202, 203). Thus the cartridge is less liable to shift its place than if the whole chamber were conical; there is however a greater difficulty in introducing it; and the French, as well as ourselves, feel all the inconvenience of chambered ordnance; but our neighbours, ever attentive to adopt processes which may tend to promote rapidity in loading and firing, have found means of overcoming this difficulty by giving to the cartridge the form of the chamber itself.[a] When diminished charges are employed, the cartridges are made smaller in diameter while their lengths are the same, and thus one extremity of the charge is always in contact with the shot.

218. Strong objections attach to guns for the naval service being chambered, particularly if they are to be employed in broadside batteries. In close action every means which can be obtained to facilitate quick firing with safety and accuracy is indispensable; but, with

---

[a] For this purpose the cartridge is made on a wooden plug (mandrin) of a figure corresponding to the chamber; in one-half of its length it is cylindrical, while the other half has the form of a truncated cone. The parts are of strong paper (*papier parchemin*), and the extremity, which is hemispherical, is of parchment; the latter, having been steeped in water, was stretched over the spherical part of the plug, and kept on by sail-twine; when dry, it was taken off, and glued to the smaller extremity of the conical part.

The cartridge thus prepared and filled with 2 kilogrammes (4 lbs. 6.4 oz.) of powder was 23 centimetres (9.06 inches) long, from the bottom of the hemisphere to the nearest end of the cylinder. It was several times successively pressed into the howitzer without experiencing any impediment, and was afterwards withdrawn without exhibiting any appearance of the friction which the old cartridges experienced at the enlargement of the chamber. This mode of forming cartridges may appear tedious, but it is easy to see that it would in time be simplified; while, by adopting it, the process of loading the howitzer is greatly facilitated, and is besides assimilated to that of loading a gun or carronade.

PART III.   GUNS FOR SHELLS AND HOLLOW SHOT.   201

chambered guns, great care must be taken (see Arts. 123, 217) in loading, from the difficulty, in quick firing, of getting a small cartridge set properly home, and from the risk of its place being shifted in running the gun up; to prevent this it becomes necessary to place a grommet-wad over the cartridge.[a]

219. The necessity of forming chambers for shell-guns, as well as for carronades, arises from the weight of metal in those guns being comparatively small, considering the weights of the bodies which are projected from them; so that a much greater thickness of metal is requisite about the seat of the charge than elsewhere, in order to withstand the great expansive force of the ignited gunpowder and render the gun safe; this excess of thickness is obtained by contracting the cylindrical bore of the gun near the lower extremity, and thus forming what is called a chamber.

220. The canon-obusier originally constructed by Colonel Paixhans for the French Naval Service was 9 feet 4 inches long, and weighed 74 cwt. nearly; and it was intended to project either solid shot weighing 80 livres ($86\frac{1}{3}$ lbs. English), or hollow shot weighing 56 livres ($60\frac{1}{2}$ lbs. English); it was subsequently designated the canon-obusier of 80, No. 1, of 1841. The charge for this gun varied from 10 lbs. 12 oz. to 18 lbs. of powder, or from $\frac{1}{8}$ to $\frac{1}{5}$ of the weight of the solid shot; and the diameter of the bore is 22 centimetres (8.65 inches). The diameter of the cylindrical part of the chamber was made about equal to that of the bore of a 24-pounder gun (French), consequently the contraction was considerable; and this was the cause that a difficulty was experienced in getting the cartridge into its place. Another inconvenience was that the *sight* piece on the swell of the muzzle projected too much, so that, in the recoil, it caught the upper sill of the gun-port when the gun had a small elevation. These inconveniences were remedied in the construction of a new

---

[a] This practical inconvenience of chambered guns is strongly set forth in the reports of some recent trials of what the French call *la charge simultanée* (see Art. 220, Note.)

canon-obusier (No. 1, of 1842), Fig. 14, about equal in calibre to the former; the diameter of its chamber was equal to that of a 30-pounder (French), and the contraction of the bore was consequently less than in the other; this piece had also, in order to receive the *sight*, a support on the re-inforce at the commencement of the chase, whereas, previously, the sight had been fastened by bands passing round the gun.[a]

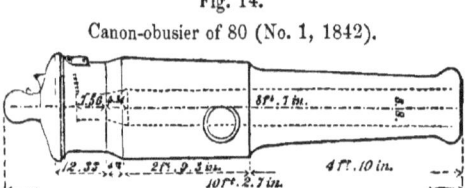

Fig. 14.
Canon-obusier of 80 (No. 1, 1842).

At a former period (1838) a different construction of the canon-obusier of 80 was tried, but the recoil of the piece in firing was found to be too great, on which account it was rejected. A modification of No. 1 (of 1842), designated No. 2 (Fig. 15), has been since executed for the armament of frigates of the second and third classes. Its charge is 3 kilogr. (6 lbs. 10 oz.), and it was adapted for moderate ranges, the tangent scales being graduated only to 1200 yards. Of this description are the 18 canons-obusiers of 80, which have been mounted for experiment on the main deck of the frigate "Psyche." (See the section on Foreign Armaments in a forthcoming work "on Shell Firing and Steam Warfare.")

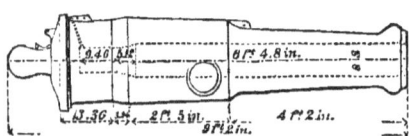

Fig. 15.
Canon-obusier of 80 (No. 2, 1842).

---

[a] On account of the great contraction at the entrance of the chamber in the gun first mentioned, it was found that the practice of ramming home the cartridge and shot at the same time, called by French writers "la charge simultanée" (which M. Charpentier acknowledges to have been first proposed by the author of this work), could not be depended on with that piece, a charge having jammed short at the second round of the trials; with the chamber enlarged as above mentioned it succeeded perfectly. It is stated, as the result of some trials made recently by the commander of the British Squadron of Exercise, that although some advantage accrues from simultaneous loading with ordinary guns, it is not desirable to attempt it with any of the new pattern chambered guns. (See the articles on Quick Loading, Part IV.)

Part III.   GUNS FOR SHELLS AND HOLLOW SHOT.   203

Another variety of the canons-obusiers of 80 consisted in placing the trunnions farther back, in order to give greater projection to the muzzle. This gun, which was lighter than the first, was tried on board the "Ocean" in 1843, but it has not been adopted in the service.

In 1848 a new canon-obusier (No. 3) of 80 was added to the other variety of this species of ordnance for the French navy. Its charge is 2.5 kil. (5 lbs. 8 oz.) for hollow shot, and 2.6 kil. (5 lbs. 12 oz.) for solid shot. (See Art. 201, Note.)

A canon-obusier of 150 (liv.), whose bore was 27 centimetres (10.6 inches) in diameter, was introduced, but its vast weight, as well as the weight and magnitude of the projectile, which render it difficult to be transported and to be worked on board a ship, are objections which will probably prevent its adoption. Its charge was 5 kilog. (11 lbs.)

In the French naval service there are also canons-obusiers of 30 (livres), Fig. 16, whose charges are 2 kilog. (4 lbs. 6 oz.) and 1.5 kilog. (3 lbs. 5 oz.); the diameter of the bore being 16 centimetres, or 6.4 inches: it is regulated that they shall fire either hollow or solid shot, and they are provisioned accordingly.

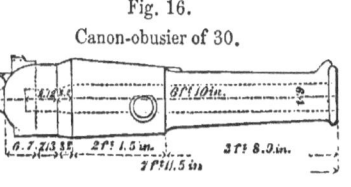

Fig. 16.
Canon-obusier of 30.

Besides the above *obusiers*, guns of 24, 30, 36, and 50 livres, and carronades, are specially destined, in French ships of the line, for firing live shells.

221. Extensive experiments were in 1850 made in France on the important questions as to the description of carriage and mode of installing the bow and stern guns of steam-vessels of war, with a view to decide between the traversing and pivot slide, and the ordinary carriage with trucks, which may be shifted from port to port as occasion may require. For the prosecution of these experiments several vessels have been fitted with guns mounted conformably to these two different principles. Several varieties of traversing

slides are also under trial in the French service, but they are all, more or less, analogous to those in use in the British service. The "Cuvier" and the "Cassini" are fitted with the affût à double pivot (Fig. 17), as

Fig. 17.

Double Pivot-Carriage with its support, for a Canon-obusier of 22 centimetres (8.65 inches).

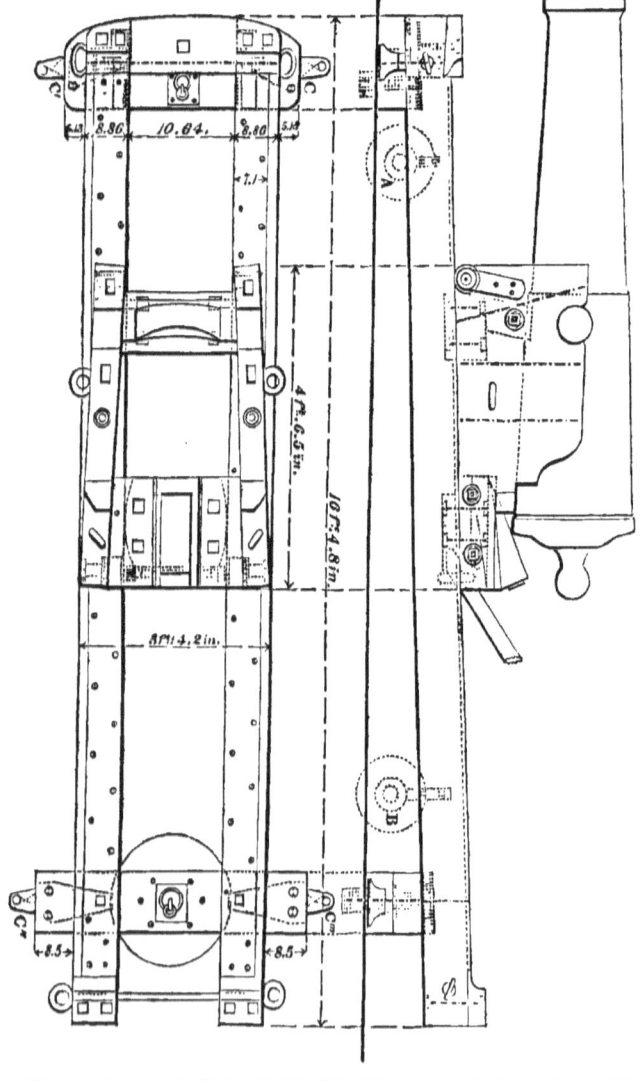

N.B.—The dotted parts A, B, C, C', C'', C''' of the support indicate the modifications made in those of the like kind which are mounted on board the "Cassini" and "Cuvier."

previously tried on board of the "Infernal;" whilst the "Caméléon" and "Pluton"[a] have been fitted with the affût à échantignole (Fig. 18) The ships are armed

Fig. 18.
Carriage of the Canon-obusier of 80 (No. 1 of 1841), in the "Caméléon" and "Pluton."

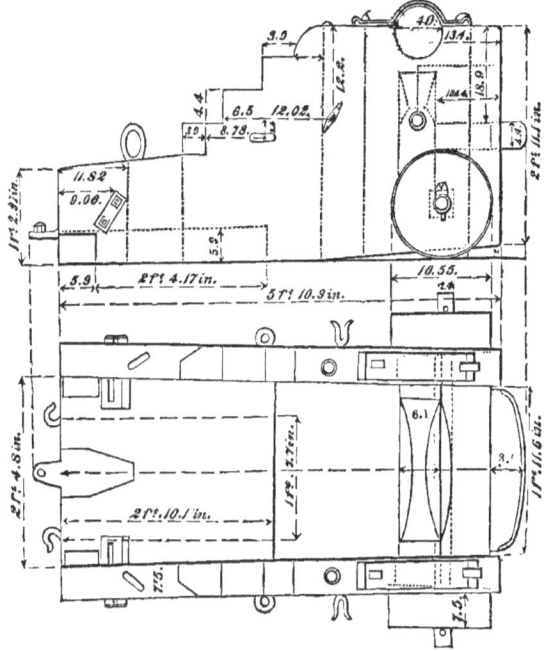

with two canons-obusiers of 80 and two 30-pounder guns, installed in broadside ports in the ordinary way.

---

[a] The state of the Sailing and Steam Navies of France in the year 1851 is exhibited in Appendix (C) at the end of the work. The total number of vessels composing the Steam Navy of England was, in March, 1854, one hundred and fourteen (Official Navy List); and though, in our merchant steamers, we have a vast resource (Report on the Steam Navy, Parl. Papers, March 27, 1849), this cannot be made available at the commencement, that is at the most critical moment of a sudden rupture. When we consider how much our Steam Navy must be dispersed throughout the dominions of the British Empire, and the concentration of that of France on the coasts of the country, the comparison appears very unsatisfactory. It results from these statements, that the number, rather than the force, of their steam ships is the object chiefly contemplated by the French authorities. This is an indication that their purpose will rather be to act in a desultory manner against our trade and coasts, than in regular operations against our ships of war. This subject will be considered hereafter, at large.

Shipbuilding is now being carried on with great activity at Cherbourg, where the "Phlegeton," 450 horse-power, and two other steamers, besides a first-class brig, are in progress of construction. A first-rate ship of the line, the "Dessaix," is to be laid down on the slip lately vacated by the "Henri Quatre."

The great disadvantage of the pivot traversing principle, as heretofore applied, is that the bulwarks must be struck for action, when the gun being what is called *en barbette*, the men working the piece are exposed to be swept off, if within the range of musketry, grape, or case shot; whilst the absence of all cover, however slight, is, besides the physical disadvantage, most prejudicial in its moral effect. On the other hand, as in the " Caméléon" and " Pluton," the installation of heavy guns in ports or embrasures, the sweep or open of which is limited to about 14° or 15° on each side of the directrix, renders frequent shifting of the guns from port to port necessary—an operation which must always be difficult and objectionable in action, on account of the time it requires, and which, when there is much motion, may be almost impracticable: no decision, to the author's knowledge, has yet been made on these important trials. In the British service, the two principles appear to be very happily combined by the following simple contrivance :—the slide is made to traverse on shifting centres, and to take up fighting points which are established on the deck correspondently with suitable points in the several ports which each gun thus mounted is intended to serve.[a] Thus the serious disadvantage of

---

[a] This simple method of changing the centres on which traversing platforms turn, so that they may be readily shifted from the *turning* to the *fighting* point, which was applied by the late General Millar to the bow and stern guns of steam-ships, was, as the General acknowledged in a letter to the author, borrowed from his invention of that principle, in 1805, for mounting guns on round towers and circular batteries (Fig. 19). A bolt passing through the hole *h* in the rear of the traversing platform, being inserted in the socket at A, the gun is turned upon that point in the direction of a radius passing through any fighting point *a*, *a'*, *c*, *c'*, &c., in the circle in which those points are established. The bolt at A being then taken out, and that of the

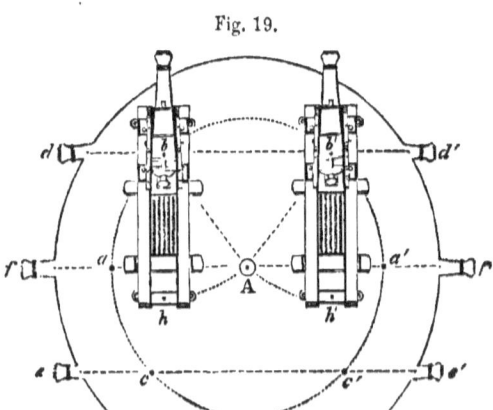

Fig. 19.

striking the bulwarks is avoided, and the advantages of the pivot carriage are retained.

In the first application of this principle, ports 4 feet wide, embrasured with sloping sides, were opened in the bulwarks, the height of which was 5 feet; the upper sill of the ports was made to unship when great elevations were required; whilst in close action the men were sheltered to this extent. The embrasures admitted of the guns being trained three points forward, and as many aft; but the ports are at present very much enlarged, in order to increase the sector of fire, and the height of the bulwarks is lowered: the expediency of this alteration may be doubted.

A very ingenious contrivance, invented by Colonel Colquhoun of the carriage department, a skilful and scientific officer, has been adopted for facilitating the operation of shifting the centres upon which the slides traverse, and for establishing in the deck, housing points or centres upon which the guns are with great facility turned.

222. In 1842 an extensive course of experiments was carried on at Gavre with canons-obusiers of 80; and the following are the results of some of the experiments:—

With two boulets creux, having no wad between them; the projectiles always issued from the piece in a multitude of fragments.

With a hollow shot and a live shell, having a wad between them; when the shell was next to the charge the fuze was invariably crushed or broken; when outward, the charge being high, the shell was frequently,

---

front pivot b or b' dropped into the corresponding socket underneath the gun (which, not being seen in the figure, is indicated by a dot perpendicularly over it, on the top of the gun), the platform is turned on this new centre into the position shown in the figure. The other gun is traversed in like manner, and thus the two guns may be pointed at the same object; or both to the right in the directions d' and e', or to the left in the directions d, e, or one on each side in the directions d' and e, or d and e'.*

* Extract of a Report of the Select Committee, dated Royal Arsenal, July 12, 1806:—
"The Committee are of opinion that Major Douglas has completed very fully the object which he wished to obtain; viz., that he can traverse guns, in a circular battery, in any direction, with facility, upon different centres within a given area."

and sometimes both the shot and shell were reduced to fragments.

Trials were likewise made with a solid shot of 88 lbs. and a hollow shot of 57 lbs.; then with a solid shot and a charge of large grape (20 balls, each of 4 lbs. weight); then with a live shell and a charge of grape: in all these cases comparatively little effect was produced; and it appeared that shells, with whatever load combined, were frequently broken in the gun.

Two solid shot were likewise fired together, but the recoil was so great, and the effect on the carriage so severe, that with the canon-obusier of 80 double shotting was peremptorily interdicted. Upon the whole it appeared that double loading could only, in some cases, be advantageous; and therefore, as a general rule, it was decided that such ordnance should be restricted to the discharge of single hollow shot.

223. The ravages occasioned by the bursting of guns on board the French frigates "Provence," "Vénus," "Triton," and others, the first before Algiers and the second at Brest, on which occasions great numbers of men were killed and wounded, while terror and demoralization were spread among the crews, have caused the naval authorities of France to enforce the utmost attention to the proof of the ordnance used on board ships of war. With respect to the canon-obusier of 80, on a first trial the chamber of the piece is quite filled with powder, and a cylindrical shot weighing 53 kilogrammes (116 lbs. avoirdupois) is projected; on a second trial, an equal charge is made to propel two cylindrical shot; and on a third, three such shot, a junk wad being rammed down on them by four blows.

Cylindrical or oblong shot are reserved for this extraordinary proof of the canon-obusier; but by a regulation, dated April, 1837, it is directed that no piece which has actually been subject to such proof shall be admitted into the service. It is often found that a gun, though it may have stood this severe proof apparently uninjured, is so strained as to give way on service when the ordinary charges are used; and the practice in

PART III.   GUNS FOR SHELLS AND HOLLOW SHOT.   209

France is to subject one gun only of each batch of metal to the extraordinary proof, all the other guns of the batch being made to undergo the ordinary proof; these are deemed capable of resisting the extreme charges, and passed accordingly.[a]

This method of proving the strength of guns rather by great loads of metal projected, than by extraordinary charges of powder, is worthy of imitation in proving naval ordnance in general. No such error can be committed, as that of putting double charges of powder—two cartridges—into a gun in action; but after guns have been loaded with the full, or distant charges and single shot, in running into action, and getting quickly within 300 or 400 yards, it may suddenly be necessary to put in a second shot, and upon such an occasion, seamen, in their ardour, may put in a third. Whatever may happen to bolts and breechings, the gun at least should stand the most severe trial; and on this account all naval guns should be proved so as to ascertain what loads of shot they will bear, with the full service charge of powder; and though, as in French proofs, every individual gun should not be subjected to the extraordinary proof, yet some one or two of every batch of metal, and from every contractor, should be tried *à outrance*.

224. On comparing together the ranges of the French canon-obusier of 80 (No. 1), weighing 74 cwt., with the

---

[a] Proof charges of the Paixhans guns are—with two shells, weighing together 132.72 lbs., 10 lbs. 12 oz. of powder; with two solid shots, weighing together 172.69 lbs., 21 lbs. 8 oz. of powder; and the maximum charge with two solid shots is 28 lbs. 1 oz. The greatest proof charge of the British 8-inch shell gun of 65 cwt. was, previously to 1848, 20 lbs. of powder with one solid shot, and a single hollow shot only was to be fired at one time; but it has since been found that this gun is capable of resisting far greater charges of powder, and of firing two hollow shots at once (see Article 246). The full service charge of the French canon-obusier of 80, No. 1, is 10 lbs. 12 oz.; that of the English 8-inch shell gun of 65 cwt. is 10 lbs. Though, as has been just said, the Paixhans gun is restricted to the use of the hollow shot, grape and case shot in general service; this is not on account of any inability to bear firing with solid shot and efficient charges, but on account of the inconvenience and difficulty in handling and loading, in quick firing; and the great proportion of the tonnage or displacement of the ship which that great weight would absorb; using low charges, the canon-obusier of 80 is capable of discharging at once, without danger or damage to the carriage, 200 or 300 pounds' weight of case-shot (*mitraille roulante ou plongeante*).

P

English 8-inch shell gun weighing 65 cwt., both of which are considered as the best of their class, the following conclusions are obtained. Using, at first, the experiments carried on at Brest in 1821 and 1824, with shells weighing 60½ lbs. and solid shot weighing 86½ lbs., the diameter of the bore being 8.95 inches, windage 0.09 inch, and the charge 10 lbs. 6 oz.; and the experiments made on board the "Excellent" in 1839 with hollow shot weighing 56 lbs. and solid shot weighing 68 lbs., windage 0.125 inch, and the charge 10 lbs.; it was found that, at an elevation of 3°, the range of the French solid shot exceeded that of the English solid shot by 651 yards, and the range of the French hollow shot exceeded that of the English hollow shot by 574 yards. At an elevation of 16°, the excesses of the French solid and hollow shot over those of the like English shot were 336 yards and 400 yards respectively. If, therefore, the Report on the French experiments above alluded to were alone considered, it would appear that the French canon-obusier is, in respect of range, far superior to the English shell gun; and an English writer of great ability has, in fact, so esteemed it; but a comparison of many ranges more recently obtained with the French and English guns has proved, beyond a doubt, that the advantages of superior range, with equal charges and elevations, are decidedly on the side of the latter.

225. In making this comparison, the ranges of the French canon-obusier (of 80), weighing 74 cwt., were taken from a general table formed from the experiments made at Gavre between 1830 and 1840, the inclinations in that table being reduced to elevations above the horizon. The ranges were reduced to English yards, and afterwards, by interpolations, reduced to those due to the particular charges and elevations for which the ranges of the English 8-inch shell gun of 65 cwt. are given in the tables of the experiments carried on on board the "Excellent" in 1839 (Table V., Appendix D.). The French and English shot were hollow, and the weights of the guns and shot were as above specified.

The windage of the English gun was 0.125 inch, and that of the French gun 0.1378 inch.

Of fourteen ranges obtained from each gun, the charges being 10 lbs. and the elevations varying from 22' 30" to 13°, the mean of the ranges of the English gun exceeded that of the French gun by 114 yards; while, of seven ranges from each gun, the charges being 8 lbs. and the elevations varying from 37' 30" to 2° 7' 30", the mean of the ranges of the English gun exceeded that of the French gun by 12 yards. In three of these ranges, the elevations being the lowest, the advantage was in favour of the French gun; but, in the four others, the advantage was on the side of the English gun. The ranges obtained from the 8-inch shell gun, weighing 60 cwt., being then compared with the canon-obusier of 80, weighing as above, the charges being 8 lbs. and the elevations varying from 22' 30" to 10°, it was found that, in eleven ranges obtained from each, the mean range of the English gun exceeded that of the French gun by 11 yards; but, in four cases, the advantage was in favour of the latter gun, the mean of the excesses being 25 yards.

226. From experiments which have been made on board the "Excellent" with an 8-inch shell gun weighing 65 cwt., and double shotted, it appears that such practice cannot be made with good effect beyond 200 yards.[a] In the experiments alluded to there was set up a butt of sound ship timber, consisting of two planks crossed, each 6 inches thick, and bolted to timbers 12 inches thick; when, at the distance of 260 yards, three rounds being fired with two solid shots and a charge of 5 lbs. of powder, 5 of the shot passed through one side of the butt, striking the opposite side and indenting it one inch; at the distance of 100 yards, of two shots fired at once, one struck the bull's eye, and the other a point about 1 foot from it. Of 30 solid shots fired two together,

---

[a] This practice is permitted in action at or within 200 yards, and the maximum charge is 5 lbs. of powder; but double-shotting is not allowed with 8-inch guns weighing less than 60 cwt.

with a junk wad between them, 11 broke; and of 30 others so fired, with no wad, only 1 broke. In 11 rounds, a shot and a shell being fired together, with wooden bottoms, 3 of the shells broke; and in 3 rounds, with two shots separated from each other from 2 to 4 inches, 3 of the shots broke.

In 1849 double shot from 8-inch guns, weighing 65 and 60 cwt., were fired in a series of experiments made on board the "Excellent" against a butt representing a section of the "Princess Charlotte" at the bends and lower deck. The shots were hollow, and weighed 56 lbs.; the charge was 5 lbs., and the distance of the butt 200 yards. At the first round both shots struck, 18 inches asunder (one in the centre of a timber), and, after passing through, ricochetted in the water at 300 yards beyond the butt.

V. ON RIFLE GUNS LOADED AT THE BREECH.

227. The practice of loading guns at the breech is not of modern date. In a work, by an Italian named Moretti, which was printed early in the seventeenth century, it is stated that the Venetians had many guns which were so loaded, and which carried shot of 4 lbs. weight.[a] And the readers of works on Military Antiquities will find in them frequent mention of that primitive method of loading ordnance which is now reproduced.

Many curious specimens of such guns may be seen in the Royal Military Repository and in the United Service Institution. The most ancient one of this description is that which was recovered from the wreck of the "Mary Rose," sunk at Spithead in an action with the French in 1545. It is made of wrought-iron bars, secured with iron hoops and fixed in a solid bed of elm 9 ft. 8 in. long: it was loaded at the breech by a detached chamber, which was kept in its place by a chock of elm. Several ancient chambers of guns of smaller sizes, of wrought iron, of the time of Henry VII., have been

---

[a] This work was translated into English by Moore about 1650.

found at Dover; and likewise the chase or body of a gun which was loaded in like manner. The diameter of its bore was 1½ inch. Artillery loaded at the breech with a detached chamber is still used in China; and the brass jingal, or swivel gun $1\frac{7}{8}$ inch diameter, is of that nature; several specimens of these may be seen in the United Service Institution. A small brass 4-pounder, with a detached chamber for loading at the breech, and bearing a cypher of the Dutch East India Company, was found on an islet on the coast of Australia, upon which a Dutch ship, named "Zeevyk," was wrecked in 1727; and a piece of brass ordnance of Dutch make, and bearing the date 1650, has recently been brought from the Gambia by H. M. Steam vessel "Teazer." The gun is made to load at the breech; and the charge and shot were kept in their places in the bore by a quoin or wedge in a manner nearly similar to the plan of Major Cavalli. (See the following Article.)

The method of loading at the breech does not seem to have been much in use for small arms. There is, however, in the United Service Institution, a small petronel of this description, of the time of Charles I., and two carbines of about the year 1740 or 1750.

228. It has been said, Art. 179, that, in 1846, iron rifled-cannon, capable of being loaded at the breech, were invented by Major Cavalli and Baron Wahrendorff, for the purpose of firing cylindro-conical and cylindro-conoidal shot (see Figs. 8 and 9 in that Article). In these guns the mechanical contrivances for securing the breech are very superior to the rude processes of earlier times; yet it appears doubtful whether or not, even now, they are sufficiently strong to ensure safety when high charges are used in long continued firing.

229. The length of the Cavalli gun (see Fig. 15, Plate II.) is 8 feet 10.3 inches; it weighs 66 cwt., and its calibre is 6½ inches. Two grooves are cut spirally along the bore, each of them making about half a turn in the length, which is 6 feet 9 inches. The chamber, which is cylindrical, is 11.8 inches long and 7.008 inches diameter. With respect to windage, it must be observed

that in all rifles with *forced* leaden shot of any shape there is practically no windage and, accordingly, no waste of the charge : but it is not so with iron shot fired from rifled cannon, since the iron cannot be made to expand so as to fill the bore and enter into the grooves; there must, consequently, be some windage; and, in fact, if there were not some, or if the charge were not greatly reduced, the blowing off the breech, an accident which happened to M. Cavalli's own gun, would be of frequent occurrence.

Immediately behind the chamber there is a rectangular perforation in a horizontal direction and perpendicular to the axis of the bore; its breadth vertically is $9\frac{1}{2}$ inches, while horizontally, it is 5.24 inches on the left side and 3.78 inches on the right side. This perforation is to receive a wrought-iron case-hardened quoin or wedge which, when in its place, covers the extremity of the chamber which is nearest the breech. The projectile, which has been described in Art. 179, being introduced through the breech and chamber into the bore of the gun, and the cartridge placed behind it, a *culot* or false breech of cast iron is made to enter $2\frac{1}{2}$ inches into the bottom of the chamber behind the cartridge; and a copper ring, which also enters the chamber, is placed over it. The iron wedge is then drawn towards the right hand till it completely covers the chamber. After being fired, the gun can be reloaded without entirely taking out the wedge; for the latter, which is shorter than the rectangular cavity in which it moves, can be withdrawn far enough to allow the new load to be introduced.

To explain at length the objects of the copper ring and the culot would require numerous drawings and occupy a space that would not be altogether convenient or consistent with the object of this work; for all such details the reader is therefore referred to the work of M. Cavalli, quoted in Art. 179.

230. The Cavalli gun is mounted on a cast iron bed, placed on a platform constructed of strong beams of timber. When this platform is horizontal the maximum

Part III.   RIFLE GUNS LOADED AT THE BREECH.   215

elevation that can be given to the gun is 15°. The elevation is produced by means of graduated quoins or wedges, which are worked very ingeniously in and out by an horizontal screw turned easily by the hand (see Fig. 7, Plate V., Cavalli's ' *Mémoire* ') : the gun is mounted on a non-recoil principle, which is effected by a strong iron trunnion attached to the front of the bed underneath, and working in a socket firmly fixed in front of the platform ; the length of the trunnion being sufficient to allow some play for the jump of the gun when fired, while it is prevented from getting out of the socket by a forelock underneath. The trunnion performs the double office of preventing recoil, and serving as a pivot on which the bed may be traversed on the platform. The shock occasioned to the platform by the jump and rebound of the immense weight of the gun and bed is softened by the elasticity of the platform, which consists of nine strong beams (Plate IV. of the '*Mémoire*') laid transversely on strong sleepers. The ends of the beams, though placed very near, are not united to each other, so that the elasticity of each may be separately exerted ; and the whole forms a highly elastic surface, by which alone the descent of the gun with its carriage, in the jump which it makes on being fired, could be sufficiently cushioned and resisted.

231. From experiments made at A°ker, in Sweden, in September, 1846, it was found that, with charges of 6 lbs. 9.8 oz. and an elevation of 14° 45', the mean of the first grazes of the hollow cylindro-conical shot weighing from 68 lbs. 13 oz. to 69 lbs. 15 oz. was at 3329 yards, and the mean deflection 85 yards to the right hand. With charges of 8 lbs. 13 oz., and an elevation of 13°, the mean of the first grazes was at 3668 yards, and the mean deflection 100 yards, to the right hand. With charges of 6 lbs. 10 oz., and an elevation of 14° 30', the mean of the first grazes of the solid cylindro-conical shot weighing 101 lbs. 8 oz. was at 2592 yards, and the mean deflection 28 yards to the right. Lastly, with charges of 8 lbs. 13 oz. and an elevation of 13°, the mean first graze of the hollow cylindro-conoidal shot was

at 3818 yards, and the mean deflection 99 yards, to the right.

The superiority, in extent of range, afforded by the hollow over the solid shot, amounts, from these experiments, to above 1000 yards; but the deflections of the solid shot are much less than those of the hollow shot. The mean range of the cylindro-conoidal shot exceeds that of the cylindro-conical shot by 150 yards; and the deflections are nearly the same. The projectiles are said to have, invariably, struck the ground with the point foremost.

232. The rifled-gun constructed by Baron Wahrendorff differs in some respects from that of Major Cavalli. (See Fig. 16, Plate IL.) Its whole length is 8 feet 10.9 inches, and its greatest diameter AB 2 feet 3.2 inches. The diameter $a\ b$ of the bore is 6.37 inches from the muzzle to within 6 inches of the chamber, in which space $c\ d\ e\ f$ it has a conical form, the diameter at $c\ d$ being 6.95 in.; the diameter of the chamber $c\ d\ g\ h$ is 7.5 inches. A rectangular wedge, 12.2 inches long, 8.1 inch broad, and 4.25 inches thick (a face of which is shown at Fig. 17), is made to slide, towards the right or left hand, in a perforation, formed transversely through the breech, for the purpose of covering, after the gun is loaded, the aperture by which the charge is admitted into the bore. A notch 7.2 inches long and 0.7 inch broad is made longitudinally in the wedge, and through this passes the stem, or bar, of a cylindrical plug, by which the charge is kept in its place. This plug (Fig. 18) is 7.4 inches diameter and 4.7 inches long, and it is provided with a stem or bar, 15.75 inches long, at the extremity of which is a screw nut having two handles. The plug is introduced in a direction parallel to the axis of the gun, through an orifice in the breech; and its stem passes through a perforation made in an iron door which closes the orifice. When the gun is loaded the door is closed; the plug is pushed forward, to the rear of the charge, by means of its stem, and the wedge is made to slide into its place : a turn of the screw nut at the end of the stem is then taken, when the whole is

PART III.   RIFLE GUNS LOADED AT THE BREECH.   217

drawn tightly together and is ready for firing. After firing, the wedge is drawn out as far as a pin which fits into a groove in its top will permit; and this just allows the plug to be drawn back close to the door, which is hollowed, as at k, to receive it; the door will then open so that the plug may be withdrawn from the breech of the gun, preparatory to a re-loading being made.

233. Some important trials were, during the year 1850, made at Shoebury Ness with the Cavalli and Wahrendorff guns, in conjunction with the British 32-pounder gun weighing 56 cwt.; cylindro-conoidal shot weighing 64 lbs. being fired from the former, and round shot from the latter, and, from those trials, the relative values of the three natures of ordnance may be said to have been in a great measure determined. The foreign guns are those which were cast at A°ker; and their construction has been carefully and ably described by Colonel Palisier of the Royal Artillery.

At the efficient service-elevation of 5 degrees, with charges of 8 lbs., the ranges, and also the deflections of the different projectiles were nearly equal to one another: and the like is true with charges of 10 lbs. At elevations of 10 degrees, the ranges of the foreign guns exceeded those of the English 32-pounder, with charges of 8 lbs., by 380 yards; and with charges of 10 lbs., by 690 yards; and at elevations of 15 degrees, the excess was, with charges of 8 lbs., about 790 yards; and with charges of 10 lbs., about 1100 yards. The ranges obtained from the Cavalli and Wahrendorff guns, with equal charges and elevations, agree very nearly with those which were obtained at A°ker in 1846. The deviations were always in the direction of the rotation of the projectiles; but they were so variable in amount that no allowance could be made for them in laying the gun with respect to the object.[a]

It must be admitted that the Wahrendorff gun has considerable advantage, in respect of range, over the English 32-pounder at a high elevation; but it ought to

---

[a] For comparisons of round with conoidal shot, fired from 32-pounder and 8-inch guns, see Arts. 191, 192.

be observed that the practice is, in that case, very uncertain. Such guns cannot enter into the armament of ships of war, but they may be used with advantage for flanking defences in casemates, or in coast-batteries for firing at objects at great distances.

234. With respect to ricochet firing, it is obvious that the rapid rotation of the shot upon its axis, and the projection of the oblique wings by which the rotation is produced in the gun, must, on the shot grazing, cause a forcible action on the surface of the plane struck, whether land or water. A remarkable proof of this appeared in the recent experiments at Shoebury Ness; when, at every graze of the cylindro-conical or cylindro-conoidal shot, it took a fresh direction, deviating more and more to the right—an important circumstance, which will no doubt be found to prevail in the ricochet of all cylindro-conical rifle shot from muskets as well as from cannon. When fired à ricochet over sea, it is probable that the shot would penetrate into the water so far that the impulsive force, resolved in the vertical direction, would not be sufficient to allow it to rise from thence.

235. The Cavalli gun became unserviceable after having fired four rounds, by the copper ring or bush imbedded in the metal of the gun at the bottom of the bore being damaged. This rendered it necessary to remove the gun to the foundery, in order to have a new copper ring put in; for this purpose it was necessary to cut away some of the metal of the gun, in order to set in the bush afresh. But however nicely this was done, it did not succeed, and at the next trial the whole of the breech was blown off.[a] The Wahrendorff gun stood well, the wedge resisting more effectually the

---

[a] Of three guns which had been constructed by Major Cavalli, one broke at the first trial in Sweden, in consequence, as he alleges, of the badness of the metal; and the third broke at the fourth round of the practice at Shoebury Ness. M. Cavalli considers that the mechanical arrangements for closing the breech were all but perfect, and he regrets that the failure in this country prevented the effect of his arrangement from being made evident. It is said that M. Cavalli seeks compensation from this Government for his invention; but having, before the experiments took place, made it known to the world by a detailed publication of his whole scheme, and this having failed, he can have no other claim than to a remuneration for his expenses in attending the trials.

force of the discharge than that of the Cavalli gun. If the latter, when it failed, had been on board a ship, the breech would have passed through, or have made a prodigious fracture in the opposite side; and consequently, besides the physical injury, it must have produced the worst moral effect on the crew.

236. Baron Wahrendorff has invented a 24-pounder gun, which is also to be loaded at the breech. It is mounted on a cast-iron traversing carriage; and, taking little room, it appears to be very fit for casemates. The upper part of the carriage has, on each side, the form of an inclined plane, which rises towards the breech, and terminates near either extremity in a curve whose concavity is upwards. Previously to the gun being fired the trunnions rest near the lower extremity; and on the discharge taking place, the gun recoils on the trunnions, along the ascending plane, when its motion is presently stopped. After the recoil, the gun descends on the plane to its former position, where it rests after a few short vibrations. The axis of the gun constantly retains a parallel position, so that the pointing does not require readjustment after each round.

The gun was worked easily by eight men, apparently without any strain on the carriage. With a charge of 8 lbs., and with solid shot, the recoil was about 3 feet, and the trunnions did not reach the upper extremity of the inclined plane, though the surface was greased.

VI. RELATIVE VALUES OF SOLID AND HOLLOW SHOT.

237. The British shell guns having been compared (Arts. 224, 225) with the French canons obusiers, it is now intended to make a like comparison between the former and the solid shot-guns, which are at present employed in the Naval Service of this country, in order to ascertain whether or not the shell-guns do really possess such qualities, with respect to extent of range, accuracy of fire and penetrating force, as to warrant their appropriation as the pivot guns of steamers: which assuredly should, with equal or inferior weight, possess

those qualities in the highest degree. It is proposed at the same time to ascertain whether or not shell-guns are better adapted than others for the broadside batteries of ships, in which situations rapid firing and extensive perforations are the essential conditions of their action.

238. From the formulæ for V, the velocity (Arts. 61, 63, and 68), it is evident that if shot, of equal diameters, but different weights or densities, be projected with charges whose weights bear the same proportion to the weights of the shot, as $\frac{1}{3}$ or $\frac{1}{4}$, the initial velocities will be very nearly equal, any difference which may exist arising only from the small difference in the lengths of the charges. Now, since the quantities of motion lost by shot in passing through a short extent of space in air are very small, as may be shown from the formulæ, Art. 60, it follows that, when the gun is laid point blank, or with a small elevation, the ranges of solid and hollow shot of equal diameters, and charged proportionally to their weights, are very nearly equal. But when the elevation is increased, so that the extent of the range and the time of flight are considerable, the solid shot, from its superior momentum, retains greater velocities at the end of equal times, and thus ranges farther than the hollow shot.[a] Since, also, a hollow shot issues from the gun with a greater velocity than a solid shot of equal diameter, when the charges are equal, and even when the charge of the former is rather less than that of the latter; it follows that, when laid point blank or with a small elevation, the hollow shot will range farther than the other: yet it will happen, for the reason above mentioned, that by increasing the elevations, so as to increase the ranges and times of flight, the less rapid diminution of the velocity of the

---

[a] This has been distinctly proved in experiments made on board the "Excellent," with 68-pounder guns, weighing 91 cwt. and 87 cwt., and 8-inch shell-guns weighing 65 cwt. and 60 cwt. The lateral deviations of hollow shot were also much greater than those of solid shot when the ranges exceeded 3000 yards. In firing at a target resembling a ship's side, at the distance of 3000 yards, it was found that the different ranges of hollow shot varied from 300 to 400 yards, while the different ranges of solid shot did not vary more than 200 yards.

heavier shot will cause the ranges to approximate; and, at length, the elevations and times of flight being farther increased, while the charges remain as at first, the range of the solid shot will exceed that of the hollow shot.

239. When the velocity with which a hollow shot strikes an object, as a ship's side, is less than that with which a solid shot strikes such object, provided it penetrate, the magnitude of the fracture, as well as the splintering and shattering effects produced by the hollow shot, will be greater than those produced by the solid shot. (See Arts. 158, 159.) But when, in consequence of the greater distance of the object from the gun, it becomes necessary to give to the latter a considerable elevation, in order to obtain the required range, the accuracy of the firing and the probability of striking the object are diminished; and, in this case, the solid shot has advantages over one which is hollow; the latter, even if accurately formed, is much more liable to lateral deviation than the other, particularly when fired across the direction of the wind; the deflections are very great in long ranges, and take place principally near the farther extremity of the trajectory. Hollow shot, and especially shells, are more liable than solid shot to irregular rotations during their flight, from their centres of gravity and of magnitude not coinciding. (See Arts. 180 to 189.) Even when empty, this will arise from the protrusion of the fuze, and the removal of metal to form the fuze hole, as well as from the unequal thickness of the metal in different places, in which respect shells are seldom perfect; and this irregularity is much greater when the shells are partially loaded with lead, sand, and even with the bursting powder, any of which is liable to change its place in the shell during its flight. The disadvantage produced by partial loading is manifest from the fact that shells so loaded neither range so correctly nor so far as when they have been entirely filled with the loading material.

240. In confirmation of what has been stated above, the following results of experiment are adduced:—

On comparing together the 8-inch shell gun of 65 cwt. with the 32-pounder weighing 56 cwt., the former projecting hollow shot weighing 56 lbs., and the latter solid shot, the charges being equal (10 lbs.) for both; it will be found, from the tables of practice on board the "Excellent" (see Tables V. and VI., Appendix D), that, at equal elevations, from 1° to 10°, the ranges of the 32-pounder invariably exceeded those of the shell-gun, the differences increasing with the increase of elevation, and amounting, at 10°, to 300 yards. Similarly, with charges equal to 10 lbs., the ranges of the 32-pounder gun exceeded those of the 8-inch gun, when loaded shells weighing 51 lbs. were fired from the latter, the difference at 10° amounting to 460 yards. The ranges of the 56-pounder guns weighing 98 cwt. and 87 cwt., the former having a charge of 16 lbs., and the latter of 14 lbs., in a still higher degree exceeded, at equal elevations from 1° to 15°, the ranges of the 8-inch gun of 65 cwt., with its charge of 10 lbs., and projecting hollow shot weighing 56 lbs. and 51 lbs., the differences gradually increasing with the elevations, and amounting at 15° to above 800 yards for the 51 lbs. shot, and 447 yards for the 56 lbs. shot.

The following table, extracted from the practice made at Deal, in 1839, exhibits the relative values, with respect to range, of the 8-inch gun projecting hollow shot, and other natures of iron ordnance projecting solid shot.

Nature of Ordnance.	Weight of Gun.	Length.	Windage.	Charge.	Ranges in Yards to the first Graze, below the corresponding Elevations.				
					P. B.	1°	2°	3°	4°	5°
	cwts.	ft. in.	inch.	lbs.						
8-in. Gun.	65	9 0	0.125	10	474	805	1133	1323	1602	1920
32-Pr.	56	9 0	0.233	10	475	877	1311	1467		
Ditto	63	9 7	0.233	12	..	..	1366	1581	1832	1998
42-Pr.	80	10 6	0.175	14	..	..	1346	1605	1842	2086
56-Pr.	97	11 0	0.175	16	..	..	1350	1600		

241. The 8-inch shell-gun (65 cwt.) is moreover inferior in respect of range to the 32-pounder gun, solid shot being projected from both. For, from experiments

Part III.   VALUES OF SOLID AND HOLLOW SHOT.   223

made on board the "Excellent," the charges (10 lbs.) and the elevations (from 3° to 8°) being equal, the ranges of the 32-pounder exceeded those of the shell-gun, by distances varying from 250 to 360 yards. The ranges of the shell-gun on board the "Excellent" compared with those of a 42-pounder gun weighing 67 cwt., charge 10 lbs. 8 oz.; and a 68-pounder gun, weighing 95 cwt., charge 16 lbs. on board the same ship; also with those of a 32-pounder (56 cwt.) at Deal, solid shot being fired from all, presented similar results in proof of the inferiority of the shell-gun. But the most striking instance occurs in the comparison of the above shell-gun with Mr. Monk's 56-pounder, weighing $97\frac{1}{2}$ cwt., charge 17 lbs. (solid shots being fired from each), the ranges given by the latter at Deal (in 1839) exceeding those of the shell-gun at equal elevations by distances varying from 400 to 590 yards.

242. With respect to the velocities and penetrating forces of shot at different distances from the gun, the following table will show that hollow shot from an English 8-inch shell gun, with greater initial velocity, preserves at equal distances greater velocity and penetrating power than hollow shot from the French canon-obusier of 80; and that solid shot from a 68-pounder gun, with less initial velocity, has at equal distances beyond 800 yards greater velocity and penetrating force than the hollow shot from the 8-inch gun. Again, it appears that the 42 lbs. and 32 lbs. shot have, at equal distances, less penetrating forces than the shot from the last-mentioned gun; and that the shot from the 56-pounder gun has both greater initial velocity and greater penetrating power than those from the 8-inch gun, and indeed from any of the guns above named:—

Nature of Ordnance.	Windage.	Weight of Gun.	Weight of Shot.	Charge.	Initial Velocities in Feet per Second.	Distances in Yards from the Gun.	Velocities at the corresponding Distances.	Ratios of penetrating Forces at the different Distances.
	in.	cwt. qrs. lbs.	lbs.	lbs.				
Canon-Obusier (80)	0.1378	73 3 14	58.8	10	1323	100	1225	6492
						500	915	3620
						1000	615	1879
8-in. Shell gun, English	0.125	Cwt. { 65 60 }	56	10	1418	100	1324	6945
						500	1019	4114
						1000	753	2250
68-Pr.	0.2	95	68	16	1280	100	1211	5812
						500	979	3799
						1000	765	2318
56-Pr.	0.175	98	56	16	1646	100	1546	8934
						500	1213	5499
						1000	913	3115
42-Pr.	9.2	67	42	10½	1360	100	1273	5489
						500	988	3304
						1000	739	1849
32-Pr.[a]	0.233	56	32	10	1600	100	1486	6819
						500	1116	3846
						1000	803	1991

243. In some trials which were made on board the "Excellent," with solid and hollow shot from a 68-pounder gun, with equal charges and elevations, the range was found to be in favour of the hollow shot from the point blank direction, to an elevation of $1\tfrac{3}{4}°$ inclusive; at $3\tfrac{1}{8}°$ the ranges were nearly equal; after which, up to $6\tfrac{7}{8}°$, the difference was slightly in favour of the solid shot: also the number of times in which the shots struck the object were nearly in the proportion of $5\tfrac{1}{2}$ to 4 in favour of the denser projectile. In the remarks on these trials it is stated that on the day of the experiment the wind was very fresh, blowing across the range, and that the solid shot were more true in their flight than the hollow shot; partly because the windage of the former was less, and partly because the hollow shot, being lighter than the others, were more affected by the wind.

---

[a] The windage is to be reduced to 0.2 inch, in a new 32-pounder of 58 cwt. From experiments made on board the "Excellent" in 1847, it is concluded that with the 32-pounder gun, 56 cwt., whose reduced charge is 6 lbs., double shotting may be employed with a certainty of penetrating, up to 400 yards; with the 32-pounder, 42 cwt., and charge of 4 lbs., up to 300 yards; and with the 32-pounder, 25 cwt., charge 2¼ lbs., up to 200 yards.

244. With respect to the relative accuracy in the ranges of solid and hollow shot from an 8-inch shell gun, it was found that, at the distance of 3000 yards, the differences between the ranges of hollow shot amounted to between 300 and 400 yards, while the differences between the ranges of the solid shots did not exceed 200 yards. Experiments were likewise made with hollow shot and loaded shells, from an 8-inch gun; when it was found that, beyond 1500 yards, the fire with the shells was more uncertain than with the shot; the number of times in which the object was struck by the former being, to the number struck by the latter, as $3\frac{1}{2}$ to 5. It was found too that the shells required half a degree more elevation than the hollow shot to attain the same range; and that, when the shells were quite filled, they ranged farther and more accurately than when not fully charged, on account of the loose powder revolving within them during their flight. (See Arts. 189, 239.)

The subjoined table of experiments is introduced in order to show the different elevations at which, with the given charges, solid and hollow shot projected from four different natures of guns attained the same range, 1250 yards; the object fired at was a target in the Marshes at Woolwich, at that distance from the guns.

Nature of Ordnance.	Length.		Weight of Gun.	Nature and Weight of Shot.		Windage.	Charge.	Elevation.
	ft.	in.	cwt.		lbs.		lbs.	
56-Pr.	{ 11	0	97	} Solid	56	.175	{ 16	$2\frac{3}{8}$
	{ 10	0	86				14	$2\frac{1}{4}$
32-Pr.	9	6	56	,,	32	.233	10	$2\frac{3}{8}$
10-Inch	9	4	84	Hollow	84	.16	12	$3\frac{1}{4}$ full.
8-Inch	9	0	65	,,	56	.125	10	3

From this table, it appears that the 8-inch gun required $\frac{5}{8}$ of a degree more elevation than the 32-pounder gun; and it is said to have made much worse practice, notwithstanding the greater windage of the latter.

245. In the absence of experiments carried on expressly for the purpose of determining the relative deviations of solid and hollow shot from the object of aim, at different

distances from the gun, the following table, computed from the formula, Art. 89, using the values of the coefficient k for the two kinds of shot, is introduced. It is extracted from the table in the "*Suite des Expériences à Gavre,*" Paris, 1844, (page 42,) and exhibits the horizontal and vertical deviations (supposed to be equal) for a 30-pounder long gun, (French,) and a canon-obusier of 80.

Nature of Ordnance.	Nature of Shot.	Diameter of Shot.	Charges.	Distances in Yards.			
				656	1094	1750	2188
				Horizontal and Vertical Deviations.			
		inches.	lbs. oz.	ft. in.	ft. in.	ft. in.	ft. in.
30-Pr. Long Gun.	Solid	6.28	8 4.4	3 7	10 10	31 6	53 9
			5 8.2	3 11	12 2	39 9	59 4
	Hollow	6.33	8 4.4	4 11	15 1	46 10	83 8
			5 8.2	5 3	16 9	51 6	91 10
Canon-Obusier 80, No. 1	Solid	8.61	7 11.5	2 3	8 2	22 4	38 4
	Hollow	8.67	7 11.5	4 11	14 5	42 4	72 6

From this table it appears, that, with the canon-obusier of 80, the deviation of hollow shot is, at all distances, about twice as great as that of solid shot.

246. Till recently, as has been observed in Art. 215, the English shell guns were restricted to the use of single hollow shot, grape and case-shot; but this restriction being manifestly a disadvantage to the service, and its necessity being doubted, at least with respect to the 8-inch gun weighing 65 cwt., experiments were carried on at Woolwich, in 1848 and 1849, in order to ascertain how many rounds the last mentioned nature of gun would stand with double shot; and what charge of powder it would bear up to the bursting quantity. The experiments decided against the necessity of the restriction, at least for ordnance cast like the gun with which the experiments were made. This was an 8-inch gun, weighing 65 cwt. 3 qrs. 14 lbs., 9 feet long, cast at Low Moor Foundry, which is distinguished for the strength of its constructions in iron. On many other occasions when guns have been burst on trial, a high charge was

PART III.   VALUES OF SOLID AND HOLLOW SHOT.   227

at once used, but, in the present instance, the first charges were low, and the rest consisted of gradually augmented quantities of powder. Thus two hollow shots, each weighing 56 lbs., and one wad, being used throughout, 60 rounds were fired with 5 lbs. of powder, (half the service charge for one shot,) then 10 rounds with 6 lbs., afterwards 10 rounds with 7 lbs., and so on, every succeeding 10 rounds having the charge increased by 1 lb., till 230 rounds in the whole were fired; the last ten rounds were consequently fired with the great charge of 23 lbs. These charges were resisted admirably, but the recoil was very great, though the platform was inclined in an angle of $2\frac{1}{2}$ degrees: with a charge of 10 lbs. of powder and one hollow shot, the recoil was $8\frac{1}{2}$ feet; with 5 lbs. and two shots, it was 14 feet 4 inches; with a 10 lb. charge, and a single solid shot, 14 feet 10 inches; but, with 10 lbs. and 2 shots, it was not less than 24 feet. The gun was mounted on a ship carriage. In January, 1849, the same piece of ordnance was used for the purpose of trying its efficiency in the discharge of two solid shots of 68 lbs. and a junk wad, when it bore well the fire of two rounds, each with a charge of 20 lbs.; but, at the third round with an equal charge, (the 243rd round,) and two solid shots, the gun burst, its fragments as well as those of the carriage being driven in all directions. On examining the vent after the 220th round, it was found to have enlarged to .28 inch at top, and 1.13 inch at bottom. This severe trial seems to have established the safety of the 8-inch gun of 65 cwt. when double shotted, with moderate charges.

247. Experiments have also been made with a new 32-pounder gun of 42 cwt. from the same Foundry; its service charge with one shot (solid) being 6 lbs.; and the following results have been obtained, solid shot and double wadding being used throughout. At first 40 rounds with two shots and a charge of 6 lbs. were fired; then 20 rounds with 3 shots and the same charge; next 20 rounds with 3 shots and a charge of 7 lbs.; again 20 rounds with 3 shots and a charge of 8 lbs., and so on, the charge being increased by 1 lb. of powder in every 20

rounds, till it amounted to 11 lbs., when at the eighth round, with this charge and three shots, the gun burst. In these trials 404 shots were fired, and 1128 lbs. of powder consumed.

Two 32-pounder guns which had been cast in Belgium, were also tried at Woolwich; and the following table contains a statement of the number of rounds which each gun bore :—

Charge.	Number of Shot.	Number of Wads.	Number of Rounds.
lbs.			
8	2	2	40
8	3	2	20
9	3	2	20
10	3	2	20
11	3	2	9

The last round burst one of the guns, which, in the explosion, struck its neighbour, and damaged it so much, that it only stood one more round with the same charge; at the next round it burst.

In the trials, 287 shots were fired from the first gun, and 959 lbs. of powder were expended.

248. In addition to the considerations of extent of range and precision in firing it is necessary to contemplate the effects of shot and shells with respect to their powers of impact—a subject of no less importance than either of the others. The penetration of shot has been already treated of in Arts. 79 and 98; and it is intended in this place to notice the effects arising from the impact of shot or shells of different diameters. A solid or a hollow shot striking a mass of timber, as the side of a ship, crushes, fractures, and splits the wood to an extent which depends, in a great measure, on the superficies of the shot, or upon the area of a section through its centre; and, consequently, is proportional to the square of its diameter. Thus, a large hollow shot will make a greater fracture, in consequence of a greater separation of the fibres of the material, than a solid ball of equal weight; though the latter may penetrate deeper,

and thereby produce, in other respects, a greater effect on and in a ship. It is evident, moreover, that the destructiveness of shot in action must also be proportional to the number of shots which strike the object at a given distance from the gun. We have no certain evidence from experiment by which we may determine the average number of hits made by shot on an object at different distances, up to the greatest at which shot can be projected with any chance of striking, suppose 2000 yards; and though, according to M. Piobert (see Art. 136), large shot struck a small object within 600 yards more frequently than smaller shot, yet it appears very probable that at 2000 yards the causes of deflection (particularly wind blowing across the range) would act more powerfully on large than on small shot, and certainly on hollow than on solid shot, and thus cause a smaller number of hits to take place in an equal number of discharges. However, in the absence of sufficient data to determine this point, let it be assumed that the number of hits of shot of different natures and descriptions are equal; and then it may be inferred that the splintering effects of shot are proportional to the squares of their diameter.

This is an important advantage in favour of the 8-inch shell-gun individually; but, on the other hand, it must in fairness be stated that magnitude of fracture is not the only thing to be considered in selecting ordnance for the general armament of ships.

249. It has already been stated (Art. 97) that it may be doubted whether the appropriation of 8-inch shell guns to the broadside batteries of ships of all classes has not been carried too far, some ships being armed chiefly with that description of ordnance; and whether it would not be more advantageous to limit, in all cases, the number of shell-guns to a lower proportion, and to combine them with others by a judicious selection of those best adapted to the circumstances of each case.

All vessels, according to their displacement, can only carry a certain weight of metal, of which their armament is to consist, and can afford only a limited tonnage for the stowage of their ammunition, projectiles and stores. The weights of the 8-inch shell-gun and of the 32-pounder gun are respectively 65 cwt. and 50 cwt., while the weights of the shot for the 8-inch and the 32-pounder gun, supposing an equal number to be distributed to each, are as 56 to 32, or as 7 to 4 : hence it will be found that eleven 8-inch guns are nearly equal in weight to fourteen 32-pounders, their complements of shot being included. Thus a vessel which could carry on her broadside only eleven 8-inch guns might be armed with fourteen 32-pounders.

In engaging an enemy's ship the aggregate magnitude of the fractures made in her side by the shot from the eleven 8-inch guns of her antagonist will be to that produced by the shot from the fourteen 32-pounder guns, (assuming that the magnitudes of the fractures made by the shot from the two natures of ordnance, supposing all to hit, are proportional to the number of shots fired and to the squares of their diameters,) as 704 to 546 ; which no doubt is an important advantage in favour of the 8-inch shot. But is not the greater number of discharges made, in equal times, by the fourteen 32-pounders relatively with the discharges of the eleven 8-inch guns, and therefore the greater probability of damage to the opponent, a very important consideration, which should enter into the question of armament ? In this case the number of discharges and the proportion of hits, supposing equal skill in gunnery and that both ships fire equally quick, will be as 14 to 11 ; a disparity which could scarcely be compensated by the greater magnitude of the fractures.

In this proportion, nearly, have the number of guns in many of the ships of the British Navy, new as well as old, frigates, corvettes, and small vessels in particular (the Thetis, Inconstant, Castor, Cambrian, and Dædalus, for instance), been reduced, in order that they may carry the heavy 32-pounders of 56 or 58 cwt. with the 8-inch guns. Some frigates have been reduced from 42 guns to 24 ; others from 46 guns to 24, 26, and even to 19 guns (see *Return to an Order of the House of Commons,* April 29, 1850).

250. The author ventures, with great deference, to think that in many of the combinations of 8-inch guns with 32-pounders in the armament of ships and vessels, too much consideration has been shown to the weight, and too little to the number of guns. He thinks that the number of shell-guns has been made too great in the broadside batteries of some ships and vessels, several of which, indeed, have the whole of the batteries on one deck so armed—the "Rodney," 26 8-inch guns; "Prince Regent," 32; "Albion," 26; "Indefatigable," 28. These, he believes, are exceptions, permitted for experiment, as in the case of the French frigate "Psyche,"[a] in consequence of special applications made by their captains. All experiments with such ordnance are, no doubt, extremely desirable, and cannot but be beneficial.

251. Judging from the armament of their ships, the problem concerning the relation between weight of gun and number of guns is solved by the naval artillerists of foreign nations with greater regard to number than by

---

[a] The "Psyche" frigate is an exception to this regulation. Her armament, as she was fitted out at Brest, in 1845, was as follows:—

Main deck	. .	18 80-pounders, Paixhans Howitzers No. 2.
,,	. .	2 30-pounders long guns forward.
,,	. .	2 30-pounders long guns aft.
Quarter-deck	. .	4 30-pounders carronades.
Forecastle	. . .	4 30-pounders long guns.

This frigate, although rated in the Budget as a 40-gun ship, was built to carry but 32 guns: she had, in 1845, but 30. Her complement of crew was the same as that of a 40-gun ship, viz. 326 men, on the war establishment. Her scantling is much stronger than that of any 46-gun frigate.

The eighteen 80-pounders on the main deck of the "Psyche" are the Paixhans canons-obusiers of 80 No. 2 (see Art. 221), weight only 53 cwt. 2 qrs. 14 lbs. (the Paixhans gun No. 1 is 71 cwt. 2 qrs. 15 lbs.), charge 6 lbs. 9 oz.; the chamber which, in No. 1 gun, is a cylinder of the diameter of a 24-pounder gun, being in gun No. 2 enlarged to that of a 30-pounder; so that the contraction being less, the operation of sponging is more readily performed, the cartridge more easily got into its place, and the *charge simultanée* facilitated. The canons-obusiers of 80 No. 2 are deficient in powers of range; they are intended for moderate ranges; their largest scales are only graduated to 1300 yards, which admits that they are not efficient beyond that distance, and consequently that the "Psyche" should avoid action at greater distances.

The observation made in Art. 97 on the reduction of the number of guns in a ship's armament, relatively to the displacement, is verified in recent practice by what is related of the "Pique" frigate, which was long under repair in Pembroke Dockyard, when she was rendered capable of carrying 40 guns, her former rating being 36 only.

those of this country. The problem is one that has been much discussed; and practical limitations have been assigned to the number of shell-guns which should be allowed to a ship of war in most of the foreign services; but in none of them has the limit been extended to the armament of whole decks. By a regulation of April, 1838, first-rates in the French navy were to carry 34 canons-obusiers of 30 on the upper decks; but, by a recent regulation (1849), all these were to be replaced by 30-pounder guns No. 3, which shows that, calibre for calibre, guns are preferred by the French to chambered ordnance. By the decree of 1838 the number of canons-obusiers in line-of-battle ships was limited to four of 80, and these were ordered to be placed on the lower deck;* but the number of canons-obusiers was subsequently increased by introducing six others, of the pattern No. 2, in the middle decks of first-rates and in the upper decks of second-rates, and four of the same pattern in the upper decks of third-rates.

According to the decree of the 27th of July, 1849, the numbers of canons-obusiers, and of solid shot guns on board of French ships of war, are to be as follow :—ships of 112 guns will carry four canons-obusiers of 80 No. 1 on the lower deck, six ditto No. 2 on the middle deck, and six 50-pounder guns; the rest of the armament, from the lower deck upwards, being 30-pounder guns Nos. 1, 2, 3, and 4. The 90-gun ships will carry four canons-obusiers of 80 No. 1 on the lower deck, six of No. 2 on the upper deck, the rest of the armament being six 50-pounder guns and 30-pounders Nos. 1, 2, and 3. The 82-gun ships, new model, will be armed in a similar manner with ten canons-obusiers Nos. 1 and 2. The 80-gun ship, old model, called 86, will carry four canons-obusiers of 80 No. 1 on the lower deck, and four of the pattern No. 2 on the upper deck. The 70-gun ships,

---

* In conformity with a resolution of the commissions composed of the chief authorities of the navy, maritime engineers and marine artillery (see Paixhans *Sur une Arme Nouvelle*, also *Expériences faites par la Marine*, pp. 44 and 58), and in compliance with the opinion and advice of the Comité consultatif de la Marine of the 17th June, 1824, ibid., p. 49.

old model, called 74, will carry four canons-obusiers of 80 No. 1 and twenty-four 36-pounders on the lower deck. No canons-obusiers will be carried on the upper deck. Frigates of the first class will carry only two canons-obusiers of 80 No. 1. The second and third class frigates and rasées will be armed in a similar manner, each carrying two 50-pounder guns.

252. The proportion of shell-guns varies very much in the ships of the British navy, even in those of the same class. See Parliamentary paper, May, 1849. The first-rates carry twelve 8-inch guns of 65 cwt. and 52 cwt.; ships of the "Rodney," "Albion," and "Prince Regent" class, twenty-six 8-inch guns each; the "Prince Regent," thirty-two ditto, which for experiment are all placed on the lower deck. By the Parliamentary paper, No. 126, (Return dated the 10th of May, 1849, Storekeeper's Department), it appears that the total number of 8-inch shell-guns, weighing from 65 to 52 cwt., in the sailing ships and vessels of the British navy, was at that date 2238, and the total number of 32-pounders, weighing from 56 cwt. to 42 cwt., was 8418, of which 3320 were the 56 cwt. gun. In line-of-battle ships the total number of 8-inch shell-guns was 1136; and of 32-pounder guns 6196, of which 2498 were the 56 cwt. gun. The Return does not specify how many of the 8-inch guns stated in the table were of 52, 60 and 65 cwt.

The extent to which shell guns should enter into the armament of a ship of war is a subject deserving much consideration, and will be examined in a more advanced part of this work.

253. The disadvantage of shell guns, compared with solid shot guns, in respect of the number of blows given, if all the shot of both kinds are fired with equal precision, holds good from the commencement of an action, when the firing is distant, till the crisis approaches, when double shotting is used. Now, 32-pounder guns, of 56 and 50 cwt., charged with 6 lbs. and 5 lbs. of powder, may commence double shotting at 400 yards with a certainty of penetration; though, at 300 yards, double shot firing with that nature of ordnance is most efficient

(Art. 107); whereas the 8-inch guns, of 65 and 60 cwt., being limited to 5 lbs. charges when firing two hollow shot, cannot commence double shotting with any effect at a greater distance than 200 yards: therefore, a ship armed with 32-pounder English, or 30-pounder French guns, should never approach so near to a ship very extensively armed with shell guns or canons-obusiers as to be within reach of her double shotted ordnance. At distances, therefore, between 200 and 400 yards double shotting can be used with effect from the 32-pounders only; and between these limits the ratio of the magnitudes of the fractures produced by the two natures of ordnance would be as $(8)^2$ to $2 (6.3)^2$, or as 64 to 79.4 nearly; but as the rapidity of loading with single shot is greater than with double shot in a certain proportion, therefore giving to the 8-inch gun, when using single shot, the benefit of this advantage, the above ratio of the effects of impact becomes, according to the best estimate that can be formed, as 7 to $8\frac{1}{3}$. With respect to shell-firing, the 8-inch guns are restricted to the firing of only single shells or shot at distances beyond 200 yards, whilst the 32-pounders may fire two solid shot, or other projectiles, at 400 yards; but as the destructive effects of shells of different diameters increase in a much higher ratio than the squares of their diameters (see Art. 249), it may be presumed that the 8-inch gun will have, in shell-firing, a great superiority over any 32-pounders at that distance: it must be considered, however, that solid projectiles from 32-pounder guns, fired in greater numbers than the shells, in equal times, and with their high degree of penetrating power, may produce destructive effects equal, if not superior, to those which are expected from the 8-inch ordnance.

It must be remarked that the recoil of the 8-inch gun exceeds that of the 32-pounder gun; with the charges above-mentioned, the velocities of recoil are 16.2 feet, and 15 feet per second, respectively; and, the weights of the guns with their carriages being 78 cwt. and 68 cwt., the momentum of the recoil of the 8-inch gun exceeds that of the 32-pounder gun in the ratio of 7 to

5.7 nearly. In comparing together the practice with two solid shots and that with two hollow shots, it must be borne in mind that there is always some risk of failure with the latter, since one or the other of the two is frequently found to be broken in pieces in the gun when fired with considerable charges, although, by experiments recently made, it appears that this very seldom occurs when, suppressing the wad, the shot are placed in contact with each other. With respect to the lighter 8-inch guns, they are incapable of being fired, either with one solid or with two hollow shot.

254. It is, in general, only in direct broadside action at close quarters that double-shotting should be applied; when, if on either side any errors should be committed in such practice, serious consequences may arise from want of sufficient penetrating power. In all oblique and raking fire, single shot and heavy charges, from all natures of ordnance, are required, particularly in raking a ship by the head, in order to break through the great masses of timber at her bow, and penetrate into and throughout the interior, for which purpose the single shot is most efficient. This observation on the importance of great penetrating power is again touched upon here, in order to introduce the very valuable and instructive Reports to which reference has already been made in Art. 106. The reader is requested to refer to that and the previous article and then to the statements which are given at length, in the note below.[a] These

---

[a] An account of shot which entered His Majesty's ship "Shannon," in the action with the "Chesapeake," June 1st, 1813:—

Shot, round 32-pounder, cut away in the wake outer gammoning, 2 in. deep; grape, 7 in number, between 4 in. and 7 in. deep.

Foremast.—Shot, round 32-pounder, 4 in. deep, 15 feet above deck.

Mainmast.—32-pounder, main-deck, 1, depth $1\frac{1}{4}$ inch; 18-pounder, 10 feet above deck, $1\frac{1}{4}$ in. deep; grape, 10 feet above deck, 2 in. deep; grape, $1\frac{1}{4}$ in. deep in different places; chain, 1, depth $1\frac{1}{2}$ in., 15 feet above deck, 3 feet deep; grazes in several places, 1 in. deep.

Mizenmast.—32-pounder, round, 16 feet above deck, 3 feet split away in breadth, 5 feet up and down, and 6 in. deep; grape, 4 do. abreast, 3 in. deep.

Cutwater and Knee of the Head.—32-pounder, round, 2 in number, through; 18-pounder, round, near the same place; 32-pounder, round, in the

documents contain detailed accounts of the number of shots, of different descriptions, which penetrated, struck, entered, or stuck in, the sides of H.M.S. "Shannon" and

---

hawse holes; hawse pieces split away under the knee; several grape 2 in. deep.
Bows.—32-pounder, round, through bows; knight-heads shot away; several grape through bulwark.
Between 1st and 2nd Guns.—18, round, through; 18-pounder, round, 2 in. deep.
Under Fore Chains, between 3rd and 4th Guns.—32-pounder, round, through forecastle; 6 grape, 2 in. deep.
Water's Edge, Fore Chains.—Shot, 18-pounder, through; chain, depth 3 in.; grape, 2, 3 in. deep; 1 bolt, iron, 10 in. deep, and 6 chain-plates.
Between 5th and 6th Guns in the Wale.—32-pounder, round, 2, 10 in. deep; 18-pounder, round, through; grape, 2, 4 in. deep; 7 grape, 5 in. deep.
Water's Edge, 7th Gun.—Chain, 1, 5 in. deep; 6 grape, 4 in. deep.
Main Chains, 8th and 9th Guns.—32-pounder, round, through; 1 canister, 6 in. deep; 18-pounder, round, 4 in. deep; 4 bar-shot, 8 in. deep.
Water's Edge, 10th Gun.—18-pounder, through; 12 grape, 3 in. deep; 4 chain-plates; 1 bolt, 3 in. deep; 6 grape, 3 in. deep; 32-pounder, round, through larboard side.
12*th Gun, Water's Edge, Channel Wale.*—18-pounder, round, through; 3 grape, 3 in. deep; 32-pounder, round, through; 4 grape, 2 in. deep.
13*th Gun, Mizen Chains, Water's Edge.*—Iron bolts, 2 in number, through 2 chain-plates; 5 grape, 2 in. deep.
14*th Gun in the Wale.*—18-pounder, round, through; 1 grape, 3 in. deep.
Quarter Gallery in the Wale.—32-pounder, round, through; 6 grape through.
Forecastle.—Grape shot, 20 through; larboard bumpin shot away; fore channel shot away; forecastle and waist hammock stanchions shot away.
Bulwarks—Quarterdeck.—Main chains or above, 32-pounder, round, through; 10 grape, 3 in. deep. Main chains much shot away; 18-pounder, round, through, 3 grape through. Mizen chains or above, 32-pounder, round, through; grape, 10 through. Mizen chains, much shot away.

H.M.S. "Shannon."
(Signed)   P. W. P. WALLIS, Commanding Officer;
Captain at Sick Quarters.

An account of shot which entered the sides of the American frigate "Chesapeake," in the action with His Majesty's ship "Shannon," June 1st, 1813:—
Bowsprit.—32-pounder, round, inside gammoning, 3 in. deep; outside do., 7 grape, 3 in. deep.
Foremast.—2 grape, 10 in. deep, 6 feet above deck; 4 grape, 4 in. deep, 20 feet above deck; 18-pounder, round, 4 in. deep, 20 feet above deck.
Mainmast.—18-pounder, round, 5 in. deep, 5 feet from main-deck; 9-pounder, round, 10 in. deep, 12 feet above quarter-deck; 2 grape, 10 in. deep, 20 feet above deck.
Mizenmast.—7 grape, 3 in. deep, 10 feet above quarter-deck; 18-pounder, round, 6 in. deep, 12 feet above deck; 4 grape, 3 in. deep, 15 feet above deck.
1 *Fife Rail, Larboard Head.*—8 grape through.
   under   ,,   4 grape through.

the American frigate " Chesapeake," also of those which hit or wounded their masts and spars respectively, in the action of the 1st of June, 1813 : the statements, there-

Bluff of Bows.—18-pounder, round, 6 in. deep.
　Do.　4 18-pounders, round, 8 in. deep; 30 grape, 8 in. deep.
2nd and 3rd Guns under Fore Chains—in the Wales.—4 grape, 3 in. deep ; 2 grape, 2 in. deep.
3rd and 4th Guns.—6 grape through ; 8 grape, 3 in. deep.
4th Gun under Channel.—6 grape, 4 in. deep.
Fore-chain Plates.—Shot away.
6th and 7th Guns, between.—9 grape-shot, 3 in. deep.
7th Gun in the Wale.—18-pounder, round, through ; 11 grape, 2 in. deep.
9th Gun under Main Chains.—5 grape, 2 in. deep.
13th Gun, side.—2 18-pounders, round, through.
Do., water's edge.—32-pounder, round, 2 through ; 2 grape, 2 in. deep.
Aftermost Port, water's edge.—2 32-pounders, round, through ; 18-pounder, round, through ; 1 pump-bolt through ; 3 grape, 2 in. deep.
Quarter Gallery, Foremast Wale.—3 32-pounders, round, through ; 10 grape through ; frame-work all carried away.
4 feet abaft After Port.—2 32-pounders, round, through ; 3 grape, 2 in. deep.
9 feet abaft after Port.—2 32-pounders, round, through.
3 feet above Water's Edge, in the Run.—32-pounder, round, through.
2 feet from Stern Ports—2 feet above Water's Edge.—18-pounder, round, 14 in. deep ; 32-pounder, round, through ; 12 grape, 3 in. deep ; stern-ports, 2 18-pounders, round, 3 in. deep.
Starboardside Counter.—7 planks, through ; stern ports carried away ; grape, 10, through.
Upper Counter.—18-pounder, round, through.
Bulwarks :—
　Forecastle.—60 grape, through ; 2 18-pounders, round, through.
　Above Chains.—5 grape, through 3 chain-plates.
　Main Chains.—5 grape, through.
　Mizen Chains.—12 grape, 3 in. deep.
　Abaft Ditto.—18-pounder, round, through ; 6 grape, 3 in. deep.
　　　　　　United States Frigate " Chesapeake."
　　　　(Signed)　　P. W. P. WALLIS, Commanding Officer.

[It is stated in James's 'Naval History,' vol. vi. p. 201, that the " Shannon's" main-deck guns were loaded alternately throughout the broadside with two round shot and a keg containing 150 musket balls, and with one round and one double-headed shot. Seeing that no vestiges or impressions of double-headed shot were found in the " Chesapeake," the author applied to the gallant and distinguished officer who instituted and certified the above surveys, for information which might account for this discrepancy. Captain Wallis states, in reply, that no double-headed shot were fired from the " Shannon ;" that the main-deck guns were loaded alternately with two round shots, and with one round shot and grape ; that kegs of musket balls were not fired as stated ; that only a few of these were provided for the boat's 12-pounder carronades. Whether any musket-balls were put into the quarter-deck guns, he (Captain Wallis) could not say ; but he thinks not. The " Chesapeake" fired round shot, grape, canister, double-headed, star, chain-shot, and other missiles, such as iron bolts and bars, about three feet long, bound together, and buck-shot in her musket cartridges.—AUTHOR.]

fore, present service-target proofs of facts, which cannot but be highly instructive, and show that even in that celebrated action there were several instances of a deficiency of penetrating power. Whether the charges of powder were too small, the weight of metal projected too great, or the powder of the "Shannon" was deteriorated, is not known; the last circumstance is highly probable, for this was undoubtedly the case in the unfortunate but gallant action of the Guerrière. These documents thus convey warnings of the extreme importance of insuring abundant penetrating power in all cases, and afford notices of many other useful facts to which reference will hereafter be made, particularly in the article on the effects of Grape Shot, Section IV., Part IV., and in the account of the action between the "Shannon" and "Chesapeake" in Part V.

255. The destructive effects of shells on ships are assumed to be proportional to the cubes of the diameters, upon the principle that the effects produced by the explosions of live shells depend on the quantities of bursting powder they contain, which, being as the volumes of the shells, are evidently in that proportion. In strictness, however, this law holds good only for shells which are fired into, and explode in earth, where they act as mines. In such cases, live shells are used with great advantage; for example, in the attack of fortresses, when they serve to destroy ramparts of earth,[a] make breaches in scarps of masonry, or for the purpose of rendering them practicable for assault.

By analogy, however, a shell lodged in a ship's timbers, and there exploding, is supposed to act as a mine, with a force depending on its charge; that is, proportional to the cube of its diameter. The greatest effect which a shell can produce in and on a ship, in action, is evidently

---

[a] "Les bombes lancées horizontalement produiront en crevant des entonnoirs proportionnés aux quantités de poudre dont elles seront remplis."—Bousmard, *L'Effet des Bombes horizontales sur un Ouvrage en Terre*, vol. i. p. 303, pl. 30.

Shells so fired were applied with great skill and science at the siege of Mooltan.

that which takes place when it penetrates into and lodges in the side or body of the ship, and then explodes.[a]

A shell suspended between decks, amid-ships, and then exploded, acts with equal force in every direction; and in such circumstances, the effect upon the crew would unquestionably be more destructive than if the shell were moving with great horizontal velocity when it bursts. Again, a shell placed in contact with the deck above it, and then exploded, would blow up the deck, and likewise produce destructive effects in lateral directions. Also a shell placed on the deck, and then exploded, would, no doubt, make a prodigious breach in the deck below it, drive up the deck above, and act severely in lateral directions. The shell which exploded by accident on board the " Medea" whilst cruising with the squadron blockading Alexandria in 1840, took place on the lower deck just above the shell-room; killed the Bombardier who was unscrewing the cap, wounded several of the crew, knocked down all the bulkheads from the captain's cabin to the boilers, broke three of the beams on the lower deck; forced up some of the planks of the upper deck, and produced, for a considerable time, much panic among the crew, and threw the whole squadron into consternation. These are portentous proofs of the terrific effects, physical and moral, produced upon a ship by the explosions of shells at rest within her, even though not imbedded in the mass of her sides or body; and the like effects must be expected to ensue should an enemy's shell be planted or lodged in the ship before the explosion takes place.[b]

256. It is this faculty of shells by which they act as mines that renders them most destructive to ships. In the experiments carried on at Brest during the years 1821 and 1824 (Paixhans, *Sur une Arme Nouvelle*, pp. 38, 62), at Portsmouth in 1838, and at Woolwich in 1850, the terrific effects of shell-firing on and in a ship,

---

[a] " Le tir de ces projectiles n'aura toute l'efficacité dont il est susceptible, qu'autant qu'ils conserveront assez de vitesse pour se loger dans les murailles des navires."—*Expériences exécutées à Gavre*, 1844.

[b] Experiments were made in the marshes at Woolwich, Nov. 19, 1850,

when the shells, having penetrated sufficiently into the timbers to lodge and explode there, took full effect, were strikingly exemplified. But the numerous failures of fuzes in those experiments[a] show that time-fuzes, which are essential to enable a shell to act as a mine, are very inefficient in horizontal shell firing. It is found that four fuzes out of five are extinguished on striking the water, and about one in three on striking a ship; if the shell strike with the fuze end forward, which is generally the case, it is found that the timber, by its resistance, forces itself into and effectually plugs the fuze.

In the experiments made at Portsmouth, with shells buried in earth,[b] and in others with shells embedded in masses of timber,[c] the explosions which took place afforded also the strongest proofs of the prodigious power of shells acting like mines. In these circumstances, the comparative effects produced by the 6.3

---

with a view of testing the effects of metal and wooden fuzes; and, on this occasion, shells were fired from 32-pounders and 8-inch guns against a strong bulkhead. The quantities of bursting powder were 1 lb. to each of the 32-pounders and 2¼ lbs. to each of the 8-inch shells; the charge to each gun being 8 lbs. Several of the shells exploded on striking the bulkhead. One of the 8-inch shells struck the ground short, and afterwards buried itself at the bottom of the bulkhead: it did not burst till after the next shell had been fired, when it exploded with tremendous effect, tearing the massive timbers into hundreds of fragments and scattering them about, besides breaking and twisting the numerous bolts with which the wood was held together.

[a] Analysing the shell experiments of 1838 against the "Prince George" hulk, at 1200 yards' distance, it appears that, on the 22nd of November, 7 shells were fired, of which 3 did not explode.
Nov. 23rd.—4 shells were fired; 2 did not explode.
Nov. 29th.—10 shells were fired; 7 did not explode, and none that grazed survived.
Nov. 30th.—12 shells were fired; 6 passed over the hulk, and of the other 6, 2 did not explode.
Dec. 1st.—31 shells were fired; 16 did not explode, and all that grazed were extinguished.
Dec. 3rd.—22 shells were fired; 8 did not explode, 1 burst short, 1 passed over the hulk, grazed, and was extinguished, 1 went over, made three grazes, and then exploded.
Thus, out of 80 shells that struck, 38 did not explode.
In the ricochet experiments at Southsea, in 1838—
Oct. 27th.—10 shells were fired; of which 2 burst.
Oct. 30th.—14 shells were fired; 3 burst.
Oct. 31st.—8 shells were fired; none burst.
So that only 5, out of 32 shells fired, burst in ricochetting.
[b] Experiment No. 13, 1847—"Excellent."
[c] Experiment No. 26, 1849—"Excellent."

and 8-inch shells respectively, were proportional to the charges, or as the cubes of the diameters of the shells.

257. In the experiments, No. 13 of 1847, and No. 26 of 1849, the shells were buried in earth at considerable depths, or embedded in masses of timber. Such as were lodged in earth formed regular mines, whose charges in exploding compressed the earth in every direction, according to well-known laws, forming what is technically called a "globe of compression," the craters having their axes in the lines of least resistance; while the shells lodged in timber, from the fibrous nature of that material, formed irregular wounds of considerable length, and detached large masses of splinters, but in a manner to show little analogy between the effects of shells embedded in the two different mediums: yet the shell which explodes when lodged in the side or body of a ship is, in fact, a mine; and, as such, is capable of producing the most destructive effects. But to obtain this effect on so small an area as a ship occupies on a horizontal plane, from shells projected at a considerable elevation, as in what is called vertical firing, is a very difficult matter, and the practice forms a very different case from that of shells projected horizontally (Art. 260). Concussion fuzes,[a] or percussion shells, are best adapted to the latter, but are inapplicable to the former. Time-fuzes, therefore, must continue to be employed in horizontal shell-firing, though, for this purpose, they are not always to be depended on.

Even when good and safe percussion or concussion

---

[a] It may here be necessary to explain to some readers wherein consists the difference between the terms "concussion" and "percussion," applied to fuzes and shells. A concussion fuze is provided with an internal mechanism, so nicely adjusted as to withstand the first shock which the shell receives—viz. that occasioned by the explosion of the charge—and resist any other that may be occasioned by grazing short, while it shall yield to the concussion occasioned by the impact of the shell on the body struck; this concussion, by shaking the burning composition of the fuze into the loaded cavity of the shell, instantly causes the latter to explode. A "percussion" fuze or shell is one fitted or filled with a chemical composition of highly explosive character, which bursts the shell at the moment of striking, without being previously ignited.

shells shall have been obtained,[a] time-fuzes will, nevertheless, be required in the naval service for shell-firing at troops on shore,[b] at open boats crowded with troops, ships filled with men on their upper decks; also for the bombardment of towns and fortresses, and the destruction of storehouses, barracks, or magazines (Art. 194); for all which purposes mortar, and not howitzer, shells are required, with time-fuzes to cause the shells to explode after having penetrated to their basements. Without time-fuzes, what would Lord John Hay have

---

[a] In the meantime it is satisfactory to know that a very near approach to this has been attained: a short-range fuze having been invented which is capable of acting by the concussion arising from the impact in striking a resisting body, as a ship; while it is enabled to resist the shock of the discharge.

[b] The answer to question 45, page 11, of the 'Catechism on Naval Gunnery,' used on board the "Excellent," states, "that fuzes, not being always driven with the same weight or force of blow (that is, by hand), do not burn equal parts in equal times; and it is recommended to ascertain experimentally the rate of burning of the batch of fuzes intended to be used, by trying one or more with a pendulum, before the practice commences"—a caution highly necessary from the deterioration of fuzes, in long services on foreign stations, from heat and damp. This observation carries the author back to a time when, in the command of an extensive brigade of 8-inch howitzers, he had ample experience of this, and of the difficulties, uncertainties, and intricacies involved in the whole question of fuzes.

It then appeared to the author that the operation of driving fuzes might best be performed by mechanical agency, which alone can give that perfectly equable compression to the composition which is requisite to ensure equality in burning.

For this a machine was contrived, by which each addition of composition received an equal degree of compression throughout, by hammers or mallets, of equal weight, falling freely through equal spaces, with nice adjustments by which the spaces of descent of the hammers might be accurately maintained throughout the operation.

The fuzes thus driven were found to burn much more equably than those driven by hand; and by this machine a great many fuzes might be driven at the same time, with the same moving power. The machine was of a very rude and imperfect construction, having been made for the trial. Perhaps the author was not wrong in principle in proposing such a machine; and having often since witnessed much error and uncertainty in shell-firing on service, he has looked back with regret that some such process had not been adopted. Having read the very just observation quoted from the "Excellent," and considering that the description of shell-firing to which this note relates demands the utmost perfection that can be given to this little implement, the author conceives that it may not be unworthy the consideration of the able and distinguished officer who now presides over the Laboratory Department to apply his skill and ingenuity to devise some instrument more constant and unerring than the arms and hands of men, which exert very different degrees of physical force, however long habituated they may be to that particular occupation.

done at Bilboa,[a] or Sir R. Stopford at Acre? (See Art. 195.)

258. From experiments carried on at Gavre, in France, in 1836, it is found that a projectile will not lodge in a mass of timber unless it penetrate to a depth nearly equal to its diameter; the strength or elasticity of the fibres forcing the shell out,[b] if the penetration be not deep enough to allow the fibres to close behind the projectile, and thus keep it embedded in the wood. It is from this circumstance that, in ordinary shell-firing from ships which are continually varying their distances, charges sufficient to obtain the necessary penetration and ensure the lodgment of the shell, must always be used, and fuzes prepared for corresponding times of flight; but the percussion or concussion fuze obviates all these complications.

---

[a] The practice of H. M. S. "Phœnix" in May, 1837, at St. Sebastian, was extremely efficient. All accounts concur in attributing to the unexpected but timely arrival of the steamers, and the extraordinary effects of live-shells, the success of the attack made by the Queen's troops on the entrenched position of the Carlists. Shell-firing against troops on shore had been previously tried, nearly in the same place, from some of the British frigates acting on the north-west coast of Spain in 1811, and in the naval operations of 1812. These were undertaken before the campaign was opened on shore, to distract the attention of the French forces in the north of Spain, and thus prevent their detaching, as was intended, a large portion of those troops to reinforce Marshal Marmont; also to intercept and break up the communication which the French still carried on by sea with Bayonne; and further, to succour and supply the guerilla corps with arms, ammunition, and stores. For these desultory enterprises, the author, then serving in the north of Spain, directed the issue of 5½-inch shells to the British frigate the "Surveillante," and some other 24-pounder frigates, to be used against any French "colonnes mobiles" which might be endeavouring to oppose or prevent the landing of arms and supplies to the guerillas, should those columns expose themselves to the fire of the ships at distances which neither case nor grape shot could reach, nor round shot be used with advantage. The shells were applied by the late gallant Sir George Collier, his first lieutenant O'Reilly, and other meritorious officers, with great skill and effect, in a manner to furnish strong proof that, for such purposes, live-shells with time fuzes might always be used with great advantage.

In attacking a redoubt, under the protection of which the Turkish ships had run on shore, in the operations of 1807, a considerable body of Asiatic troops, together with a part of the crews of the ships, having remained on the beach, were dispersed by a few shells from the "Pompée."—James's 'Naval History,' vol. iv., p. 181.

[b] This frequently occurred in the shell experiments of 1838, by which it appears that many shells rebounded on striking, from not having penetrated sufficiently.

In horizontal shell-firing considerable velocity is required, that the shell may have sufficient penetrating power to act as a shot,[a] if from any defect in the fuze it fail as a shell; for, if the striking velocity and penetrating power be small, the explosive action is rather towards the exterior, on which side the line of resistance is the least, than towards the interior of the ship; and this is the reason why shells that do not strike with considerable horizontal velocity so frequently explode outwards, and throw their fragments back. (See shell experiments of 1838 against the "Prince George" hulk, at 1,200 yards.) With a little more penetrating power the explosion may take place both ways. If an 8-inch or a 10-inch shell, either of which constitutes a powerful mine, should hit and lodge in a vessel anywhere below the water-line, and explode outwards, it might prove most serious, and even fatal to her; whereas, if it should hit anywhere above the water-line, and act outwardly, the effect would not prove very injurious to the ship, and not at all so to the crew.

259. The maximum effect of horizontal shell-firing with time-fuzes is when the shell perforates the near side of an enemy's ship, crosses the deck as a shot, penetrates into the other side just so far as that the line of least resistance may be inwards, and then exploding, scatters its fragments back into the ship (fig. 20.)[b] It thus combines, in one and the same projectile, the effects of both shot and shell; but this requires a rare and almost miraculous concurrence of the con-

---

[a] A remarkable proof of the importance of this occurred in the shell experiments of the 22nd of November, 1838—round No. 5, when a 32-pounder shell penetrated the side of the hulk "Prince George," below the water, lodged behind a rider, and made an opening in the side, which allowed the water to rush in with considerable force: the perforation was in such a position that the carpenters who were present stated that it would have been impossible for them to have plugged or stopped the leak from the inside: the shell did not explode. Had the fuze acted concussively, no such serious penetrating effects could have been produced.

[b] The targets, intended to represent parts of the sections of ships, as in Figs. 20 and 21, pages 245, 250, consisted of piles driven into a mud-bank at low-water: the bank, when not covered by the sea, exhibited on its surface the grazes of the fragments which passed beyond the surfaces of the targets.

ditions and circumstances under which the firing is made.

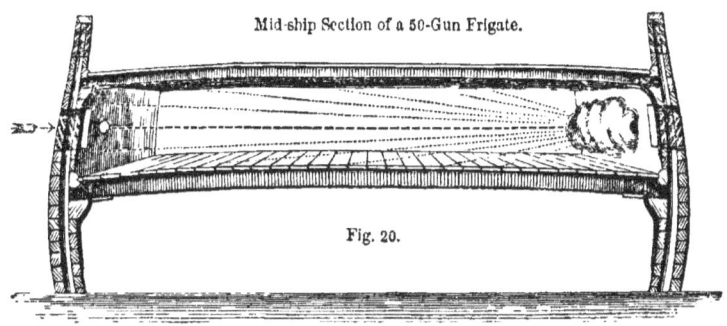

Fig. 20.

260. In vertical shell-firing against a ship, the great difficulty and uncertainty is to hit the object,—a mere speck in the amplitude of a long range; but if a shell should chance to fall upon a ship, it must either lodge in her, or pass through her bottom and sink her. In this practice the fuze must burn long enough to exceed somewhat the time of flight. In horizontal shell-firing the conditions are very different: the probability of hitting is very great; and, in close action, in a smooth sea, may be considered certain; but the difficulties and uncertainties relating to the fuze become then very great. The horizontal velocity is such that the shell passes over large portions of the range in times so short, that the smallest error in the estimated time during which the fuze will burn, destroys the effect. The shell may either explode prematurely, or may pass through both sides of the enemy's ship without exploding. The latter will be most frequently the case from the great care which must be taken that the shell should act as a shot in the event of the fuze being extinguished on grazing the water or striking the ship. On an average the probabilities of the former event happening and not happening are as 4 to 5, and of the latter as 1 to 3. (See Art. 256.)

261. From what has been stated in the preceding Articles (256 to 260), it is evidently essential to the efficiency and simplicity of horizontal shell-firing, that

the shells should be of a nature to explode with the utmost certainty on striking the object. It became, therefore, of vast importance to endeavour to obtain either an infallible concussion fuze, or a safe and efficient percussion shell (see the definition, Art. 257, Note), which may be free from the danger arising from unscrewing the cap of a metal fuze (Arts. 255, 274), and may not require to be introduced into the gun before the cap is taken off; so that the gun may be loaded with shell as quickly as with shot. It is of equal importance that the vital principle of the shell should not be destroyed in its flight by being immersed in water, in grazing, which so frequently extinguishes the fuzes of the common and concussion shells, nor be liable, like them, to be choked by the fibres of the medium struck.

Should such weapons of destruction be obtained, their effects upon and in ships will depend mainly on the number of impacts, in a given number of rounds, on such solid parts of a vessel as present sufficient resistance to the blows to occasion the explosion of the shell, by which, together with the horizontal force of the latter, the fragments may be driven through the side and into the ship with power sufficient to produce an effect similar to that which is depicted in fig. 21, p. 250.

The explosive agents contained in a percussion shell being necessarily so fortified as to withstand the resistance of a graze and the reaction which produces the rebound, and being capable of exploding only when a more violent shock of impact takes place, it follows that a percussion shell might perforate a slender or unrigid body without exploding, and even enter a port and pass across the deck to the opposite side of the ship, without producing any effect as a shell.

It may, therefore, here be remarked that neither a concussion fuze nor a percussion shell are suitable weapons to be used for dismantling purposes (see Sect. IV., Part IV.); for although a shell may be capable of bringing down a mast by a direct hit, whether it burst or not, and would infallibly fell a lower mast, did it explode on striking, yet this is but a remote probability; and therefore shells whose explosion depends upon striking a resisting body with great percussive force, would be wasted if fired at the rigging; and in general it may be said that although shells which do not explode will produce as much damage to masts, sails, and rigging as shot of the same diameter, yet this equality of effect would be obtained by a projectile which costs two or three times as much as a shot: shell-firing with such an object must therefore be enormously expensive. Shells may, no doubt, be used with considerable advantage for dismantling purposes at considerable distances, particularly in chasing or when chased; but for this they should have time-fuzes and should burst at a suitable distance short of the ship (Sect. IV., Part IV.).

262. It has been stated (see Articles 136, 248) that large projectiles are found to strike objects more frequently than small projectiles of equal density, fired with proportional charges;[a] but that some of the causes of deflection (as a strong wind blowing across the range) act more powerfully on large projectiles, especially those that are hollow, than on solid shot; and that the probability of large hollow projectiles striking an object as frequently as solid shot, diminishes as the distance of the object increases; also that the deflection is most apparent towards the end of a long range (Art. 239.)

When large and small projectiles are, both, shells or hollow shot, and are fired with charges which are to each other as the densities of the projectiles, the same law obtains: the inferiority of precision in the smaller shell, with respect to the proportion of hits, is manifest, and increases with the distance of the object, on account of the greater surface of the smaller sphere, compared with its weight, and consequently with its momentum. The smaller shell has, in fact, less power

---

[a] Piobert, *Traité d'Artillerie*, tom. ii. p. 270. Ward, p. 28, Metz experiments, 1816 to 1825; and *Suite des Expér. à Gavre*, 1844.

to overcome the resistance of the atmosphere, and therefore does not retain its velocity so long as the larger shell.

But when large and small projectiles are not of equal density—the former being a shell or hollow shot, and the latter a solid ball—the case is materially altered, and even reversed.

Lieutenant Ward, in his excellent treatise, already so frequently noticed, as well as all other naval authorities, lays it down as a principle that, at long ranges and with guns of different calibres, the probabilities of hitting a given object are as the squares of the diameters of the *balls*, supposing all to be of equal density, and to be fired with proportional charges.[a] But when the densities are so unequal as those of hollow shot, or shells, and solid shot, whether of the same or of different diameters, the probability of hitting, in distant firing, is in favour of the denser projectile; and it appears from the experiments of 1838, made on board the "Excellent," that this was found to be the case at short ranges likewise.[b]

---

[a] This maxim teaches an important lesson in favour of solid shot guns of large calibre, which appears not to have been without profit to the service to which Lieut. Ward belongs.

[b] On the 18th of October, 1838, on board the "Excellent," 11 rounds of single shot were fired from a 32-pounder, 9 ft. 6 in. long, and weighing 56 cwt., against a target set up at the distance of 400 yards; the charge was 8 lbs. and the elevation half a degree. At every round the target was struck; but at the fifth its top was just grazed. The gun was worked by 13 men, and the time in which all the 11 rounds were fired was 7 mins. 10 seconds. On the same day 11 rounds of single shot (hollow) were fired against the target, at the same distance, from a 68-pounder gun, 9 feet long and weighing 65 cwt.; the charge was 10 lbs., and the elevation $\frac{7}{8}$ths of a degree. At the first, second, and fourth rounds the target was struck; but at all the others the shot fell short, except at the eleventh round, when it fell close to the right staff; and at the sixth and 8th rounds the shot was deflected to the right hand. The gun was worked by 15 men, and all the eleven rounds were fired in 7 mins. 40 seconds.

On the same day also 11 other rounds of solid shot being fired from the same 32-pounder gun, with an equal charge and elevation, there were three hits, one of which was in the bull's eye, and three other shot touched the target: while 11 hollow shot being fired from an 8-inch gun, weighing 68 cwt., with a charge of 10 lbs. and an elevation equal to half a degree, two only hit.

Experiment, 17th of October, 1838.—15 rounds of hollow shot being fired from an 8-inch gun, of 65 cwt., charge 12 lbs., elevation $\frac{3}{4}$', at 400 yards, and

The author has been at great pains to examine very closely the important problem of the probability of hitting small objects of equal and different surfaces, under all varieties of circumstance; and he would refer the reader to the subjects of probabilities and deviations which are stated at Articles 136, 137, and 143. Also to Articles 158, 159, 160, 241, and 242, for the results relating to ranges and penetration which have been obtained from experiments made on board the "Excellent." Reference may also be made to Articles 242 and 243 for the relative accuracy of solid and hollow shot, and of solid shot and shells.

263. Although, as has already been shown (Art. 256), the destructive effects of an exploding shell are the greatest when it is embedded in a material whose fragments may be torn and scattered in every direction, and consequently that the maximum effect which a shell is capable of producing on and in a ship is when the shell lodges in some part of her mass and then explodes; yet so great are the difficulties of obtaining such a result in practice, that it has become a highly important object to endeavour to obtain some safe method of causing shells to explode, whilst moving with great horizontal velocity, on striking a ship; and the effects of horizontal shell-firing we are now to explain.

The actions of shells which explode on striking some solid part of a ship, though extremely formidable both to the vessel and to her crew, are far less so to the ship than those of an embedded shell; and the two cases are

---

the same number of rounds of solid shot fired from a 32-pounder gun, of 56 cwt., charge 10 lbs. 11 oz., elevation $\frac{3}{8}°$, the last five rounds double-shotted, same distance, the 32-pounder gun made the best practice.

With respect to the accuracy and penetrating power of solid shot fired from 32-pounder guns at 1200 yards, the following results were obtained from the Experiments on the 17th of October, 1838.—Six rounds of solid shot being fired from a 32-pounder gun, 50 cwt., charge 8 lbs., elevation $2\frac{5}{8}°$, there were five direct hits; and one shot ricocheted and struck the hulk; the penetrations were 22, 25, 36, and 48 inches. The two first penetrated directly or diagonally into perfectly sound wood; the last diagonally into solid but unsound wood.

Five solid shot being fired from a 32-pounder gun, 46 cwt., charge 6 lbs., elevation $2\frac{3}{4}°$, all struck the object directly. Penetration 39 inches diagonally, through sound plank and into solid timbers.

essentially different, since, if the percussive agent, whatever it be, perform its peculiar office, the shell to which it is attached cannot lodge in the wood. When a shell explodes in passing through a ship's side, the effect resembles that which accompanies the explosion of a spherical case-shot after being projected from a gun: the splinters, usually fifteen or sixteen in number, are carried forward into or across the ship in diverging paths, whose directions are compounded of that of the original projectile force and of those which the splinters would take from the action of the bursting charge alone. The cone of dispersion formed by the splinters

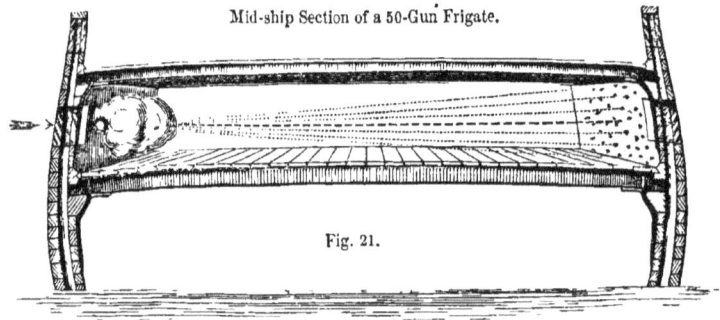

Mid-ship Section of a 50-Gun Frigate.

Fig. 21.

(fig. 21) has frequently been observed by the author when standing in a line perpendicular to the axis of the cone, or in the direction of the target. The paths of the splinters, being thus in the directions of the resultants of the two forces just mentioned, have not a great lateral divergency; and therefore the risk of casualties, at some distance from the original direction of the shell, is small compared with that which attends the explosion of an embedded shell. Whilst it is admitted that the force and effects of the explosion of an embedded shell on a ship are in general more destructive than those of a shell which bursts in passing rapidly through the side, it is equally true that the latter are more destructive than the former upon the gun crews of the adjoining, and still more upon the opposite, guns, when both batteries are manned, and upon all the people on the fighting decks within the limits of the cone of dispersion.

The 32-pounder gun most frequently used in the experiments referred to in the note, Art. 262, in the comparison with other ordnance, was the 56-cwt. gun, the windage of which (see Tables VI. and XVII., Appendix D., and Art. 198, Note) is excessive, being .233 inch, whereas that of the gun against which its range and precision were tried is only .125 (Table XVII., Appendix D.). This old pattern gun is the only specimen of naval ordnance of former times remaining in use. It is no doubt a powerful gun, from its length and service-charge, but it is defective in precision on account of its excessive windage: it is therefore very much to be desired that the new 32-pounder of 58 cwt., length 9 feet 6 inches, charges 10, 8, and 5 lbs., windage .2, proposed by Colonel Dundas, should be sanctioned, and forthwith provided, at least for trial. It may then be determined what more can be done towards the improvement of that valuable class of gun, and particularly whether the windage might not be reduced to .175. The 32-pounders of 45 and 42 cwt. (Monk's guns, B. and C.) have only .173 windage, and all the guns bored up from 24-pounders only .123; and there can be no reason why the new 32-pounder of 58 cwt. should have more. The pernicious and discreditable anomaly which still exists with respect to the windage of 32-pounders (Art. 198, Note) will then in a great measure disappear.

264. From all the preceding statements of the relative practice with shell-guns and the new ordnance for projecting heavy solid shot, it appears that, for power of range (see Arts. 240, 241, 243), for penetrating force (Arts. 158, 242), and particularly for accuracy of practice (Art. 262, with the Note), the solid shots fired from the long and powerful guns recently constructed are more efficient than hollow shots from shell-guns. By many experiments (see Arts. 243, 244) it is found that the deviations of hollow shot from the point of aim are greater than those of solid shot. The

inferiority of shell-guns would, however, be amply compensated by their great capacity for throwing projectiles of superior magnitude, either hollow shot or shells, if, as was generally the case during the last war, a ship could always attain a position for action within 1500 yards (Art. 244) without previously suffering serious damage: yet one disadvantage to which a ship carrying shell-guns would still be subject is that, being chambered, those guns are not so favourable as others for quick loading (Art. 218).

Shell-guns and shell-firing are, as yet, untried in actual combat, in broadside batteries; but there can be no doubt that, in a future war, a ship armed chiefly with shell-guns, and not carrying with them a due proportion of solid shot guns, by which it may be adapted for action at great distances and at close quarters, will have to endure the serious ordeal of much distant firing, in which the superior accuracy of solid shot, in their flight, will give to the opposing ship, if more abundantly provided with such projectiles, and if its crew have equal skill in gunnery,[a] a decided advantage over the other (Art. 249).

It appears therefore, to the author, to be a great mistake to suppose that the propulsive power of steam, and the effects anticipated from shell-guns, will cause actions at sea to take place at close quarters only, and to be decided in a few minutes. Steam gives the means of avoiding, as well as of closing to action, and when that giant power shall have been applied to line-of-battle ships, as well as to other vessels of war, as a locomotive battle power, naval evolutions will be capable of being executed with the utmost precision; and hence the principles of naval warfare with steam as well as wind will require to be well studied and practised by naval officers, in order that they may qualify themselves for the various

---

[a] "L'artillerie est une science très compliquée. L'officier de marine qui possède cette science est le maître de celui qui l'ignore: c'est une assertion qui deviendra pour vous une vérité évidente, en proportion des connaissances réelles que vous acquériez dans la science de l'artillerie."—Général Du Bourg, *L'Organisation de la Marine*, p. 358. Paris, 1849.

operations to which the new power in alliance with the sail will give rise. The destructive effects of shell-firing at short distances, and the extension that has been given to the projectile powers of artillery for very distant firing, far from tending to render close action unavoidable or preferable, and naval battles of very short duration, will, it may rather be presumed, cause actions at sea in general to be commenced at great distances, and be conducted with the utmost circumspection, tactical skill, and practical science.

265. The final struggle at close quarters will, no doubt, be preceded by efforts even more strenuous than those employed on former occasions to cripple the enemy at a considerable distance, whilst endeavouring to close, as in the actions between the "Macedonian" and the "United States" (see Part V.). No longer will a ship be permitted, with impunity, to run down, end-on, against her opponent; nor, without serious damage, will divisions of fleets in line a-head, as at Trafalgar, be permitted to advance, almost unscathed, against the broadside batteries of heavily armed ships waiting to receive, and, with improved gunnery, well prepared to meet that mode of attack. Of this we find distinct warning in all the commentaries that have been written on the naval actions of the late war. (See 'De la Gravière,' vol. ii., pp. 185, 188.) Seeing the progress which is now being made by foreign nations in introducing into their ships new and more powerful guns, with the doubts which begin to be entertained of the superiority of canons-obusiers and the incendiary system ('De la Gravière,' vol. i. pp. 98, 99, and other writers; see also Arts. 249, 253), we cannot fail to be convinced that distant firing will be the ruling practice on which success will principally depend in future war. This will require a corresponding armament of long and powerful guns in our ships of war, and particularly in steam-vessels. At present the heaviest solid-shot gun in line-of-battle ships of the British navy is the 32-pounder! while shell-guns are preferred for the bow and stern armament of steamers. It is important to remark here that the weight of the

French 30-pounder shot is about 32½ lbs. English, and that the French 36-pounder shot weighs nearly 39 lbs. English.[a] The new French 60-pounder (weight of shot 66 lbs. English), tried at Gavre in 1848, though not yet adopted as a naval gun, is nearly equal to our 68-pounder, which the author has no doubt will hereafter be applied to the naval service.

266. With respect to the United States, the naval authorities of that country are slow to imitate, and cautious not to commit themselves by any extensive adoption of shell-guns; they have accordingly incurred very little expense in providing new guns of that description for the broadside batteries of their ships of the line and frigates; and the naval officers of the United States, in general, do not appear to view with much favour the appropriation of shell-guns for the bow and stern armament of steamers. The Bureau of Ordnance know too well what they owe to long and powerful guns when these were opposed to carronades and short guns in the actions on the lakes of Canada (Art. 147), as well as on the ocean (see Part V.), to be induced, without mature consideration, to provide shell-guns for general service, more particularly for steam warfare, which they think requires guns possessing, in the highest attainable degree, power of range, accuracy in distant firing, and penetrating force. The United States' shell-gun, called the Paixhans' gun, of 63 cwt., is only the old 42-pounder of 70 cwt., bored up to the calibre of a 64-pounder to fire hollow shots of 43 lbs., which, when filled with sand, are increased to 46 lbs. These shell-guns are incapable of firing solid shot: four only of them are placed on each deck of line-of-battle ships, and two on the main decks of frigates.[b] They have been found greatly wanting, both in range and accuracy, two points most essential to a steamer's battery (Ward,

---

[a] "*Aide Mémoire Navale,*" p. 664. A new 50-pounder gun, which, by a regulation of the 27th July, 1849, was to have entered largely into the armament of the French navy, has since been withdrawn from the batteries of ships.

[b] Present armament of the ships and vessels of the United States' Navy. Bureau of Artillery, May 29, 1845.

p. 32); and accordingly we learn that new ordnance is being prepared, which there is every reason to believe will be guns for solid shot, combining accuracy with extent of range so as to enable a steamer to reach a line-of-battle ship with effect, at great distance.

In the work just quoted we find the following important and very sage question propounded :—Might not the weight of metal required to make a shell-gun be much more usefully employed if cast into a doubly-fortified, long and efficient gun, even though of less calibre? Such a gun would be far more accurate and efficient in distant firing; and at close quarters, when double-shotted, would do powerful execution. Since the publication of the third edition of this work a 64-pounder unchambered gun, fulfilling in all respects these conditions and expectations, has been introduced into the naval service of the United States, as mentioned in Art. 201.

There has lately been tried in the United States a new shell-gun, called a Columbiad; its calibre 12 inches, and weight of shell 172 lbs. With a charge of 20 lbs., and an elevation of 10°, it gave a range of 2770 yards; and, with its maximum elevation and charge, produced a range of 5761 yards, or about $3\frac{1}{2}$ miles. (*United States' Ordnance Manual*, 1850, p. 364.) But the weight of this gun, 25,000 lbs., was found to render it inapplicable to sea service, though the results of the experiments are said to have been highly satisfactory.

New 32 and 64-pounder long guns, of various weights and dimensions, are now under trial in the United States, for the naval service, and the best of these will no doubt be adopted.

267. Reverting to foreign navies, it is important to state that the Russian navy during the war with Turkey in 1829 consisted of five divisions, each comprehending nine line-of-battle ships, six frigates, eight corvettes and brigs, with eight steamers. This force has since been augmented, in each division, to twelve line-of-battle ships with a corresponding number of frigates and smaller vessels. The total establishment of the Russian fleet now consists of sixty ships of the line, armed with from 70 to

120 guns; thirty-seven frigates of from 40 to 60 guns; seventy corvettes and brigs; forty steamers, and two hundred gun-boats. The system of manning is by establishments of *Equipages de Ligne*, as in France (Art. 35, Note). Of this vast naval force three-fifths are stationed in the Baltic, and two-fifths in the Black Sea. In the former there were, in March, 1854, thirty-two sail of the line, armed and ready for sea, in the arsenal or harbour at Cronstadt, with the full proportion of frigates and smaller vessels; and in the Black Sea, at Sevastopol, thirteen sail of the line, ready for sea, with a large proportion of steam ships and small vessels.

The Russian three-decked ships carry a mixed armament of canons-obusiers, French pattern, and heavy solid shot guns[a] on the lower deck, 36-pounder long guns on the middle deck, 36-pounders No. 2 on the main deck, and 36-pounders No. 3 on the quarter deck and forecastle. Two-decked ships have the same natures of guns on their lower, main, and upper decks, as the three-decked ships have on their lower, middle, and upper decks.

The armament of the "Pallas" frigate, of 1600 tons measurement, which was pierced for 54 guns, but carried during the peace 50 guns, may be taken as a correct specimen of the general armament of the Russian frigates. The "Pallas" is provided with 30 guns on the main deck; viz. twenty-six 24-pounders (9 feet long), weighing 50 cwt., and four 8-inch shell-guns (65 cwt.), British pattern; she carries on the forecastle two 24-pounders similar to those on the main deck, and six 24-pounders ($6\frac{1}{2}$ feet long) of 30 cwt. On the quarter deck she has twelve 24-pounders of 30 cwt.

The Russians have no percussion shells, nor have they adopted the French Billette shell (Art. 299, and Note). They prefer time fuzes, and use the metal screw fuze. It was with these that they set fire to, and burned the Turkish squadron in Sinope Bay (Art. 302).

---

[a] Judging from the fact stated in note, p. 300, communicated to the author by Vice-Admiral Deans Dundas, concerning the nature of the shot picked up at Sinope, these guns must have been 42-pounders (equivalent to guns firing shot of 38 lbs. weight, English). Such guns are known to be in use in the Russian Navy.

The Russian 36-pounder is superior to our 32-pounder, and their 24 lbs. shot is equal in weight to 22 lbs. English. A new 50-pounder (the weight of its shot nearly equal to 55 lbs. English) is being prepared for the Dutch navy: this is an excellent gun, weighing 4624 kilog. (equal to 91 cwt.); its preponderance of breech $\frac{1}{21}$, and the length, from the base ring to the muzzle, 3638 millim. (equal to 12 feet nearly). The Danish gun-boats are armed with long and powerful 60-pounder guns (the weight of the shot equal to about 67½ lbs. English). The weight of a Danish 36-pounder shot is 39 lbs. 11½ oz., and of a 30-pounder shot, 33¾ lbs. English.[a]

These facts, and others which will be stated hereafter, are distinct indications that in the foreign naval services professional opinion is undergoing a considerable re-action with respect to the value of shell-guns and shell-firing. They also demonstrate the prevalence of an opinion that distant firing with powerful solid shot guns will be the most effectual means of avoiding or counter-acting the destructive effects which hollow shot and shells would unavoidably produce if the ship which uses them were by any chance to gain the requisite proximity.

It will not indeed be always in the power of the commander of a ship, however desirous he may be of avoiding close action, to be able to accomplish that pur-pose; for in thick weather, or in a dark night, vessels may unexpectedly fall in with and be very near before they discern each other. These contingencies may hap-pen, and a close action may be thus suddenly brought on; but it may be presumed that the occurrence of such

---

[a] To convert the weights of foreign ordnance and shot, in pounds, into English pounds, the following table may be used:—

Add to the number of { Danish pounds $\frac{1}{13}$ of that number.
Dutch ditto $\frac{1}{11}$ ditto.
French ditto $\frac{1}{12}$ ditto.
Spanish ditto $\frac{1}{11}$ ditto.

Subtract from the number of { Swedish ditto $\frac{1}{15}$ ditto.
Russian ditto $\frac{1}{11}$ ditto.

James's 'Naval History,' vol. i. p. 42.

surprises will be as rare as they have been in former wars between ships armed in the usual way. Either the ships may come together by accident, and then the surprise will be mutual; or one of the two, previously knowing the position of the other, may take advantage of the obscurity to approach her unseen; but this implies a difference in the degrees of vigilance on board the two ships which is scarcely within the limits of probability. Should the surprise be mutual, the disadvantage will certainly be on the side of the ship which is not armed with shell-guns; if both are so armed, and of equal strength, it is evident that they may then engage on equal terms.

268. The 10-inch shell-gun is no doubt a formidable piece of ordnance; its shell in exploding is a powerful mine; but it is inferior in range to the heavy guns now in use in the British navy, excepting the lower classes of 32-pounders. The weight of the 10-inch shell-gun, when first introduced, was 84 cwt. (see Table XVII., App. D); but this was increased to 86 cwt., by surrounding the breech and charging cylinder with an additional 4 cwt. of metal, and taking away about 2 cwt. from the forepart, by which means it has more "preponderance," and is made to approximate nearer to the new principle stated in Art. 208. The small proportion of windage given to the 10-inch shell-guns (.16 of an inch—the same as a common howitzer—the 68-pounder gun having .2 of an inch) is also an improvement. A 68-pounder gun (Art. 210) may be used for firing 8-inch shells as well as 68 lb. shot. For shell-firing, indeed, it is not superior to the 10 or 8-inch guns, because shells will not bear higher charges than those established for shell-guns; but a 10-inch gun for firing hollow shot is inferior in range to a 68-pounder gun with solid shot. The difference of weight between these two pieces of ordnance is only 9 cwt., but the superiority of the latter in power of range, accuracy, and penetrating force, is of vast moment in steam warfare.

This superiority of 68-pounder unchambered guns, to the 10-inch chambered shell-guns, for the pivot guns of steamers and sailing-ships, so strongly asserted by the author in the third edition of this work, is confirmed by the introduction of the new 64-pounder gun, unchambered (Art. 201), which has been adopted in the naval service of the United States, to be exclusively the pivot-gun for steamers, and indeed for all classes of vessels in which guns so mounted are employed. The reasons assigned for this substitution are precisely those which have been expressed by the author in the preceding passage; to which we may add that the unchambered gun can fire shells as well as solid shot, whereas the chambered guns cannot fire solid shot at all, nor even shells so far, so accurately, nor with so much penetrating power as the unchambered gun can project solid shot. It may be added that the principle upon which is founded the new field artillery proposed by the Emperor Louis Napoleon, and lately adopted universally in the French service, is that an unchambered 12-pounder gun will do the work of a howitzer as well as its own, and will thus render the last nature of arm unnecessary. See Appendix B.

Comparing a 10-inch shell-gun of 84 cwt., charge 12 lbs., and a hollow shot of 84 lbs., with a 68-pounder gun of 95 cwt., charge 16 lbs. and solid shot, in respect of their range, we see that at an elevation of 1° the difference of range is 142 yards, at 2° it is about 190 yards, in both cases in favour of the latter; at 3° the 68-pounder gun ranges about as far as the 10-inch gun does at 4°; the range of the former at 4° is greater than that of the other at 5°; and at higher elevations the differences in favour of the 68-pounder increase considerably. At 15° the range of the 10-inch gun is 3050 yards, and that of the 68-pounder gun 3673 yards (Tables V., VI., Appendix D). If we compare the trajectories of a hollow shot from a 10-inch gun, and of a solid shot from a 68-pounder, the elevation of the former gun being 5° and that of the latter 4°; also the ranges of the 10-inch shot and the 68-pounder shot being, by the experiments, 1670 yards and 1737 yards respectively, the angle of descent for the solid shot is much less than for the hollow shot; or the former, near the end of its flight, approaches much nearer than the other to a horizontal direction; and the chance of hitting a small object near the extremity of the range of a shot is manifestly greater in proportion as the trajectory[a] approaches, in that part, nearer to a horizontal direction; that is, as the angle of descent is less. It should be added, that hollow shot, being greater than solid shot of equal weight, are more liable than the latter to suffer lateral deflection from the action of wind and other causes. (Arts. 141, 239, 243.)

---

[a] The practice of representing the trajectories of shot may be very advantageously employed to exhibit the comparative curves in which projectiles move; and, from thence, the circumstances which determine the relative efficiency of shot and shells, of different diameters and weights, when projected from ordnance of equal or different dimensions and weights, at various elevations, and with various charges of powder. The author has laid down, by means of the first equation for y, Art. 80 (the second, or third equation for y, on page 57, is more simple but rather less accurate), the trajectories, among others, of a solid shot from a 68-pounder gun, and a hollow shot from a 10-inch gun, the ranges 1737 and 1670 yards, at the respective elevations of 4° and 5°, having been taken from the results of experiment; the inconvenient magnitude of the scale has, however, deterred him from having them engraved for this work. In order to obviate any misapprehension which may arise on the part of the practical reader, with respect to the correctness of such tracings, from an opinion that, because they are made by means of a theoretical formula, they may differ materially from the figures of the paths in nature—the author thinks it right to observe that such difference cannot exist. The ranges having been given by experiment, the initial velocity, or the height due to it, may be obtained from the formula itself without the uncertainty which may attend the direct computation of that element by the formula for V., Art. 61, or otherwise. The ordinates of the curve, corresponding to any assumed values of x, may therefore be determined with as much correctness as simplicity, the value of c, for the particular nature of projectile, being determined as in Arts. 59, 60. In the examples alluded to, the range of the 68-pounder shot is, by experiment, greater than that of the 10-inch shell, while its elevation is less; therefore the curves must intersect one another near the end of the range, and the angle of descent of the hollow shot be necessarily greater than that of the solid shot. On this depend the limits within which an object may be struck, and it is capable of being ascertained with all requisite accuracy.

The computation of the ordinates from the formula, and the manner of tracing the curve, should be taught in all schools of scientific gunnery; the practice of doing so is capable, if the work is laid down on a scale of sufficient magnitude, of giving very correct notions concerning the various circumstances on which the probability of a shot or shell taking good effect, at a given distance, depends. This subject is of particular importance where the comparative trajectories of solid and hollow shot are required, on account of the different degrees of rapidity with which the projectiles lose their velocities in passing through the air, and the consequent increase in the inclinations of the descending branches to the horizon, as the projectiles approach the end of their flight.

269. Table XXVII., Appendix D, exhibits the strength of the British Warlike Navy at the present time, March 1854.[a]

For these there have been provided 112 10-inch guns.[b] So far as these have been appropriated as bow and stern-guns to steamers, the expediency of the measure is very much doubted, and it seems to be deserving of reconsideration whether or not it might be better to replace them with 68-pounder guns of 95 cwt. in large vessels, and by the 88 cwt. 68-pounder guns for those that cannot well carry the heavier ordnance; also whether or not vessels that cannot carry these should be armed with 32-pounders of 56 cwt., or of 58 cwt., in preference to the 8-inch shell-gun. Vessels, however, which are unable to carry these guns at bow and stern, cannot be considered as efficient war steamers.

VII.—THE STOWAGE OF SHELLS AND THE PRECAUTIONS NECESSARY TO PREVENT ACCIDENTS IN SHELL-FIRING.

270. Great alterations have been made in the armament of the British Navy since the regulation of July, 1848. Shell guns have been introduced to a much greater extent, and the complements of shells for all rates and classes of H.M. ships and vessels[c] have been considerably augmented. See the statements in the following table:—

[a] OFFICIAL NAVY LIST.

Sailing Ships, and Ships propelled by Screws, in Commission, in Ordinary, and being Built.							
Rates.						Sloops.	Guard, Troop, Store Ships, &c.
First.	Second.	Third.	Fourth.	Fifth.	Sixth.		
17	50	24	48	44	34	84	38

Ships propelled by Paddles.		
Frigates.	Sloops.	Guard, Troop, and Store Ships.
18	32	2

Besides 97 Tenders.

[b] Parliamentary Paper, No. 127.
[c] See statement, No. 128, in the Appendix to the Second Report of the Select Committee on Ordnance Expenditure, 1849.

NAVAL GUNNERY.

Number of Shells per Gun or Carronade, supplied to all Classes of H.M. Ships or Vessels of War.

	Per Gun, or Carronade.	
SHELLS filled with Powder and fired to Wood Bottoms { All Bow and Stern Pivot Guns	60	
10-In. Side Guns	40	
8-In. Side Guns and Carronades { For the first six	40	No more than 500 Shells 8 In. to be supplied to any one Ship (the difference to be made up in Round Shot*).
For the second six	20	
For the remainder	10	
Steamers, Side Guns, and Carronades	10	

	Total Numbers.	
6-In. or 32-Pounder, Sailing Ships { First Rates	450	
Second Rates (Princess Charlotte Class)	450	
Other Second Rates	350	No more than 500 Shells 6 In. to be supplied to any one Ship (the difference to be made up in Round Shot*).
Third Rates	300	
Fourth Rates	200	
Fifth Rates	160	
Sixth Rates	120	
All Sloops	100	
Brigs, "Espoir," and "Dolphin" Classes, with 3 In. Fuzes only	50	

* When the proportion of Shells amounts to more than the maximum allowance of 500 of each nature, Round Shot should be substituted for the Shells withheld.

NOTE 1.—The following numbers of empty Shells with White Metal Fuzes will be supplied for Drill in addition to the above proportion:—

1st, 2nd, and 3rd Rates	6
4th, 5th, and 6th Rates	4
All other Ships and Vessels	2

NOTE 2.—When Ships cannot stow all the filled Shells allowed, empty ones are to be supplied to complete the proportion without Boxes, but with the proper number of Fuzes and quantity of Powder. The quantity of Powder required for filling Naval Shells is:—

10-Inches	5 lbs. 8 oz.
8-Inches	2 lbs. 4 oz.
56-Pounder	1 lb. 12 oz.
32-Pounder	1 lb.

NOTE 3.—No Shells for Guns or Carronades of a lower calibre than 32-Pounders of 25 cwt. and 17 cwt. respectively.

PART III.   COMPLEMENTS OF SHOT AND SHELLS.   263

PROPORTIONS OF FUZES, of different kinds, supplied to H.M. Ships or Vessels of War.

FUZES		
Moorsom's	. . . .	Three to every four Shells, 10 and 8 In., for Pivot and Side Guns. Two to every three Shells, 6 In., for Sailing Ships. Three to every four Shells, 6 In., for Steamers.
Graduated or Time, 3 In.	. . . .	One to every four Shells for Pivot and Side Guns. One to every three Shells, 6 In., for Sailing Ships. One to every four Shells, 6 In., for Steamers.
Spare 4 In.	. . . .	One to every two Shells with 3 In. Fuzes.
Metal White, for Drill	. . . .	One to every empty Shell supplied for Drill.
		But, until the store of Moorsom's Fuzes admits of a larger supply, Fuzes will be issued as under :—
Moorsom's*	. . . .	One to every four Shells, 10 and 8 In., for Pivot and Side Guns. One to every three Shells, 6 In., for Sailing Ships. One to every four Shells, 6 In., for Steamers.
Graduated or Time, 3 In.	. . . .	Three to every four Shells, 10 and 8 In., for Pivot and Side Guns. One to every three Shells, 6 In., for Sailing Ships. Three to every four Shells, 6 In., for Steamers.
Spare, 4 In.	. . . .	One to every two Shells with 3 In. Fuzes.
Short Range	. . . .	One to every three Shells, 6 In., for Sailing Ships.

NOTE.—No Vessel under a Sloop, commanded by a Commander, is to be supplied with Moorsom's Fuzes.

* The Board's Order of 2nd May, 1853, $\frac{A}{1185}$, directs Moorsom's Fuzes to be supplied to one half the established number of 10 and 8 In. Shells instead of one fourth, and 3 In. Fuzes to the other half.

To provide space for the stowage of these increased complements of shells, great alterations have necessarily been made in the internal fittings of all ships and vessels.

271. In line-of-battle ships the shell-rooms for 6-inch shells have been obtained by converting the upper part of the shot-lockers before the main-masts into rooms for their reception. A space about 4 ft. 6 in. deep has thus been taken from the shot-lockers, and appropriated to the stowage of 6-inch shells instead of the shot formerly contained in that space. This arrangement is sufficient to enable two-decked ships to receive their additional complement of 200 6-inch shells; but in three-deckers it is found necessary to make more room by unfixing, emptying, and removing as many of the 8-inch shells as may be necessary—one 8-inch shell occupying nearly as much space as two of the others.

The two 8-inch shell-rooms, in line-of-battle ships, are abaft the main-mast, one on each side of the passage under the scuttle. Over the loaded shells there is space sufficient for a considerable number of empty shells, which may accordingly be placed in compartments on the crown. Shells fitted with 3-inch fuzes are placed in one of these shell-rooms, and those fitted with short-range fuzes in the other. The 4-inch fuzes are considered as "spare fuzes," and kept for distant firing.

DIMENSIONS of the 8-Inch SHELL-ROOMS in SHIPS and VESSELS of different RATES, according to Regulation. No two Shell-Rooms are, however, exactly of these dimensions.

Rates.	Guns.	Complement of Shells, distributed equally in two Rooms.	Athwart Ships.		Dimensions Fore and Aft.		Height.	
			ft.	in.	ft.	in.	ft.	in.
First Rates . .	110 to 120	420	9	5	5	2	6	0
Second ditto .	{ 92 [a]	600	12	3	5	2	5	10
	{ 84	360	9	3	5	1	5	10
Third ditto .	{ 74	200	7	0	5	1	6	1
	{ 70	200	8	3	4	3	5	7
Fourth ditto .	50	240	8	3	4	3	5	7
Fifth ditto . .	{ 44	100	4	1	5	0	5	4
	{ 36	100	6	0	5	0	5	7
Sixth ditto . .	26	80 { In one Room.	7	2	4	7	6	3½

[a] The 92-gun ship is armed with twenty-four 8-inch guns—the 84 has only eight: hence the difference in the complement of shells, and the capacity of the shell-rooms.

STOWAGE OF SHELLS.

DIMENSIONS of the 6-Inch SHELL-ROOMS in SIXTH RATES, CORVETTES, and SLOOPS.

Rates.	Guns.	Complement of Shells.	Athwart-ships.		Fore and Aft.		Height.	
			ft.	in.	ft.	in.	ft.	in.
Sixth Rates	22 [a]	80 ⎫ Stowed in ⎫	2	9	4	7	4	3
Corvette	18	80 ⎬ two Shell- ⎬	4	3	3	0	5	6
Brig	16	80 ⎭ Rooms. ⎭	4	3	2	6	4	7
Brig	12	60 { In one Room.	7	10	2	1	4	9

The dimensions of the shell-rooms in the "London," 92, and "Formidable," 84, are:—

	"London."		"Formidable."	
	ft.	in.	ft.	in.
Height	5	0	5	10¼
Fore and Aft	2	11½	2	8
Athwart-ships	9	0	7	10

The two last dimensions being dependent on the size of the shot-locker, no fixed quantity can be assigned to them for other classes of vessels; but it appears that room sufficient will be found in ships of inferior classes, to stow the additional shells in the present shell-rooms, and in the spaces allowed for passages.

The DISTANCES between the Load Water-Line, and the tops or crowns of the Spaces allotted for the 8-Inch Shell-Rooms, in the undermentioned Ships and Vessels, are—

	Class.	ft.	in.
Caledonia	120	6	8
London	92	8	0
Formidable	84	5	8
Benbow	74	4	0
Cumberland	70	4	6
Vernon	50	7	4
Resistance	44	7	0
Pique	36	5	6
Alarm [b]	26	3	6
Dido, Corvette	18	2	10
Bittern	16	2	3
Espiègle	12	2	10

The water-tanks intervene between the 8-inch shell-room and the ship's side, and should therefore always be kept filled. The chain-cables, shot-lockers, and spare hawsers, as well as water-tanks, lie between the 6-inch

---

[a] Old class, carrying only two 32-pounder guns, and twenty 32-pounder carronades.

[b] Besides the "Alarm," 26, we still have a small class of sixth-rates (formerly the 28-gun frigates), which now carry only two 50 cwt. 32-pounder guns, and twenty-two 32-pounders of 40 cwt., and which therefore take a complement of eighty 6-inch shells, instead of 8-inch, as in the cases of the "Alarm," "Vestal," and "Trincomalee" classes.

shell-rooms and the ship's side in these classes, to render the rooms inaccessible to shot.

The shell-rooms in frigates being sufficiently large to receive their complements of 6-inch shells in addition to those of 8-inch, there is no necessity for emptying and removing any of the latter.

The operation of passing shells through the tiers to the main hatchway, in frigates, requires so many hands, and is so slow, that it has been found necessary to make a new scuttle, immediately over the present one, to communicate with the main deck, as in Ships of the Line, to admit of two shells being passed up at the same time.[a]

Sloops are provided with 6-inch shells only; and, of the established number, one-fourth are fitted with " Short Range Fuzes;" as many are stowed away in the shell-room as it will hold, and the remainder are placed empty in the wings of the bread-room.

The shell-rooms of smaller brigs are constructed in the wings, or spaces on each side of the bulk-heads of the magazine.

DISTANCES between the Load Water-Lines and the tops or crowns of the Magazines (column 1) in the undermentioned Classes of Ships and Vessels, and the difference in feet and inches between the *Heavy* Service and *Light* Service Water-Lines (column 2).[b]

Class.		1.		2.	
		ft.	in.	ft.	in.
Caledonia	120	5	6	1	6
London	92	5	6	1	4
Formidable	84	6	0	1	3
Benbow	74	5	6	1	3
Cumberland	70	6	0	1	2
Vernon	50	4	10	1	0
Resistance	44	3	6	1	0
Pique	36	2	4	0	10
Alarm	26	3	4	1	0
Dido, Corvette	18	2	0	0	8
Bittern	16	1	9	0	6
Espiègle	12	1	7	0	6

[a] In the French frigate "Psyche" (see 'French Armament,' in the forthcoming work on 'Shell Warfare') a scuttle is opened in the deck between the guns, through which the shells are handed up, for the service of every piece. (See the new method of handing up cartridges for the supply of the gun-decks, Sec. V., Part IV.)

[b] The figures in the last column denote the "rise" of the respective ships on the supposition that they have exhausted two months' water and provisions —as much, probably, as any British man-of-war consumes before replenishing either water or provisions, unless extraordinary circumstances have prevented the replenishment from taking place.

STOWAGE OF SHELLS.

DISTANCE from the LOAD WATER-LINE to the Crown of the MAGAZINES and SHELL-ROOMS of the undermentioned Screw-Propelled SHIPS.

Ship's Name and No. of Guns.	Fitted with Dell's Hexagonal Powder Cases.		Shell-Rooms.
	After Magazine.	Fore Magazine.	
	ft. in.	ft. in.	ft. in.
Royal Albert, 121 .	4 10	6 3	5 3
Agamemnon, 91 .	4 9	5 9	2 3
Majestic, 80 . .	6 6	5 0	6 5
Cressy, 80 . . .	4 9	{ 3 6 } { 5 6* }	4 3
Tribune, 31 . .	3 3	2 5	5 2

* Before the Ship was adapted for the reception of a Screw Propeller.

In the Prince Regent, which has lately been armed and fitted to a great extent for shell firing,[a] there is space sufficient for the stowage of 740 8-inch shells.

The 6-inch shell-room is before the main-mast, having two doors which open into the main-hold, lined throughout with copper, and 9 feet 6 inches wide, 5 feet 3 inches long, and 6 feet high. The top of the space for loaded 8-inch shells is about 5 feet below the load water-line, or 2 feet below the orlop-deck.

272. To avoid as much as possible the inconvenience and danger of filling and fuzing shells on board of ship they are in general sent on board in boxes, each containing one shell.

The space required for stowage is computed from the dimensions of the boxes (Fig. 22), which are as follow :—

	For the 10-inch Shells.	8-inch Shells.	32-Pounders.
	Inches.	Inches.	Inches.
Length . . .	12	10	8.2
Width . . .	12	10	8.2
Depth . . .	12½	11	8.8

The bulk of 100 10-inch Shells in boxes is about 87 cubic feet.
,,   8-inch Shells   ,,   64   ,,
,,   32-lb. Shells   ,,   38   ,,

---

[a] Thirty-two 8-inch guns of 65 cwt. on the lower deck; thirty-four long 32-pounders of 56 cwt. on the main deck; and twenty-six 32-pounders of 42 cwt. on the quarter deck and forecastle. The lower deck of the "Queen" was, in 1849, armed with 8-inch shell guns, displacing the 32-pounders of 56 or 58 cwt.—Parliamentary Paper, No. 128, 1849. The whole of the lower deck of the "Royal Albert" is armed with 8-inch shell guns of 65 cwt.

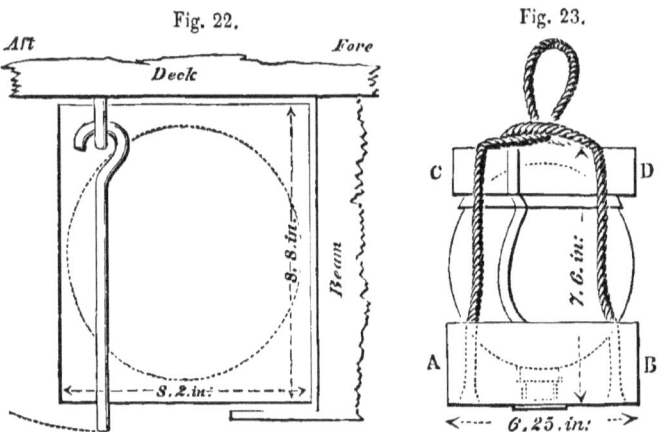

Fig. 22. Fig. 23.

But such is the difficulty of finding stowage for the large equipment of shells in boxes of the above dimensions that it is proposed to substitute for the boxes a wooden fuze cover, as represented at AB in Fig. 23. For the purpose of carrying or suspending the shell, the cover AB is attached by a rope to the wooden bottom CD; and between these plates the shell is confined, but so that it may be easily detached from them when it is to be introduced in the gun. The height for stowage would be thus reduced from 8.8 inches, which is that of the box, to 7.6 inches; and the breadth from 8.2 inches to 6.25 inches. The French, experiencing the like want of room, propose to suppress the box[a] and to trice the shells to the beams.

273. Wood fuzes having been found perishable, or to deteriorate by damp or heat in the vicissitudes of naval service, and more liable to be accidentally ignited than metal fuzes,[b] all sea-service shells, before they are shipped,

---

[a] See French system of Stowage, Sect. V. Part IV.
[b] The following awful catastrophe occurred on board H. M. S. the "Theseus" in 1799, from the explosion of French shells, fitted, as shells are in that service, with wooden fuzes:—
Captain Miller, when in command of the "Theseus," having a great desire to try the use of shells fired from ships; and having obtained seventy 36-pounder and 24-pounder shells from a French store-ship which had been captured, directed the gunner of the ship to stow them in a secure place; and some time afterwards, to examine and see that the shells were fit for use. For this purpose the shells were brought up and placed on the quarter deck.

Whilst

are now fitted with metal screw fuzes, whose lengths are 4 inches, 3 inches, and 1¼ inch (see Figs. 24, 25, 26, p. 271), and these are protected by metallic caps.

Exclusive of the advantage of these metallic fuzes, on the ground of safety and that they suffer no deterioration from long custody on board of ship, shells fitted with them burst with greater violence than those which have wooden fuzes. The diameter of the fuze hole for the former being only .9 inch, whilst for the other it is 1.2 inch, the larger tap permits the escape of the charge in some degree; and thus, either the shell is not broken or it bursts with comparatively little effect. An 8-inch shell with a wooden fuze requires a charge of 22 oz. of powder to burst it, but a shell with a metal fuze bursts with a charge of 16 oz.; a 6-inch shell with a wooden fuze requires 14 oz. of bursting powder, whilst a metal fuze shell of the same nature requires only 5 oz. A shell with a metal fuze is therefore a much more powerful mine when it explodes on being fired into an enemy's ship than a wood-fuzed shell of the same nature. On all these accounts, therefore, wood fuzes have been abolished in the naval service.

274. To avoid danger in supplying guns with shells, the latter should be brought up in their boxes, the fasten-

---

Whilst the gunner and his mate were so employed, the fuze of one of the shells took fire, the shell exploded, and set off all the rest in quick succession! The ship was instantly in flames, in the main rigging and mizen top, in the cock-pit, in the tiers, in several places about the main deck; the whole of the poop and after part of the quarter-deck were blown to pieces, all the booms destroyed, and eight main-deck beams broken. Captain Miller, the gunner and his mate, and 32 men were killed; 42 men were wounded; 9 men jumped overboard and were drowned. The ship was reduced to a perfect wreck, and nothing but the great exertions of the surviving officers and crew saved her from total destruction.

Shells fitted with metallic fuzes are not liable to such a deflagration as this; yet it appears (Charpentier, '*Essai sur l'Artillerie de nos Navires de Guerre,*' p. 167) that the French, notwithstanding the fragility of wooden fuzes, their liability to break in the gun, the perishable nature of wood, the chemical action of the composition, and the difficulty of preserving them from humidity on board a ship, prefer wood to metallic fuzes; their argument being that these are conductors of heat, which wood is not, and are raised to such a temperature by the inflammation of the charge, and the ignition of the priming and contents of the fuze, as to set fire to the powder with which the fuze is in contact, before the shell arrives at its destination.

ings of which should not even be loosened till the shell is required to be put into the gun, and the cap of the fuze should not be taken off or unscrewed until the shell is actually introduced into the gun; the cap is then removed, and the shell is pushed home, the rammer head having a cavity in it to receive the head of the fuze, and thus protect the priming from contact with the rammer.

The precaution of not unscrewing the cap of the fuze till the shell is in the gun is of the utmost consequence. A terrific accident occurred in the shell-room, or on deck, on board H.M.S. the "Medea," by the ignition of a fuze in unscrewing the cap (Arts. 255, 261), and the like accident occurred on one other occasion.[a] The screws have since, as a preventive, been fitted on the outside of the fuze (see Fig. 27), and this method has been found to succeed.

275. As it may not be practicable to bring up shells from the shell-room rapidly enough for quick broadside firing after an action shall have commenced, two or three shells per gun, enclosed in their boxes, Fig. 21, or fitted with fuze-covers, Fig. 22, are, before going into action, placed on shelves behind, or triced to the beams of the fighting decks amid-ships; the reason given for this arrangement is that shells so placed are less exposed to the danger of being struck by the enemy's shot, than in any other position on the deck.[b]

---

[a] While the crew of the "Hogue" were being exercised at Plymouth, one of the men, in taking the cap from off the fuze of an 8-inch shell, turned the screw the wrong way, by which the fuze itself became unscrewed; the consequence was that, when the shell was fired, it split the gun (one of 50 cwt.) near the muzzle in three places. Such an accident having occurred in a mere experiment, and with trained seamen gunners, what dangers may not be apprehended in a naval action, and, through our insufficiency of skilful bombardiers, with men imperfectly trained?

[b] By a recent general order, issued to the fleet in the Baltic, the number of shells to be placed on the fighting decks of ships going into action is increased from four to five; the whole to be ranged on the opposite side to that engaged; and cases of reserve ammunition hung up amid-ships ready at quarters. This increase in the number of loaded shells and cartridge cases will multiply, in proportion, the chances of shells being struck and exploded, and of ammunition being ignited. Nor does this arrangement appear to be preferable to that of stowing shells amid-ships on shelves behind the beams;

PART III. LENGTH OF FUZES. 271

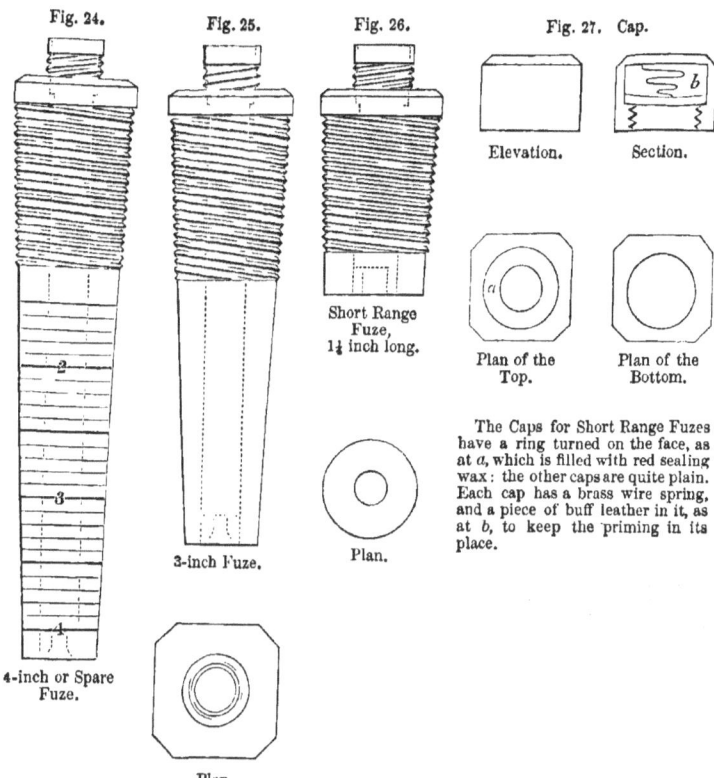

Fig. 24. Fig. 25. Fig. 26. Fig. 27. Cap.

Elevation.   Section.

Short Range
Fuze,
1¼ inch long.

Plan of the   Plan of the
Top.   Bottom.

3-inch Fuze.   Plan.

The Caps for Short Range Fuzes
have a ring turned on the face, as
at a, which is filled with red sealing
wax: the other caps are quite plain.
Each cap has a brass wire spring,
and a piece of buff leather in it, as
at b, to keep the priming in its
place.

4-inch or Spare
Fuze.

Plan.

276. Quick broadside shell-firing will not in general commence till the distance is within that which admits of the short range fuze being used, viz. 600 yards, for if the firing commence at great distances, it would be necessary to begin with the 3-inch fuze, and then change to the short-range fuze; or, if the distance be much greater, to commence with the 4-inch fuze, then change to the 3-inch fuze, and ultimately to the short-range fuze;

---

for the side not engaged at the commencement of the battle may suddenly require to be manned for action on that side alone, or on both sides at the same time; which, in either case, will necessitate the immediate removal of the shells from the side not originally engaged. It may therefore be inferred that the action thus to be prepared for, is such that one side only will be required for action; and this can only be in the attack of fortresses or other stationary objects, when each of the attacking ships shall have taken a fixed position in which to bring the guns of a single broadside to bear on the object of attack; but, in so far as the attack may be against walls of stone, shells, as has been already observed, had better remain in the shell-room.

but any change made in the class of fuze in action is highly inexpedient, and should be avoided, if possible. If, however, ships are furnished with so large a supply of shells as not to be obliged to reserve them for close action, shell-firing may commence with 8-inch guns as soon as the ship gets within the distance corresponding to the time of the 3-inch fuze, viz. at 1,800 or 1,900 yards. In such cases, officers must bear in mind that the necessity of a change in shell-firing, from one class of fuze to another, is as great as that of the change from one charge of powder to another, and may be made with equal facility.

277. From the difficulty and inconvenience there would be, in action, in reducing the lengths of fuzes or the times of their burning, in order to suit the distances at which shell-firing may be carried on as vessels approach each other, time-fuzes have been divided into three classes, as stated in Art. 273. By this very necessary arrangement the inaptitude of time-fuzes to horizontal shell-firing is in some degree remedied; for although, in changing the class of the fuze, it is theoretically true that a jump is made from one class to another in respect to time, without relation to other circumstances, and that there is obviously a large margin for error in taking any one time-fuze as constant when all other circumstances, as distance, charge, elevation, and actual time of flight, to all of which the fuze should be nicely adapted, are variable, yet practically this classification is an important element in the use of time-fuzes. Where the 4-inch fuze of 20 seconds ceases to be suitable, the class of 3-inch fuzes of seven seconds and a half come into use; and when the distance of 600 yards is attained, the short-range fuze takes place of the other. Shells fitted with any one of these different fuzes are calculated to range the corresponding distance with the lowest charge before exploding. At shorter distances shells so fitted might penetrate both sides without exploding. In Table XV., Appendix D., the times of flight have been made intentionally longer than are strictly necessary, by $\frac{1}{4}$ of a second, to obviate any risk of the explosion taking

place before the ship fired at is struck. Should the explosion take place prematurely, it is obvious that the projectile would be lost both as a shot and a shell. This quarter of a second would allow for any little error that might exist in the fuze from a defect of length. If the fuze be too long the error is of less moment, since there would be at least a tolerable certainty of the shell exploding soon after striking, from the shortness of the column of composition then remaining unconsumed.

278. The 4-inch fuzes, fig. 24, are driven with fuze composition, and are intended for distant firing; their full time of burning being twenty seconds. They may be cut off, or bored into, or reduced with the slitting saw, or bored up with the fuze auger as occasion may require; but it must be observed that if the 4-inch fuzes are cut into with a slitting saw very deeply, the lower part of the fuze is liable to be broken off by the concussion of the charge; the composition may then be disturbed, and, consequently, premature explosion may take place.

The 3-inch fuzes, fig. 25, are driven with mealed powder; their time of burning is seven seconds and a half; they are not to be cut off, but may be bored into, or down, from the top, when to be used for distances under 1900 yards. The 1¼ inch, or short-range fuze of .35 of an inch of fuze composition, fig. 26, is, with the priming, intended to burn a short two seconds.

Metal screw fuzes possess this great advantage, that they do not protrude so much as common fuzes above the surface of the shell, and, being firmly screwed into it, they are not so liable to be broken or knocked out, either in the gun or on striking and passing through a ship's side.

279. To ascertain the comparative capability of shells having metallic screw fuzes, with metallic caps, to resist the explosion of a contiguous shell, nine shells so fitted were placed in a pile, and one of those most covered by the others exploded. None of the fuzes of the other eight shells were ignited by the explosion, as undoubtedly would be the case of shells having common wooden fuzes with canvas caps.

From the above experiment it was concluded, that shells fitted with metallic fuzes would be proof in all cases against the explosion of a contiguous shell; but it appears by subsequent experiments, that this is not an invariable fact. It is now known that any loaded shell at rest, when struck by a solid shot, fired with even a moderate charge, from a sea-service gun, will be exploded, if not with as much violence as when burst by its own fuze, yet with force sufficient to scatter in every direction, and to considerable distances, any other shells that may be placed together.

Sixteen shells were piled on the top of each other, on the mud, about 40 yards from the "Excellent," and fired at with a solid shot from a 32-pounder gun of 56 cwt., charge 6 lbs.; the result was, that the shot, on striking the shell, broke it, exploded the powder it contained, and scattered the other shells in every direction. Nine were driven into the sea, the rest into the mud, and could not be found. It was "*thought*," but this is not certain, that only the shell that had been struck exploded. This experiment was repeated with the like result.

"It follows," states the Report on this experiment, "that a shot will explode a shell on striking it; and although with less force than in the ordinary manner, yet sufficiently so to render such an explosion disastrous; consequently great care should be taken in action to expose shells as little as possible."

280. To account for the explosion of a shell on being struck by a shot, it has been surmised, that the shot first breaks the shell, and in doing so, elicits a spark which fires the bursting charge contained in it.

There is reason to doubt this. The powder which the shell contained, exploded, according to this supposition, when in a state of dispersion; it could not therefore have strength sufficient to scatter the other shells in every direction, so far and so forcibly as in the experiments by which that effect was produced. The only rational way of accounting for this fact is, that the powder contained in the shell was ignited contemporaneously with, or at least instantaneously after the breaking of the shell by the blow; and this could only be by the powder in the shell having been exploded by the percussion. This will also account for the detonation being less than usual, while the force of the explosion was still so great as to produce the above-mentioned effects.

281. The direct and secondary effects produced by firing a solid shot into a pile or mass of shells, were first exhibited in Holland by the following experiments, to ascertain whether the effects that might be produced in a ship in action, by a shot striking a rank or pile of shells, were as destructive and serious as the author of the Naval Gunnery had stated in Art. 252 in the third edition of this work.

For this a pile of 430 shells in their boxes, ranged three deep, was fired at with solid shot at a considerable distance. At every round that hit, 3 or 4 of the shells were broken, and their contents exploded, without the usual "report," but with sufficient violence to derange the whole mass, and to set fire to the fuze of a shell which had not been struck: by which *contre-coup*, the shell exploded with its full force!

282. These results being far more formidable and alarming, than those obtained from the previous experiments (Art. 279), by which the explosion of a shell on being struck by a shot was fully proved,* the author deemed it his duty to communicate to the first Lord of the Admiralty this confirmation of the apprehensions which he had long entertained of the extreme danger of stowing ranks and multiplicities of live shells on the fighting decks of ships in action; and he strongly recommended that a course of experiments, similar to those made in Holland in 1852, should be undertaken at Portsmouth, to try whether it is safe and prudent to leave in force, the regulation which prescribes, that all ships on going into close and the closest action, should have two shells (increased to five by a late regulation, Art. 275, Note) per gun of every nature and description, placed, ready for

---

* It was thought at the time that this explosion produced the secondary effect of exploding a shell which had not been struck directly; but the explosion of the latter shell was caused by the ignition of its own fuze.

use, on their fighting decks. This was accordingly done. The author did not attend the experiments, but the following are the results as communicated to him by several naval and military officers his friends, who were present; and for the accuracy of which he can vouch.

A mass consisting of sixteen 32-pounder 6-inch shells, in their boxes (Fig. 22), some fitted with 3-inch, others with short range metal fuzes (Figs. 25, 26), were piled in three tiers, one above another, but only one tier deep (so that no more than one shell could be struck by the same shot) : the mass was then fired at with a solid shot from a 32-pounder gun of 56 cwt., with a charge of 10 lbs., at about 60 or 70 yards distance. The shell struck was broken to pieces, the powder it contained exploded as in the preceding experiments, without much detonation, but with force sufficient to demolish or injure many of the adjoining boxes, to break the strappings by which the shells were fixed to their wooden bottoms, and to scatter the shells in various directions to different distances. These were the immediate and direct effects. The secondary effects were, that the fuzes of two shells which had not been struck were ignited, though capped, by the explosion of the shell which had been hit, and both burst with full force! A large splinter passed over the ship from which the gun was fired, 60 yards distant from the mud bank on which the shells were piled, and fell within the dockyard. Thus the blow of one shot caused, directly and indirectly, the explosion of three shells! The experiment was repeated with the same results upon the pile; the secondary effect was the explosion of one shell by the ignition of its fuze.

To try whether 8-inch shells or hollow shot are as liable to be broken by the blow of a solid 32-pounder shot as the 6-inch (32-pounder) shells in the previous experiment, a number of 8-inch empty hollow shot (unloaded shells), in boxes, were piled in three tiers, and fired at with a solid shot from a 32-pounder gun of 56 cwt. with a charge of 10 lbs. Two of the hollow shot were broken into pieces, some of the splinters were driven forward to a distance of 500 yards; the remaining hollow shot scattered in every direction, though to no great distance, but all the boxes were broken or much injured.

Had the empty hollow shots been loaded shells, the powder contained in those which were struck would have exploded with greater force than the 6-inch shells, in proportion to the cubes of their respective diameters, and the effect upon the other shells, in driving them out of their boxes, scattering them about the deck, and igniting the fuzes of shells not struck, would evidently have been far more serious than in the preceding case. The object of the experiments being to ascertain the amount of danger produced by solid shot striking shells, the greater as well as the minor effects should have been tested; but this not having been done, the experiments were imperfect with respect to the former object: this may, however, be pretty accurately deduced from the results obtained in the case of the 6-inch shells, which show how much more formidable and disastrous the like trial against the same number of 8-inch live shells would have been.

283. Shells fitted with short-time fuzes frequently act by concussion, from the column of composition being dislocated on the shell striking the side of a ship; in this case the explosion is almost instantaneous, and the effect very destructive. The displacement of the fuze composition in the gun by the shock of the discharge is prevented with considerable certainty by an ingenious contrivance, adverted to in Art. 257, note, which gives to the short-range fuze a high degree of value in horizontal shell-firing.

284. To obtain the advantages specially, which the short-timed fuze gives incidentally, the production of a good and efficient percussion fuze, which should fulfil all the conditions stated in Art. 261, would, it was considered, be a most important discovery for horizontal shell-firing. For the attainment of this end, efforts displaying *great ingenuity* have long been making with every prospect of success by an able and accomplished

officer, Captain Moorsom, of the Royal Navy, in his percussion fuze, which being intended to explode on striking a sufficiently resisting body, may be used at any distance, and the fuze having no other cap or cover than a kitt plaster, which need not be removed when the shell is put into the gun, the loading is performed as simply as with a shot, care being taken, however, to enter the shell with the fuze outwards, lest the explosion of the charge should damage it; and that the rammer head should be made slightly concave with a hole in the centre, to receive the head of the fuze (for it does protrude a little, at some sacrifice of accuracy in the flight of the shell), to protect the fuze from injury in ramming home.

285. It was generally imagined, and such was the author's opinion, that percussion or concussion shells would be more liable to explode by the action of their fuzes, if struck by shot, than common shells; on the principle that a body made explosive would, on striking a resisting medium with a sufficient amount of force, be as likely to explode as if, being at rest, it were to receive the blow of a shot moving against it with the same momentum. But this does not appear to be the case; whether it arises from the circumstance that a percussion shell is so contrived as to resist the shock produced by the firing of the charge, and to require, in order to be exploded, a second shock—that of impact; or that a shell, when struck by a shot, is broken before there is time for the percussion principle to act, is doubtful. This, at least, is proved, that percussion shells are not less liable to have their contents ignited or exploded by the blow of a shot than common shells; for all loaded shells, however they may be fuzed, or if not fuzed at all (the fuze-tap being stopped by an iron plug), will be broken by the blow of a solid 32-pounder shot, fired even with a reduced charge of 6 lbs., as in the experiments of 1847; and in every case their contents will be exploded, as was fully proved by the experiments of 1853, in which a solid 32-pound shot, fired into a mass of 70 or 80 percussion shells, broke

four of them and ignited the powder they contained, deranged the whole mass down to the lowest tier, and displaced many of the shells from their cases. It was also found that when a percussion shell struck a percussion shell, both were broken, and the charges which they contained exploded.

286. Were shells stowed according to regulation, or distributed throughout the fighting decks in positions deemed least accessible to shot, the extreme cases stated in the preceding account of the experiments, in which the shells were piled or massed together, could scarcely happen. The chances of shells being struck by shot will depend, first, upon their number and the positions in which they are placed on going into action; secondly, on the number, which must be brought up in quick succession after the action has commenced, and, again, upon where they are placed. The complements of shells for all ships and vessels have been so largely increased since the publication of the third edition of this work (see Art. 270 of the present edition), that great difficulty has been experienced in providing shell rooms sufficiently capacious to contain them, and in distributing throughout the fighting decks the large number required for immediate use, in positions the least liable to be struck by shot.

The method proposed in the Report on the experiments of 1847, Art. 279, is to trice the shells close up to the deck between the beams amid-ship, inside a fore-and-aft carling, in which position it is stated that the shells can only be struck by shot coming in at right angles to the side of the ship.

By the method now in use (*Admiralty order of the 8th December*, 1849) two shells for each gun are placed in boxes on shelves between the beams (Fig. 22, page 268), in which position, it is stated in the instructions for shell firing, these can only be struck by shot coming in at right angles to the keel—that is, as admitted in both methods, when the antagonist ship is a-beam, which is precisely the position for action in which the shells are most exposed to be struck.

287. The figures 28, 29, 30, below, show the method of stowing filled shells, two for each gun, on the lower, middle, and upper decks of the Royal Albert.

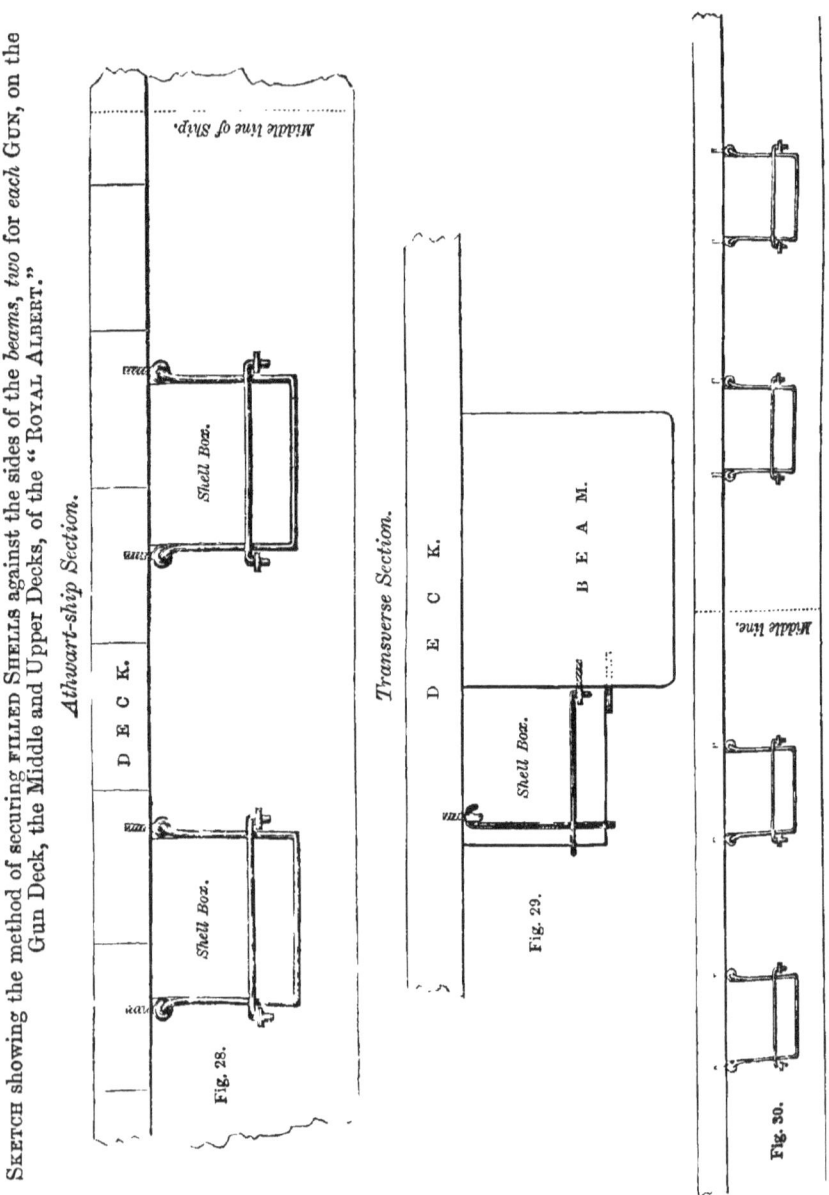

Fig. 28 is half of the athwart-ship section, showing half the beam from the middle line of the ship; the live shells there represented being two for each gun on that side. The like supply of shells for the guns on the other side being in like manner placed on shelves behind the other half of the beam.

Fig. 29 is a transverse section of the beam, the shell being placed behind it.

Fig. 30 represents on a smaller scale an entire beam with four shells behind it, from which it is plain that there are as many ranks of shells, consisting of four in each rank, as there are guns on a side, all exposed to be struck by broadside fire from an antagonist ship.

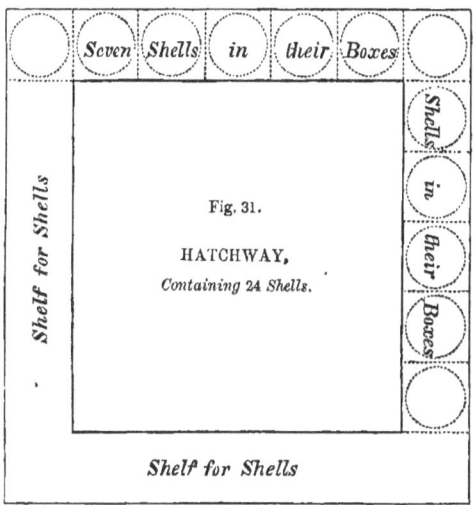

In addition to this supply of shells, the fore and main hatchways are fitted with shelves inside, by, and above the *combings*, upon which twenty-four shells in their boxes are placed. The main hatchway, containing the shells for the aftermost guns; the fore hatchway, those for the foremost quarters; the shells being, it is said, somewhat protected from shots coming in *a-beam*, by the sides of the hatchway, and from raking shot, either before or abaft, by the beams between which the hatchways are placed. In the directions for thus filling the hatchways it is enjoined that the boxes should be placed as low as possible, the better to protect the shells; but it is found that this alone is not sufficient to protect the tops of the shells, at least 6 inches of the boxes being exposed.

288. In a recent regulation drawn up by Admiral Chads, and adopted by a General Order, the complement of shells on the fighting decks, on going into action, is increased from 4 to 5 per gun; so that in a ship's battery of 15 guns on each side, there will be 75 shells.

289. When the action commences, the shells, triced to, or placed behind, the beams, are used first; then the shells in the hatchways; and, in proportion as they are expended, supplies must be brought up in quick succession, to enable the battery to keep up quick shell firing; for if this be interrupted, from want of shells, the guns must be loaded with shot, which requiring an alteration in the elevation used in firing shells, and a different charge of powder, would be extremely disadvantageous. The supplies of shells brought up from the shell room, during the action will neither be triced up to, nor be placed on the shelves behind the beams; but will be placed on the deck, behind each gun. This circulation of the shells, and manner of placing them, is a new condition of the case, in which it is impossible to compute, by the doctrine of chances, how many may be struck by shot, but every shell put into the guns will have to undergo that ordeal; and it appears to be impossible that they should all escape. The complements placed on the decks are so large, that they cannot be dispersed and isolated: some must be grouped together. Those in the hatchways are in mass; those behind the beams in ranks: and it is vain to suppose that either a "carling," or a "combing," or a beam, can protect the shells from the impact of solid shot, capable of perforating any beams, or of penetrating from 22 to 48 inches into sound oak. (Art. 158.) The expedient of stowing twenty-four live shells in the fore and main hatchways, is, in fact, to establish two *expense* shell-rooms on every fighting deck! Surely that is a desperate measure. And is not the con-

traction of such an important thoroughfare extremely objectionable? It was an invariable rule of the service, and for very sufficient reasons, to batten down the hatchways before going into action, leaving only the scuttles, through which ammunition was handed up.

The question tried and solved by these experiments is one of fact, and not of chance. It is no doubt possible, but very improbable, that in a ship in action, or a target hulk, having on her fighting decks the regulated number of live shells, &c., no shell may be hit, and consequently no explosion of this description take place; but it does not follow that there is no danger of such a catastrophe: one ship in a fleet might escape the danger, whilst another might have twenty of her shells struck, and exploded. The author founds his conviction upon no theory, but upon well established facts: he thinks he will be considered, from this, and his other practical works, to be a practical man; or, if a theorist, that it is on matters of exact science, that admit of no mistake. Those are theorists who would disregard the facts of the case, and stake the results upon the most uncertain of all theories—the doctrine of chances.

290. The preceding experiments are sufficient to account for the frequent explosions of grenades collected in the tops of French vessels in action, to throw down upon the decks of their enemy's ship:—" In the action with the 'Hoche' and her consorts (1798) a shot from the 'Foudroyant' struck the 'Bellone's' mizen-top, and ignited some hand-grenades which had been placed there for use against the enemy; these set fire to her rigging and sails, and, but for the prompt exertions of the crew, would have burnt the ship."—*James's Naval History*, vol. ii. p. 130.

291. When the shell system was first introduced in the French navy, and imitated in that of Great Britain, it was asserted there, and believed here, that loaded shells were liable to no such accidents as those demonstrated in the preceding experiments; for that gunpowder contained in iron shells, *globes de fer*, must, it was said, be less dangerous than powder enveloped in bags of

serge;[a] that no danger was to be apprehended from the accidental ignition of a fuze well luted, the shell to which it is attached being, moreover, shut up in its case, and deposited in the shell-room; and that hundreds of shells might be placed there with far less danger than as many barrels of gunpowder, which we are habituated to place in ships' magazines.[b]

But it has since been abundantly proved, that these opinions and assumptions—for such they are, never having been verified by actual experiment—are utterly erroneous; for powder contained in an iron shell placed upon the deck of a ship in action, is by no means so safe as a cartridge. In the one case the blow of a shot explodes the contents of the shell with great force; whereas, in the other, it only breaks the cartridge and scatters the powder. It appears, too, that fuzes, however well capped and luted, may be ignited by the explosion of a contiguous shell when struck by a shot. One should, therefore, have thought that strict attention ought to have been paid to the regulations, enjoined in the French navy, and likewise in ours, viz. : that shells should be stowed away in the shell-room, as soon as they are shipped, and there remain untouched until required to be put into the gun : that no shells should upon any account be placed anywhere else, and that shell-firing should be confined to shell-guns (Art. 274). But the reverse has been done. Vast numbers of live shells, increased by a late regulation from 4 to 5 per gun, are accumulated on the fighting decks of ships in action. Shell-firing, at first the exception, is now the rule in the British navy; and in all other respects the precautions deemed indispensable, before these dangers were known, are now either relaxed or abandoned, notwithstanding the proofs which have been given that those dangers exist.

[a] "La poudre, enfermée dans des globes de fer, sera évidemment moins exposée aux accidens que celle qu'on emploie dans des barils de bois ou des gargousses de serge."—Paixhans, p. 84.

[b] "En résumé : quelque centaines de bombes à bord, bien lutées et encaissées dans leur soute, seront infiniment moins dangereuses que les centaines de barils de poudre auxquels on est habitué."—Paixhans, p. 86.

292. The moral effect produced upon the crew of a ship in action, by the unexpected explosion of one of her own shells, would naturally be far greater than that occasioned by the explosion of an enemy's shell; a contingency which the crew must be prepared to expect. It has been well said by a gallant naval officer, that the panic occasioned by the explosion of one of her own shells on the fighting-deck of a ship in action, would be greater than that produced by the bursting of one of her guns, and what *that* is, we know by the effects produced by the bursting of a gun on board the " Redoubtable " at Trafalgar ; the " Isis " at Copenhagen ; the " Cæsar " on the 1st of June, 1794 ; the " Ambuscade " in action with the " Bayonnaise ;" the " Princess Royal " at Toulon ; the French gunnery-ship " l'Uranie " in exercise in 1852, &c. &c. The effects upon the material of a ship produced by the bursting of a shell are shown in the case of the " Medea " (Art. 274), and the experiments mentioned in Arts. 279, 282.

293. The worst enemy to be dreaded in naval action is internal explosions ; panic is the invariable accompaniment of them. The great object sought to be attained in all the improvements in the material and service of naval artillery, has ever been the prevention of accidental explosions of powder in ships during action, and security against fire. For this the flannel cartridge was introduced, in order to avoid the danger of pieces of a paper cartridge, lurking in the gun, or flickering about like touch-paper between decks, being brought in by the wind after the gun is fired, or by indraughts of air through the ports. For this, also, was introduced the quill-tube, in order to get rid of the priming of loose powder ; and the lock to banish the lighted match. In former times no ship was allowed to fire a salute, far less to go into action, until all the fires had been extinguished.

By a wise and resolute disuse of hand-grenades in the tops ; the avoidance of all firing from aloft among the sails ; by careful inspection on the part of the officers ; and, in general, a wholesome dread of all incendiary projectiles from which the French suffered so severely during the late war, our ships have been more free from accidental fires, than those of any other nation, and there is no instance on record of an English ship of war blowing up in action, during that protracted period.

294. But now receptacles (shells) containing gunpowder, infinitely more dangerous than cartridge-cases, and these, to boot, are accumulated on the fighting-decks of ships; and, with respect to fire, actions will now be carried on in the midst of heat and smoke mingled with sparks, in steam-propelled ships, particularly on the upper decks; and this, with every respect and value for the introduction of that giant power so successfully applied to the propulsion of ships of war, adds unquestionably a new element of danger arising from fire, which demands more than ever the utmost precaution to expose as little as possible any explosive or ignitable bodies upon which it might act. From this, together with other dangers and contingencies relating to their machinery, it has been truly said, "that steam-vessels, compared with sailing ships, have vastly more vulnerable or vital points than they, and far more to dread from fire."—*Report on the National Defences of the United States*, 1852.

295. The author has considered attentively the numerous instances which are to be found in naval history of the panic which is invariably created by any unexpected catastrophe of the description above alluded to. Suppose a ship having vast numbers of live shells on her decks to take fire, either in action or not; what is to be done? The magazine may be flooded, and the powder immediately rendered innocuous; but the shells, whether in the shell-rooms or on the decks, are impervious to water. Those in the shell-room cannot be got at, or at least got out in a hurry; those on its decks should be immediately thrown overboard; but all this can scarcely be expected of a crew in a state of panic. In the case of the "Ajax," even without the presence of these terrible agents, on the alarm of the ship

on fire being given, 380 men jumped overboard, and 250 of the crew perished. (*James's Naval History,* vol. iv. p. 301.)

296. There can be no doubt, that if a solid shot were to penetrate into a shell-room with much force, and strike a row of shells, as in the Dutch experiment mentioned in Art. 281, and in those of 1853, there would be an end of that ship. Is it quite certain that the shell-rooms of ships in general are sufficiently below the water-line to be perfectly secure against such accidents? Their crowns should be, at least, six feet below the load water-line (" Excellent" experiments, No. 10, pp. 22, 23). Are they so in all cases? It appears from the table given in Art. 271 that, excepting in first and second-rate ships, the crowns of shell-rooms are not so much as 6 feet below the load water-line, and still less are they below light water-line: in all smaller vessels, down to the lowest class, the shell-rooms are less and less safe. It must be remarked, likewise, that when a ship in action heels off from her enemy, as when fighting her weather-guns, she exposes much of her side between the actual and general water-line; also when she heels towards her enemy, her deck may be penetrated by shot; and again, when engaged at a considerable distance, a shell, fired at a high elevation, may descend upon her deck. The author saw a shell fall upon the deck of " L'Aigle," British frigate, when in action in the Scheldt in 1809; it penetrated into her bread-room and there exploded within a mass of bread in bulk. Had the bread-room been a shell-room there would have been an end of her.

297. Nor can the crowns of shell-rooms or magazines be lowered in steam-propelled ships in particular; for so much space is occupied by the engine (the great object being to keep this entirely below the water-line), that it is not possible to stow the large complement of shells, under a less height. In order, therefore, to provide sufficient space for the stowage of shells, their rooms must be made proportionally longer. Hence, the necessity of making ships of greater length is

mainly attributable to the vast capacity required for the stowage of shells, each of which being placed in a rectangular box, absorbs a very large portion of the interior of the ship. Powder, packed in hexagonal cases, is stowed compactly and with no loss of space.

298. It was first enjoined peremptorily, by regulation, that shells should always be stowed as far as possible from the sides of a ship; but live shells are now stowed in increased numbers, on the fighting-decks of ships, either on shelves between and near to both sides of the deck, or massed on the side opposite to that engaged.[a] So that in the first case, a shot might strike and explode a whole rank of several shells; and, in the latter case, strike a mass of shells so placed. These regulations cannot be all right or rational. The discrepancy between the former rule (Art. 274) and the recent regulations is most alarming as to the future, and the author, being certain that the original precaution was well founded, while the relaxation is perilous in this respect, as well as in others, strongly urges the necessity of a return to the first method, confirmed as its advantages have been, by the results of experiments, which concur in proving the danger of stowing live shells anywhere but in rooms appropriated for their reception.

299. The French, as well as ourselves, have long been experimenting on percussion fuzes, and appear, from what is stated by M. Charpentier and other authors, to have brought an instrument of that description to perfection in the Billette shell.[b]

---

[a] See New General Order, June, 1854.

[b] "Les boulets creux employés sur la flotte doivent désormais être assujettis au mécanisme percutant, de l'invention de M. le capitaine de corvette Billette. Nous avons dit les raisons que ne nous permettent pas de nous étendre sur ces projectiles, d'invention nouvelle, raisons que le lecteur saisira aisément."—*Essai sur le Matériel de l'Artillerie de nos Navires de Guerre*, p. 164. M. Billette had previously applied the fulminating principle to the ignition of the fuzes of sea-service grenades, to be thrown by hand, or tossed by *bracelets* (a sort of sling, one end of which is strapped to the right arm of the user) from the tops of French ships on the decks of the enemy's vessel; and all French ships of war are now largely supplied with these incendiaries, viz., ships of the line with 300, and frigates with 250 and 170, according to their classes.—*Charpentier*, p. 174.

Thus it appears that the naval forces of the two great maritime Powers are both provided with appalling weapons of mutual destruction, and were prepared, if unhappily there had been occasion, to use them, *à outrance*, against each other's ships, in a barbarous and ignoble strife, in which it seems the only question is, which shall be burnt first! We know the dangers of the shell system, and it is our duty to provide against them. The French, it should be observed, experienced very numerous terrific proofs of the treacherous and suicidal effects of incendiary and combustible projectiles in action with our ships and fleets in the course of the war of the Revolution; a detailed account of these will be given on a future occasion. It is sufficient for our present purpose to state, that, exclusive of numerous comparatively trifling explosions, " Qui s'est trop souvent renouvelé dans la longue et funeste guerre de la révolution,"[a] four or five ships of the line, six frigates and smaller vessels, were either burnt, blown-up, or so terribly damaged by their own incendiary and combustible weapons, as to be incapable of further resistance, without injuring or destroying one of ours; and that in those awful catastrophes many hundreds of Frenchmen perished; whilst numbers of those who threw themselves into the sea to escape the fury of a more relentless element were rescued from a watery grave by the humanity and intrepidity of British seamen,[b] who in the heat of action, at great risk to themselves, and with some loss in killed and wounded, succeeded in their humane attempts.

---

[a] De la Gravière, vol. i. p. 97.
[b] In the action of the 13th July, 1795, the Alcide, 74, set fire to herself with her own grenades. Of the 615 men on board, 300 were saved by the boats of the British ships. (De la Gravière, vol. i. p. 97. James's Naval History, vol. i. p. 271.) The Achille, at Trafalgar, set herself on fire in the same manner, when closely engaged with the Prince. " As soon as Captain Grindall perceived that his opponent was on fire, and her crew jumping overboard, he sent his boats to the rescue, which, together with those of the Swiftsure, Captain Rutherford, soon after aided by those of the Pickle schooner, and the Entreprenante cutter, effected their noble and generous purpose."—James's Naval History, vol. iv. pp. 74-77.

300. We feel deeply sensible of the atrocious character which such a system of warfare must assume, and in which we may be involved; but the adoption of that system by us will cast no slur upon our national character, for self-defence is the first law of nature, and the first duty of nations; and we are provided with abundant means to try the shell system in war, if unhappily it should be forced upon us. But are we quite prepared, in all respects, to carry out the incendiary system on equal terms with those from whom we have to a great extent already imitated it? The French naval shells contain incendiary bodies,[a] which, when ignited by the bursting of the shell, are scattered about in every direction, burn with far greater intensity than *la roche à feu*, develop more heat, and give out dense smoke during the combustion, which must interrupt for a considerable time the working and aiming of the guns. The incendiary faculties of these *cylindres incendiaires* must increase prodigiously the chance of setting fire to any antagonist ship in close action, with comparative safety to that from which they are discharged, unless the opponent possess and use the like means. If he do, then must the issue be tried by the practices and artifices of the *brûlotier*, and both combatants perhaps, or at least one, will surely be burnt. If we disdain to descend to such a practice—and this,

---

[a] " Ces cylindres sont les mêmes pour tous les projectiles creux; et ils ne diffèrent entre eux que par leurs dimensions, qui varient suivant le calibre des projectiles.

"La nouvelle composition dont on se sert pour garnir les cylindres, brûle avec beaucoup d'intensité, et donne un grand développement de chaleur ainsi que beaucoup de fumée pendant sa combustion; en sorte que cet incendiaire remplace avec avantage la roche à feu et les mèches que l'on employait pour obtenir le même effet.[*]

"Le département de la Marine a adopté ce perfectionnement; et un tableau, indiquant le mode de chargement des projectiles creux et la composition du nouvel incendiaire, a été envoyé dans tous les Arsenaux Maritimes pour que les artificiers aient à s'y conformer."—*Essai sur le Matériel de l'Artillerie de nos Navires de Guerre*, par F. E. A. Charpentier, Colonel d'Artillerie Marine, p. 165; *Aide Mémoire Navale*, p. 270; and *Gassendi*, p. 179.

---

[*] The quantity of this composition put into the shell of the canon-obusier of 80 is 0·270 k. (9·53 oz.); and into the canon-obusier of 30, 0·150 k. (5·3 oz.).

so far, appears to be the case—then we must endeavour to act in some other way which shall not expose us so disadvantageously to the terrific effects of a destructive weapon which we scorn to use. What would Nelson have said to this incendiary warfare?[a]

Red-hot shot, too, a *projectile incendiaire* which Napoleon denounced ('Mémoires,' tom. i.) as a dangerous, troublesome, and *difficile* weapon, so repugnant to the feelings of Frenchmen as to have been renounced, will again come into use, if not in ships, at least in coast batteries (witness the fate of the Christian VIII., Danish ship-of-the-line, Art. 330), and, no doubt, shell-equipped ships will, much more than others, invite a return to that practice. Notwithstanding the assurances contained in the 'Aide-Mémoire,' p. 438, that "Le tir à boulets rouges est aujourd'hui presque inusité," and Charpentier's observation (*Essai sur l'Artillerie Navale*, p. 183), "que l'usage de faire rougir au feu les boulets des batteries de côte a beaucoup perdu de son ancienne faveur," several vessels of the combined fleets in the Black and Baltic seas have been struck with red-hot shot. Two lodged in the "Vauban" at Odessa, which obliged her to retire out of action, in order to have them extracted, which with the aid of some carpenters sent from a ship of the line, was effected, when she rejoined the squadron in action. The practice of firing red-hot shot from land batteries is not without its danger. A 32-pounder of 56 cwt. burst at Gibraltar in October, 1852, in using hot shot, by which fatal accident three officers and ten men were either wounded or killed.

---

[a] *Translation by the Hon. Capt. Plunket, from De la Gravière* :—" Having witnessed the destruction of L'Alcide and L'Orient, Nelson considered *l'incendie* as the greatest danger of a naval action. Before the battle of Trafalgar commenced, he ordered all the hammock covers on the sides of the Victory to be well watered, and to throw into the sea or remove everything that could serve to aliment fire. It was to this pre-occupation, above all, that the absence of musketry in the tops of the Victory may be attributed. Nelson thought that a careless discharge, a fortuitous explosion, might set fire to the tops and rigging, and become the cause of a frightful accident. This, in fact, happened in that very battle to the French ship L'Achille."—[He might have added, the Redoutable.]—vol. ii. p. 224.

The gun broke into many fragments, some of which were thrown to distances of 140 yards. Another melancholy accident of this description, for which no satisfactory cause has been assigned, took place at Malta in May, 1854. The gun burst into eight pieces, one of which was thrown to a distance of 100 yards: one man was killed, one lost his leg, and three were wounded; three of the officers present miraculously escaped imminent danger; the carriage on which the gun burst was broken to pieces; the carriage of the gun next on the right was broken; and the gun thrown into the ditch. The expansion of the shot by being raised to red-heat is less likely to have occasioned such an accident with this gun than with any other; for its windage is greater. However this may be, —were such a catastrophe to happen in a ship, it might be fatal to her. This resumption of the use of red-hot shot in land batteries cannot be revived in ships, without extreme danger. It may be said with respect to steam-ships, that as they have furnaces in which shot may be heated without any difficulty, so the case should be exceptional with respect to them; and this appears to have been so at Odessa, where red-hot shot were fired from the "Terrible;" but there is already so much danger of fire in all ships by the incendiary system now in use, that it were better not to add another element to those perils.

We find in the chapter on Projectiles, in M. Charpentier's work '*L'Artillerie de la Marine*,' that experiments have been made at L'Orient and Brest, with a new description of missile, denominated "*asphixiants*,"

---

* " Une nouvelle espèce de projectiles, dits asphixiants, parce qu'ils ont en effet la propriété de produire par le développement du gaz délétère l'asphixie immédiate des êtres organisés, ce qui les rendrait surtout redoutable pour les navires ennemis, sur lesquels l'agglomération d'un grand nombre d'hommes dans un espace resserré en favoriserait puissamment l'effet suffocant."
—*Essai sur l'Artillerie de nos Navires de Guerre*, p. 185. In 1851 experiments were made secretly at Brest with boulets asphixiants, under the direction of a special committee appointed for the purpose.

The author learns, with great regret, that some awful experiments have been made, with fearful success, in the Royal Arsenal, with projectiles asphixiants, combining in a frightful degree incendiary with suffocating effects!

because it develops a deleterious gas, which produces the immediate suffocation of all organized beings." Whether those truly diabolical weapons, as M. de la Gravière denominates all such means (vol. i. p. 98), have been adopted in the French navy, is not known; but they stand in type as one of the weapons of the new French system of warfare.

Rockets are included in the category of projectiles adopted in the French navy for incendiary purposes (Charpentier, p. 200). They are said to be "particulièrement propres pour les navires à vapeur, qui ont la faculté de s'approcher aussi près que possible des côtes, et compenseraient d'ailleurs avantageusement sur ces navires le petit nombre de bouches à feu que leur nature permit d'y placer." The Rocket system is explained further on (Art. 315 et seq.).

301. The "Vengeur" was sunk in the battle of June 1, 1794, by the fire of the "Brunswick," 74, Captain Harvey. When in a sinking state, all the boats of the "Alfred," "Culloden," and "Rattler" cutter, that could swim, were sent to save as many as they could of her people. Thus 213 men were saved by the boats of the "Alfred," whilst those of the "Culloden" and "Rattler" rescued nearly as many more.[a]

But these noble and generous impulses—these humane exertions—far from being cherished and practised, will be smothered and repressed in that merciless, ruthless, and inglorious system of warfare for which we have been compelled, with the utmost repugnance and at enormous cost, to prepare. The black flag displayed over the depository of the sick, the wounded, and the dying, in a besieged fortress, is ever held sacred by the usages of war, as marking a locality appropriated to purposes of humanity. There the medical officers—non-combatants—perform their mournful duties in safety; the sick and the wounded are no longer exposed to the casualties of war; and the dying depart in peace. But what shall be said of that inhuman system prepared for

---

[a] James's Naval History, vol. i. p. 164, Chamier's edition.

naval warfare, in this age of enlightened humanity, which would advisedly, purposely, and deliberately consign the whole of these, and all other survivors, to indiscriminate and instant death or mutilation? A ship may be sunk in action; yet, as we have seen, there is always time to remove the sick and wounded, and save the survivors; but who shall approach a ship on fire to rescue her crew from the sudden and awful effects of that merciless and barbarous system, the object of which is to set fire to her at heart, and if possible blow her up?

302. An awful and portentous realization of the horrors depicted in the previous article has marked the commencement of the present naval war, at Sinope, where the Russians have proved that they well understand, and can effectually use, the weapons which they, too, have imitated from France; and, from what they effected, they may learn what they have to expect in return, when, in combat with ships armed in a similar manner, they shall be exposed to the incendiary projectiles of their opponents, as well as to the suicidal effects of those which they carry on board of their own ships.

" —— neque enim lex æquior ulla,
Quam necis artifices arte perire suâ."
—*Artis Amatoriæ*, lib. i. v. 655, 656.

The burning of the Turkish squadron, which, when related in our journals, produced such a burst of popular horror, hatred, and detestation in this country —but in which the Russians only did that which our noble ships are expressly designed, and fully equipped, to do—was not effected by percussion or concussion shells; neither of which, as has been already observed (Art. 257), are so destructive on a ship, nor so likely to set her on fire, as shells coming in with fuzes blazing, and especially if they lodge before they burst. The Russians have not adopted either, but use shells having screw-metal time-fuzes attached; and with these the Turkish ships were either blown up by the shells, which having lodged, then exploded in their interior; or which ignited, by their fuzes as well as by their explosion, the powder circulating on the fighting decks, and thus occasioned so much panic among the crew, that they were unable to extinguish the fire; in consequence of which the whole of the Turkish squadron was either burnt or blown up; thus verifying the prognostication of General Paixhans, expressed in his original work:— " Enfin la grande faculté incendiaire des canons-à-bombes sera de mettre de feu aux poudres qui circulent pendant le combat; effet terrible, qui causera une déflagration totale, que rien ne saura empêcher"[a]—the very catastrophe which occurred at Sinope, as reported by the Commission of naval officers sent by the French Government, at the request of General Paixhans, to inquire into the facts and circumstances of the destruction of the Turkish frigates.[b] The Turks have great cause to dread incendiary warfare in general, and Russian shells in particular. A large Turkish squadron was wholly burnt in that manner by the Russians in the bay of Tschesmé in 1770. The Greek Brûlotier " Canaris" and others destroyed a Turkish squadron and many Turkish ships in the war of the Greek revolution by setting them on fire; and the services of the " Karteria," under the command of Captain Hastings, during that war, further inspired the Turks with terror of the incendiary warfare, which that commander prosecuted with red-hot shot and shells; by the former of these principally he obtained his successes. We may, however, rely upon it, and take warning accordingly, that this mode of naval warfare, which the Russians were the first to practise, and which proved so destructive at Sinope, will be used from land batteries with still greater effect against ships, in proportion to their magnitude, and with more safety to themselves than when employed

---

[a] Paixhans, ' *La Nouvelle Force Maritime de la France*,' p. 78.
[b] It is very probable that these shells may have contained *cylindres incendiaires*, described Art. 300, Note, as in the French naval service; but that the projectiles used by the Russians on that occasion were not boulets asphixiants, as has been stated, nor percussion shells, the author has it in his power to state on the authority of the Commission mentioned above.

ship against ship, fraught as it is with danger to the users, as stated by Paixhans, Jeffers, and others.

Shells fitted with time fuzes combine the incendiary with the explosive quality, whereas percussive shells are purely explosive. The Russians prefer the former shells; and in order that there may be time for them, before bursting, to set fire to the ship in which they lodge, they use rather long fuzes. It is to this circumstance probably that the gallant young Lucas owed his life, and his comrades their preservation, while, by his heroic conduct, the ship was saved. Percussion shells are useless for an attack on stone walls; and in the incendiary warfare carried on by British ships against works which may be burnt, they are not so efficient as time-fuzed shells.

303. The affair of Sinope, in which the Paixhans shell system was first used in war, forms so important an epoch in the history of the new system of naval warfare, that the reader will naturally expect to find some account of that conflict in these pages; and much interest is given to that subject by an important paper written by General Paixhans, and entitled ' *Observations sur l'Incendie, dans la Mer Noire, des Bâtiments Turcs par la Flotte Russe.*' The observations are dated 15th of February, 1854, and were published by authority in the *Moniteur* a few days afterwards. In that document it appears to have been his principal object to point out the errors committed, as he thinks, in France—and this applies to us in a greater degree — of constructing ships of greatly increased magnitudes, and concentrating in them multiplicities of explosive and ignitable bodies, instead of the reverse, as proposed by him and approved of at the time (pp. 20, 22 of the work above quoted). He states in his original work, and in his recent publication, that, as a few of his canons-obusiers in a comparatively small vessel are capable of destroying any ship however vast, costly, and well manned, so no ship should be armed with more than four or six canons-obusiers; no shells should be exposed on, or be conveyed to great distances along her fighting decks;

he adds that horizontal shells fired with skill from a ship so equipped will destroy any vessel, and this with the greater certainty in proportion as she is large ; because the circulation of powder, and the accumulation of projectiles for the service of a greater number of guns, must multiply, in that proportion, the chances of an entire explosion of the vessel ; he therefore strongly recommends that, instead of investing in a single vessel, the vast sum of money required to construct a ship of 80 or 130 guns, armed and fitted for shell-firing, and so risk the total loss of the ship, and the lives of a thousand men, on one fatal chance, it would be better to lay out the same sum of money in constructing two or three smaller vessels, which might together carry that amount of armament and the same number of men. But before entering further upon that important document, the author will endeavour succinctly to explain, as he has often been recommended to do, for the information of the general reader, and of naval and military officers who may not have acquired a competent knowledge of the Paixhans system, from better sources, the principles, effects, objects, and dangers of that new system of naval warfare which has now become an object of intense interest and vast importance to the country.

304. The reader, if an adept in the practice of artillery, should remember—if not, he should be informed—that the shells which produced those very extensive ravages upon the " Pacificator" hulk in the experiments made at Brest, in 1821 and 1824,[a] upon the evidences of

---

[a] The Pacificator hulk, of 80 guns, was moored in the roadstead of Brest, at 300 toises (639 yards) from a small *ponton de service* (a lighter), on which the gun was placed. The first shell struck the vessel low, made a fracture 8 inches diameter, perforating the side, which was there upwards of 28 inches thick, and *exploded in-board,* committing very extensive ravages. It set the ship on fire ; and she would have been consumed, had not men been sent to work the pumps, which had been previously prepared. The second shell perforated the quarter-deck, and carried away two planks ; after this it struck the mainmast, taking off a splinter 3 or 4 feet long, and 9 inches deep, *and then burst*, carrying off an iron hoop 10 feet in circumference, weighing 130 lbs., a large fragment of which was driven with great force against the opposite side of the hulk. The splinters of the shell did vast damage to the materials, and must

which the French naval shell system was founded, were loaded shells, having fuzes attached, which, ignited by the explosion of the discharge in the gun, continued to burn for a time somewhat greater than that of the estimated flight, and then exploded; thus producing the maximum effect which any shell is capable of producing on a ship. But, as has been already explained in Article 259, so much difficulty has been experienced in getting a *time fuze* to fulfil the conditions of the case, that it appeared preferable to our naval authorities to sacrifice that precarious advantage to the minor effect, which may be obtained with greater certainty, by causing the shell to explode on striking any resisting part of the body of a ship.

But General Paixhans, considering that the maximum effect of one of his shells is far more destructive on a ship, than any percussion or concussion shell can possibly be; and that, should the former take its fullest effect, it would prove more fatal to the ship than perhaps all the percussion shells that might be fired at her, adheres to that description of fuze to which his system is in-

---

have carried off a great many of the crew, had the vessel been manned. The third shell perforated the side between her ports, broke off an oaken knee, 7 feet long, and from 6 to 13 inches thick, which, with its iron bolts, weighed upwards of 200 lbs.; it *then exploded*, knocking down upwards of forty planks, placed to represent men at quarters, and doing very extensive damage to the material.—*Paixhans*, pp. 38, 39, 40.

In the experiments made at Brest in September, 1824, a shell penetrated into the hulk at 3 feet above the water-line, and then exploded, making an opening of nearly 3 square feet, ravaging, tearing, and separating the planks above and below. The Report states, p. 54, that had that effect taken place at the line of floatation it would have very soon sunk the ship. Another shell exploded in the hulk, and set her on fire, and would have burnt her had not prompt measures been taken to prevent it.

So at Portsmouth, in 1838, in the shell experiments against the Prince George, and in others more recently made, the most destructive effects in the hulk were produced by shells which lodged and then exploded. Assuredly no such formidable effects could have been produced by concussion or percussion shells; and it must be observed that the latter have never, or rarely, been found, in the experiments made with them, to set the hulk target on fire. They pass too quickly, and burst too soon, to communicate their incendiary qualities to the medium through which they pass. They may blow up cartridges and powder-bags, but not directly set fire to the ship. The extremely destructive effects of Paixhans' shells in war is a great fact, which has been proved at Sinope. The performance of percussion-shells is a subject which has yet to be tested.

debted for its success experimentally, and for its ultimate adoption. We have seen in the affair at Sinope that the Russians have much reason to be satisfied with the destructiveness of such shells; and we may be sure that they will be used in any naval action, ship against ship, and likewise from fortresses and other land defences against fleets, if these should be so indiscreet as to attack in that manner such formidable positions: and, in truth, it must be said, that in any encounter, large ships, particularly three-decked ships of 120, 131, or 140 guns, are marks so large as scarcely to be missed, even at considerable distances; that their frames are so thick and stout that no shell could pass entirely through them; and, consequently, that any shell which hits the hull will lodge in her body, and, embedded there, will, by its explosion, produce the destructive and terrific effects of a mine sprung in her interior.

This is the Paixhans system, as stated in the General's original work, and in the following exposition; in both of which the essence is declared to be, to avoid the error of building large ships, and the equipment of any ship so extensively for shell firing as necessarily to expose her to the greater risk she would thereby run of being set fire to, or blown up, by her own weapons. It is added that the effects of this system upon an enemy consist rather in the destructive powers of a few canons-obusiers, in comparatively small and swift vessels drawing little water, than in accumulating many in the same huge ship. Into these extremes, however, both France and England have rushed, but England furthest,[a] in rivalry to each other: out of so expensive, dangerous, and extravagant an extreme, both may hereafter have occasion, if the General is right, to retrace their steps.

---

[a] There are only seven ships of three decks afloat in the French navy, and none of more than 120 guns. The 100-gun ship on three decks is to be suppressed, and no three-decked ships are being built. The disadvantage of ships of three decks, as having a ballistic force not much more considerable than that of a 90-gun ship, the expense of their construction, and the large surface they present to the enemy's projectiles, are strongly urged by the French authorities.—*L'Enquête Parlementaire.*

305. *Observations on the Burning of the Turkish Frigates by the Russian Fleet in the Black Sea.*

[In a letter written by General Paixhans, and published in the *Moniteur*, February, 1854.]

The Russian ships, in their iniquitous and violent attack[a] at Sinope, not only had a superiority, two or three fold, of guns, but, moreover, the advantages of an armament of a different nature from that of the Turks. The Russian admiral, after having explained in his official Report the manner in which he commenced the combat, states, " Our vessels, having entered the bay and " taken their positions, cannonaded the enemy with a well-directed " fire. In less than five minutes, a ship-of-the-line, Grand-Duc- " Constantin, destroyed the battery under her fire ; and the enemy's " frigate anchored near to that battery, on which was directed " principally shells fired from the Paixhans guns on the lower " deck of the Russian ship, blew up. A short time afterwards " the Ville de Paris, Russian ship-of-the-line, blew up by her shells " another Turkish frigate." Another Russian account sent to Berlin states, that in the second tack of the Russian ship Constantin the Turkish frigate carrying the admiral's flag blew up also.

Every one will allow that, in seeing my name thus placed, it was natural for me to feel the necessity of making some observations on this occasion ; the publication of these, however, I deferred that I might affirm nothing until I should have other information than that given in the Russian bulletins.

To obtain these I addressed myself to the French Ambassador in Turkey, in consequence of which an investigation was made at Sinope, and I select the following passages from the reply to my questions :—" None of the Turkish ships destroyed had any Paix- " hans guns ; their largest pieces of ordnance were of the calibre " of 24, and there were but few of these."—" The land batteries " were not armed with Paixhans guns, but with cannon of small " calibre."—" Those batteries were in a very defective (pitoyable) " state of defence." Then comes a description of the projectiles

---

[a] The gallant General designates the Russian attack of the Turkish squadron at Sinope as iniquitous, on the ground, probably, that the attack ought not to have been made under the political circumstances then existing. It may, however, be considered as iniquitous from the fact that the Russians used shells in the action ; for such was their superiority in number and calibre of guns (they were heavily armed : many 68 and 42-pounder solid shot fired by them were picked up at Sinope), that, instead of burning the Turkish ships and exterminating the crews, they might have taken the whole squadron without firing a shell, and have brought away the surviving men as prisoners of war. The ships might have served to increase their own effective force.

used by the Russians, some of which not having exploded, enabled the commission to explain them exactly. After having described them the Report proceeds, " One of these projectiles had " lodged in the earth near one of the batteries, and, on exploding, " projected its splinters to distances of about 300 or 400 metres." Then the Report states that the Turkish officers interrogated as to the effect of these projectiles, were unanimous in attributing to them the setting fire to the greater number of the Turkish ships, and in stating that the fire (incendie) was followed by their explosion or blowing up. The Report concludes with these words, " The Turkish Colonel of Artillery sent from Con-" stantinople to inspect the state of the forts is so much impressed " with this truth that he urgently demands that Paixhans howitzers " be supplied."

The facts of the case having been thus certified, the following are my observations:—

The canons-obusiers named as above were tried first at Brest in 1823 (qy. 1821 and 1824?), and from these experiments, as well as others made elsewhere, the great destructive effects of that piece of ordnance have been generally inferred. Several distant facts of war have since added to those proofs ; and recently the English Admiral P. said to M. B. that, if two vessels armed with these guns were to fight, it might happen that in a few instants one would disappear in the air and the other under the water. In short, in the disastrous combat of Sinope the effects produced on the Turkish frigates, all of which perished, and even on the town, have but too distinctly proved what will in future be the issue of maritime combats, every navy in Europe and in America having adopted that nature of ordnance.

Whatever be the sad impression resulting from these facts, it is impossible to throw the least blame on the introduction of that new arm. For, if blameable, why should Governments the most enlightened have hastened to adopt it? Improvements in arms have never been rejected any more than other improvements. Is it not established in history, that the more powerful the arms, the less bloody have been the wars? The truth is, that with naval artillery whose effect will be prompt and decisive, a comparatively small vessel, well using this powerful artillery, may make herself feared by the largest ship ; but, when the weak is put in a state to be no longer despised by the strong, the means employed should not be unfavourably judged. Will it not be, on the contrary, fortunate for all maritime nations to be able to arm their ships, their coast batteries, and their seaports, with ordnance the most capable of repelling aggressions, against which the ancient artillery was often comparatively ineffectual?

As to the deplorable disaster of Sinope, the Turks displayed the most admirable energy, but foresight had prepared no chance for the success of their courage. They exposed themselves to a most unequal combat, in opposition to the soundest counsel. They placed themselves in a roadstead under the protection of batteries to which they had only given the semblance of an armament. And it may be affirmed that, if a part of their naval armament, and, above all, their land batteries, had consisted of *canons à bombes* properly served, the Russian ships could not have approached without receiving a severe lesson; land batteries firing much more accurately than those of ships, which are always more or less agitated by wind and water.

That which has lately happened in the Black Sea, and that which may soon take place elsewhere, is that which will always happen in favour of any power which may first use effectually and ably a new weapon of war; and this truth, if unfortunately war should ensue, is about to appear in another manner, by the musket, which is now receiving such remarkable improvements in its range and accuracy of fire.

To these observations I add two others, which, if just, will perhaps at no remote period prove useful—one relating to our navy, the other to our artillery:—

1st. Guns which fire shells horizontally will destroy any vessel, and will do this with a greater certainty in proportion as the vessels are large, because the circulation of powder and projectiles during an action, being more multiplied for the service of a greater number of these guns, will multiply the chances of an entire explosion of the ship. From this fact results the important question whether, instead of concentrating in a single ship of 80 or 130 guns and 1000 men, and exposing that large quantity of military and financial power, and that amount of lives, to perish suddenly, it would not be better, from motives of humanity and considerations of economy, to lay out the same sum of money in constructing two or three much smaller vessels, which might together carry the same amount of armament and the same number of men? Our principal ships, being then far less enormous and drawing less water, may enter a greater number of our ports, which at present are limited to five, accessible to large ships. The construction of three smaller vessels would neither require so much time, nor timber, nor be so costly. Our fleets would then find at home, and in our colonies, more ports of refuge accessible to them, and they would find more points accessible to attack on the coasts of the enemy. The battery of a frigate may, as well as the battery of a large ship, carry the means of keeping at a distance or of destroying an enemy. In the combat of two or three such ships against one adversary of colossal magnitude, the latter may doubtless, if near, be able to destroy either of the others singly, but these might concentrate upon him, at a distance, mortal blows, and remain masters of a field of battle from which the greater ship will have disappeared.

Without entering into the details, I shall add, that, in considering the different maritime and commercial conditions of other nations, it will be perceived that the mode of warfare best adapted to our country should rather be a system of desultory and unexpected expeditions and offensive cruises, than the heavy and disastrous system of regular battles and general actions. A hundred powerful and rapid frigates will evidently be more useful to France, than 40 or 50 ships-of-the-line; and this change of system is supported by two authorities of great weight—that of the Emperor Napoleon I. towards the end of his career, and the decisive facts of our recent history. Other powers may indeed adopt the same change, but, if it were not to give this notice too wide a circulation, it would not be difficult to demonstrate that, if a general abandonment of the expensive system of fleets were to take place, France would have little to lose and much to gain.

2nd. With an arm the effect of which is very destructive, the advantage will evidently be in favour of those who know best how to give it length of range and accuracy; thus, both in our actual armaments and in the progress to be made, these two conditions, together with a superiority of calibre, should above all others be satisfied; to this I shall add, that if the same effects could be produced by lighter pieces of artillery of the same description, and which do not require vessels of such great draught of water, nor expose so many men, we should have resolved a problem which, together with great speed in our steamers and greater numbers of them, would give to France a system of naval economy which suits her in the highest degree.

In general, on the land as well as on the sea, to enable combatants to keep distant from each other by increasing the range of their arms, is to reproduce that effect which happened when firearms superseded the pike and the club, and gave to science and address an invincible superiority over brute force.

One cannot tell whether nations will ever obtain from their governments the abandonment of warlike conflicts by sea and land, in which flows the best and most valiant blood of the peoples; but, until that period arrive (if ever it arrive!), those who increase the power and the range of arms, will at least constrain combatants to keep distant from each other, and will therefore have acted in a sense favourable to civilisation and to humanity.

     (Signed) General Paixhans.
Paris, 10th February, 1854.

306. Adverting to what General Paixhans has stated

in his original work, and in his late remarkable publication, there can be no doubt that there will now, more than ever, be reason to dread the ignition of ammunition circulated or circulating on the fighting decks of ships in action. Numerous instances may be found in the naval history of the late war (*James's Naval History*, vol. ii. p. 245,[a] and in many other places) of the explosions of powder on the fighting decks of ships in action, and of the panic into which those explosions threw the crews, before the agency of live shells was introduced.

The danger incurred in bringing up and distributing throughout the decks, ammunition and shells for the supply of all the guns of a ship, must hereafter, irrespective of other perils, be greatly increased, and is, in the opinion of many naval officers whom the author has consulted on these matters, perhaps more to be dreaded than the accidental explosion of shells. This dictates the absolute necessity of adopting additional precautions to prevent as far as practicable such fearful accidents; and assuredly there can be none so effectual as to forbid, as in the United States' service, any accumulation of shells or of inflammable or combustible bodies on the fighting decks of ships, which may be exposed to the chances of being ignited or exploded by shells or shot striking them.[b] Paixhans observes, page 18 :—" C'est là surtout qu'il devait y avoir opposition, parce que l'objection principale consistait à dire, qu'il sera dangereux d'employer, à la fois, beaucoup de projectiles chargés, au milieu d'un équipage nombreux." But far from listening to the proposition that this accumulation should be forbidden or reduced, it has by a late regulation been greatly increased (Art. 271, Note).

---

[a] " In the action between the Ambuscade and the Baionnaise, an explosion of some cartridges took place on board the British frigate, which, besides the panic it created, badly wounded every man at the adjoining gun, and occasioned so much confusion, that she was immediately boarded, and, after a very short struggle, captured."—*James's Naval History*, vol. ii. p. 245.

[b] See the letter of General Paixhans above. The *Times*, with its usual vigilance, has thrown out the subject of that letter for consideration; and certainly it most nearly concerns us, who have such immense ships as the Duke of Wellington, Prince Albert, &c.

Lieutenant Jeffers, of the United States' Navy, says, in his treatise on the "*Theory and Practice of Naval Gunnery*," New York, 1850, p. 174, that "No accumulation of shells in the batteries should be allowed, since, if struck by a shot they will explode, and cause great damage; for that the shock occasioned by the stroke of a shot on iron develops a heat so great, that the pieces cannot be held in the hand, and their blue colour indicates that the temperature has been raised to about 600°, the point at which powder ignites. If a loaded shell be struck by a shot, the heat developed may ignite the charge, and cause a most disastrous explosion"—an awful fact, which has since been fully established.

It is enjoined in our regulations for conducting shell firing, that the greatest possible care should be taken to expose shells as little as possible to the chance of being struck by shot; for which reasons the shells which are placed on the fighting decks for the service of the guns are directed to be stowed on shelves behind the beams, because in that position they can only be hit by shot coming in at right angles to the keel! (Art. 275, and the *Regulations for Shell-firing*.) Should a shell be struck, though it might not explode with full force, yet the ignition of the charge which is the immediate result, must produce, as Lieutenant Jeffers says, the most disastrous results. Can it be detrimental to state these facts and opinions? Are they known to the officers of the British Navy? Is it not right that these warnings should be given? To conceal from ships' companies, and from the country, any danger which seamen must face, in the new and peculiar system of warfare in which they are engaged, but have not yet experienced, and in which the people of this country, of every class, are so deeply concerned, with a view to prevent seamen from being panic-stricken when the case occurs on service, and to avoid alarming the country, by suppressing the truth, will assuredly produce effects the very reverse of that intended, and inevitably give ground for the outbreak of panic, at a moment when it is most to be feared—*in action*. Prepare British seamen for what they have to expect, and must face, and they will meet it like brave men. To conceal danger from them is to treat them like children whose nerves are not to be depended upon; and when the truth which had been concealed or suppressed, is practically disclosed, the very mystery that has been attempted will magnify the danger, and surprise the men into a panic which it will not be easy then to allay. This will act disadvantageously, not only immediately on the ship's company actually serving, but on the people of this country in general, who appear to be kept, by this policy of concealment, totally ignorant of the true character of the new system of naval warfare, and for which they likewise should be prepared.

It is no doubt wise, prudent, and even necessary, to conceal the destructive effects which may be produced on a foe, by the discovery and introduction into the service of any new weapon of war, so to reserve to ourselves the initiative, and the exclusive use of the new arm (although the French did not so with respect to the Paixhans' system, nor can anything of a practical description be kept secret in these times); but when the subject of experiment is to ascertain how far the new weapon or system may be used with safety to ourselves, or the reverse, the results should be divulged to the profession at large, to enable them to understand the nature and extent of the danger, and if, upon a full consideration of the facts of the case, it should be decided that the danger is unavoidable and must be encountered, brave men will meet it with the advantage of being well prepared to face it.

Had the facts which are now fully proven, here and elsewhere, been established and as well known before the shell system was carried to its present dangerous extent, it is impossible that it should have been permitted to go so far, at least without a positive prohibition against placing live shells anywhere on the fighting-decks of ships in action. Let not the reader imagine that this is the first announcement of danger to the users! It was first declared before a select committee of the House of Commons in 1849,[a] as the present First Lord of the Admiralty well knows, in answer to some questions propounded by Sir James Graham, namely, that "in order to conduct with safety the dangerous service of shell-firing on board ships, which has of course greatly increased the chances of a ship being set fire to in a general action," it was necessary to send every filled shell on board in a wooden box, and to provide stores and implements (metal screw-fuzes) for shellfiring. This fearful announcement greatly strengthened the apprehensions entertained by the author and by many British officers and seamen, of the perils incidental to shell-firing—perils which it is neither possible nor prudent to attempt to deny.

The dangers incidental to shell-firing having thus been declared by authority five years ago, but the causes and amount of the danger never having been fully examined and ascertained, the author, rather tardily perhaps than otherwise, proposes to examine in this edition of his work, in what the danger consists;—whether " sending shells on board in separate boxes, with the expensive *implements* " above alluded to, which the select committee were led to believe would be found effectual safeguards against the accidental explosions of shells, have proved so—what is the amount of danger to the users—what the chances of accidental explosions taking place—the effects produced by the bursting of a shell on the fighting decks of a ship, its consequences, —and whether by any, and what means, those dangers might be obviated, abated, or avoided.

The danger of the entire explosion of a ship—and par-

---

[a] " Question 5075.—The change of the armament introduced a variety of expenses, which have added annually to the Ordnance vote; amongst them are the stores necessary for firing shells horizontally, and the implements necessary for conducting *that dangerous service* safely on board ship. We are obliged to send every filled shell on board in a wooden box. Of course *this employment of shells has greatly increased the chances of a ship being set fire to in a general action.*

"It has been necessary to substitute leather boxes for that purpose, having a cap fitting so tightly that, in the event of the powder boy being killed, the box might roll about the deck without a chance of the powder being ignited by the bursting of shells near it."—*Second Report on Army and Ordnance Expenditure*, 1849, p. 325.

ticularly of a large ship—by the accidental ignition of her own ammunition, or by the explosion of her own projectiles, having been publicly declared and widely circulated, on the authority of the great originator of the system of horizontal shell-firing (which we have imitated in principle, and far surpassed in degree), as a warning of danger to his countrymen, to which no doubt they will listen, and endeavour as much as possible to avoid—inclines the author to think that General Paixhans' opinion and advice demand the most serious consideration from all who are interested in the British Naval Service. The multiplicity and accumulations of incendiary and explosive projectiles on the fighting decks of our ships, the system of shell-firing from all natures and descriptions of ordnance, and the inordinate magnitudes of ships, not only exceed what General Paixhans proposed in his original work (*La Nouvelle Force Maritime*), but even surpass the extent to which subsequent French naval administrations have assumed the responsibility of transgressing his proposition; and, consequently, we are more likely than they, or any other naval power, to incur the risk of such a catastrophe as that which General Paixhans has darkly foreshadowed.

The author has stated in the previous edition of this work (page 270 and the Note), that M. Paixhans' original proposition was to apply the canon-obusier to a small extent only in the naval service. The commission appointed by the Minister of Marine, to witness and report upon the experiments of 1821, approved of Colonel Paixhans' new and powerful canon-obusier of 80, and reported that it was well adapted to become the armament of small vessels (as a pivot-gun), but that the introduction of shell-firing into the naval service would require the utmost precautions to prevent the many fearful accidents which might arise from the use of live shells on board a ship, and therefore recommended that *if* introduced into ships of the line, it should be in very limited numbers.

In 1824, the Minister of Marine, unwilling to carry into effect, on his sole responsibility, the recommendation of the Commission of 1821, nominated another, the members of which were to revise its Report, and to state their own

PART III.   DANGER INCIDENTAL TO SHELL FIRING.   309

opinion and decision.[a] Both reports were then referred to the Comité Consultatif de la Marine, with a series of questions upon them; to these the Comité replied upon the subject of danger,[b] that no possible precaution could altogether suffice to prevent effectually the most dangerous consequences if the use of *canons à bombes* and live shells should be extended generally to every deck of a ship; but, being of opinion[c] that, if the number of those guns were limited to two, or at most four, placed only on the lower deck, and near to the shell room, the dangers would be comparatively small, the Comité consented to this limited arrangement. Upon the question of magnitude of vessels, they stated that, as large ships in which great numbers of men, guns, shells, and quantities of ammunition were concentrated, would be exposed, in proportion to their size, to the chances of a *déflagration général*, or *une explosion entière*, therefore the Committee recommended that comparatively small vessels should be introduced in place of large ones. The reader will have perceived that the adoption of a contrary measure in France has produced General Paixhans' severe condemnation of the error. These limitations and regulations for the safe custody of shells, were (as may be seen by referring to Section No. 16 of General Paixhans' work, page 85,[d] where the danger

---

[a] Rapport fait sur les premières expériences de Brest, en Janvier, 1824, par une commission composée des chefs supérieurs de la marine, du génie maritime et de l'artillerie.—*Paixhans*, p. 41.
  Signé de MM. le Vice-Amiral Comte de Gourdon, commandant la marine à Brest; De Kerlerec, major-général de la marine; Geoffroi, directeur des constructions navales; D'Herli, directeur du port; Godebert, directeur de l'artillerie; Lemaraut, Lahalle, Courcy et Touffet, capitaines de vaisseau; Simon, sous-directeur des constructions; Gérodias, sous-directeur de l'artillerie; Gicquet des Touches, sous-directeur du port; Lettré, De Rossy, Couhitte, et Demaré, capitaines de frégate.—*Paixhans*, p. 44.
[b] Avis du Comité Consultatif de la Marine; Nouvelles expériences ordonnées, plus étendues que les précédentes.—*Paixhans*, p. 47.
[c] Rapport fait sur les secondes expériences de Brest, en Septembre et Octobre 1824, par une commission composée des chefs supérieurs de la marine, du génie maritime et de l'artillerie.—*Paixhans*, p. 57.
  Signé de MM. le Contre-Amiral Bergeret, président de la commission; le Colonel Godebert, directeur de l'artillerie navale; Lassale, Russel, et Béhic, capitaines de vaisseau; Simon, sous-directeur des constructions navales; Gérodias, sous-directeur de l'artillerie navale; Longueville et Pasquier, capitaines de frégates.—*Paixhans*, p. 59.
[d] Observations sur le danger que pourront courir nos vaisseaux par les bombes qu'ils emploieront eux-mêmes.—*Paixhans*, p. 84.

which shell-armed vessels will incur from the explosion of their own shells is pointed out) framed expressly to prevent the recurrence of such awful catastrophes as those which occurred on board the "Theseus" (Art. 273, Note), which produced a prodigious sensation at the time, though she was no longer a French ship: such also as the explosion which took place in a French ship at Aboukir: the burning of the "Formidable," which took fire in the poop in the combat off the island of Groix, in 1795, from the explosion of shells, by which she was disabled and obliged to surrender, with the loss of 220 men out of a crew of 666; and lastly, the blowing-up of a French ordnance store-ship, containing quantities of live shells. All these accidents were ascribed to want of prudence in the safe custody and stowing of shells, and particularly to exposing them on deck, "*autour des canons;*" which accidents, a strict observance of the foregoing regulations and precautions, it is asserted, would have effectually prevented: relying on the efficacy of those regulations, we adopted them; but they have since been either rescinded or violated, both in the French and in the British service, and thereby the dangers said to have been extinguished, are reproduced and reactivated.

The Academy of Sciences was consulted on the whole of those proceedings, and reported their concurrence in the views taken, by the Commissions and Comités Consultatifs, to which the professional questions had been referred.[a]

By decree of 1838, in conformity with the Reports and recommendations of the commissions and Comités Consultatifs, as stated in the previous pages, the introduction of canons-obusiers into line-of-battle ships was limited to, at most, four; these to be placed on the lower deck, and near to the shell-room; and the like

---

[a] Rapport fait à l'Académie des Sciences et approuvé par elle le 10 Mai, 1824.—*Paixhans*, p. 45.

Signé de MM. le Baron Vane, ancien inspecteur général des constructions navales; De Rosse, contre-amiral; De Roni; Marquis de La Place; le Maréchal Duc de Raguse, rapporteur; et Baron Fourier, secrétaire perpétuel de l'Académie pour les Sciences Mathématiques.—*Paixhans*, p. 46.

limitation was long observed in our service, with respect to the analogous and equivalent pieces of ordnance, the 8-inch shell guns. But since 1838, successive French naval administrations, regardless of the limitations prescribed by the commissions of 1821, 1824 —in defiance of the injunctions of the Comité Consultatif, and of the decision of the Minister of Marine, by whom those recommendations were approved and confirmed, have sanctioned the great augmentations decreed by the regulations of 1848, 1849, and subsequently, in the number of canons-obusiers assigned for the armament of the French navy.[a] These augmentations, with the increased magnitude of ships, and other extensions which General Paixhans disowns having had any share in promoting, and against which he now remonstrates, we on this side of the water have far exceeded. See Armament of the British Navy, as regulated in July, 1848, Appendix F, compared with what it is now, Art. 271, and Note, p. 267.

It was not the author's intention to open, or even to touch upon, this delicate subject on the present occasion, farther than he has already done in the third edition of this work, (Article 246 et seq.), but the whole of this question having been opened by the distinguished originator of the shell system, in the important document which he has recently published; and concurring entirely, as the author does, with him as to the great perils which he apprehends from such an extension of the armament, and magnitude of ships : seeing also that the fleets of England and of France are now acting together upon a system, the dangers of which, from its excesses, may be perilous to both, particularly if they should attack powerful land batteries or fortresses, he has been induced to express himself strongly, in the hope that the threatened danger may be diminished

---

[a] It will be perceived, on referring to the regulation for the armament of the French navy (Appendix C), that this limitation was strictly observed in the decrees of 1824, 1838 ; relaxed in that of 1848 ; further relaxed by the decree of 1849, and subsequently still more so by extending shell-firing to 30 and 36-pounder guns as well as canons-obusiers.

by attention in high quarters to the advice of General Paixhans, and the suggestions which the author has long since ventured respectfully to offer.

With respect to the remark of General Paixhans, in his recent paper, that the more the weapons of war are rendered powerful, the less are battles destructive of human life; the author wishes to observe that, however this may be true of the number of men killed or wounded in actual conflict, it is not so of incendiary warfare, whether practised against ships or towns: it was not so at Sinope, and will not be so in like circumstances hereafter. The loss of life which will take place when incendiary projectiles are employed cannot but be far greater than on any occasion in ancient warfare. War may be made sufficiently costly to deter nations from entering into it, unless when forced to do so by circumstances similar to those in which this nation is now placed; but it may be rendered effectual without being entirely merciless. The great Nelson prayed that humanity, after victory, might be the predominant characteristic of British seamen.

307. If General Paixhans be right in stating that the danger of a ship being blown up, increases in proportion to the number of explosive projectiles and the quantity of inflammable matter which may be accumulated on, or in circulation throughout, her fighting decks, the author cannot be wrong in asserting that the danger will be lessened in proportion as the causes are diminished, and consequently that the danger must disappear when no such cause is present. This being so, it must follow, when a ship has no live shells stowed on her fighting decks (*mines*, in fact, ready laid to her enemy's hands, and ever ready to be sprung by the blow of a solid shot or the explosion of a shell), and is so armed, in the main, as to be able to drive into and through her opponent the greatest number possible of solid shot, in the least time, from guns double shotted, at distances beyond the reach of shell guns similarly charged; that such a ship, all else being equal, must carry the day over any vessel armed with much heavier,

PART III. DANGER INCIDENTAL TO SHELL FIRING. 313

and therefore, in proportion, fewer pieces of ordnance for shell-firing: the latter vessel must, moreover, have vast numbers of live shells stowed on her decks; she will, therefore, have to contend with the rapid inpourings of her enemy's solid shot, and at the same time be exposed to the treacherous effects of her own projectiles.

Upon a full and very serious consideration of the danger, admitted by General Paixhans in his original work, and the limitations under which only his system was adopted (as stated in the preceding article), the author cannot forbear to express his opinion that no shells should be permitted on the fighting decks of a ship under any circumstances; and the time for making this alteration is apt; because the 'reason for placing complements of shells on the fighting decks of ships was, to enable them to keep up quick shell-firing when in close action with ships. But this is not likely to occur in the present war, at least not so suddenly as not to have time to bring shells up as quickly as necessary. But in action against fortresses, forts, and land-defences faced with hard-stone walls, neither quick shell-firing nor any shell-firing will be required from the broadside batteries of ships (see Art. 331, *et seq.*), though howitzer or horizontal shell-firing from the pivot-guns of steamers, against troops, or for enfilading, or for incendiary purposes, and mortar-shells from bomb-ships, will always be important adjuncts in the attack of fortresses and arsenals.

308. It appears from what is stated in the proceedings of, and the evidence taken before the French Parliamentary Committee (*l'Enquête Parlementaire*) that French naval officers are becoming very sensible of the fact, that they have overloaded their vessels with armaments which exceed, in the aggregate, the just proportion which should be maintained between the total weight of metal and other material forming the armament, and the displacement of the ship; and consequently affecting prejudicially her stability and speed; that they have augmented the weight of armament for these ships, to

extents disproportioned to the "développements relatifs de leur carène." That they have not only overloaded their ships with too much weight of metal in the aggregate, but have introduced ordnance of such surpassing weight, individually, as to have reduced the number of guns beneath that which ships should carry, in relation to their tonnage. That displacing a superior number of guns, (30-pounders No. 1), by a smaller number of *canons-obusiers*, far from adding to the real battle-power of a ship, detracts from it. The number of hits or impacts, all else being equal, will be proportionally greater when the armament consists of 30-pounder guns, than when *canons-obusiers* are used instead of those; that 30-pounder shot have force sufficient to penetrate the side of any ship, even when two shots are fired; that firing two shots from *canons-obusiers* of 80 is interdicted; that the principal and most decisive battle-power in close action, is the faculty of pouring into an enemy's ship a superior number of solid shot in the least possible time; that the use of solid shot should therefore be the rule, shell-firing the exception; that a single man cannot easily, and quickly, put into a gun, a projectile larger or heavier than at most a 36-pounder; and that the time required for loading heavier guns, even with increased gun crews, becomes greater in proportion as the weight and magnitude of a projectile increase: these disadvantages are exclusive of the objections to all chambered guns in quick firing. De la Gravière declares: " C'est à cette supériorité dans le tir que nous eussions dû attribuer les pluspart de nos reverses depuis 1793 : c'est à cette grêle de boulets, comme l'écrivait Nelson, que l'Angleterre devait alors l'empire absolu des mers, qu'il devait lui-même la victoire d'Aboukir, et qu'il allait devoir celle de Trafalgar. . . . . . Ils jonchaient nos ponts de cadavres." (vol. ii. pp. 175, 176.) And in another place he adds: " Aujourd'hui même, en effet, où la science pyrotechnique a fait d'immenses progrès, on peut se demander encore si les boulets creux méritent bien réellement l'effrayante réputation qu'on leur a faite, et si le tir plus rapide et plus sûr des projectiles pleins n'est point toujours celui dont l'efficacité demeure le mieux établie." (vol. i. p. 99.)

These views and opinions coincide so much with what the author had previously stated in the former edition of this work, and has now more expressly enunciated in the foregoing article, that he may be excused for thus availing himself of the important weight of opinions coming from so experienced and enlightened a body of men of science and experience, on the other side of the Channel, concurrently with those of many persons at home. But before quitting this part of his subject, he is anxious to state explicitly, that though perhaps it may be generally regretted that shell-firing should ever have been thought of for the naval service, yet, having been introduced in other navies, it is no longer a question whether or not we should adopt it likewise. The author is far from thinking that it ought to be abolished in our service so long as it is retained in others. No doubt the burning of the Turkish squadron at Sinope is an awful proof of the destructive effects of the Paixhans' shell practice, and may well be brought against any proposition for the rejection of that practice; but the author does not propose to relinquish or to enfeeble that system: he affirms it in principle, but condemns the excesses to which we have carried it. Its originator disowns the excesses, and the author, agreeing with him, condemns them, from a thorough conviction of the dangers to which they give rise.

309. If the armament of the British Navy consisted of a judicious combination of shell-guns and solid-shot guns, each confined to the use of their legitimate and appropriate projectiles, our ships would be deprived of none of their real battle-power, by restoring to solid-shot guns the exclusive use of their own missiles. The shell-power of a ship would, by such a wise combination, suffice for a special use of shells; there would be no need to place any shells on the fighting decks, and there would be no difficulty in providing ample space for stowing them in shell-rooms; whilst by restoring to solid-shot-guns the use of projectiles suited to their very natures,

the danger of placing multiplicities of shells on the fighting-deck for their use, would be entirely avoided. "If it were proposed," observes the Comité consultatif de la Marine, pp. 49, 50, " to give hollow projectiles to a hundred pieces of ordnance on board a ship, the precautions to be taken would be numerous, and the danger would be difficult to avoid; but it is proposed to have only four or six *canons à bombes*, which being placed on the lower deck, can be supplied with their shells by conveying these last a short distance only ; and, though powerful guns, being few, they will require but a moderate quantity of ammunition."[a] Yet we have vastly multiplied the number of shell guns, and thus we persist in braving the dangers against which these warnings are pronounced, and have hurried on the French to do the like. The difficulty of providing safe stowage, for many hundred of shells, in some cases, in shell-rooms whose heights are limited, by the necessity of keeping their tops or crowns sufficiently below the water-line, renders it necessary to add to their length, and hence has arisen, in a great degree, the costly lengthening of ships, to enable them to stow their vast complements of shells. The real battle-power that would be restored to ships, by this emancipation of solid-shot-guns from the obligation to use shells, consists in the superior range, accuracy, and penetrating force of solid-shot-guns, in their superiority for firing double-shot (Arts. 253, 264), and in the greater number of guns which vessels so armed would carry in relation to their tonnage; and the reader, pursuing the argument briefly stated in the third edition of this work (Arts. 246 to 248 of that edition), and given at greater length above, will not fail to arrive at a just conclusion.

---

[a] " S'il s'agissait," observes the Comité consultatif de la Marine, pp. 49, 50, " de donner des projectiles creux à cent bouches à feu du haut en bas d'un vaisseau, les précautions seraient infinies et le danger difficile à éviter ; mais il ne s'agit ici que de quatre ou six canons à bombes, qui, étant placés dans la batterie basse, recevront leurs projectiles, sans longs trajets, et qui, ayant une très-grande puissance, n'auront qu'un approvisionnement peu nombreux."

The number of shots that were found sticking in the sides of line-of-battle ships after actions in the late war has since been used as an argument against the efficiency of solid shot, and in favour of shells: but Simmons, in his excellent work on the effects of heavy ordnance, pp. 69, 70, attributes this most justly to the custom of double and treble shotting then so general; and the author has shown in Art. 254 that this arose in some cases from carrying a little too far the well-known fact that by reducing the charge to what is just sufficient to produce penetration through the side of a ship, the ravaging and splintering effects are increased: these errors will probably exist no longer; but however this may be, if solid shots have not power to penetrate under the circumstances, still less could shells or hollow shots do so.

310. When it is unavoidably necessary to fill shells on board ship, the metal fuze should be carefully brushed and luted before it is put in; the shell should be filled by means of a funnel, taking care to insert its orifice below the screw in the tap of the shell, so that no grains of powder may get into the thread. To guard the more effectually against this accident, the female screw in the tap of the shell should likewise be carefully brushed out before the fuze is introduced, and a *washer* of parchment or some other material should be placed under the collar or head of the fuze, that there may be no contact between the metals of the fuze and the shell.

All sea-service shells are fixed to wooden bottoms by straps or bands of tin, or snaked by yarns to grommets of rope, to prevent the shell from turning in setting home; the latter is the more certain expedient, on account of the tin straps being liable to be broken, or injured by damp.

311. Fuzes should be frequently examined to see that their caps have not become so corroded as to be immoveable, but it must be remarked that this can be safely done only by putting the shell whose fuze is to be so examined into a gun previously loaded with a few ounces of powder—the range to the front, right

and left, being clear. The cap may then be screwed off safely with the instrument proposed by Captain Nott, a brass wrench supplied for that purpose; the same instrument is used for fixing the fuze in, and withdrawing it from the shell; the men who turn the instrument keeping well within it and the shell; and, in the event of the fuze igniting, the shell would be blown out of the gun.

The examination of percussion fuzes will be still more perilous, but at the same time more necessary; for the deteriorating effects produced by vicissitudes of climate and temperature, and by sea-damp on the chemical compounds contained in those shells when long stowed in a ship, have not yet been ascertained. To tamper with such explosive bodies, by opening, examining, renewing, and replacing them on board of ship, should not be permitted on any account;[a] and there only remains to be tried the effect of firing them against rocks or other resisting masses. This may not always be convenient or possible, and at best will only show whether or not the shells fired have been efficient: it will not show that all those in the ship's store rooms will prove so thereafter; and failures not unfrequently occur, even with new made percussion fuzes.

312. In the experiments of 1853, several 8-inch shells were found to have struck the object without exploding; some stuck in the side of the hulk; others passed through the wood, which was unsound, the shock not having been sufficient to produce the explosion. An 8 or 10-inch shell passed through the near-side, and struck an iron bolt on the opposite side, yet did not explode, the percussion fuze being found entire. Thus those projectiles failed as shells, and did not even succeed as shots. Had those shells which stuck in the sides or lodged in the body of the hulk been fitted with time fuzes, the effect of their explosion would have been the greatest possible. Many

---

[a] "An awful explosion took place in the laboratory at Rochefort, in 1851, in preparing the implements for shell-firing, in which a maître-d'artificiers and several men were killed; and a similar accident at Mauvillon, by which seven seamen gunners perished."—*L'Enquête Parlementaire*, vol. i. p. 328.

of Moorsom's fuzes were picked up entire and uninjured amongst numerous splinters of shells after the experiments had been made. The old hulk the "York" was no doubt very much riddled and ravaged by these experiments; but she was not burnt, which she must have been by even one such shell as that mentioned, or by one or two of the shells, having time-fuzes, which were planted in the "Prince George" in the experiments of 1838.

313. It is now of vast importance to the naval service of this country that a series of experiments should be undertaken in order to establish, if possible, the following points :—1st. What are the effects produced on the sides of a ship by firing against them shells with different degrees of velocity? for example, at short ranges, with full charges; at medium ranges and with medium charges; and at long ranges with reduced charges. 2ndly. At what time, and where, would the bursting take place? 3rdly. At what ranges and with what velocities would the shell explode by the concussion without penetrating at all into the timber? and must the blow be very slight to have this effect; that is, to render the shell harmless? The following facts may be considered as affording answers to some of the above inquiries; and may be of use in regulating the charges of powder in shell-firing against ships. At 1250 yards, 10 lbs. of powder drove an 8-inch shell through one side and lodged it in the further side of a line-of-battle ship at the lower-deck; with 8 lbs. of powder an 8-inch shell perforated the first side and rebounded from the other. At 900 yards, with charges of 8 and 10 lbs. of powder, 8-inch shells passed through the first side and lodged in the second. At 600 yards, with charges of 7 and 8 lbs. of powder, 8-inch shells just perforated both sides and then dropped. At 600 yards, an 8-inch shell, with a charge of 5 lbs. of powder, passed through the first and lodged in the second side. At 300 yards, with charges of 7 and 8 lbs. of powder, 8-inch shells perforated both sides and buried themselves in the butt. With hollow shot, plugged, of 56 lbs., and a charge

of 10 lbs. of powder, at 1250 yards, the shot passed through the first side and buried itself in the second. With 12 lb. charges (now abandoned) shells frequently burst in the gun, or at its muzzle.

In May, 1853, at the request of the Special Committee of Artillery Officers, six 32-pounder shells, and six 8-inch shells, all fitted with Freeburn's concussion fuzes, were fired from the "Excellent" at the "York" hulk, at 1200 yards distance, when the following results were obtained :—Every shell struck the hulk : of these 10 burst immediately ; one struck an iron knee and broke, and one did not burst. There was no premature explosion ; the trial was therefore very satisfactory, and it was recommended that some fuzes of the same kind should be made at the Royal Laboratory, in order to be preserved there as patterns for future constructions.

314. It has hitherto been generally supposed that the frequent bursting of shells in or near the muzzles of guns, when impelled by large charges,[a] arose, in common shells, from the dislocation of the composition contained in the fuze, and particularly in short-time fuzes; and with respect to Shrapnel shells, in which it is most essential that large charges should be used, the premature explosions which frequently take place were supposed to be due to the same cause. It was to determine this point that a committee was appointed to make experiments and report upon those failures. The committee recommended that the service charges for spherical case shot should be considerably reduced; this measure, having been approved and adopted, has, by so much, deprived that description of shell-firing of the essential cause of its efficiency. But it was surmised by many, and firmly believed by the author, that those premature explosions neither arose from dislocations of the short column of composition in the fuze, nor from the friction of the bullets on the bursting charge

---

[a] "In firing Shrapnel shells from heavy 32-pounders and 8-inch guns, a charge of 6 lbs. is not to be exceeded, and is the most effective that can be used with these guns; with a greater charge the shells will very frequently burst in the gun from the concussion on firing."—"*Excellent*" *Instructions.*

contained in the shell, in the interstices between the balls; for it has been recently proved that shells without fuzes, the taps stopped with iron plugs, explode at or near the muzzle of the gun when fired with large charges; and the friction of the bullets on the powder (which is effectually prevented in Captain Boxer's fuzes) can scarcely be the cause of the explosion, because friction cannot have had time to produce the heat required to ignite the bursting charge on the shell issuing from the gun, although it might do so in the course of, and near the end of its flight, from the grinding of the powder by the balls. It may therefore be surmised that the explosion of the shell at the instant of firing can arise from no other cause than that gunpowder is explosive by percussion, and that the bursting charge is exploded by the shock of discharge, just as when a loaded shell is struck by a shot it is exploded, not by a spark elicited on the previous breaking of the shell, but, as already stated, by the ignition of its contents contemporaneously with, or instantaneously after, it is broken by the blow.

The fact, incontestably, is that gunpowder is, in a considerable and dangerous degree, explosive by percussion, concussion, compression, and friction, and though not instantaneously, yet gradually, by even a small degree of heat. To one or other of these latent dangers in gunpowder, and the more easily ignited compounds for fuze-primings, &c., may be ascribed the unaccountable explosions which so frequently occur in the powder-mill, the laboratory, the fireworkers' manufactory, and in ships, as well as the frequent explosions of ammunition in tumbrels and on the field waggons, and of which no good account has been given, because in general all the persons near and about are killed. A very terrific case of this description occurred at Portsmouth, in 1851, where a large concourse of naval and military officers and many other persons assembled, at the Fire Barn, to witness the trials of a 10-inch shell, filled with a slow-burning composition, which, on being ignited by its fuze, should emit a dense smoke or vapour, having moreover suffocating effects. But instead of burning slowly, the composition, as soon as ignited, exploded the shell, scattered its splinters with great force in every direction, and, besides the casualties it occasioned to the bystanders, had well-nigh put an end to the valuable life of a gallant and distinguished general officer, one of the author's most valued friends.[a] After the most careful investigation, this accident has never been accounted for, and remains one of numerous instances of the treacherous character of all such artifices.

It was proved by Le Roi that gunpowder is explosive by percussion, for this chemist found that a few grains strewed on an iron anvil, and struck by an iron hammer, exploded. Jeffers, in the work above quoted, says, p. 174, that gunpowder likewise explodes by the collision of iron on brass, brass against brass, copper against copper, though less readily; likewise by the shock of bronze on copper, iron on marble, iron on lead, and lead on lead. The latter was abundantly proved in an experiment made by Professor Faraday, who is in possession of a large fragment of a shell containing a mass of lead formed by the bullets it had contained, which were transformed, by the shock of impact, into a mass of prisms; this was made use of to try whether the action of lead on lead would ignite gunpowder; for this purpose a small quantity was placed in the hollow of the shell, which, when struck by a leaden bullet, was ignited. A stone shot, of 770 lbs. weight, entered the lower deck of the "Windsor Castle" in 1807, set fire to some powder, and produced a terrible explosion, by which 46 men were killed and wounded, and such a panic was created that some men jumped overboard and were drowned. Why, then, might not the blow of a 32 lb. shot, fired with a full service charge, and striking a shell at rest, explode the shell by percussion? But whether the shell be broken and its contents ignited thereupon, or whether it be exploded by its charge on

---

[a] Major-General Simpson, Governor of Portsmouth.

receiving the blow, this at least we know, that such an ignition of powder may produce, as it did at Sinope, the most disastrous effects in a ship in action.

The expense of shell equipment and shell-firing, to the extent to which it has been carried in the British navy, is enormous. The cost of every 8-inch shell shipped in its box, including the fuze, is 11s. 6d. Every 8-inch shell fired costs 17s. $4\frac{3}{4}d$.[a] The cost of the fuze for a 10-inch shell is the same as that for an 8-inch, but the expense of the shell itself is greater than that of an 8-inch, in proportion to the greater weight of the larger shell. Fully admitting the advantages of pivot-guns to steamers and other vessels, and of a limited number of shell-guns for special shell-firing, there is nothing to object to the expense, though large, of guns, carriages, slides, and appurtenances of the former,[b] nor to the supply of 8-inch shells for a limited number of shell-guns. But with respect to the vast expense of arming whole decks of line-of-battle and other ships with shell-guns; and of shell-firing from all natures and descriptions of ordnance, great and growing objections are made, and making their way. The vast expense of our shell system consists not only in providing great numbers of new shell-guns, shells, fuzes, &c., but, moreover, of constructing ships and vessels of vastly increased and enormous magnitudes, capable of carrying such heavy armaments and of providing stowage in their interior for the

---

[a] Cost per Round of 8-inch Naval Shells fired.

	s.	d.
Shell	5	$4\frac{1}{2}$
Metal fuze and its cap, about	5	0
Charge, 10 lbs. powder at 7d.	5	10
Bursting powder, 2 lbs.	1	2
Box	1	11
Total, including the box	19	$3\frac{1}{2}$

Moorsom's fuzes cost about 6s. each; Boxes, 6-inch, 1s. 6d.

[b] The expense of the gun, carriage, slide, and appurtenances
of a 68-pounder gun £225
Ditto ditto of a 10-inch gun 176
Ditto ditto of an 8-inch gun 172

greatly increased complements of shells. For this the relation formerly existing between the displacement of ships and the number of guns they are to carry (Arts. 249–252) has been greatly altered, in the construction of new ships, and in re-modelling those formerly built. If General Paixhans, the great originator of the shell system, is right, and unless the author and many experienced first-rate naval officers be egregiously mistaken, fresh expense of no small amount will have to be incurred merely to undo much that has been already done.

The author can scarcely doubt that he will be considered by the reader to have established, in the preceding pages, a case of very great peril, deserving the most serious consideration:—this, at least, is so strongly impressed upon him in the course of the laborious, protracted, and most anxious attention which he has given to this important subject, that the author feels it impossible to repress the avowal of his apprehension, now that he is called upon by the demand for his work to resume his labours. Thus he disburthens his mind of the load he would have to bear, and clears his conscience of the responsibility which, in certain cases, he might incur, did he conceal the convictions of his own mind. The reader will pronounce whether, in his judgment, a case of peril to the profession and to the country has, or has not, been established in these pages; and, if the former, whether by any, and what means, the danger may be prevented, abated, or avoided.

VIII. ON MILITARY ROCKETS.

315. A rocket for military purposes consists of an inflammable composition contained in a cylindrical case of stout paper or of iron, the head, or anterior part, having usually the form of a cone, but, occasionally, that of a hemisphere or paraboloid. The composition for the cylindrical part consists of nitre, sulphur and charcoal, in the following proportions:—Nitre, 4 lbs. 4 oz.; sulphur, 12 oz.; and charcoal, 2 lbs. When the

rocket is used merely for making signals, the head, which is then conical, is filled with a composition for producing, at the explosion, the stars of light which constitute the signal.

A rod of wood is attached, at one end, to the base, the neck or choke of the rocket;[*] its length being equal to about 60 times the diameter of the cylindrical part of the rocket, and its thickness equal to about half that diameter. Signal rockets weigh from half a pound to 2 lbs.: the diameter of a one pound rocket is $1\frac{2}{3}$ inch; the length of the cylindrical part is $12\frac{2}{3}$ inches, and of the conical part $3\frac{1}{3}$ inches.

The composition is driven into the rocket case till its density is equal to about twice that of gunpowder; but, in the interior, about the axis of the case, is left a void space of a conical form, its base coinciding with the neck of the rocket; and, in this neck several apertures are formed for the admission of air: at one of these, in which is left a piece of *quick match*, the fire is applied to the composition. The rocket, when about to be fired, is fitted in a tube, which is attached in a given position, to a rest; when, on applying the match, the whole surface of the conical space is put in a state of slow combustion and the rocket is propelled: the combustion continues till the composition is entirely consumed; the elastic gas generated by the combustion escaping through the apertures.

316. The propelling power is produced by the expansion of the gas generated in the burning composition: the force thus originated causes a pressure, outwards, against the sides and ends of the rocket; but the apertures in the neck allowing the gas to escape there, (being resisted only by the pressure of the atmosphere at the apertures), the pressure against that end is consequently less than that which is exerted by the gas against the head, or anterior part of the rocket; and the

---

[*] At first the rod was attached to one side of the rocket; but this caused great irregularities in the flight, and the late Sir William Congreve placed it in the direction of the axis of the rocket: this disposition in a great measure remedied the evil without interfering with the escape of the gas.

difference between the pressures at the opposite ends is a resultant force acting against the head during all the time that the composition is burning; this constitutes, therefore, a pressive force by which the rocket moves onwards with a motion continually accelerated till the resistance of the air against the head becomes equal to that force, or till the composition is burnt out (about 20 seconds), when the rocket falls to the ground. This action of the gas is quite analogous to that which produces the recoil of a suspended gun, when fired without shot or wadding.

317. The rod serves to guide the rocket steadily in its flight, the lateral resistance of the air about it preventing, in some measure, its vibrations. In a 1 lb. rocket, before combustion begins, the common centre of gravity of the rocket and rod is about 2 feet from the head of the former, and about 7 feet from the opposite extremity of the latter; and then the resistance of the air, in checking the vibrations of the rocket from accidental causes, acts with considerable effect, like a power applied at the end of the longer arm of a lever; but, in proportion as the composition is burnt out, the centre of gravity approaches the middle of the length of the whole missile; the resistance of the air is then less able to counteract the accidental deviations of the rocket itself; the head at the same time begins to droop, and at length the whole comes obliquely to the ground. It has happened, even, when the angle of elevation was small, that the weight of the rocket preponderated so far over that of the rod as to cause the missile to come to the ground in a direction tending towards the spot from whence it was fired.

Signal rockets, whose diameters vary from 1 to 2 inches, will ascend vertically to a height of 500 or 600 yards; and those whose diameters vary from 2 to 3 inches, to a height of 1200 yards. A 12-pounder rocket fired at an elevation of 16°; and a 6-pounder rocket, at an elevation of $14\frac{1}{2}°$, range about 1200 yards. The distances at which the explosions of rockets have been seen vary from 40 to 50 miles.

318. The use of rockets was first introduced into the military service by Sir William Congreve. This scientific officer caused them to be made to serve as shells or carcasses; and their weights, for these purposes, are 3, 6, 12, 24 and 32 pounds. When fired against timber or earth they penetrate to considerable depths. A 12-pounder rocket, after a range of 1260 yards, has entered to the depth of 22 feet into earth.

Every shell-rocket is fitted with a fuze, screwed into the base of the shell. The fuze is as long as the size of the shell will admit of, so as to leave sufficient space between the end of it and the inner surface of the shell, for putting in the bursting powder; and the end of the fuze is cupped, to serve as a guide in the insertion of the boring bit. There is a hole in the end or apex of the shell, secured by a screw metal plug, for putting in the bursting powder, and for boring, according to the different ranges at which it may be required to burst the shell. The following table shows the dimensions of the parts of the rocket which relate to the fuze :—

Nature of Rocket.	Distance from the surface of the Shell to the end of the Fuze, in Inches and Tenths.	Length of the Fuze.	Distance from the surface of the Shell to the top of the Cone in the interior of the Rocket, in Inches and Tenths.	Diameter of Fuze Composition, in Tenths of an Inch.	Diameter of Fuze-hole, in Tenths of an Inch.	Diameter of Plug-hole, in Tenths of an Inch.	Shell contains, of fine grain Powder, in Ounces	Thickness of the Rocket Composition above the Cone.
24-Pounder	1.6	3.3	9.3	.25	.75	.4	8¾	3.3
12-Pounder	1.	2.5	7.2	.25	.75	.4	3¼	2.8
6-Pounder	.9	2.	5.7	.2	.55	.25	1½	1.8
3-Pounder	.7	1.3	4.	.2	.55	.25	¾	1.5

If the rocket is to be used as a shot-rocket, the only thing to be attended to, is to take care that there is no powder in the shell, and that the plug is secured in the plug-hole.

If the rocket is to be used as a shell-rocket, at the longest range, the plug is to be taken out, and the shell filled, the fuze left at its full length, and the plug replaced.

If at the shortest range, the fuze is to be entirely bored through and the rocket composition bored into, to within one inch and a half of the top of the cone, in the 24-pounder rocket, and to within one inch in the 12, 6, and 3-pounder rockets. The distances from the surface of the shell to the top of the cone, and from the surface of the shell to the end of the fuze, and also the length of the fuze, being fixed and known, the place on the boring bit at which to screw the stopper, whether for various lengths of fuzes or length of rocket composition to be left over the cone, is easily determined; these distances are marked on the brass scales for each nature of rocket, and the length of rocket composition available for boring into, and the lengths of fuze, are also set off, and subdivided into tenths of an inch.[a]

---

[a] The following rules concerning the lengths of rocket fuzes, the ranges and elevations, may be useful, though they have not been confirmed by an extensive course of practice :—

For 24-pounder rockets ; if the whole length of the fuze is left in the shell of the 24-pounder rocket, it may be expected to burst at about 3700 yards, elevation 47 degrees.

If the whole of the fuze composition be bored out, and the rocket composition left entire, the shell may be expected to burst at about 2000 yards, elevation 27 degrees.

If the rocket composition be bored into, to within 1.5 inch of the top of the cone, the shell may be expected to burst at about 700 yards, elevation 17 degrees.

For 12-pounder rockets ; if the whole length of fuze be left in the shell of the 12-pounder rocket, it may be expected to burst at about 3000 yards, elevation 40 degrees.

If the whole of the fuze composition be bored out, and the rocket composition left entire, the shell may be expected to burst at about 1500 yards, elevation 20 degrees.

If the rocket composition be bored into, to within one inch of the top of the cone, the shell may be expected to burst at about 420 yards, 10 degrees elevation.

For 6-pounder rockets ; if the whole length of fuze be left in the shell of the 6-pounder rocket, it may be expected to burst at about 2300 yards, 37 degrees elevation.

If the whole of the fuze composition be bored out, and the rocket composition be left entire, the shell may be expected to burst at about 1100 yards, elevation 15 degrees.

If the rocket composition be bored into within one inch of the top of the cone, the shell may be expected to burst at about 420 yards, elevation 10 degrees.

For 3-pounder rockets ; if the whole length of the fuze be left in the shell

319. The very vague observations on shell-rocket practice given in the Note, above, are sufficient to show the great uncertainty of that practice against troops in the field; and to this uncertainty must be added the liability of the sticks to be broken on grazing the ground, when fired at low angles.

The forward motion of a rocket, besides being impeded by the resistance of the air at the head, is further impeded by the action of gravity when the missile is fired at an elevation. Again, in firing across the wind, the action of the air upon the stick causes the rocket to come up more to the wind instead of being driven bodily to leeward, and the stronger the current of air is, the more the rocket points towards the quarter from whence the wind comes. When the rocket is fired against the wind the range is considerably shortened, and when fired with the wind, it is lengthened. Thus, in firing across the wind, some allowance must be made for its effects, and the rocket must be pointed by so much to leeward of the object; in firing against the wind, greater elevation than that which the distance requires must be given, and in firing with the wind, less elevation must be given; but the amount of these allowances can only be assumed approximatively according to an estimate of the strength of the wind; and therefore the practice must be uncertain.

320. The author has seen sufficient of rocket practice on service to convince him of the uncertainty and inefficiency of that weapon in firing at small objects. When used against large towns, which can scarcely be missed, in order to set fire to habitations and other structures formed of combustible materials, rockets may answer well as incendiary weapons; but they are far

---

of the 3-pounder rocket, it may be expected to burst at about 1800 yards, elevation 25 degrees.

If the whole of the fuze composition be bored out, and the rocket composition be left entire, the shell may be expected to burst at about 850 yards, elevation 12 degrees.

If the rocket composition be bored into within one inch of the top of the cone, the shell may be expected to burst at about 420 yards, elevation 8 degrees.

more formidable to private dwellings and their unfortunate inmates than destructive of military defences and the lodgments of the troops. If a rocket strike the roof of a house it will there stick and set it on fire; but it has not penetrating power to produce any serious effect on the defences of a place.[a] Flushing was set on fire in many places, at the bombardment of that place in 1809, but no mark was left of the rockets having done any material injury either to the defences or the defenders. Rockets may be used with considerable advantage against cavalry, from the scaring effects of that blazing projectile upon horses: they may also be employed efficaciously against squares or masses of infantry, and in dislodging an enemy from villages or towns, which could not otherwise be approached by infantry alone. At the battle of Leipsic, in 1813, the British rocket troop under the command of Captain Bogue, is said to have rendered essential service; and at the passage of the Adour some discharges of rockets fired across the river checked a French column that was advancing to attack the lodgment which had been effected on the right bank of the river by a body of 600 British troops; but it appears upon the whole that the effects produced by these first uses of this apparently fearful weapon were rather moral than real. Rockets may also be fired from ships against troops or towns with considerable effect as incendiary projectiles; and were used as such by the French in 1844, against the cities of Tangier and Sucrah; they have also been found very useful for incendiary purposes in Algiers. The portability of rockets, when great numbers are to be conveyed in countries impracticable for wheel carriages, renders them very desirable weapons with troops in the field, and even with the artillery of an army, to be used on special occasions, as when guns cannot be brought up.

---

[a] It must be remarked that what is called a shot-rocket is one whose shell is not loaded with powder, and which has the top plugged. It is therefore only a thin hollow shot, having consequently comparatively small penetrating power, and which, like other hollow shot, would break to pieces on striking any very hard material.

321. Sir William Congreve enthusiastically believed that his rockets would entirely supersede the use of artillery in the land service, and many exaggerated opinions were once entertained of the efficiency of that weapon ; but these have long since sobered down to the idea, now very generally prevalent, that they are only substitutes for field guns, when these cannot be brought up; they may easily be carried by men or drawn by horses, and may commence firing upon an enemy before artillery could be brought into position. The most efficient use, however, that can be made of rockets is, as an incendiary projectile, to set fire to towns or single buildings; but it may be doubted whether the rocket system is not carried too far for field service in the organization of rocket troops. These consist of mounted men, as in the horse-artillery, with a number of carriages for transporting the rockets and their appurtenances. The carriages cannot be brought up unless the country in which they are to act is practicable for them ; and if so, would it not be better that a troop consisting of so many men, carriages, and horses, should have guns rather than rockets ? According to the present organization of rocket troops, these very uncertain weapons are made substitutes for artillery which the same number of carriages might transport and the same number of men might serve. M. Charpentier in his '*Essay on Artillery*,' p. 199, makes the following very just observation respecting extreme opinions on the value of rockets: " Quelques militaires regardent les fusées de guerre comme des projectiles insignifiants. D'autres, s'exagérant leur puissance destructive, en font une sorte d'invention infernale, dont ils voudraient, dans des vues philanthropiques, voir l'usage interdit. Il y a erreur de part et d'autre. Les fusées incendiaires peuvent avoir un effet très-utile dans certaines circonstances de la guerre, ne fut-ce que pour porter la démoralisation chez l'ennemi."[a]—

---

[a] " There are military men who consider rockets as weapons of small value ; and there are others who, exaggerating their destructive powers, consider them as a sort of *infernal* invention, the employment of which they would, from motives of philanthropy, interdict. Both parties are in error : rockets can be employed with advantage in war only as means of producing disorder among an enemy's troops."

Thus in the affair at the Adour, in 1812, the French soldiers were certainly very much scared, but not much hurt; one of them, who was made prisoner, had the skirts of his coat set on fire by a rocket, and on surrendering exclaimed in great consternation—" Sacré Dieu! j'ai vingt ans de service et je n'ai jamais vu des armes à feu comme celles-là."—But, continues M. Charpentier, if rockets are powerless against the strong materials of ships of war, they may be used with great efficacy against places on a sea-coast, to protect landings, and against crowds there assembled; and, accordingly, they should be plentifully supplied to steam ships, which, having small draught of water, may approach close to an enemy's coast and by using these incendiary projectiles compensate advantageously for the small number of guns to which their armament must be limited. Thus large proportions of rockets are issued to steam ships in the French navy, and even to other vessels.

322. But rockets are dangerous inmates in ships: there is no space to stow them in shell-rooms or magazines, nor would it be proper to place them there. A terrific exemplification of the danger of these incendiary bodies, in ships overmuch provided with weapons of this description, is given in the subjoined note.[a] Rocket firing from ships is a very dangerous practice. The first rush of back-fire before the rocket starts is capable of igniting any combustible body upon which the tongues of the flame act. An expedient, we believe, has been proposed, to protect the ship from this back-fire, but it does not appear to have been successful.

---

[a] "A terrible misfortune has occurred to the ship of the line 'Valmy,' while at sea, and in the passage from Torbay to Brest, owing to the imprudence of a gunner, who had in his possession some boxes of rockets, one of which exploded. About five o'clock in the morning of the 8th a tremendous explosion was heard, like a clap of thunder, on board. At the same moment the rappel was beaten, minute guns were fired, and orders given to get out the boats. The shock was so great that the whole of the lights were extinguished; darkness the most complete prevailed; and the crew ran a risk of being suffocated by the smoke of the powder. Nothing was heard but the cries and moanings of the wounded, the greater part of whom were as if buried under the timbers. Twenty seamen, whose forms had lost all human appearance, were found amongst the ruins. Ten of them died in half an hour after, and it is feared that very few can be saved. The 'Valmy,' damaged completely in her inside, is to put into Brest to be repaired."

323. When, to all the other aberrations to which a projectile is subject in its flight, we add that the trajectory which a rocket describes is made up of two portions produced by very different causes and governed by different laws, the very great uncertainty of rocket practice will be obvious. When it first starts from the tube the velocity is so small that it is not sufficient to prevent the fore part of the rocket from drooping or dipping below the axis of the tube; the actual angle of departure is, therefore, less than that at which the tube is set, and allowance for the error can only be made by a vague estimation. As the rocket proceeds its velocity increases, and is supposed to be greatest at one third or one half the range. The common centre of gravity of a rocket and its stick on starting is situated near the propelling power, and the vibrations of the rocket during its flight take place about that point; this point is, however, continually changing its place in proportion as the composition is consumed, and this change causes continual irregularities in the deviations of the rocket during its flight. When the composition is entirely burnt out, the rocket proceeds under new and very different conditions: so that upon the whole it is utterly impossible to lay down the trajectory of a rocket, or to obtain good and sufficient rules for conducting the practice with that arm.

324. A very ingenious method of dispensing with the stick of the rocket has recently been proposed by Mr. Hale. This consists in causing the rocket to rotate on its axis during its flight; and, as in the case of an elongated shot, move steadily with the point foremost. For this purpose, instead of permitting the rush of flame to escape from the bottom orifice in a line with the axis of the tube, by which the flame acts directly against the air, the burning material issues from five orifices made near the neck, obliquely to the axis of the tube; the effect of which is that the body of the rocket is made to rotate while it is also propelled. This is certainly a very ingenious contrivance which may be expected to produce advantageous results. In the experiments made with

these rockets several modes of directing them have been tried : First, by firing them from a small trough formed of wood in two inclined planes: Secondly, from a frame carrying two portions of rings, which grasp the body of the rocket and retain it in one position till it has acquired, after ignition, sufficient force to overcome the pressure of a spring below it ; this force, suddenly releasing the body from the rings, permits the rocket to escape with a velocity sufficient to prevent the usual droop or dip above mentioned. This droop was supposed to be the cause of the failures in some previous experiments which had been made at low angles of elevation without the rings. Thirdly, the rockets were directed by a circular machine consisting of three hoops made of iron bars, between which the rocket was introduced ; on being ignited, it proceeded round the circle between the bars with increasing velocity, and escaped at the lower part of the machine with force sufficient to prevent any droop as well as to carry it to a very great distance.

325. These rockets were not recommended by the select committee, on account chiefly of their liability to failure at low angles ; but whatever may be their present defects they appeared to the committee to be capable, when further improved, of being made very valuable weapons.

Unless the cause of the failure of the Hale rocket, when used at low angles, can be removed, it will be of little use against troops in the field ; and it is in horizontal firing, on plane battle fields, that rockets are most formidable. In other respects also the success of the Hale-rocket may be doubted : the stick-rocket continues its flight, directed by the stick, after the composition is burnt out ; but the Hale-rocket loses its directing power as soon as the composition is consumed, because the rotation then ceases, and nothing can be expected from the rocket, beyond the distance it has reached when the composition ceases to burn.

PART III.   ATTACK OF MARITIME FORTRESSES.   335

IX. ON THE ATTACK OF MARITIME FORTRESSES.

326. The most recent cases of vertical shell-firing which have occurred since the termination (1815) of the general war, are those carried on by the French squadrons which bombarded the fortresses of San Juan d'Ulloa in 1838, and Vera Cruz in 1839; and these may be taken as proofs of the uncertainty of vertical shell-practice against castles or other small places, as well as of the inability of fleets to contend with fortresses and other powerful land batteries, unless the ships be very close to them. Whether the attacking ships or squadrons should, under any circumstances, be permitted to approach to such proximity unopposed, are questions which will be surely solved in the negative, whenever it may happen that ships, advancing with such temerity, shall be properly *cannonaded*, as soon as they come within the reach of well-placed, powerful and well-served long-range guns. This not having been done at Algiers and Acre, and we may add Navarino, the daring and success of the operations against those places have tended to create the erroneous notion that land-service batteries cannot under any circumstances withstand the concentrated fire of ships of the line.

327. If, indeed, ships be permitted to approach, with impunity, to measure well their distances from a fortress, and then deliberately open their fire, the torrent of iron which they may throw in must be irresistible and overwhelming, particularly if the batteries are placed *à fleur d'eau*, and consequently commanded by the upper decks of large ships. Batteries placed nearly on a level with the water are far more subject to the fire of ships, and are much less formidable to them, than batteries elevated somewhat above the surface of the sea: these last command the upper deck of a ship by a direct, though depressed fire, which will penetrate obliquely into, and through her sides, and possibly come out below the water-line; whereas shot, even the best directed, from the ship, will, except such as may chance to enter the embrasures, or graze the crest of the parapet, pass over the heads of

the defenders in the battery without doing any material harm. Should any guns be mounted *en barbette*, which ought never to be the case in sea-batteries intended for close or flanking defences, they would inevitably be dismounted. No ricochet from a ship can touch a battery in a commanding position; whilst, unless the battery be situated so high that its shot would strike the water under angles exceeding 3 or 3½ degrees (Art. 139; see also Arts. 160-164), shot ricocheting on the water will strike the ship.

In the *Aide Mémoire Navale* it is stated, p. 404, that the height of a coast or land battery above the level of the sea should be from 10 to 15 mètres (yards), because that height will permit the ricochet to take effect as far from the battery as about 200 mètres (yards), and will avoid the effect of the ricochet of shot fired from ships' decks, which are only from 5 to 6 mètres (yards) above the water-line.

It is not easy to assign any general rule for the most advantageous height of a battery above the level of the sea, because that level alters with the tide: the height should also depend on the degree of proximity to which vessels may, from their draught of water, approach to attack. It may be stated in general, that all batteries should have some command over the body fired at. The most favourable situation for a gun battery in the field service is about one hundredth part of the range above the position of the enemy.[a] But this low command is sufficient only in firing against troops: when the fire of one battery is directed against another, a more considerable command is of great importance. At the siege of Burgos, in 1812, the batteries of the place were 50 feet above the breaching battery, at the distance of 150 yards only; and their effect was irresistible, the shot plunging on the very platforms of the besiegers' battery.

M. Sar, in his '*Cours d'Etudes Militaires*,' pp. 349, 350, 359, *et seq.*, has treated this subject with great perspicuity and intelligence: his words are to the following

---

[a] Practice Cards by Lieut.-Colonel Burns.

effect—" The experiments which, at different times, have been made prove that shots *ricochet* more perfectly on water than on land; and, according to Gassendi, all ricochets with elevations of 2, 3, and even 4 degrees cause large shots to lose very little of their force."—pp. 348, 349.

" A battery of 10 pieces, served by skilful gunners, firing in succession, would soon overpower a ship, whatever might be her force, especially if the guns are of high calibre, or if red-hot shot are projected."—p. 350.

" The shot from ships whose decks are 6, 12, or 18 feet above the water cannot, in ricocheting, rise up to the battery, while the latter can employ both a direct and a ricocheting fire against the whole body of the ship. On the other hand, only those shots from the ship can take effect which pass 18 inches above the parapet of the battery, since the guns in the latter are only so far exposed, and the gun itself covers the head of the man who points it; all the rest of the service is performed behind the parapet. Thus the ship, for every 18 feet length of gunwale, has no other object to aim at than the muzzle of a gun presenting only about 2 square feet of surface, while the battery has before it an object presenting 2000 square feet of surface, independently of the masts, ropes, and sails."—pp. 350, 351.

328. Land batteries properly placed, well armed and skilfully served, may open with great effect on an enemy's ships at great distances, and keep up, as the latter approach, a continued and deliberate fire, the effect of which will become still more formidable in proportion as the ships are nearer. This fire the ships cannot return but by their bow guns, until they shall have taken their position for attack; and during all that time they will have been severely maltreated.[a]

---

[a] It results from experience, that a battery of four heavy guns, well placed and served, has a superiority over a vessel even of 120 guns.—*Aide-Mémoire Navale*, p. 404.

It is generally admitted that four guns of the calibre of 18 or 24, protected by a wall and properly served, are equivalent to a ship of the line; and that it is almost impossible to destroy a fort by fire from ships, even if at anchor.—*Concise Treatise of Naval Gunnery*, by William Jeffers, U. S. Navy, page 175.

While ships are approaching, under fire of the heavy ordnance with which coast batteries should ever be armed, a few well-directed shells, having time-fuzes, thrown in at suitable distances, to act against the whole expanse of a ship—masts, sails, and body—can scarcely fail to produce very severe dismantling effects, which will very much interfere with, and impede the operations they have yet to execute, before they can open their fire with any safety or effect; and Shrapnel shells well applied during the operation of furling sails, would be extremely deadly to the crowds of hands then aloft.

329. It has, indeed, been observed that steam-tugs may be used to tow sailing-ships to the positions necessary for enabling them to attack batteries and fortresses with the advantages which proximity will give to a fleet or squadron, and the Prince de Joinville's attack of Tangiers is quoted as a proof of what steam may effect by traction. But if well opposed, this mode of approach would, perhaps, be more dangerous and uncertain than any other, from the difficulty of passing a tow-rope, and the probability of its breaking or being cut by a shot, besides the risk of the steam-tug or ship being disabled, as was the case with the ship the "Christian the Eighth," which occasioned her destruction (see Colonel Stevens's account of that catastrophe. See also Art. 330, p. 343).

This is a sufficient warning against trusting to traction by steam-vessels. The application of steam-power for battle purposes should be by propulsion inherent in the vessel, which should therefore be provided with adequate locomotive powers.

330. Lord Exmouth, in the "Queen Charlotte," was allowed to approach and anchor with impunity within 50 yards of the Mole of Algiers! He then opened his fire, and poured in such a torrent of projectiles as to silence every gun opposed to him, with the loss of only 8 men killed and 131 wounded. But the Impregnable, which was fired upon at 1200 or 1500 yards distance, had 50 men killed and 138 wounded, and was greatly damaged in her material; thus the ship engaged at a distance sustained greater loss and damage than that

which fought in close action. And it appears that only those ships which were very near the enemy silenced the batteries with which they were engaged.

The British fleet was permitted to approach Acre almost without any opposition; to sound as they advanced, to buoy the positions which the several ships were to take up, and then to open a most destructive fire upon the place, which was very inefficiently returned.

It did not escape the sagacity and vigilance of the Duke of Wellington, in voting the thanks of the House of Lords to the admiral, officers, and seamen engaged in that successful operation, that wrong and perilous impressions might be created as to the ability of fleets to contend with fortresses in general; that the achievement at Acre was an exceptional case, and that it would not be safe or practicable to do the like against fortresses or land batteries well armed and skilfully defended. Thus His Grace, after expressing his cordial approbation of the services performed by the navy in the Mediterranean, and of those who were engaged in this glorious expedition, goes on to say:—

"He had a little experience in services of this nature, and he thought it his duty to warn their Lordships on this occasion that they must not always expect that ships, however well commanded or gallant their seamen might be, were capable of commonly engaging successfully with stone walls.

"He would repeat that this was a singular instance, in the achievement of which great skill was undoubtedly manifested; but which was also connected with peculiar circumstances, which they could not hope always to occur. It must not, therefore, be expected, as a matter of course, that all such attempts in future must necessarily succeed."[a]

The victory of Copenhagen in 1801 was dearly purchased—the loss in killed and wounded was far greater, and the ships more severely damaged, than in the great battle of Aboukir, especially in their hulls; most of them

---

[a] Hansard's Debates, vol. lvi. p. 254.

had several of their guns rendered useless, whilst the land batteries were comparatively little damaged.

The "Agamemnon," "Bellona," and "Russel," having run aground, occasioned gaps in the British line, which exposed the van-ships to a greater share of fire from the enemy's land and floating batteries than was intended; on perceiving which the gallant Riou in the Amazon, with the frigates Blanche and Alcmene, and two sloops, bravely attacked the Crown batteries, but suffered so severely that they were obliged to haul off, by which they were probably saved from destruction. If the Crown Prince of Denmark had refused to listen to Nelson's overtures for a cessation of fire,[a] Nelson could neither have withdrawn his crippled ships nor effected his own retreat, in compliance with the signal of recall, if he had been disposed to obey it; for the Crown batteries, which had driven off the frigates, effectually stopped that outlet. But the British fleet on that occasion was provided with powerful bomb-ships, which had taken position behind Nelson's line, and which continued, throughout the action, to throw their shells into Copenhagen, over the ships in line of battle, and might, if hostilities continued, bombard, and in great part destroy, the city on the morrow. This the Crown Prince well knew, and to save the capital from the horrors of a bombardment, ordered the fire to cease.

The Crown batteries were the great difficulties with which Nelson had to contend: to attack Copenhagen from the south, it was necessary to pass under the fire of those powerful works, in order to get into the King's Channel, an attempt which was at first meditated, but afterwards abandoned for the purpose of avoiding their fire; and the attack was made through the passage to the north of the Middle Ground, the ships entering the Royal Channel by the south. When the crippled state of many of Nelson's ships rendered it

---

[a] The circumstances under which wax was used instead of a wafer, prove an intention to avoid giving ground for an opinion that the letter was sent off in haste.

advisable to endeavour to withdraw them as soon as possible from the intricate channel in which they had gallantly fought, but severely suffered, it was clear that the Crown batteries, which had not been directly engaged since the defeat of the frigates under the gallant Riou, and had subsequently been reinforced with fresh men, would effectually prevent the passage of any ships through the outlet which those batteries commanded: but Nelson was happily extricated from this very painful, and perhaps perilous predicament, by the acceptance of his overtures for a cessation of fire, or the result of the battle might have been somewhat different.[a] However this may have been, this, at least, is certain, that it would not be prudent, with fleets only, to repeat such an attempt against fortresses or powerful batteries armed as all now are, and provided with expert gunners and skilful bombardiers.

When Copenhagen was attacked in 1807, with a large military force, combined with a powerful fleet, the place and all its sea defences were first invested on the land side, then besieged, bombarded, and captured; and thus the naval objects of the expedition were accomplished without much difficulty, and with the loss of only 56 killed and 179 wounded in both services. The loss of the Danes in killed and wounded in the naval and military operations, external to the city, was much greater: in the subsequent bombardment, 305 houses were destroyed, many more much injured, and 2000 inhabitants, men, women, and children, perished: not a man was hurt in the attacking batteries, but an armed transport, in which there were nine mortars, was blown up by a mortar-shell fired from the Crown batteries, and by which the master of the transport, 2 officers, and 28 men were killed and wounded.[b]

At Navarino the combined fleet was permitted to pass under the guns of commanding and powerful batteries; which, had they opened, as they ought, on the ships of

---

[a] James's Naval History, vol. iii. p. 74; De la Gravière, vol. ii. p. 14 *et seq.*
[b] James's Naval History, vol. iv. p. 290.

a fleet entering in this equivocal if not hostile manner, would have severely crippled these ships before they could have got into a position to engage the Turkish fleet.

In 1814 a French 80-gun ship, in attacking at anchor a 2-gun battery in the Scheldt, mounting one long 18-pounder gun and one $5\frac{1}{2}$-inch howitzer, at a distance of 600 yards, was beaten off with the loss of 41 men killed and wounded, besides being severely damaged in her hull. The battery lost only 1 man killed and 2 wounded. No doubt the howitzer shells fired from the battery contributed greatly to its success, but shell-firing from a $5\frac{1}{2}$-inch howitzer, at 600 yards, could scarcely have penetrated the side of an 80-gun ship; it must have been chiefly by the solid shot of the long gun that the vessel was so severely damaged as to be in a sinking state. The ship might, it is said, have been sunk or captured had the position of the battery been such as to have given it a more oblique command of the ship.

Sir Sidney Smith, in the "Pompée" of 80 guns, with the "Hydra" of 38 guns, and the "Aurora" of 28 guns, cannonaded, at 600 or 700 yards distance, a 2-gun battery protected by a tower; both battery and tower being placed considerably above the level of the sea. The two frigates remained under weigh, and occasionally fired at the fort, but without silencing it; at length the marines landed in the rear of the battery; and on their approach the serjeant in command of the post immediately surrendered. This shows in principle, though on a small scale, the advantages of a combined attack by land and sea. The "Pompée," having remained stationary at anchor, had 35 men killed and wounded, and received 40 shot in her hull.

In the engagement between the "Loire" frigate and a fort armed with 12 long 18-pounders, the fort being placed in a commanding situation, the disadvantage of an attack by sea was strongly exemplified; the defenders of the fort were so well covered that the frigate's fire, though accurately directed, was comparatively ineffectual; whilst almost every shot from the fort struck and penetrated the frigate, so that in a very few minutes of this unequal warfare, the "Loire" sustained considerable loss (James's History, vol. iv. p. 135).

The batteries which the "Christian the Eighth" so indiscreetly engaged, were placed 18 feet and 12 feet above the level of the sea. The first was armed with two 8-inch guns (French) and two brass 24-pounder siege-guns; and the second with four 18-pounders.

The Danish commander in his official Report states that, finding, after much expenditure of ammunition, the ships were suffering greatly from the fire of the most elevated battery, while that of the "Christian the Eighth" and other vessels made no material impression upon it, he withdrew from his position and engaged exclusively the lower but weaker battery, mounting four 18-pounders. His ship soon afterwards blew up, having been set on fire by shells and red-hot shot fired from the commanding battery.

331. From all that has been said respecting horizontal shell-firing against ships, it is plain, that the inability of fleets or squadrons to contend with fortresses and land batteries, if they are properly armed and their guns well served, is much greater now than it was in the cases to which we have referred, in the late war. Nearly half the armament of our ships consists of ordnance neither designed for, adapted to, nor capable of encountering heavy-armed land batteries. The smashing effects of hollow shot[a] of large diameter, and the ravaging effects of horizontal shells on ships, in close and the closest action, were the objects for which shell-guns have been so largely introduced into the broadside batteries of our ships. But, in the remarkable naval war in which we are now engaged, there will, to all appearance, be no fights on the open sea, ship against

---

[a] Hollow shot ought only to be used (if ever without being loaded) at very limited ranges. At such ranges their velocity is sufficient, their *smashing* effects, from their volume, very great; and they possess that great advantage of being more readily handled than the solid shot of equal weight; but at considerable ranges, solid shot are efficient, where hollow shot would be useless.—*Simmons*, p. 28.

ship, and fleet against fleet, as of old, but attacks of fleets against fortresses and other powerful land defences. And we must observe that the heavy guns and solid shot, whose battering and penetrating powers were so great, and which formerly stood on the lower decks of our ships, and the main decks of our frigates, have been displaced by ordnance incapable, at any distance, of contending with fortresses (the reader is requested to refer to the table of relative penetrations of solid and hollow shot, Art. 242, p. 224). In fact, the force of our wooden walls, applied against stone walls, is reduced to the number of solid-shot guns which may yet remain on the upper decks.[a] Hollow shot and shells break to pieces, and solid shot frequently split, on striking walls of granite or scarps of hard stone.[b]

---

[a] It has been already stated (Art. 209, p. 191) that, since 1838, when the 42-pounder ceased to be a naval gun, the largest solid shot gun in the British Navy is the 32-pounder, whilst in other navies guns of larger calibre form the armament of the lower decks of line-of-battle ships. Thus it appears that the real battering power of our line-of-battle ships in respect to solid shot guns is not in general so great as it was during the late war, when many of our ships carried 42-pounders, and, as James states in several places, did good service with them (the "Britannia" for example, vol. i. p. 262). This reduction in the battering power of our ships is now further, very materially, diminished by displacing solid shot guns on the lower decks and arming the ships, many wholly, and others chiefly, with 8-inch shell guns.

The advocates for retaining the 42-pounder in the U. S. Navy argue that its superiority to the 32-pounder, in respect of accuracy, penetration, and the magnitude of the fractures it makes, is nearly as 2 to 1; while, for the same weight of battery, the 32-pounder outnumbers the 42-pounder only as 4 to 3: hence the relative values of the 42-pounder and 32-pounder are as 3 to 2. A higher ratio than that of the weights of the two natures of shot.

[b] L'effet des obus contre la maçonnerie est à peu près nul; ils se brisent au moment du choc, ou bien, tirés à de très-petites charges, ils ne produisent que des impressions très-faibles.—*Metz Experiments*, 1834. *Aide Mémoire*, p. 434. The like results were obtained from the Experiments of 1839 at Fort Monroe.—*U. S. Ordnance Manual*, p. 372.

From a summary of experimental practice against a four-foot square wrought-iron plate ⅜-inch thick, fixed against a solid-granite block placed oblique to the line of fire from a 32-pounder of 56 cwt. with solid shot, and charge 10 lbs., in order to ascertain up to what angle it would deflect, it appears that every shot broke or split into pieces. Thus neither hollow nor solid iron round shot will avail for destroying such a hard material. (See the Table below.) Perhaps 8-inch hollow shot, filled with lead and fired from 68-pounder guns, with charges increased in the ratio of the weight of the solid shot of 68 lbs. to that of the hollow shot filled with lead, might produce some impression on granite. The iron or shell would no doubt break, but the lead would not, and, the velocities being equal, the momentum of the blow would

ATTACK OF MARITIME FORTRESSES.

Horizontal shells, which have been shown to be so destructive when fired from ships, against ships, will be very generally used, with greater relative advantage, and more safety to the users, from *land batteries*, against ships; whilst shell firing from ships against batteries, and with percussion shells in particular, will produce little or no effect on the material of which the batteries are formed, whether earth, or blocks of granite, and still less against the *personnel* posted behind parapets which cannot, like timber, be penetrated, ravaged, or set on fire by shells.

In promulgating these opinions, and giving these cautions, upon a matter of such vast, and perhaps immediate importance, the author is naturally desirous of availing himself of the support of any influential and competent judgment expressed in other works which may inspire confidence in his conclusions; and finding in a recent publication—*Report on the National Defences of the United States*—some very remarkable coincidences of that description, the author extracts the following:—

" So far as the new projectiles are concerned, these have, relatively to ships, strengthened forts; for hollow shot crumble into fragments, and strike harmless, when be greater than that of a solid 68 lb. shot in the ratio of their respective weights.

Angle of the surface with the line of fire.	Charge.	Deflection of the fragments.
10°	10 lbs.; shot broke	4°½
12°½	,, ,,	17°½
15°	,, ,,	13°
17°¼	,, ,,	30°
20°	,, ,,	21°
22°½	,, ,,	18°½
25°	4 lbs. ,,	21°¼
27°	10 lbs.; shot broke into 4 pieces	11°
27°	4 lbs.; shot broke into 4 pieces, and many small ones	21°¼

Two rounds were fired at an angle of 30°, formed by the line of fire, with the surface of a 5⅜-inch wrought-iron plate, 4 ft. square, fixed against a 4 ft. cube mass of oak, built of squared logs.

30° Charge 10 lbs. The shot broke; some pieces went through the plate, and some deflected.

30° Charge 4 lbs. The shot broke; one half went through the plate, and upset the whole mass of oak; the other half deflected in broken pieces.

directed against stone walls. It takes solid shot, and plenty of them, rapidly discharged and concentrated upon or near one spot, to batter walls and make breaches. On the other hand, a few 8 or 10-inch shells fired from forts at ships, pass through the side of any line-of-battle ship into the main or lower deck, and, there exploding amidst the dense crowds at the batteries, every fragment multiplies itself in countless splinters of wood and iron; or, if a shell enter the orlop-deck among the men passing the powder, or, lower still, strike the water-line, large irregular splinters will be torn out, leaving openings which will defy all shot-plugs. Changing the scene to a steamer, it has been said that, compared with a sailing ship, a steamer has many more vulnerable and vital points. No! ships armed with hollow-shot ordnance will do well to prefer contending with something similarly constructed. No ship or ships can, at this day, lie under a fort having furnaces for heating shot, in addition to its murderous shells."[a]

The principle laid down in this admirable work on the national defences of the United States is that all assailable points should be guarded by forts, so as to leave the naval forces free. Forts can be made impregnable against any naval force that could be brought against them, and are needed for the protection of our fleets while preparing for hostilities on the ocean. The government and people of the United States view not with favour the substitution of floating batteries for permanent land defences, on account of the perishable nature of the former, and the inefficient state in which they may be when sudden danger menaces. The value which they might have, if in perfect order at the

---

[a] Among the various propositions made for the defence of naval arsenals and maritime places in general, floating batteries made of iron so thick as to be shot-proof have been recommended; and, in order to test the value of such constructions, a target, representing the side of an iron floating battery, was formed with seven thicknesses of boiler-iron, well bolted and riveted together. A shot from a heavy gun passed, without difficulty, through the target, and tore out large fragments.—*Report on the National Defences of the United States*, 1852, page 6.

moment of being wanted, ceases as soon as the occasion which called them forth no longer exists; and their speedy decay is certain. To leave the defence of harbours and other permanent establishments to temporary constructions so costly as ships, which are formed of perishable materials, would be to expend enormous sums in a manner which would invite attack by sea. If we rely for our defence on our naval force, no portion of it should be permitted to leave our coasts for the protection of our foreign commerce, in the event of an alarm of war occurring. To employ our active navy, in whole or in part, for defence, instead of strengthening our fortifications and raising new ones, would be to supplant impregnable bulwarks by perishable ones—a fixed security by a changeable one: it would be to expose ourselves to the chances of being suddenly left for a time without adequate defence. In so doing we should resign our sense of security, and our confidence of safety; we should divert our navy from its highest duty, deprive it of its chief honour and its chief claim to the respect and support of the people: we should lose the power of vindicating the national honour and independence, and of asserting the freedom of the seas. The navy is not a defensive, but a protecting force.

332. The attack of fortresses and powerful land batteries with a naval force only, must ever be a hazardous, and perhaps desperate, undertaking. But if skilfully combined with a military force sufficiently strong to make good its landing, to invest the place or the batteries on the land side, to take the defences in reverse, and so open the way to the attack by sea, the object of the attack will in general be successful. But this mode of proceeding can only be applied when the place to be attacked occupies a position, insular or otherwise, of such extent as to admit of being attacked by land as well as by sea.

In combining military and naval operations of this description, the first and main difficulty to be encountered is to effect a landing and establish a lodgment on the enemy's coast, in the face of a large military force, which ought always to make the most determined efforts to oppose a debarkation or prevent a lodgment from being made good; for, as in the assault of a breach, and in forcing the passage of a river, if a solid lodgment be once established on the crest of the one, or on the further side of the other, a fulcrum is obtained which, if skilfully used, and supported with sufficient means, will ensure the success of the enterprise.

A very large army may now be transported with great speed and convenience in a very few large steamships to any seat of war, however remote; but to transfer 1000, 1500, or 2000 men from the transports to the shore is a work of considerable time, and requires great numbers of boats, specially constructed, for that purpose. This preparatory operation cannot be attempted or executed under fire from the enemy; and, therefore, the troops intended to force a landing must be embarked in the boats which are to take them to the shore, whilst the transports are anchored at a safe distance. The success of the operation will mainly depend upon the nature of the locality that may be chosen. It should not be too near to the fortress or stronghold to be attacked, because, in this case, the garrisons of the forts or fortresses might safely co-operate with the force in the field, to oppose the landing and attack the lodgment. Nor should the point of debarkation be too distant from the great objective of the expedition, because that would necessitate a long march to invest the place, and much difficulty in getting up the siege-train and stores.

If the enemy (exclusive of the force in his garrisons) is not strong in the field, it might be advantageous to endeavour to seize some capacious bay or inlet capable of affording shelter to the numerous ships, vessels, and small craft, and near which a fort might be constructed to serve as an entrepôt and base of operations; but these great objects can rarely be effected immediately; indeed, if the enemy has occupied and strengthened the localities, and if he is, moreover, strong in the field, it would not be prudent to attempt a landing there. In this case some point, deemed apparently by the enemy of minor importance, should be sought for—some promontory, with a nearly level surface, and remote from high lands, having also water about it of sufficient depth to permit the boats to arrive at the beach, and to enable bomb-ships, steamers, and gun-boats to cover the advance of the flotilla containing the troops, support their landing, and protect the lodgment they may form. Having thus obtained a footing, and received such increase of strength as may be deemed necessary, including field-artillery, the whole force should move forward to meet the enemy in the field, and conquer for itself some position which may afford shelter to the fleet, and become a tête de débarquement and base of operations to the invading army. In forcing the passage of a river, the operation is undertaken, if possible, in a sinuosity re-entering, with respect to the invaders, and a lodgment is made upon the opposite salient in the enemy's position; the whole interior of that position is commanded from the points in the possession of the assailants, and consequently the lodgment to be made is capable of being supported and protected. (*An Essay on the Principles, &c., of Military Bridges*, by Gen. Sir Howard Douglas, Arts. 157, 159, 3rd edit.) In like manner, in order to obtain a footing on an enemy's coast, a low level promontory or salient should be chosen, because ships on each side of it may perform the same office (commanding the opposite ground) as, in forcing the passage of a river, is performed by the batteries placed at the two salient points which contain between them the re-entering sinuosity. The ships are thus enabled to support the lodgment on the coast and protect the flanks of the troops which have gained the shore. To attempt to force a landing in a bay, reverses these conditions, for the shore of a bay, unless it be very extensive, cannot be held, nor even approached, until both the promontories which contain it are occupied.

332 a. When the Duke of Wellington, then Sir Arthur Wellesley, invaded Portugal in 1808, it was a favourite object with the ministry that the descent should be made at the mouth of the Tagus. Wellesley

decided otherwise, and made choice of a landing-place remote from Lisbon, in order to avoid the danger of a debarkation in face of a large force. He effected his landing at a part deemed by the enemy of minor importance, the mouth of the Mondego River : he moved forward as soon as he could, fought a general action, gained a complete victory, and obtained possession of Lisbon. (Napier, *History of the War in the Peninsula*, Book I., ch. 4.)

332 *b*. For great operations of this description any want of mortar-ships, gun-boats having small draught of water, and flat-bottomed boats for landing the troops, would be seriously felt. All the landings of troops in the face of an enemy, in the course of the late war, at some of which the author served, were conducted in the following manner. The troops intended for debarkation being placed in the boats out of fire of the shore, were directed by signal to form line abreast on points marked by men-of-war's boats, carrying distinguishing pennants, and containing the naval officers charged with the direction of the several divisions of the flotilla, and the whole was placed under the superintendence and command of a naval officer of rank. When the line was formed, the whole moved forward by signal, rowing easily, the better to keep in line, until within the reach of musketry from the shore, when orders were given to row out. The whole of the operation, from its commencement, was covered by bomb-ships carrying 10 and 13-inch mortars, and these protected the advance of the troops by firing shells, when necessary, over the line of boats, in order to reach the beach ; a like firing, with increased charges, being directed against the enemy's supports in rear of the troops disputing the landing : at the same time gun-boats drawing little water, placed on the flanks of the operation, scoured the beach upon which the troops were to land. Whilst these operations were being executed, the fleet of line-of-battle ships remained at a distance in reserve, unscathed and ready to take their part in the ulterior operation when the proper time arrived.

PART III.   ATTACK OF MARITIME FORTRESSES.   351

When, in 1801, the British army under Sir Ralph Abercrombie arrived in Aboukir Bay, and the weather, at first tempestuous, became calm enough to permit the troops to land, General Abercrombie, who had himself reconnoitred the coast in a small vessel, gave orders for the first division, consisting of 6000 men, to prepare for landing early on the following morning (March 8). The preparations could not be made, however, without attracting the notice of the French; and these disposed themselves, with a numerous force of infantry, cavalry, and artillery, to prevent the invaders, if possible, from gaining the shore. The "Fury" and the "Tartarus" bomb-vessels, with sloops and gun-boats, were appointed to protect the landing of the force, and, though they suffered severely from the fire of the French, the troops succeeded, though with difficulty, and only in detached parties, in making good their landing. The enemy retired, and, on the 21st of the same month, the battle of Alexandria, in which Sir Ralph Abercrombie fell, took place.

332 c. When the place, fortress, or arsenal to be attacked is covered and protected by isolated points of defence, mutually protecting each other, and when no previous military operation can be made, those points or outposts should be attacked in detail and successively reduced; after which the fleet may arrive at, and attack the main position. This must evidently be a protracted and difficult process even with such means: with ships alone it cannot be effected without severe loss and damage; and it should always be remembered, that many of the attacking ships would be severely injured, probably disabled, in the attempt, whilst the enemy's fleet would remain untouched and in reserve. It would therefore follow, that the attacking fleet must be exposed to a very disadvantageous action with the enemy in the event of the latter subsequently leaving his place of shelter.

In the Walcheren campaign—ill-fated, unfortunate expedition! yet in this respect instructive—it was at first intended that the fleet, consisting of 10 sail of the line and 11 first-class frigates, should at once force the passage of the West Scheldt, and cannonade Flushing. But Sir Richard Strachan very judiciously determined to remain in the offing until the place should be invested on the land side, and the batteries of attack ready to open. Thus the place, not having been invested on the sea-side, became a *tête* to Cadsand, from whence reinforcements to a large amount were sent over to Flushing, until the frigates had forced the passage and intercepted the water communication between Cadsand and that town. When all was ready, a powerful and irresistible combined attack was made upon the place by the naval and military forces, and it surrendered the next day. Neither of the forces, however considerable, could have succeeded singly in reducing the place. No military force, without the co-operation of the fleet, could have reduced the *tête*, which was capable of being supported to any extent by sea; and the naval forces, without the co-operation of the army, would have been crushed by the artillery of the place, exclusively directed against them.

When the fortress or arsenal to be attacked is situated on a coast which may be approached from the open sea in any direction, steam-ships may avoid the danger of a direct attack, end-on or oblique, by approaching the place on either, or perhaps on both sides; and, having gained the proper proximity, clear of raking or diagonal fire, may range quickly up in parallel order, to attack the place in line or lines. So, in steam warfare, ship against ship or fleet against fleet, direct advances upon the broadside batteries of ships, may, upon the same principle, be avoided, and the enemy be attacked in parallel order, by ranging up to him, if the attacking ships are superior in speed, and forcing him to fight.

But, when the fortress, arsenal, or place to be attacked, is only approachable by a narrow and intricate channel, through which ships can only pass singly, or nearly so, there can be no manœuvring for position. There is no way of avoiding being met, first by direct, then oblique, and ultimately by raking fire from the batteries that defend the channel; and steam can only perform its office of propulsion into or through the intricacies, and under disadvantageous and hazardous circumstances. Steam-ships might, indeed, run past any advanced or covering batteries at full speed, without being much damaged; but it would be extremely perilous to leave such forts unsilenced in their rear; and, unless the daring enterprise should succeed, like Nelson's at Copenhagen, to produce a cessation of hostilities, the fleet, or at least any disabled ships, could never get out again.

However successful a naval attack of a fortress or arsenal may be, the work of destruction can never be effectually accomplished by ships. The sea defences may be silenced, guns dismounted, parapets ruined, magazines blown up by mortar shells, and habitations devastated by the cruel process of bombardment; but no substantial demolition of the defences, or material destruction of public works and property, can be effected unless the damages inflicted by the attacks of ships be followed up and completed, by having actual possession of the captured place for a sufficient time to ruin it entirely. No naval operation, however skilfully planned and gallantly executed, can, alone, reap the fruits of its victory.

In the desultory operations of small active steamers, employed, with their pivot guns, to shell open towns, roadsteads, harbours, and slender buildings, magazines, stores, &c. &c., or to shell bodies of troops on shore, the attacking vessels should never anchor, but, having given their end-on fire, go off at speed to reload, and prepare to take up the fire in turn with others, whenever they regain a favourable position for a good effect. To hit a steamer running with speed across a line of fire is no easy matter; and, when in the end-on position, she presents but a small target to be hit at a long range.

But the attack of such comparatively insignificant places by horizontal or howitzer shell-firing, is a very different affair from the regular attacks of formidable fortresses, naval or military stations, or other defences solidly constructed, and efficiently armed and manned, by the broadside batteries of line-of-battle ships, or

2 A

from their bombardment by sea-service mortars which still figure in the list (Table XVII., Appendix D) of ordnance in use in the Naval service, though, for using them, no bomb-ships exist. There can be no skirmishing with huge line-of-battle ships. When these are employed to attack fortresses, they must do so at anchor, after they shall have attained favourable positions in sufficient proximity. They then become stationary floating batteries, large targets to fire at, and scarcely can be missed. This will assuredly invite retrospect to the defence of Gibraltar, and the destruction of the Spanish floating batteries, by red-hot shot; nor will that example be lost on the great power with which we are now at war. The Russians will, doubtless, have recourse to red-hot shot as well as shells, in defending their great naval stations.

333. The disuse of mortars in the British navy is founded upon the belief that sea-service howitzers will supersede them in naval bombardments; while mortars are retained in the land service, as indispensably necessary in sieges and bombardments, for important purposes which cannot be effected without them, whether the attack be made by land or by sea.

We have seen, p. 166, that, in 1851, a range of 4866 yards was obtained with an excentric shell fired from a 10-inch gun at an elevation of 28°; and that in 1852, p. 168, a range of 5860 yards was obtained with an elevation of 32° from a new 10-inch gun of 116 cwt. cast for the purpose of those experiments. But these great elevations are so trying to the gun (at the round next after that which produced the maximum range the gun burst), and so straining to the ship, unless she is so well trussed in her framework as to give her longitudinal strength to bear the pressure of so heavy a gun placed over a fine bow, or at the stern over a clean run, that these high elevations cannot easily and safely be used.[a] These guns may be found useful in

---

[a] This has been amply verified by the bursting of one of these guns at Shoebury Ness, 12th October, 1854.

" shelling " towns, roadsteads, open batteries, and troops on shore at great distances; but how far they are fit to supersede mortars in the attack and bombardment of large fortresses and arsenals, is another question. In this case small elevation is a disadvantage; there is too much horizontality in the trajectory to produce the crushing effects of bomb-shells. The shell wants that tremendous force which can only be acquired by the descent of mortar-shells fired at the elevations which produce their greatest range (42° or 45°), when they fall with force sufficient to penetrate into any building, and finish their course with the destructive effects of a mine. Shells fired horizontally from any ordnance, however useful in shelling places of minor importance, are not efficient substitutes for mortar-shells for purposes of bombardment. A bomb-ship may, without much exposure, do great damage to an extensive fortress or arsenal, which, being a large object, ought to be struck at every discharge at upwards of 4000 yards, whilst she is a mere speck on the sea at that great distance. (*Ward*, p. 48.) The French mortars did good service in the attack of Bomarsund by land: the powerful effect of the descent of their shells into the forts was particularly noticed. Sir Charles Napier reports (19th Aug.) that they never missed, and contributed largely to the reduction of the place. Then why did not the French send some of the *Bombardes*, which they wisely retain in their Naval service, to the attack of Cronstadt and Sevastopol? In the projected bombardment of Sevastopol mortar-ships would be of more use than the Despatch gun-boats, by throwing bomb-shells of 100 lbs. weight into the arsenal over surrounding heights, which cover the dockyard from any horizontal firing, and screen the arsenal from being seen.

334. The dimensions of the vessels now being built as *Despatch-Boats*, each of which is to be armed with two Lancaster guns of 95 cwt., are, in length 165 feet; extreme breadth 25 feet; their tonnage is 450, and their draught of water 11 feet 4 inches. They are propelled by screws, and their horse-power is 160. The guns are

mounted on pivot carriages and slides, in the body of the ship; one abaft, and the other before the funnel. In voyaging, the guns are housed, longitudinally, in the middle of the deck.[a] For action on either side, the slide carrying the gun is made to traverse upon a rear pivot, to take up the fighting bolt on the side to be engaged, and then turned into the position for broadside action. The ports are sufficiently wide to admit of the guns being traversed upon their fighting points, to about an angle of 56° before, or abaft the beam, forming altogether a sector of 112° on the horizontal plane, upon which the guns may fire; so that there remains a sector of 34° on each side, on which there is no fire, which leaves a dead sector of 68° a-head and a-stern, upon which the guns cannot be brought to bear, a defect which would not be at all compensated by installing 4⅜ inch brass howitzers in the bow ports. The arrangements for turning the guns into their fighting positions appear to be very complicated. To turn the gun into action, on either side, the rear housing-bolt is first disengaged from the socket, and the traversing platform turned upon its front housing-bolt, to take up a rear turning point established on the deck, on that side of the middle line of the ship on which she is to be engaged, and so placed as to be equi-distant from the front housing-bolt and the fighting point on that side: the former centre being then disengaged, the platform is turned upon the rear point, to take up the fighting bolt in the centre line of the port, when, the rear point being

---

[a] It is no longer a secret that gun-boats with heavy ordnance of Lancaster's construction are to be sent forthwith to the Baltic Sea; since we read in the *Times* of the 21st of August,—" The 'Arrow' is the first of the despatch-boats so called, four of which were lately constructed by Mare and Co., and two by Messrs. Green, of Blackwall. The 'Arrow,' 'Beagle,' 'Lynx,' and 'Snake' are by the former, and the 'Wrangler' and 'Viper' by the latter firm. They are each of 160 horse-power, mounting 10-inch and 8-inch Lancaster's new oval rifle-guns and 12-pounder brass howitzers. Their great rifle-guns are mounted on pivots amidships, and are capable of firing on a line with the keel on each bow. They are rakish-looking craft, with three masts, schooner-rigged. The 'Arrow' has been sent to Portsmouth for the inspection of the Lords of the Admiralty, now at that station, who will make some experiments with her new and peculiar armament to-morrow, in the presence of Prince Albert, who will be on board."

disengaged, the gun is ready for action. Thus two operations are required to move either gun from its housed position, into that for action, and the same operations, in the reverse, to re-house the gun. To shift the gun from one side of the ship to the other requires therefore four operations; whereas if the rear housing-bolt were placed equi-distant from the other and from both fighting points, the gun might be simply turned into position for action on either side, or wheeled on that centre from one side to the other without difficulty, and in much less time.

335. In the Report of experiments recently carried on at Shoebury Ness (see *Times*, August 5, 1854), it is stated that "Lancaster's gun (a 68-pounder, 10 feet long and weighing 95 cwt.) projected elongated shells weighing 88 lbs. at an elevation of 17° to the distance of 5600 yards, and that none of the shots fell wide or short of the target." The greatest range obtained on that occasion was only 4500 yards; a charge of 12 lbs. instead of 16 lbs. having been used, from some distrust, we believe, of the strength of the gun to resist the full charge,[a] and likewise in order to reduce the impulsive force, which might otherwise have broken the shell; but notwithstanding this diminution of charge, one of the shells broke soon after it left the gun. Though none of the shots fell wide or short of the target at the long range, none fell very near it. These experiments not having been made, however, for the ordinary purposes of practice, but for particular objects which are not disclosed, the powers of the gun, and the quality of the practice, must not be judged by the above results.

The Lancaster elongated shells are made of wrought-iron, on account of the frequent breakings of cast-iron shells which took place in the former experiments (Art. 190); but it appears, from what is stated in the Table, p. 171, and in the present article, as well as from similar failures in the trials made on board the "Arrow" Despatch gun-boat at Portsmouth, on Tuesday, August

---

[a] See Appendix H, on the recent bursting of a Lancaster gun at Shoebury Ness.

23, that wrought-iron shells are likewise extremely liable to be broken by the prodigious force with which they are driven through the spiral of the elliptical bore.

It is moreover a matter of no small difficulty to make a wrought-iron elongated elliptical shell. The body, without the bottom, is easily forged and formed; but to complete the shell, by welding the bottom firmly to the body, is not easily executed, and it appears that this is the main cause of failure. The cost of manufacturing one of these shells is at present something enormous. The formation of the elliptical bore is likewise a nice and costly operation, which enhances considerably the price of the gun over that of a 68-pounder. Wrought-iron shells do not, like those of cast iron, break into many pieces by the explosion of their bursting charge, but they are apt to open at their weakest parts. Also, the percussion fuze, being at the apex of the shell, would, it appears, act before the body of the shell is imbedded in the material which it is intended to injure or destroy; and, when a wrought-iron shell is fired against a hard body, as granite, the point yields to the blow, and is blunted or doubled back, by which the impact of the shell is deadened, and the penetrating power is diminished.

336. It is no doubt a serious, and may be a fatal defect in the armament of these vessels, that, whether chasing or chased, they cannot bring their guns to bear upon the enemy's ship but by means of the helm, which would have the effect of losing distance from the vessel pursued, or from that pursuing. Fighting-pivots are indeed established at each of the bow ports; but besides the difficulty of transporting a gun, which, together with its carriage and slides, weighs 7 or 8 tons, it is the opinion of many experienced naval constructors that the fine and lean bows of these vessels have neither displacement nor longitudinal strength, however well trussed, to bear that weight, far less to resist the shock of an elevated discharge of the gun. Sensible of this, an expedient of a novel and somewhat startling character has been proposed—possibly for serious emergencies which cannot otherwise be met—

PART III.   ATTACK OF MARITIME FORTRESSES.   359

viz. to establish other fighting points on the decks, so that the guns may be turned into positions parallel to the keel, and so fire a-head over the bows, or a-stern over the taffrail, as the case may require. But it is clear that this cannot be done unless the guns are fired at high elevations, especially the gun firing a-head, on account of the rise of the bow of the ship; and consequently this expedient is not available when the enemy is near. The pivot of the foremost gun being about 60 feet from the bow of the vessel, the blast from the discharge of the gun would sweep over all the fore part of the deck : a shot might pass close over the head-rails, without damage to the ship; but it remains to be seen whether the bulwarks would not be injured or blown down, unless the gun were fired at its highest elevation, and whether there would not be much danger of setting fire to the rigging, hammock-cloths, or sails, though brailed up or furled, by sparks being driven into the foldings or other crevices. The aftermost gun is about 100 feet from the taffrail : fired " over all" astern, its discharge would sweep with equal danger and violence over the after part of the deck—in both cases there must be no men nor ammunition placed in the way of those explosions. How then would it be with the helmsmen ? This expedient for enabling the vessels to fire a-head or a-stern in the line of the keel, cannot be practised, nor would it be prudent to attempt such firing in action. These vessels can therefore only fight in the broadside position—an utter abandonment of the peculiar advantages of steamers. By extreme training of their guns they might fire at angles of 34° with the keel; but they would then expose to the enemy a surface expressed by the area of the broadside, reduced in the ratio of unity to the sine of 34°; or a length of about $89\frac{1}{2}$ feet ($160 \times \sin. 34°$) ; and thus would form a target three and a-half times larger than the area of the transverse section.

These Despatch-boats are very beautiful, fine, and speedy-looking craft; but they are greatly overloaded, with respect to their displacement and form, by the enormous weight of their deck-load, which, consisting of two guns, each of 95 cwt., with their massive carriages and slides, cannot be less than 14 tons! With such a top-weight they must roll excessively in any sea; which will interfere much with their speed in voyaging, with their stability in action, and with their safety in a heavy sea. When tried, they will, no doubt, confirm the apprehensions which are very generally entertained, that such heavy armament will detract greatly from the essential quality of a despatch-vessel, and make a very bad gunboat. Perhaps it may be said that these vessels were not designed for the ordinary service of steamers (which in every case requires that they should be so armed at their ends as to make what would otherwise be their weak point their strongest), but are intended to use the enormous guns with which they are armed for broadside firing, to batter granite walls at great distances. It were much better to make the experiment against a granite wall, in some actual siege operations by land, than from the battery of a ship at sea. To hit a wall with any projectile fired at 18° of elevation, at the distance of 5000 yards, is a very remote probability, even when the gun is fired on solid ground; from a ship the practice must be infinitely more at random, when there is any swell; and here we would observe that the deviation to the right, occasioned by the rotation of the projectile in that direction, is largely developed in practice from the Lancaster gun, with elongated shells (the average amount of the deviation is 150 feet at 5000 yards). This ought to be allowed for, in proportion, at any range; but the deviations not being constant, they cannot be accurately estimated; and, as many readers are apt to believe that practice made on shore, or in harbour, may be taken as a criterion of what it will be at sea, we would particularly refer to what is stated in Art. 343 on that subject. But if these vessels were designed, and armed with such heavy ordnance, for battering purposes, they will rather have to serve as stationary floating batteries than as active steamers to shoot flying; and therefore, instead of being built for speed, they should have been formed for stability and steadiness. The alternate recoils and running up of guns of such vast weight, on the sides of vessels of their size and form, must occasion a rolling motion which will never cease, however still the water. The motion may be insensible to the eye, and yet such as to destroy entirely the accuracy of practice, in which deviations of half a degree in the elevation will make a vast difference in the range. A skilful seaman-gunner knows well that a rolling motion is more disturbing to good gunnery than a pitching motion; that the former does not affect the length of range from a pivot-gun at bow or stern, though if there is much of that motion it may occasion lateral deviations; that very small rolling motions have great influence on the range in broadside firing; and that a swell a-beam, or a cross sea, produces very little pitching, but a great deal of rolling motion.

The aberrations in gunnery practice arising from the floating motions of a ship, affect vertical and horizontal shell-firing very differently. If the shell of a mortar whose greatest range is at 42° of elevation, depart from the muzzle at any greater or smaller angle, the range is shorter in both cases, and equally so if the error, plus or minus, be equal; but small differences of elevation are of very little moment in a range of 4200 yards, and the shell would scarcely miss the area on which it is intended to fall; for its horizontal velocity, never very great, is nearly exhausted, and the velocity of descent is very great; but in horizontal shell-firing, errors of the same amount taking place above or below the angle of elevation at which it is intended to fire, produce very great differences in the range—there is so much horizontal velocity that the shot passes over large spaces in very short times; there is very little vertical velocity and less chance of the howitzer shell falling within a given area than a mortar shell, both being subject to the same amount of disturbance in the intended elevation.

As gun-boats these vessels are far too large, their draught of water much too great; and, in general, all our ships and vessels in the Baltic are disadvantageous in these respects. Ships drawing from 24 to 27 feet of water have no business there; while there is not a sufficient number of small ships to enter creeks or inlets, or to approach the batteries on the coasts, without getting frequently aground. No doubt it is necessary that there should be a fleet consisting of a sufficient number of line-of-battle ships to prevent the enemy's fleet from coming out, and to beat them, should they endeavour to disturb the desultory operations of the squadrons of the allied fleets which may be engaged in blockading and other detached services; but for this, huge vessels of great draught of water are rather hurtful than serviceable. The ships of a fleet, acting in shallow seas and intricate channels, imperfectly surveyed, should be as much as possible of the same draught of water: if not, and even if one or two draw some feet of water more than the others, the operations of the fleet must be as much impeded by shallow water, if they are to act together, as if all the vessels of the fleet were as deep in the water as are the one or two larger ships.

337. The Russian ships of the line—and the same may be said of those belonging to Denmark and Sweden—do not exceed in draught of water 23 English feet; that of our ships is from 25 to 27 feet, and in some still more. England should always be provided with ships suited for service in all seas. Our old 74-gun ships were just the class of line-of-battle ships required for service in the Baltic, but they are now extinct.[a]

Referring to the results of the experiments of the 22nd August, at Portsmouth, the author has not found it necessary to alter a tittle of what he had previously

---

[a] The gallant and distinguished Vice-Admiral commanding the combined fleet in the Baltic must now be feeling the full force of what he wrote in 1849 (see p. 192 of his recent publication). After extolling the old 74's, and condemning their extinction, he goes on to say,—" We seem to have quite forgotten that there are such seas as the North Sea and Baltic, where small ships on two or three decks, with a light draught of water, are indispensable. There are also many places where the like qualification is necessary; and I well recollect that, during the siege of Cadiz, we were obliged to look up all our old 64's, with a light draught of water, so that they might lie clear of the enemy's shells. Another thing we seem to have forgotten—that all the work of the war was done by the 74's, and that the large ships did not stand the blockade so well as the small ones."

PART III.   ATTACK OF MARITIME FORTRESSES.   363

written; and has to add the fact that the rolling of the "Arrow" was excessive, whilst other vessels were comparatively at rest. These vessels are universally considered to be too narrow and crank to carry such a heavy deck-load. So much of their volume is absorbed by the engine, that there does not remain sufficient available capacity for the stowage of provisions and stores. The gun maintained the character given of it in Art. 190 for length of range; but the practice, always uncertain at such distances, was extremely wild and random, from the motion of the vessel. Two shells broke soon after leaving the gun, thus affording additional proof that even wrought-iron shells cannot be depended upon; and, unless this serious defect can be effectually remedied, it must be considered a great objection to the adoption of the gun; for, considering the enormous cost of the shell and the vast expense of the system, one lost opportunity of obtaining the full effect of a discharge, from the breaking of the shell in or near the gun,—of which so many instances have recently occurred,—might be productive of the severest disappointments in action. How this defect is to be remedied, appears to be a very difficult and doubtful matter. But however this may be, the difficulty has not been overcome, as the late experiments show; and the Despatch gun-boats, constructed expressly to demolish the enemy's defences in the Baltic and Black Seas, with elongated shells of 100 lbs. weight, fired from Lancaster's elliptical guns, were actually sent to their destinations, provided, it appears, with only 25 of their own peculiar oval shells (*Times*, Oct. 16), on the much-vaunted efficiency of which the whole of the Lancaster system depends, but largely supplied with 68-pounder round shot, with which it is said the Lancaster guns make good practice! It was not for this that the enormous expense of the Lancaster system has been incurred, or can be justified; and if there be any truth in all that has been said, written, and done, upon the subject of windage, the principles upon which it is regulated, its annular uniformity, the important condition of a perfectly cylindrical bore, and the rigorous condemnation of any gun in which the smallest irregularity in the surface of the bore can be detected, by the nicest instruments, it is utterly impossible that the anomaly of firing spherical projectiles from guns having elliptical spiral bores can be precise; and this incongruity is therefore repugnant to science, and absolutely fatal to the Lancaster system, as applied to guns.

There never yet was a gun-boat without a powerful gun at the prow, or which could not fire in the direction of her keel. It was therefore a great mistake to call these vessels *Despatch gun-boats*; they are *steam-vessels* armed and fitted in a novel and peculiar manner, for experiment; and much is it to be wished that the experiment, so far unsatisfactory, may in the end be successful; but believing, for the reasons stated, that these vessels will not altogether realise the expectations which the public in general are led to entertain of their services as *gun-boats*, the author deems it his duty thus fully and frankly to state the facts and reasons which have led him to this conclusion.

The smaller class of vessels now fitting out as gun-boats come fully up to the author's notion of what a good steam gun-boat should be. Their dimensions are— length 100 feet; extreme breadth 22 feet; depth of hold 7 feet 10 inches; tonnage 212; engines, two of 30 horse power each; draught of water at load-line 6 feet 6 inches; armament two 68-pounders of 95 cwt. These vessels are lugger-rigged, without bowsprit; they are sufficiently full at each end and abundantly strong to bear that heavy gun, at either or both ends, for action. In voyaging, the guns are housed in the middle of the vessel, and may either be turned into action on the broadside or at either end, for which ports are made at the bow and at the stern. The only blemish is the fittings for the dangerous and objectionable project (Art. 336) of firing the guns from the body of the vessel "over all," a-head or a-stern; and arming them with Lancaster guns instead of 68-pounders, before efficient projectiles congenial to the principles of his system shall have been produced, and at such a moderate cost as would justify a general use of them, but which, confessedly, is not the case at present.

338. With respect to the Armament of Coast Batteries; wherever a battery is so placed as to command long sea-ranges, and, if properly armed, to open its fire upon an enemy at great distances; or, from the nature of the sinuosities of a coast, to be in a position to enfilade an enemy's fleet or squadron advancing to the point of attack which that battery is established to defend, there assuredly should be placed, exclusively, in positions which demand the service of guns possessing, in the highest degree, power of range, accuracy, and penetrating force, powerful 32-pounders, 56-pounders (Monk's), and 68-pounder guns. These have superseded shell-guns, as pivot-guns in our service; and, in the United States Navy, the new 64-pounder unchambered gun has displaced their 8-inch shell-gun.

In batteries which do not command long sea-ranges, which are not in enfilading positions with respect to an advancing force, and are so situated in the scheme of defence, as to serve for the flanking defence of collateral works, or whose seaward range is only open to distances at which shell-firing may be used with effect, there shell-guns, or other howitzers, or 68-pounder carronades, may with advantage be placed.

339. The British shell-guns, as analogous ordnance to the French canons-obusiers (Art. 215), are well adapted to the sea-service; the British 8 inch-gun of 65 cwt. is superior to the French canon-obusier No. 1 (Art. 224, and Note), and is perhaps the best chambered gun ever produced, here or elsewhere. But, as has been shown in Art. 218 and the Note, great objections attach to all chambered guns, which, though, in part, obviated by the expedient of the cork wad (Art. 411), are not removed—the defects being inherent in their construction. Chambers are, however, necessary evils in the naval service, in ordnance of such large calibre as 8 inches and 10 inches; because, if their bores were cylindrical throughout, they would be too weak

round the seat of the powder to stand the explosion of even a very diminutive charge; and it would therefore be necessary to fortify an unchambered gun of that calibre by casting it at least 30 cwt. heavier than it now is. But this additional weight, for vessels of equal "displacement," would disable them for carrying an equal weight of ordnance without greatly reducing the number of guns; and, in the vessels, there would not be space, even if they could carry the same number of the heavier gun, to stow the additional number of men required to work such heavy ordnance.

But why impose the evil of chambered ordnance on the land service? The cork wad must, in that case, be adopted in the land, as well as the sea, service, for without it there would be frequent miss-fires.[a] Objections to weight as limited by "displacement," and want of space for the accommodation of additional men required for working the guns, do not attach to the land service; and consequently the question, cleared of those conditions, rests solely upon the comparative merits, faculties, and capacities of the chambered and unchambered guns, as best adapted to the distinct purposes for which they were designed—the one to fire shells or hollow shot, within limited distances—the other to fire solid shot with superior power of range, accuracy, and penetrating force, at great distances.

340. To place in *fixed* batteries chambered 8-inch guns, in lieu of unchambered solid-shot guns, on account of the difference of weight, since neither displacement, transport by sea, nor conveyance by land enters there

---

[a] "In 8-inch guns with conical chambers, to insure the reduced cartridge being set sufficiently home it is *most essential* that spherical cork cartridge tops of $5\frac{1}{4}$ inches diameter and $2\frac{1}{4}$ deep, should be fitted inside these cartridges when sent on board, without which the guns *will frequently miss-fire.*"—*Instructions for the exercise and service of great guns and shells.*

But no cork or other tapering wads are provided for this ordnance in the land service (fig. to Art. 411), where there is more need of that expedient than in the sea service, because land batteries, which should always have some command over the object or plane in front, are necessarily often placed in such elevated positions for coast defences as to require great depression in firing at any object upon the plane of the sea, by which a reduced cartridge would be far more liable to slide from its place in a conical chamber than in horizontal firing, and thereby occasion miss-fires more frequently than on board a ship.—AUTHOR.

into the question, were a grievous error, which, much we fear, may have been committed, particularly with respect to 8-inch shell-guns of 50 and 52 cwt.—a very inferior class of that tribe—seeing that more than one-third of the ordnance appropriated to the armament of our coast defences are shell-guns, and that so large a proportion can scarcely be required for shell firing only.

Extent of range, with the least elevation, the greatest accuracy at long ranges, and penetrating force, are pre-eminently required of guns appropriated to batteries, wherever they command long sea-ranges, or are in enfilading positions.

No law of gunnery is more clearly demonstrated and irrefutable than this—that elevation is, inversely, the exponent of accuracy. No scientific or experienced artillerist will dispute this, and no practitioner of the science can deny it, until he shall first have demonstrated that this well established law is untrue, and shall cause it to be erased from the codes in which Robins, Hutton, and all professors of the science of gunnery have stereotyped it for our guidance.[a]

" The gun that makes the greatest range with the least elevation, and consequently with the greatest horizontality in the flight of its shot, is, assuredly, the most accurate in its practice, and the most destructive in its effects. Sea-service howitzers do not possess this property in so great a degree as solid-shot guns; for the former require higher elevations to attain the same ranges, and accuracy and effect are sacrificed accordingly." (Monk.)[b]

It appears by Tables V. and VI., Appendix D, that the range of a 32-pounder of 56 or 58 cwt. at an elevation of 1° is very nearly equal to that of a 10-inch gun at 2°, and about 200 yards more than the range of the

---

[a] The uncertainty of striking objects increases considerably with the increase of range, for it requires corresponding increase of elevation.—*Straith, Fortification*, vol. ii. p. 227.

[b] The accuracy of the solid shot will be much greater, both on account of its greater velocity, and because the hollow shot, to obtain the same range, must have greater elevation.—*Simmons*, pp. 26, 27.

8-inch gun (65 cwt.) at 1°; the range of the 32-pounder at 2° is greater than that of the 10-inch gun at 3°, and about equal to that of the 8-inch gun at 3°. Again, the range of the 32-pounder at 3° is fully equal to that of the 8-inch gun at 4°, and considerably greater than that of the 10-inch gun at 4°. At 4° the range of the 32-pounder gun is about equal to that of the 10-inch gun at 6°, and nearly equal to that of the 8-inch gun at 6°; the range of the 32-pounder at 5° is very nearly equal to that of the 8-inch gun at 7°, and fully equal to that of the 10-inch gun at 7°. Again, in more distant firing, the 32-pounder of 56 or 58 cwt. at 10°, ranges very nearly as far as the 8-inch gun at 13°, and quite as far as the 10-inch gun at 14°. At 1600 yards the 10-inch shell-gun would be firing shells or hollow shot at about 5° with a full-service charge of 12 lbs., the 8-inch, at about 4°, whereas the 32-pounder of 56 or 58 cwt., with a charge of 10 lbs., would be firing solid shot at the same distance at about 3°. At 2000 yards the elevations for the 10 and 8 inch guns would be respectively $7\frac{1}{2}°$ and 7°, whilst the 32-pounder would be firing its solid shot at a little more than 5°. See likewise shell and shot experiments of 1838, "Excellent" (Art. 262, Note).

The results given in those tables are conclusively in favour of the 32-pounder of 56 or the new one of 58 cwt. The 56 and 68 pounder guns have a still greater superiority over the shell guns; and, therefore, a grievous error would be committed in preferring the latter to the former for the armament of coast batteries in general.

341. The comparative values of solid and hollow shot, and consequently of unchambered and chambered ordnance, have been well laid down by Lieut.-Colonel Burns, of the Royal Artillery, on his 'Cards'—excellent codes—for the guidance of the practical artillerist.

Why has the whole tribe of carronades, which at first were so victorious (Art. 148), disappeared? because they are chambered ordnance, which, though capable of firing solid shot, are incapable of doing so with certainty and effect, excepting in close action; and when properly opposed by solid-shot guns, at long ranges, were overpowered, and ultimately beaten out of the service. It is not meant by this to disparage shell-guns by comparing them with carronades, but to show the defects of chambered ordnance as shot-guns. (Arts. 123, 218.)

The unfortunate selection of chambered instead of unchambered ordnance of the same calibre, in 1812, cost the allied army, under the Duke of Wellington, nothing less than the failure before Burgos. Had the so-called 24-pounders, which, being chambered, were only 5½-inch howitzers, been really 24-pounder guns, Burgos would have been taken.

It is commonly said that 8-inch shell-guns of 52 cwt.—an inferior and inefficient class of shell-guns, of which vast numbers have been provided, but which are rapidly and justly falling into disfavour and disuse in the naval service (Art. 215)—form a large portion of the siege-train for service in the East; not, it is hoped, to interfere with the usual proportion of the good old 24-pounder—a capital siege-gun—but to be used as howitzers; for they are incapable of serving with efficiency as battering ordnance (excepting against earthen works, *Bousmard*, vol. i. pp. 97, 98, 279, 283), or for ricochet firing, with shot; and they are very inconveniently heavy howitzers for siege service. In siege operations howitzers are not usually required till the siege is far advanced; they are used to ricochet the covered ways, to shell the places of arms, and when the place is about to be assaulted; for which purposes less weighty howitzers would be sufficient.

342. That greater degree of horizontality in the trajectory of a projectile, which is produced by superior velocity, is still more important in the armament of coast batteries, whose sphere of action is over the horizontal surface of the sea, and not over the uneven surface of land; because there is, throughout the range, more of parallelism between the plane of operation and the flight of the projectile, in proportion as the trajectory partakes of horizontality, that is, in proportion as the gun possesses the power of throwing its shot to the

greatest distance with the least elevation. The probability of striking an object, and, if missing it, the chance of hitting some other body on the battle-plain, either with cannon, rifles, or common muskets, depends more on the angle under which the projectiles strike, or graze the ground than that at which they were projected. (Art. 268.) Science shows, as stated in Article 60, that hollow shots, being less capable than solid shots of retaining their initial velocity, graze at angles with the ground which exceed those of their departure in a much greater degree than in a solid shot fired with the same initial velocity. It shows also, that the deviation of the hollow shot increases in proportion to the length of the range, and is greatest at its termination.

343. In practice on shore, or in still-water in harbour, the relative superiority of solid-shot guns may be somewhat obscured by extreme skill and nicety in target-firing, to hit a point at a well-ascertained distance, in slow and deliberate school practice; but in battle, on the open sea, where the firing is rapid, the ship in motion, the enemy not stationary, and the distance not accurately known, and ever varying, the practice cannot be precise; and it must ever be considered that hitting an object under those circumstances is rather a matter of probability than of certainty; and that probability will be greater or less, according as the elevation is less or greater.

Irrespective of the superiority of the range of an unchambered gun, firing solid shot, over a chambered shell-gun of the same calibre, firing hollow shot, at the same elevation; or the advantage of obtaining the same range with equal elevation, it must be observed that the penetrating power of the solid shot is a matter of immense importance in coast batteries.

The relative penetrating powers of shot fired from the under-mentioned heavy ordnance, proposed for the armament of coast defences, are tabulated as below by Straith (*Fortification*, vol. ii. p. 226), chiefly from the work by Captain Simmons, entitled, *A Discussion on the Present Armament of the Navy*. The penetrations are

PART III.   ARMAMENT OF COAST BATTERIES.   371

calculated by a formula equivalent to the equation for p in Art. 79. Two columns are added from Captain Simmons' work (p. 24) to show the velocities of the 32 lbs. solid shot, and the 48 lbs. hollow shot, at the further extremities of the different ranges:—

Range in Yards.	24-pounder iron gun, of 42 cwt., 7 ft. 6 in. long; charge, 8 lbs.	Velocities, in feet per second.	32-pounder iron gun, of 55¼ cwt., medium windage ·203; charge, one-third of shot.	42-pounder iron gun, medium windage ·203; charge, one-third of shot.	56-pounder iron gun, 11 ft. long, 98 cwt.; charge, 16 lbs. solid shot, medium windage ·125.	Velocities, in feet per second.	8 inch or 68-pounder sea-service howitzer, 48 lb. hollow shot. charge 12 lbs. windage .
	Relative Penetrating Force.		Relative Penetrating Force.	Relative Penetrating Force.	Relative Penetrating Force.		Relative Penetrating Force.
500	60·104	1200	68·302	85·921	104·126	1092	51·136
800	39·055	1001	47·527	66·846	74·158	877	32·983
1000	29·829	883	36·982	46·583	59·596	765	25·096
1200	23·027	783	29·080	37·076	48·304	672	19·365
1500	20·252	661	20·724	26·699	35·679	562	13·544
2000	9·559	515	12·873	16·305	22·396	436	8·152
2500	6·231	418	8·287	10·714	14·869	358	5·494
3000		354	5·944			310	4·121

The superior penetrating powers of solid shot fired from 32, 42, and 56-pounder guns, compared with that of the hollow shot fired from the best of the 8-inch shell guns, at the distances stated in the first column, are quite sufficient to establish the great inferiority of the 8-inch shell-gun to solid-shot guns in this important respect.

Shell guns should therefore, only be placed in branches of coast batteries, forts, or fortresses which have not ranges open to them on the sea beyond the limits at which shell-firing ceases to be efficient. The hits of shells are never so frequent as of shot, and beyond 1200 or 1300 yards shell-firing becomes extremely uncertain, and should therefore give place to guns that can best fire solid shot. In some recent experiments made at Portsmouth, out of 100 rounds of shells fired at a large hulk, distant 1200 yards, not less than 15 per cent. missed. No solid shot from a 32-pounder would have failed to hit at that distance.

Unchambered solid-shot guns may, after having efficiently cannonaded an enemy when distant, take up shell firing when he comes within the limits at which it is efficient; but shell-guns cannot open with much effect, either with hollow shot or shells, upon the enemy, in very distant firing, and not with more effect than unchambered guns, when shell-firing commences. They cannot fire red-hot shot, a defect which should exclude them from use in coast batteries, except as howitzers.

Simmons, in his excellent work above quoted, concludes the examination of the relative merits of solid and hollow shot, and therefore of unchambered and chambered guns, thus: "It may even be feared that, in introducing hollow shot in lieu of solid shot, if persevered in during war, very deplorable results will ensue" (p. 40). In this opinion the author unhesitatingly concurs, and we shall consider it unfortunate if these sea-service chambered guns are preferred to unchambered solid-shot guns, in the armament of any of our coast batteries, which require the superior power of solid-shot guns for long seaward ranges.

344. In consequence of the bursting of guns at Gibraltar, and more recently at Malta, in firing hot shot, the charges have been reduced to three-quarters of their present amount. We do not see how this is to prevent such accidents; but, for other reasons, the reduction of charge is advantageous. Wood is more rapidly and certainly set on fire by hot shots, when these penetrate but little, than when they lodge at great depths, because in the latter case the crushed fibres of the timber, by their elasticity, close upon the shot, and more or less contract the orifice; and the communication with the external air is not sufficiently free to excite combustion. The penetrating power of a red-hot shot is greater than that of a cold shot, fired with the same charge; because, by the expansion of the hot shot, the windage is diminished. Great care must therefore be taken so to regulate the charge according to the distance and dimensions of the body fired at, as to be sure of causing the hot shot to lodge.

345. A very important error in classification and nomenclature, and one calculated to convey wrong

notions to the uninitiated, and even to mislead those who should know better, is frequently observable even in official documents, in which the 8-inch shell guns are called 56-pounders, and sometimes 68-pounders; but when it is stated that the so-called 56-pounder is a chambered gun, the weight of whose hollow shot is 56 lbs., and that no shell gun, with one exception, can fire solid shot of 68 lbs., also that none do fire solid shot, the error of not calling things by their right names will be made manifest.[a] From this error in nomenclature our 8-inch shell gun, called a 56-pounder, led the French to believe (see *L'Enquête Parlementaire*, vol. ii. p. 414, and *Notes* thereon by the author, pp. 14 to 16) that we had introduced a 56-pounder unchambered solid-shot gun into our naval service,[b] and this was actually the reason why the French introduced their 50-pounder as an equivalent gun! This they have now withdrawn from the broadside batteries of their ships to give place to a greater number of 30-pounders, No. I.

Simmons, throughout his able and very useful work, calls shell-guns *sea-service howitzers*. So are they denominated by Straith, vol. ii. p. 226. General Miller brought them forward by that title; and by that name they were then, and still are, designated by Monk. In French, and all other foreign treatises, they are called simply *howitzers*.

The author has throughout this work called shell-guns *chambered guns*, since, by the introduction of shell-firing from all natures and descriptions of naval ordnance, the former distinction ceased. But without some such explicative, that description of ordnance cannot be correctly and scientifically named and classed; for *guns*, simply, they are not. Chambered guns are contradistinguished from *un*-chambered guns by definitions

---

[a] Thus Guildford battery at Dover is erroneously believed to be armed with 56-pounder solid-shot guns, whereas the ordnance improperly placed there are 8-inch shell guns which only fire hollow shot of that weight. Assuredly that battery should be armed with 68-pounder guns.

[b] The armament of the main deck of the "Indefatigable" is stated in a document of authentic character to consist of 28 56-*pounders*, but they are really 8-inch shell guns.

which cannot be mistaken, and must not be violated. Mortars, howitzers, and shell-guns, are designated in terms of their calibres, expressed in inches and decimals; unchambered ordnance for solid shot are contradistinguished by being denominated in terms of the weight of their solid shot, by pounds and decimals or pounds and ounces.

346. Lest erroneous notions should be entertained by the student from the use which has been made in the preceding articles of the term command, the author wishes here to observe that it is not an uncommon error to suppose that the advantages of *command* depend entirely on the height of a work or battery above the object against which its fire is directed, as if a shot would produce, practically, a greater effect against a work situated on a level below that on which the gun is mounted than it would against a work on a higher level. Such is not the case in any appreciable degree, and a work is not effectively commanded unless it can be fired into. But the prevention of this depends not so much upon the relative heights of the ground as upon the positions of the planes of site, or rather of defilade, by which the *reliefs* of the parapets are determined, so that a plane passing through their crests and the enemy's fire-arm may be parallel, or nearly so, to the terreplein of the work. Thus, though such terreplein may be considerably lower than the ground on which the enemy's battery is constructed, it can no more be seen from the guns in the latter than if the work and battery were raised on the same horizontal plane.

On the other hand, when the parapets of two works have equal heights above a horizontal plane passing through both, the terreplein of either of them would be commanded by the other if it should have a rise towards the rear; even when two works are formed on the face of a slope, the upper one may be commanded by the lower, should its terreplein have a greater inclination to the horizon, than the ground on which both works are constructed.

The reason why a difference in the vertical heights

ARMAMENT OF COAST BATTERIES.

of works gives scarcely any appreciable advantage to the higher work, and causes as little disadvantage to the lower, in respect of range, is, that all modern improvements in the construction and uses of artillery, having been introduced to render firing more direct, and consequently more accurate; it follows that, within the comparatively short ranges at which artillery can be so used, whatever be the relative heights of the battery and the object fired at, there is little difference in the form of the trajectory of the shot, whether the gun and object be on an oblique or on a horizontal plane, and there is no significant difference in the elevation required for guns of equal powers of range, whether the firing take place from below upwards, or the reverse.

POSTSCRIPT.

The capture and destruction of Bomarsund, with little loss, and in a very short time, by the skilful manner in which the military and naval forces cooperated with each other in those operations, are satisfactory illustrations—may they prove happy omens!—of the success which, as stated in Art. 332, p. 347, and Art. 332 c, p. 335,[1] will usually attend all such well-concerted undertakings. The forts were breached, and the fortress reduced to the necessity of surrendering, by a few powerful solid-shot guns landed from the ships of the combined fleet. Shell-guns and hollow shot could not have effected this, whether fired from the land or from ships' batteries. It is no disparagement to the naval forces to assert that they, alone, could not have demolished the defences of Bomarsund in so short a time, and not without much damage to the ships. The firing of shot and shells from the French and English ships, at long ranges, caused no serious injury; but the breaching of the forts, and the skilful establishment of a breaching battery within 400 yards of the rear of the fortress, rendered all further resistance vain. General Jones states, in his official despatch, that the interior of the place showed that the fire of the ships did such trifling injury, that the governor, with so strong a garrison in a well-casemated work, and without a breach having been formed, ought not to have surrendered; but the batteries, armed with 32-pounder British guns, being ready to open, and the forts destroyed, Bomarsund was no longer defensible. The British admiral exercised a wise discretion in not exposing the ships of the combined fleet to the severe damage they would have sustained by recklessly advancing under the fire of the enemy's sea-ward batteries; and having conducted with admirable skill the whole fleet through the intricacies of the channels leading to the point of attack, and arranged with his gallant allies and associates the plan of operations, he did right to leave to the land batteries to do what they so well and skilfully accomplished, and to keep the fleet in reserve, upon the principle that the sea-ward fronts of such forts being always thicker and stronger than the works, if any, which enclose them in the rear, should never be attacked in front if by any means they can be turned and battered in the reverse. Thus there can be no question

---

[1] In the articles to which these figures refer, these results are fully predicated.

that the place was taken by the land attack under Marshal Baraguay d'Hilliers, and the skilful engineers, General Harry Jones and General Niel; and by means so effectual, that, without any other assistance from the naval forces than the loan of a few of their solid-shot guns, and that of investing the place by sea, to prevent succours from being thrown in, Bomarsund must have fallen. Authentic information, for the accuracy of which the author vouches, enables him to state that, with respect to the effects of the solid shot on the granite, with which the walls were faced, the French guns made no impression on the blocks when they were struck perpendicularly in the middle of their faces; nor did the shot fired from the more powerful 32-pounder British guns split the granite when so struck; but, when the blocks were hit by the latter near the edge, or on a joint of the masonry, they were displaced, the joints penetrated, the wall shaken; and this not being backed with solid masonry, but filled-in with rubble, the mass was thrown down and a practicable breach formed. This successful operation is very generally, but erroneously, stated to have been effected by the fire of the ships, and is even strongly held up as a proof of what ships can do, and ought to attempt, alone, elsewhere. But the results of the experimental firing at the remnant of the fort which, unless the previous firing of the ships during the attack was absolutely harmless, must have been somewhat damaged, and moreover shaken, by the blowing up of the contiguous portions, do not warrant this conclusion, even should the attacking ships be permitted, like the "Edinburgh," to take up, quietly and coolly, positions within 500 yards, and then deliberately commence and continue their firing, without being fired at! (pp. 335, 338, at top.) The firing of the "Edinburgh," at 1060 yards, was unsatisfactory. 390 shot and shells were fired from the largest and most powerful guns in the British navy (viz. from the Lancaster gun of 95 cwt., with an elongated shell of 100 lbs.; from 68-pounders of 95 cwt. and 32-pounders of 56 cwt., solid-shot guns; from 10-inch shell guns of 84 cwt., with hollow shot of 84 lbs.; from 8-inch shell guns of 65 and 60 cwt., with hollow shot of 56 lbs.), but did little injury to the work. At 480 yards, 250 shot, shells, and hollow shot were fired: a small breach was formed in the facing of the outer wall, of extremely bad masonry, and considerable damage done to the embrasures and other portions of the wall; but no decisive result was obtained—no practicable breach formed by which the work might be assaulted, taken, and effectually destroyed (p. 353, line 12), although 640 shot and shells (40,000 lbs. of metal) were fired into the place, first at 1060 and then at 480 yards. Several casemates had fallen-in, the embrasures not having been made high enough to admit of giving the guns sufficient elevation for very distant firing, in attempting which the key-stones were actually blown out, the casemates ruined, and their guns rendered useless. The Lancaster shells, of which such high expectations were entertained, failed signally in precision of fire, even at 480 yards, and evidently cannot be depended upon, even when fired from a large and steady ship (pp. 358–363). What then can be expected from those guns and projectiles, at 5000 yards, when fired from the side of a crank Despatch gun-boat? (p. 363.) Shells, whether percussion or time fuzed, and hollow shot, broke on striking the granite (pp. 344, 345); and it was found that percussion shells could be of no use in such a case (p. 296, line 11; p. 313, line 20), whereas time-fuzed shells might produce their effects upon the personnel in the interior of the fort, by entering a casemate through the embrasure, and then exploding. It is of very great importance that the facts of the case should be rightly understood, lest wrong notions should induce the attack of more formidable places which do not admit of military co-operation, with ships alone; and which the author repeats, now more confidently than ever, would be a desperate and perilous experiment.

PART IV.

ON THE SERVICE OF GUNS IN ACTION.

SECTION I.—ON THE DETERMINATION OF THE DISTANCES OF OBJECTS AT SEA.

347. In all cases of gunnery an accurate knowledge of the distance is of the first importance. When considerable, it is usually estimated very vaguely; but the necessity of knowing it as correctly as possible, at long ranges, is greater than when the trajectory is nearly rectilinear, as in short ranges; elevation being given according to the distance, and inaccuracy increasing with length of range. At considerable distances, also, there is more leisure and opportunity, as well as greater necessity, for determining those distances with precision, while in closer action all that is required is, to be certain that the enemy is within point-blank range. When two vessels are opposed to each other at great distances, the effect will depend almost wholly on the skill of the gunner; and that vessel which has most correctly estimated its distance from its opponent will do most execution, supposing everything else equal. Let those who may be inclined to disregard such niceties refer to our actions with the Americans (which will be more particularly noticed hereafter), and they will perceive that, in our unsuccessful affairs with them, our vessels were in general crippled in distant cannonade before close battle commenced. This is fighting skill against skill, and shows the absolute necessity of attending, minutely, to everything that can contribute to pre-

cision of fire at great distances. In such trials the most devoted heroism will avail little, as we have seen, unless trained to precision. It is not proposed that trigonometrical calculation should be used to determine the degree of *proximity* of an enemy who will fairly close for straightforward battle; but, should the opponent prefer to keep off for a time, in order to try the effect of his guns and his skill at long ranges, the importance of what has been recommended with respect to the training of gunners and to the attainment of an accurate knowledge of the distance at which the fire is to be returned, cannot be disputed; the latter is, indeed, the fundamental requisite for efficiency of practice. If the distance be only vaguely guessed, suppose within two or three hundred yards of the truth, the elevation cannot be right to corresponding extents. The consequent error may indeed be somewhat corrected, when committed, by observing the effect; but *trying the range*, as it is called, is, in such a case, extremely objectionable—discreditable to the system, and consequently injurious in its moral effects.

348. Numerous methods have been proposed for readily estimating distances, and, of these, none appears so simple and obvious as making use of the different angles, subtended, at different distances, by the heights (when known) of the masts of the ship whose distance is desired, the heights and distances being arranged in a table; so that by simply measuring, with a sextant or quadrant, the angular height of a mast (as is commonly done in line of sailing, or in chasing, to ascertain whether the chace be gaining or losing distance), and entering the column of angles, the corresponding distance may be taken out. Table XXI., Appendix D, at the end of the work, is one in which the heights of the different parts of French ships of war, and of their masts, above the surface of the water, at load-watermark, are given, for every rate and class of vessel. From these elements Table XXII., Appendix D, has been formed, in which are given the distances, expressed in English yards and

cables' lengths, corresponding to the angles subtended by the heights from the water-line to the truck of the mainmast, and also to the main-topmast cross-trees: the eye of the observer being 20 feet above the level of the water.

349. It has been ascertained from authentic documents, that the dimensions of the masts of American ships of war are, in many cases, exactly the same as those of French ships of corresponding rates, as expressed in the tables; and, in general, they are so nearly alike, that this method may be used to estimate distances with tolerable certainty in acting either against French or American vessels. It is not, indeed, likely that established dimensions, which are in general rigidly observed, will differ much from the quantities stated in a table which has been formed from official documents; if they should vary in some cases, the difference cannot be so much as would create an error of more than a few yards in the required distance; and the results will always be much nearer the truth than the most practised eye could give: at any rate the distances found from the table will be sufficiently correct to produce good practice, if all the other essentials be well understood.

Every man-of-war should be provided with a table showing the angles subtended by the masts of foreign ships of war, with the corresponding distances in English feet or yards. This table should be painted on a board, so that the officer observing might take out the distances by inspection; and the author hopes to have it in his power to give, in another work, the necessary data for the practice of this method, with respect to the heights of the masts of foreign ships of war, in addition to Table XXI., Appendix D, relating to France.

This method of obtaining distances at sea, simply by the inspection of a table containing the angles subtended by a mast of known height at different distances, as now very generally practised in the service, is preferable to any other when it can be used.

350. Another method which has been recommended

consists in taking simultaneously, at the bow and stern of the ship, the horizontal angles between the line joining the places of the observers and lines drawn from those places to some point at the enemy's ship. This method requires either a logarithmic computation, or a double inspection in a table prepared for the purpose; it consequently presents great difficulties when the ships are under weigh, particularly if one is much before or abaft the beam of the other, and is quite inapplicable if one is on the bow or stern of the other; whereas the former method may be used in all positions of the ships, provided the height of a mast of the enemy's ship be known.

351. That method will not, however, serve to obtain the distances of steamers, which either have no masts or have them of no regular height. In this case the distance may be determined by making use of the ship's own mast as a given height, causing an observer aloft to measure the angle A B C formed by the mast A B, when vertical, and the line of sight B C from the observer to the enemy's ship at C; and then either computing the required distance A C by the formula AB tan. ABC, or obtaining

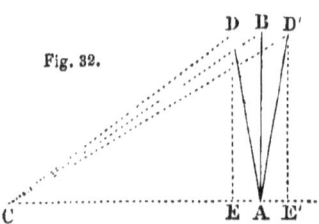

Fig. 32.

it by inspection in a table. If, during the operation, the ship should be on a *heel*, so that the mast has an inclined position as A D or A D', the angle of heel B A D or B A D' (as shown by a pendulum suspended from the centre of a graduated arc) must be taken, and either subtracted from the observed angle A D C, or added to A D'C, according as the *heel* is towards or from the enemy's ship (supposed to be on the beam). The sum is equal to the angle E' D' C, and the difference to the angle E D C; and, by trigonometry, A D cos. B A D tan. E D C = E C, the required distance: but the correction for the heel will rarely be necessary in good fighting weather.

Part IV.   DETERMINATION OF DISTANCES AT SEA.   381

In observing the angle C B A or C D A, since the image of the distant object is that which would be brought by reflexion to coincide with that which is the nearest to the observer (a mark on the *top* or on the deck), the parallax of the instrument would not, in any sensible degree, affect the accuracy of the observed angle.[a]

This, in principle, is nearly the method given in the Blue Book of H. M. S. " Excellent ;" which is to observe, with a sextant, at the cross-trees, the angle between the enemy's water-line and a point *vertically* under the observer. If this observation could be always made, it would be indifferent whether the enemy's ship were on the beam, or the bow, or the quarter.

352. An objection to this method is that, when the mast is not in a vertical position at the time of making the observation, it is difficult to have on the deck a point vertically under the observer's eye; and if the observed angle is one between a line, as D C, from the observer to the distant ship and the direction D A of the mast, it would be necessary, when such ship is not directly on the beam, to introduce a correction to that observed angle. This correction could be made either by computation, or by means of a subsidiary table, provided the angle at the foot of the mast between the direction of a fore-and-aft line on the deck and that of the enemy's ship be observed at the same time that the angle is observed on the mast. The formula for the correction is, by trigonometry, sin. h sin. $a =$ sin. C ; in which h is the angle of *heel*, a the angle between the fore and aft line and the direction of the enemy's ship, and C the

---

[a] The parallax of the sextant is the angle subtended, at the place of the object which is seen by reflexion, by a line joining the observer's eye, and the point in which the produced ray of light incident on the index glass, in coming from that object, intersects the ray from the object seen through the unquicksilvered part of the horizon-glass. This parallax is always greater as the object seen by reflexion is nearer to the eye, and as the observed angle is less; but, at no greater distance than 100 yards, and with an observed angle as small as 10 degrees, the parallax of a common sextant would not exceed two minutes of a degree.

correction.[a] This is subtractive from, or additive to, the angle observed on the mast, according as the heeling of the mast and the enemy's ship are on the same or on contrary sides of the fore and aft line.

With respect to the method in which the water-line of an enemy's ship is brought by reflexion to the foot of a plumb-line on deck, it would be, perhaps, impossible to observe the angle with accuracy. But, as the pendulum placed in the main hatchway is useful for other purposes, as in determining the proper instant for firing guns horizontally, or otherwise, so there is nothing to prevent the employment of the instrument for finding the angle of heel in combination with the observation of the angle A D C (fig. 32). An objection has been made to the method of observing such angle from the mast-head, when the ship or object is to leeward, on account of the top-gallant sail being in the way; but the top-gallant sail may not be set; if it be, and in the way, it may be clewed up, or the lee-sheet started, and the clew hauled-up during the operation.

353. An able and already distinguished young officer[b] has lately proposed an ingenious mode of determining the distance of steamers, or of ships the heights of whose masts are not known, by observing from the cross-trees,

---

[a] Let A B represent a mast in an inclined position, A E a fore and aft line on the deck, and A C the direction of the enemy's ship: let B m be a vertical line let fall from the cross-trees at B to meet a horizontal plane passing through A and C, and let B m n be a triangle in a vertical plane passing through B m perpendicularly to A C. Then A B C represents the observed angle, and A B n is the correction which is represented in the text by C; also A B m is represented by h, and E A C, or its equal, A m n, by a.

Fig. 33.

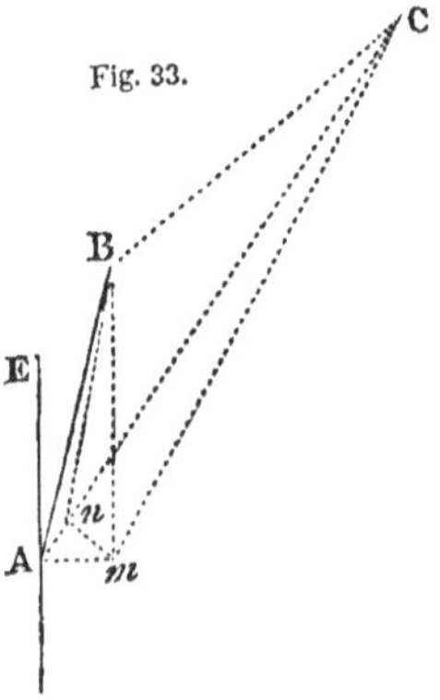

If A B be represented by unity, we have A m = sin. h: and by trigonometry, A m sin. A m n = A n; that is, sin. h sin. a = sin. C, as in the text. Thus the angle A B n, or C, is found; and being taken from the observed angle A B C, it leaves the angle n B C. Thus A B cos. C tan. n B C = n C, the required distance.

[b] Captain A. P. Ryder, R.N.

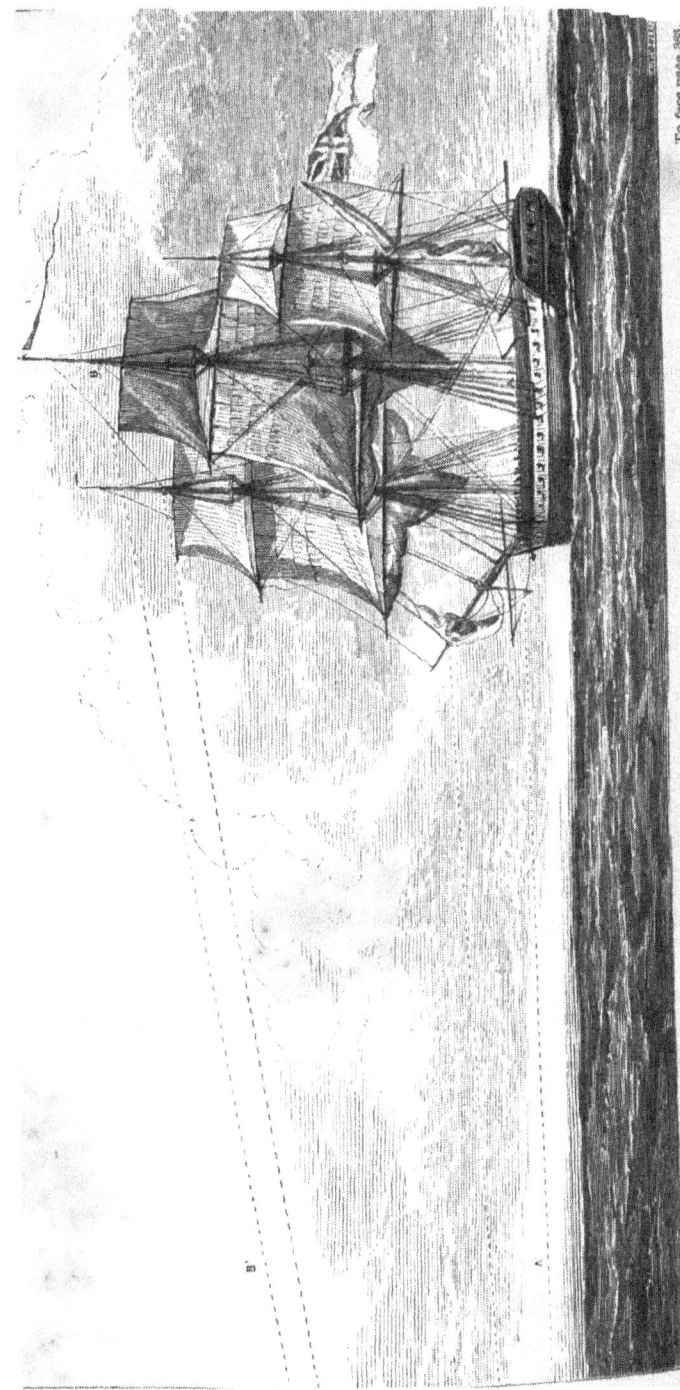

To face page 388.

top-sail yard, or any other convenient perch, the higher the better, the vertical angle between the visible horizon and the enemy's water-line. The computation, by this method, would be very laborious, requiring logarithms with seven places of decimals, but the proposer has given a double table from which the required distance may be taken by inspection. The chief objection to the method is that the angle subtended, at the observer's eye, between the edge of the horizon and the enemy's water-line is very small, being, at 2630 yards, only 33' 53''; when land, instead of the sea-horizon, is beyond the enemy's ship, the observation must be made with a sextant fitted with an artificial horizon, and consequently would be liable to great uncertainty.

354. But why might not the following application, in an oblique plane, of Sir E. Belcher's horizontal method, be adopted?

Let two observers, each visible from the other, one at B (suppose at the cross-trees of a mast), see the preceding figure, and the other at A (suppose on the deck) as far forward or aft as possible, take simultaneously (C being the ship or object whose distance is required) the angles C A B, C B A. The length of rope which is made the base A B, being once measured, may be considered as constant; and the required distance A C may be obtained by the usual process. Or, if half the sum of the two observed angles at A and B be taken as one of the angles at the base of the triangle (supposed then to be isosceles), the distance may be found with sufficient precision, on multiplying the half length of the base by the tangent of an angle equal to that half sum. At the distance of 300 yards, the error, in the most unfavourable case, would not exceed 4 feet; and it becomes less when the distance is greater. It is easy to see that a table might be made from which the required distances might be had immediately on inspection.

This method will be better understood by an inspection of the annexed plate. The maintop-gallantmast head is supposed to be the point marked B in figure 33, in the preceding note : the length of the maintop-gallant back-stay,[a] to the point to which it leads in the main chains, may be made the base A B ; and the length of that rope must be accurately measured : the point C (indicated by the convergence of the lines of observation, A A' and B B', produced) being the ship or object whose distance is required, the angles A B C and B A C are taken simultaneously to the water-line of the ship C, by observers placed at the two points. Or, as the lines traced in the plate show, the fore-top-mast or fore-top-gallant-mast stay may in some cases be more conveniently used, the observers being placed at the head of the top-mast or top-gallant rigging, and on the jib-boom respectively. The exact lengths of these ropes may easily be verified by actual measurement, as occasion may require.

The distance of steamers not being determinable by the method explained in Art. 348, and the heights of the masts of French sailing ships, though very correctly given in Table XXI., Appendix D, being subject to alteration from time to time, by variations in the immersion of the ship, and other unknown circumstances ; and, seeing that no very accurate account of the heights of masts of other foreign ships of war can be obtained, it is obvious that the method which proceeds upon an accurate knowledge of the height of a vessel's own masts, or length of any convenient rope of its rigging, possesses great advantages over that which depends

---

[a] TABLE showing the Lengths of the Fore and Maintopmast and Top-gallant Stays and Backstays of English Ships of War.

	Line-of-Battle Ship.	Frigates.		Corvette.	Brig.
	80 Guns.	50 Guns. 1st Class.	40 Guns. 2nd Class.	18 Guns.	16 Guns.
	ft. in.	ft. in.	ft. in.	ft. in.	ft. in.
Maintopmast Backstay . .	136 6	129 8	120 6	84 0	79 6
Maintop Gallant Backstay .	167 0	158 6	147 6	105 0	99 6
Foretop Gallant Stay . .	148 0	147 0	137 0	99 0	100 0

upon the assumed height of the masts of the vessel to be fired at.

The method proposed by the author (Art. 351) of using the known height of a ship's own mast above the water-line as the base of the triangle by which the distance of an enemy's ship may be computed, though liable to some difficulties, will be of use on many occasions when other means cannot be resorted to, as when it is required to ascertain the distance of a steamer, or other craft, which has either no masts or such as have not their heights in accordance with a fixed regulation; or, again, when the horizon, from the intervention of the land, cannot be seen. The difficulties alluded to are such as may arise from the ship having so much motion as to render it scarcely practicable to take the angles on the mast and at its foot, as well as the angle of *heel*, simultaneously. In smooth water there can be no difficulty. In inland seas, at anchor, in preparing to attack land-batteries, when it is of importance to determine the distance of the object from the ship, there seems to be no other method of obtaining such distance than by some triangulation in which the base is either the length of the ship, as in Sir Edward Belcher's method, or the height of the ship's own mast, as proposed by the author. In suggesting this last method it is not intended to disparage the very ingenious one proposed by Captain Ryder or that of Sir Edward Belcher, but merely to offer a means which may be employed when circumstances render it applicable; and it may be observed that the difficulties indicated would be equally felt in Sir Edward Belcher's method, which is, besides, applicable only when the object whose distance is required is well on the *beam*.

SECTION II.—ON THE POINTING OF NAVAL ORDNANCE.

355. When, without manœuvring, ships come fairly to close action, the necessity of determining the elevation due to the distances of the objects does not exist.

Superior precision, and rapidity of horizontal fire, will then determine the affair; it is important therefore to consider the best means of laying ordnance readily in horizontal or point-blank positions, whatever be the inclination of the vessel. It very frequently happens that ordnance cannot be pointed accurately by sight, particularly in general actions, on account of the smoke in which the hulls of the contending vessels are usually enveloped. In such cases, therefore, it is necessary to resort to some expedient by which each piece of ordnance may be readily laid, and correctly fired, in a horizontal direction. Various very ingenious contrivances were devised during the late war to regulate generally the position of ordnance for horizontal fire; and the most successful method of doing this was by means of a pendulum, as practised by the gallant Sir Philip Broke in his Majesty's ship "Shannon."

356. This ingenious expedient was also practised by other gallant officers, from the example of its distinguished author; and it is described in a tract by Captain Pechell, entitled 'Observations on the Defective Equipment of Ships' Guns, in a Letter addressed to Vice-Admiral Sir Harry Neale, Bart., G.C.B., dated H.M.S. "Sybille," Dec. 1st, 1825.'[a]

357. The motion of a large ship, in good engaging weather, is so easy and slow, that anything of a pendulous nature, nicely fitted, on board, will act with considerable accuracy—witness the marine barometer.

Though, at first sight, there may seem to be grave objections to an application of the pendulum for this purpose, on account of the disturbing effects of the

---

[a] The author's lamented friend Sir John Pechell, sent him a copy of this tract soon after it was published, in a letter dated Dec. 12, 1826, of which the following is an extract:—" I am sure you will approve of everything that has a tendency to bring gunnery into fashion. You will perceive I have quoted your work, and only wish the Admiralty had long since adopted your system, but it appears they were frightened at the expense; something, however, has been gained by you to begin with, and that is an ample allowance of powder and shot."

vessel's motion, yet experience has proved (see Arts. 358, 359, and 367) that the pendulum may, with advantage, be used in regulating the position in which ordnance should be laid, with reference to the angle of heel, so as to produce a fire nearly horizontal. Those, again, who object to the principle of intrusting to captains of guns the care of regulating the pointing by the position of a pendulum, can scarcely be aware of the high degree of efficiency, intelligence, and practical skill, to which, though not in sufficient numbers, the whole corps of seamen gunners has been raised by the Institution described in Part I.

358. The established practice in the British service, with respect to the use of pendulums, is to have one or two of these instruments placed in the square of the hatchway or hatchways in order to denote the inclination of the ship; and general orders, circulated by signal or otherwise, direct the elevation to be given to the lee-guns, or the depression to the weather-guns, according as the ship may be engaged, so that, in either case, her firing may be horizontal. The guns are always kept in horizontal positions, and the sights at point blank, unless contrary orders are given. The depression chock is always to be used so long as the elevation will admit of it, as the coins are then less liable to start out. This method may be used with advantage when approaching an enemy, by running down from the wind, the guns being prepared for horizontal firing by the best possible estimation of what the heel of the ship may be when brought to the wind.

359. The pendulum is very generally used in the French naval service to indicate the moment when the axis of the gun becomes horizontal; and it is justly considered so important to determine this with the utmost precision, that the indications of the pendulum are compared with, and checked by means of a reflecting instrument, so constructed as to show by the coincidence of the real and reflected horizon, the instant in which the axis of the gun is horizontal. This combination, called L'Horizon Balistique, is mentioned in

several works as an important discovery.[a] French writers observe that accuracy in aiming, and certainty in hitting, have now become matters of the utmost importance in an economical as well as in a professional sense, on account of the enormous expense of the shell system.

360. However accurately the distance of an object may have been determined, and however well a ship may be furnished with tables and rules for practice, and provided with skilful and experienced gunners, the firing cannot be efficient unless the ordnance be furnished with correct sights.

The only means provided in the ordnance arsenals during the late war for pointing naval ordnance, were the quarter-sights engraved on the sides of the base-rings, in quarter degrees from point-blank to two or three degrees. For close, horizontal fire, guns not provided with proper sights (and some such may yet be required) may be laid by the point-blank quarter-sight, with sufficient accuracy, by simply bringing the notches upon the base-ring and muzzle to bear upon the object aimed at; the elevation will then be correct, and, in close action, the line will be sufficiently true. But when the distance is such as to require any elevation, this method of pointing guns is totally inapplicable to naval service; because, unless the line be correctly taken over the top of the gun at the same time, great errors in horizontal divergence would be produced. In land service the line may be taken over the top of the piece, and the elevation afterwards regulated by the quarter-sight; but in naval practice these two operations must be executed simultaneously. It is, therefore, a matter of great importance in all naval artillery-practice, to reduce these two operations to one, and this can only be done by the tangent scale, or some other top-sight.

---

[a] The reader will find a description of these expedients in Colonel Préauxs' work, 'Sur le Canonnage à Bord' (p. 100); but no figures are given to show how they are applied. The author has not succeeded in his endeavours to procure the instruments, but hopes he may have it in his power to give drawings of them in an Appendix.

361. In our naval actions of 1812 and 1813 the disadvantages arising from not having any regulated sights or adjustments of this description were severely felt; and, had it not been for the intelligence, ability, and skill of our naval officers, who, by various expedients, supplied this great defect, the consequences might have been very serious. The guns of the United States' frigates were fitted in a manner which enabled them to be fired horizontally, or at any required inclination. In some cases this was accomplished by the dispart, in others by tubes placed on the tops of the guns, and either fixed parallel to the axis or provided with the means of being inclined to it so as to give to the gun the requisite degree of elevation.

362. It was not till after that war that the subject of sights, and scales of elevation, was taken into serious consideration by the proper authorities in this country, when instruments of various descriptions were devised to take the line of direction and give the necessary elevation to the gun at the same time. To effect this, *top-sights* (Congreve's) were first introduced and very generally used. Moveable sights of different heights had been used on board of the "Shannon" and "San Domingo" during the war, the only fixture being a confining sight on the breech, to warn the captain of the gun to keep his eye down to the level of the notch on the base-ring, in line with the other sight. Congreve's sights consisted of tubes of delicate construction, fitted, with mathematical and mechanical nicety, to minute differences of elevation; but it was soon found that instruments of this description standing high upon the top of the gun, would, on service, if attached to guns upon the quarter-deck, forecastle, and gangway, be exposed to damage at every instant by the fall of rigging, &c., from aloft; and, if on the main deck of line-of-battle ships or frigates, by the working of the tacks and sheets in manœuvring. Another serious objection to fixed top-sights standing considerably above the breech, and for which the means then in practice afforded no remedy, was that the lowness of the upper

sills of the ports prevented the use of those scales when a ship might be fighting her lee-side under pressure of a smart breeze, her opponent being so distant as to require considerable elevation to be given to the guns.

It appeared also on further trial that whenever there is so great a swell as to occasion much and rapid motion, any close sight having but a small aperture through which to catch a glimpse of the antagonist's ship, as it passed rapidly into and across the field of the tube, was not found so advantageous as an open sight; because the quick motion of a ship giving but a glimpse of the object fired at, without a distinct perception of the approach to coincidence, induced the captain of the gun either to take his eye from the tube, or by keeping his eye too long to it, prevented him from anticipating the right instant, which a skilful and experienced gunner ought to seize.

For these reasons all complicated instruments, as sights, have been abandoned, and the simple tangent scale, as long used in the land service, is now universally adopted in the navy.

363. The tangent scales, with which all naval guns are now fitted, are made of brass. For all 32-pounders, up to 50 cwt., the scales are graduated to 5°. The 56 or 58 cwt. 32-pounder, length 9 ft. 6 in., admits of the tangent scale being graduated only to 4°. When greater elevations are required than the brass scales can give, a wooden scale [a] of the same form is substituted; this is graduated from 4°, or, as the case may be, from 5°, to $6\frac{1}{2}$°; and for still greater elevations, a longer scale is put in, which is graduated up to 10°, the greatest elevation which the ports of ships in general admit. If greater elevation or depression is required than the ports will admit of, inclined planes placed, in the former case, under the rear trucks, and in the latter, under the fore trucks, will permit about two or three degrees more, of either, to be given.

---

[a] Assuredly these ought to be of brass likewise: no expense should be spared in these very important instruments.

PART IV.   ON POINTING NAVAL ORDNANCE.   391

The elevation may be increased beyond that at which the gun can come in without its muzzle striking the upper sill of the port, by placing the edge of an inclined plane beneath the after axletree, so that in the recoil of the gun the hinder part of the carriage being forced up the inclined plane, the muzzle is thereby depressed, and comes in without striking.

364. The elevating screw is coming rapidly into use in the French naval service for all natures of ordnance: by its means the gun is easily elevated or depressed by one man, whereas the adjustment by means of the coin requires two handspike-men per gun; and in this way the elevation is effected by jerks, whereas with the screw the movement is uniform and gradual. When great changes of elevation are required, it is true that there is some loss of time by the slow process of the screw, and for this and other reasons it is objected to in our service; but a great advantage of the screw, particularly in rapid firing, is, that by the very ingenious mode of preventing the screw from turning, the elevation of the gun is not altered by the firing; whereas, in the other mode of adjustment, the coins are displaced by the shock of the discharge, so that after being fired the gun has to be restored to its previous position, the distance and charge being supposed to remain the same.

The following Table contains the extreme elevations and depressions of guns, which can be given, without injuring the ports, on each deck of different ships of war in the British navy:—

Deck.	Nature of Gun.	Elevation.		Depression.	
	"PRINCESS CHARLOTTE."	°	′	°	′
Lower	. .	10	0	5	30
Middle	. .	9	0	5	0
Main	. .	9	30	7	30
Upper	. .	9	30	7	30
	"ILLUSTRIOUS."				
Lower {	32-Prs. (9 ft. 6 in. long) . .	12	0	5	0
{	68-Prs. (9 ft. 0 in. ,,) . .	10	0	6	0
Main .	32-Prs. (9 ft. 0 in. ,,) . .	12	0	8	0
Upper {	32-Prs. (8 ft.), forecastle . .	9	0	7	0
{	32-Pr. Carronades . .	12	0	3	0

Deck.	Nature of Gun.	Elevation.		Depression.		
	"Excellent."	°	′	°	′	
Lower	32-Prs. (9 ft. 6 in.), without inclined planes	10	0	5	30	Height of port, 2 ft. 9¼ in.
	Ditto, with inclined planes	13	30	8	30	Ditto of axis above the deck, 3 ft. 4 in.
Middle	32-Prs. 50 cwt. (9 ft. long), without inclined planes	10	30	5	45	Height of axis above the deck, 3ft. 1½in.
	Ditto, with inclined planes	12	0	8	30	
Main	32-Prs, 41 cwt. (8 ft.), without inclined planes	13	0	5	30	Height of axis above the deck, 3 ft.
	Ditto, with inclined planes	14	0	7	30	
	"Asia."					
Lower	32-Prs. (9 ft. 6 in.), without inclined planes	10	0	4	30	
	Ditto, with inclined planes (with toggle)ᵃ	15	30	6	30	
	8-inch Guns, 65 cwt. (9 ft.) without inclined planes	12	0	2	15	
	Ditto, with inclined planes (with toggle)	16	0	4	45	
Main	32-Prs. (7 ft. 6 in.), bored up Guns, without inclined planes	15	0	7	0	
	Ditto, with inclined planes (with toggle)	16	0	10	30	
	8-inch Gun, without inclined planes	7	30	6	0	
	Ditto, with inclined planes (with toggle)	12	30	7	30	
Upper	Carronades	10	0	8	30	
	Ditto (amount which the screws will allow)	11	30	8	30	

365. Although for extreme elevations and depressions the inclined planes may be used, yet it must be observed that, beyond the elevation which the ports admit, the sight can no longer be taken by the tangent scale, nor by any other top-sight. A gun can then only be laid by lowering the breech to an extent regulated by a wooden graduated scale, of which one end is placed

---

ᵃ The toggle, and tripping-line, which we describe for the information of the general reader, is an expedient by means of which a greater degree of depression may be given to a gun than the height of the lower port-sill admits, without being struck by the muzzle of the gun in its recoil, as in fighting the weather-side in strong breezes. This is effected by placing a wooden pin (*toggle*) vertically under the breech of the gun, and having one end of a tripping-line fastened to the centre of the toggle, and the other end tied to the breeching-bolt in the side of the ship, so that at the commencement of the recoil the toggle is tripped from its place, when, by the preponderance of the breech, the muzzle is raised sufficiently to come in clear of the port-sill.

upon the carriage, and the elevation is determined relatively with a mark or quarter-sight on the base-ring of the gun in the horizontal plane passing through its axis.[a] Graduated coins are found to answer extremely well, and are more readily used.

366. The following rules for concentrating, or directing the fire of several guns towards a given point, are extracted from the practice of the "Excellent:"[b]—

"In concentrating any number of guns on one point at a given distance, the directions are always taken from one particular gun. If the given point is on the beam, the midship gun is taken as a guide to direct the rest; if on the bow, the after gun, and, if on the

---

[a] DIMENSIONS of the Ports, in English measures, for different natures of Ordnance, on board of French Ships of War.

SHIPS of the OLD MODEL.

	Natures of Guns.					Carronades.		
	36.	24.	18.	12.	8.	36.	24.	18.
	ft. in.	ft. in.	ft. in.	ft. in.	ft. in.	ft. in.	ft. in.	ft. in.
Width . .	3 3.4	3 2.2	2 10.8	2 7.2	2 3.5	3 3.4	3 2.	2 10.2
Height . .	3 2	2 10.8	2 7.2	2 4.7	2 1.6	3 3.4	3 3.4	3 2.
Height of lower sill above the deck . .	2 3.5	2 1.6	1 10.4	1 6.1	1 4.9	1 4.9	1 2.9	1 2.1

SHIPS of the NEW MODEL.

	Canons-Obusiers.			Guns.			Carronades.
	80, No. 1.	80, No. 2.	16.	30, Long.	30, Short.	12.	30.
	ft. in.	ft. in.	ft. in.	ft. in.	ft. in.	ft. in.	ft. in.
Width . .	3 3.4	3 2.5	2 11.4	3 3.4	3 2.5	2 7.2	3 2.2
Height . .	3 2	2 11.4	2 7.2	3 2.	2 11.4	2 4.7	3 2.2
Height of lower sill above the deck . .	2 2.8	2 0.	1 8.6	2 2.8	2 0.	1 6.1	1 3.7

[b] Since this article was written the author has received a very interesting publication, entitled, 'Remarks on the Means of Directing the Fire of Ships' Broadsides in Converging Directions,' by Commander Arthur Jerningham, of the Royal Navy, and to which the reader is referred for his able and elaborate propositions, into which the author regrets he has not space to enter.

quarter, the foremost gun may be used for that purpose; but as bow and quarter concentration is only intended for chasing and being chased, it is proposed to use the midship gun as the director in each case, because its training is generally equal both forward and aft, and the subject will be rendered more simple. If this method be adopted, the midship gun will be used in the three following positions:—'Right-a-beam,' 'Extreme trained aft,' 'Extreme trained forward.' Concentration should not be attempted for distances beyond 400 yards, nor when there is much heel: it might be very useful in close broadside action, if an opponent were obscured by smoke, on the lower deck, but could be seen from the upper deck; for then a sight or picket over the gun concentrated on would give the line of direction, and pendulums for each division of the guns below would mark the time for firing, *i. e.* when the ship was upright; or it may be obtained by the clever instruments invented by Commanders Jerningham and Moorsom, each having a bar with two sight-vanes, one of which is a tangent sight, traversing on a graduated sweep, the guns below being trained and elevated to the same angles, and fired by signals from the deck, by checking lines, on tubes fitted for conveying orders from the quarter-deck to the decks below."

367. When a vessel heels much, or there is any motion, the concentrated fire of a ship's batteries must be given simultaneously in broadsides, or by divisions,[a] either of which methods is objectionable, and should be avoided as much as possible. Firing by volleys, in the naval service in particular, is extremely objectionable, and is rarely as efficacious as the independent and deliberate firing of one or two well-directed and well-served guns.

Perhaps, as stated in Colonel Stevens's interesting pamphlet,[b] it may be doubted whether the somewhat

---

[a] See Answer to Question No. 103, page 192: Questions on Naval Gunnery, "Excellent."

[b] It was stated, by more than one eye-witness, in several public prints published at the time, that the *Christian the Eighth* fired by salvos of broadsides, or by divisions of guns, and that these salvos, as indeed evidently was the case, mostly missed the battery, taking effect on or at the base of the redoubt to the eastward. The author is informed that the mode pursued was to fire the salvos by a single word of command, as volleys of infantry *were formerly* in the British service. Thus the moment to fire was not confided to the captain of each gun, its proper director.

In reflecting on the failure of these salvos, the author regards it as a useful warning against the practice of discharging whole broadsides, or divisions of guns, at a given instant, by a word or signal of command, which fatally in-

complicated system of concentration may not be carried too far, and whether it is likely to prove as efficient as the independent and quick firing of guns conducted in the ordinary manner. The moral, as well as the material, effect would probably be greater if the same number of guns were fired independently, according to the skill and judgment of well-trained captains of guns, who, each seizing the favourable moment, might spread destruction throughout the enemy's ship.

368. Naval officers and gunners should be well skilled in every available expedient for immediate substitution in all the uncertain and contingent operations of gunnery on service; it may therefore be useful to explain a method of obtaining the elevation required for striking a ship at a given distance, which may be easily practised should other means fail, as by the breaking, or want of a tangent scale, or some other contingency, and which may, moreover, be generally useful.

The method of pointing ordnance by tangent-practice, in the manner to be explained, was suggested to the author by Sir Philip Broke; and it forms the basis of the French principle for regulating elevation, which, however, by a complicated use of the line-of-metal sight, is extremely objectionable.

The elevation given to a piece of ordnance, at any range beyond point-blank, is intended to allow for the space through which the projectile falls by the action of gravity in the time of flight. Now the vertical space through which the projected body, in its flight, descends below the line of aim, is equal to the tangent of the angle of elevation multiplied by the range or horizontal distance of the object from the gun. Thus, suppose a gun to be at A, at a known height A A above

---

terferes with the perception of the British seaman, instructed as he now is in the use of the sights with which his gun is equipped.

To the captain of the gun should be confided the important trust of the trigger-line, and with it the duty to fire when, his eye being brought down to the horizontal plane of the sights, he judges that they coincide with the object.—*Account of the Destruction of the Danish Ship of the Line Christian VIII., by Colonel Stevens, Royal Marines.*

the level of the water, and at a known distance A B from a vertical object B'C, as a ship's mast upon which is a distinct mark D (suppose the main-top), whose height B'D above the level of the water is known. Then if, from the experimented or computed ranges of shot fired from a particular nature of ordnance, there be taken the corresponding elevations; and (BB' being equal to AA') with the ranges and elevations, the heights B'D be computed trigonometrically by the formula

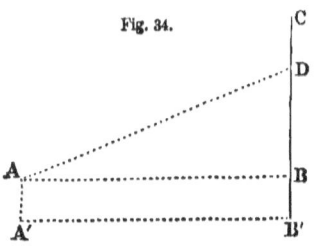

Fig. 34.

$$AB \tan. BAD + B'B = B'D;$$

a table of such heights, with the corresponding distances A B (the names of the marks D being also expressed), may be formed; from which, by inspection, there may be taken the designation of the point to be aimed at, in order that the shot may strike the object at the water-line, the dispart sight being used so that the line of aim may be parallel to the axis of the bore of the gun. In this manner Table XXIII., Appendix D, was formed for an 8-inch and a 32-pounder gun; the points to be aimed at being on the hull or masts of French ships of two classes.

The ships B and C, represented in the annexed plate, are supposed to be of 82 guns, and the distance from the ship A to the ship B is found to be 630 yards. This requires that A, in firing at a ship B of that class with 8-inch guns (charges 10 lbs.), should aim at a point about midway between the water-line and the main-top (see Table XXIII., Appendix D). The distance from the ship A to the ship C, supposed to be 1078 yards, requires that, with the same gun and charges, the aim should be taken to 1 foot below the main-topmast cross-trees, or to points equally high on the other masts. The masts being thus the aim for the line as well as for the elevation, will, with this description of practice, be more

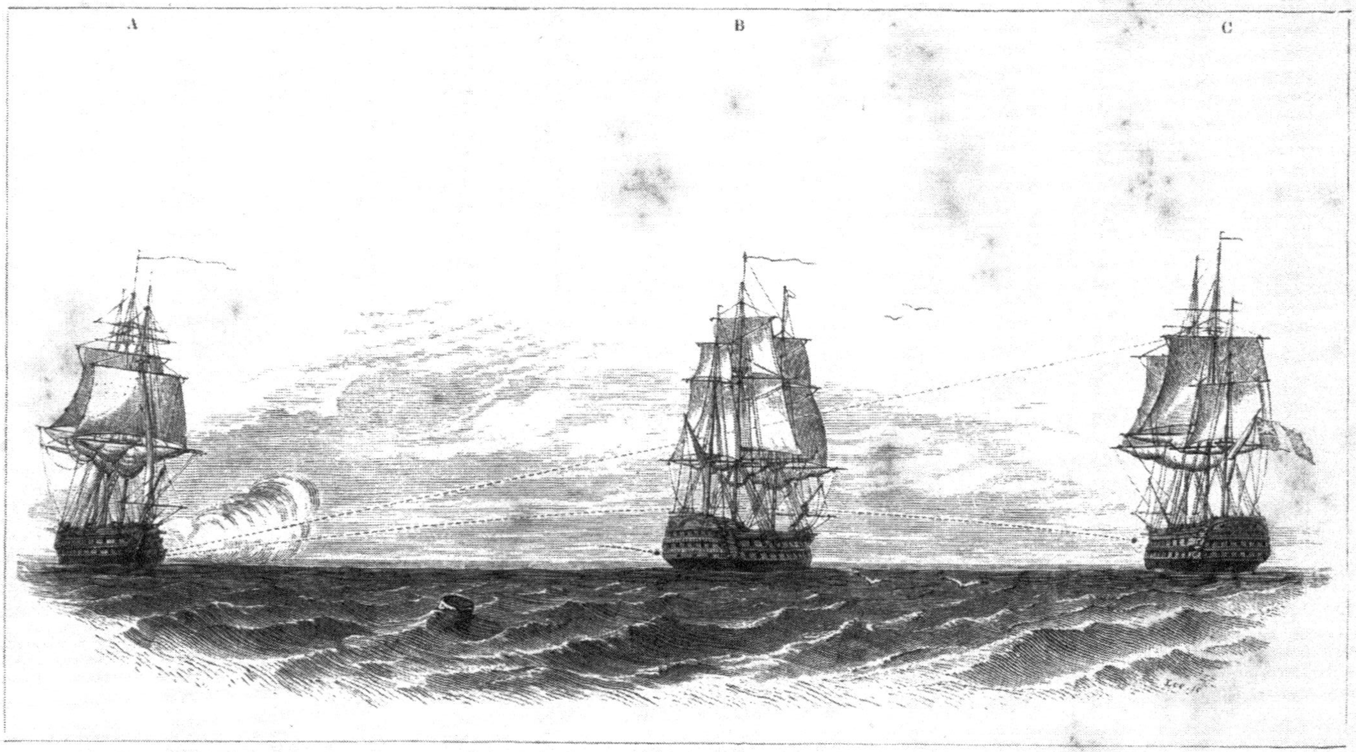

To face page 208.

perhaps, be deemed somewhat interesting by the professional reader.

On Sir Charles Douglas's appointment to the "Duke," he brought before the Admiralty and the Ordnance several propositions for improving, facilitating, and quickening the service of naval ordnance. The cartridges were at that time all made of paper, which required the operation of worming guns after every discharge, on account of the lower end of a paper-cartridge remaining generally at the bottom of the bore in a state of ignition. To obviate this, Sir Charles Douglas proposed that the cartridges should be made of flannel. He recommended, and urged repeatedly, the full equipment of his ship with flint-locks, by which the use of the slow-match and the powder-horn for priming might more or less be discontinued; and as tin tubes would manifestly be dangerous and highly objectionable on the fighting decks of a ship, he recommended that quill-tubes should be substituted for them.

371. Neither of these propositions was immediately or fully adopted; paper-cartridges, the match, and the priming-horn continued for some years in general use. It appears by official documents [a] that no locks were supplied during 1778 to the "Duke" at Plymouth, where she was commissioned in April of that year; but in the following year she was furnished, at Portsmouth, with two locks for 24-pounders, four locks for 12-pounders, and two for 6-pounders; total eight locks.[b] Yet the "Duke" was fully equipped with gun-locks in the celebrated victory of the 12th of April, 1782, to which that ship and the other three-deckers, the "Formidable" and the "Namur," so greatly contributed. By what means then had the "Duke," in the previous years, been completed with these important implements? By Sir Charles Douglas, out of his own funds and by his own energies: he bought a sufficient number of common musket-locks, which being let into pieces of wood, as

---

[a] Report of the Ordnance Storekeeper, Devonport, 5th Dec. 1850.
[b] Report of Mr. G. Stacey, Chief Clerk, Tower, 7th Dec., 1850.

into the stock of a musket, might then be fastened with iron wire to the guns.[a] He purchased flannels sufficient to make bottoms for paper-cartridges, goose-quills for tubes, and the ingredients necessary to fill and prime them.

On Sir Charles Douglas's appointment to be Captain of the Fleet he was succeeded in the command of the "Duke" by Captain, afterwards Lord Gardner; and, in the battle of the ensuing year,[b] the quick and efficient firing of that ship was so conspicuous and powerful as to enable the gallant Gardner to widen the gap which his leaders had made in the enemy's line, and so open

---

[a] The subjoined is one of the accounts of these disbursements, which the author finds among the papers of his honoured Parent:—

"AN ACCOUNT of Expenses by me incurred for the better use of the Cannon of H. M. S. the 'Duke,' under my command, between the beginning of the year 1780 and the day of the date hereof.

	£.	s.	d.
To 28 left-hand locks[1] at 10s. 6d., made by Mr. Sandwell, of Tower-hill, gunsmith	14	14	0
To 400 best black flints at 2s. 6d. per hundred . .	0	10	0
To carriage and porterage of said locks and flints from London	0	2	6
To 1004 goose-quills for tubes	3	9	4
To spirits of wine to make an inflammable paste for the tops thereof, to keep in the powder	0	15	6
To sewing silk for tying down and crossing their heads to keep in such paste	0	12	6
To 71 yards of flannel at 1s. 2d. per yard, for bottoms to paper cartridges	4	2	10
To worsted for sewing in ditto	1	4	6
	25	11	2

Given under my hand, on board H. M. S. 'Duke,' at sea, the 13th March, 1781,

(Signed)   CHARLES DOUGLAS."

"These are to certify that the foregoing articles are actually in use on board this His Majesty's Ship in my department, in quantity and quality as above-mentioned; and I do moreover certify, to the best of my knowledge and belief (*I myself having purchased sundry of the said articles for Sir Charles Douglas*), that the charges of the whole thereof are fairly stated.

Witness my hand, on board H. M. S. 'Duke,' at sea, 13th March, 1781,

(Signed)   WM. IRELAND."

[b] An analytical and tactical account of that celebrated and remarkable battle, 12th of April, 1782, will be found in the author's work entitled 'Naval Evolutions,' in which the manœuvre by which that great victory was gained is fully described. Boone, 1832.

---

[1] The locks were made to be placed on the left side of the vent, in order, it would appear, not to interfere with the stationing of the men in the exercise or manual, the man who serves the vent being always on the right, and he who fires on the left of the gun.

the way for Rodney to pass to a memorable victory. That glorious day settled the question of the locks, by bearing down all further opposition to the introduction of improvements which the prejudices of the time deemed useless and unnecessary refinements;[a] but that battle having likewise put an end to the maritime part of the war, no measures for the supply of locks to naval ordnance appear to have been taken till 1790, when "brass locks" of a new pattern were provided, and continued in general use throughout the late war. These no doubt contributed greatly to the efficiency of our practice, to the accuracy and rapidity of which all French authors attribute the superiority of our gunnery in the actions and battles of the early part of the war, the French not having adopted locks till 1800.

372. The flint-locks of the pattern of 1790 remained in general use till 1818, when the double-flinted lock, likewise adopted by the French, of the author's invention, a drawing of which was given in the former editions of this work, was ordered for general use in the navy.[b]

---

[a] "The 'Duke,' commanded by Sir Charles Douglas, was always considered one of the best ordered and best regulated ships of the fleet, consisting of forty sail of the line (1780); and, although strong prejudice and attachment to old customs prevented the general adoption of that excellent system of naval gunnery (locks, quill tubes, flannel cartridges, &c.) of which Sir Charles was the sole inventor, it was found to answer so well by Captain Gardner, who had so great a share on the 12th of April, 1782, that it afterwards became universally adopted in the navy."—*Admiral Lord Saumarez's Naval Evolutions*, p. 46.

[b] The adoption of a double-flinted lock of the author's invention was communicated to him in a letter of which the following is a copy:—

"Office of Ordnance, Jan. 16, 1818.

"SIR,—Having submitted to the Board of Ordnance your letter, dated the 15th instant, requesting to be informed of the proceedings which have taken place in consequence of your letter of the 8th November last, respecting the cannon locks of your invention, I have the honour, in reply, to acquaint you that a copy of your said letter, as well as the observations which accompanied it, has been transmitted to the Lords Commissioners of the Admiralty, and that, on a subsequent communication which has been had with their Lordships on the subject, it has been decided that the provision of locks for sea-service ordnance now in use should be discontinued, and those of your invention gradually introduced into the service.

"With respect to the measure of appropriating your locks to land-service ordnance, as suggested in your letter dated the 12th September last, I am to inform you the Board have not signified their orders on that subject.

"I have the honour, &c.,
 (Signed) "W. GRIFFEN.

"Colonel Sir Howard Douglas, Bart."

Having thus been the means of introducing improved cannon locks into the naval service, and strongly convinced of the very great importance of applying locks to all land service artillery, whether field, siege or garrison, the author addressed to the Master General, and Board of Ordnance, a recommendation to that effect in the same paper in which he proposed the improved double-flinted locks for general adoption in the navy. But the Board of Ordnance not having, as the reader will perceive, in their letter dated January 16, 1818, come to any decision on that proposition, in conveying to the author their approval of the other, he resolved to endeavour to procure support in the prosecution of his plan by consulting officers of greater experience and authority, and, in the event of their judgment coinciding with his, to press the subject on the reconsideration of the Board with all the energy in his power. He accordingly transmitted a copy of his proposition to his gallant friend and associate the late Sir Alexander Dickson, an officer eminently qualified by his great experience, knowledge, and talent to pronounce authoritatively upon all matters connected with artillery service, and received in reply the letter printed in the subjoined note.[a]

---

[a] "Valenciennes, 20th April, 1818.

"MY DEAR SIR HOWARD,—Having fully considered the matter, I feel much pleasure, in compliance with your wish, in being able to state that your proposals for the more extended use of locks coincide in the most essential points with my own ideas on the subject.

"The use of locks with heavy ordnance, particularly in the operations of a siege, presents very great advantages; for by the employment of slow-match only the fire is frequently retarded, and nothing can be more dangerous than lighted portfires in a battery. I have seen several very shocking accidents occasioned by the use of them, owing to the want of presence of mind of the gunner having the portfire lighted in his hand at the moment of a shell falling near him. In the sieges I have directed I have ever prevented, as much as in my power, the use of portfires; but Ciudad Rodrigo was the only operation in which I was fully successful in this respect; and it was to the help of about sixteen or twenty naval gun-locks, in addition to the slow-match used, that I was indebted for the vigorous fire kept up in that attack. I trust in future, therefore, that in all siege equipments each piece of ordnance will be supplied with a lock, the use of which, under every circumstance, except in heavy rain, would supersede the portfire, which in the very confined situation of a land battery, and where much powder is in circulation, is so dangerous.

"In the operation of defence, also, the same arguments, in a great degree,

Supported by that distinguished officer, the author's proposition received favourable consideration, and would then have been carried into effect had not financial and other considerations rendered it inexpedient to provide forthwith the necessary supply of locks. Thus the measure remained in a state of suspension, until at length the discovery of the percussion principle, and its application to fowling pieces, indicated that all flint-locks would ere long be superseded by percussion locks and tubes. This having been effected, the author, in 1851, reverted to his proposition of 1817—under a still stronger conviction of the greater importance, or it may well be said, of the more pressing necessity of this measure in the present improved state of artillery science and gunnery practice, than when he first wrote. This equipment is so forcibly urged in Sir Alexander Dickson's letter with respect to field and siege artillery, that nothing need, or could be said, to add to the weight of that testimony, whilst much that has been stated in the course of this work on the vast importance of the percussion lock and tube, in firing at ships in motion, applies strongly to the necessity of providing these means of procuring a more instantaneous explosion to all guns in coast batteries, which, though at rest themselves, will have to fire at moving objects,—ships

---

hold good in favour of locks; and they are truly valuable in coast batteries, and in all night firing.

" With respect to their being applied to artillery in the field, I am convinced that, with a tube such as you describe, locks would be of infinite service with field guns in ordinary duties, such as reviews, firing exercises, and salutes, and in all real service, when not too closely engaged. The effect of the lock being sufficiently certain in these situations, the saving of portfires would be great, and there would be far less risk when in action amongst ripe corn, dry grass in villages, or amongst houses; for the setting fire to a country more generally arises from cutting the portfire than from the discharge.

" In concluding, I have to observe that I do not mean to resolve the question merely into one of economy, as I deem it on most occasions an effectual simplification of manœuvre by saving the trouble of continually lighting and cutting portfires, at the same time affording the means of firing with celerity at moving objects.

" I remain, my dear Sir Howard, &c. &c.
(Signed) " A. DICKSON.
" Colonel Sir Howard Douglas, Bart.
&c. &c."

advancing, in continued change of distance and position, to that proximity which they must attain to attack batteries or fortresses with effect,—and particularly in firing at steam-vessels, running at full speed across the line of fire of a battery, instead of stopping to engage it in its strongest direction. In this case coast battery guns will be required to "shoot flying" (Art. 383), but they will have little chance of hitting active steamers at considerable distances, unless, in aiming the gun, allowance be made for the distance the steamer moves in the time elapsed between the firing of the gun and the shot striking the object; in fact, unless the action of the lock and tube be so quick as to produce the most instantaneous possible discharge, the chances of hitting in such shooting will be very remote.

373. This important manner of igniting the charge was first attempted by fixing the lock to the vent-field out of the way of the explosion from the vent. For this purpose it was necessary to make the tube of a rectangular form, so that one part might receive the percussion of the cock, and the other part convey ignition to the charge. This construction was found, however, to be so sluggish as not to accomplish the great desideratum in naval gunnery, which is, that the firing of the charge and the actual delivery of the shot from the gun shall take place as quickly as possible after pulling the trigger-line, in order that there may be little time for any alteration to take place, from the motion of the ship, in the aim of the gun. Thus it was necessary to devise some means by which the hammer, after having struck fairly upon the head of the tube placed in the vent, should instantaneously slip or be drawn aside, so as to be out of the way of the explosion through the vent. Various modes of effecting this have been devised in the British and in other naval services, but the most efficient and simple implement of this nature is that which was invented by an American named Hidden, and patented in 1842.[a] The hammer

---

[a] Ordnance Memoranda. Naval Percussion Locks, by Lieut. Dahlgsen, U. S. Navy, Assistant Inspector of Ordnance.

was, however, modified and improved so as to be stronger, and better suited to the locks in use in this country, by Colonel Dundas, Inspector of Artillery, to whom the service is so much indebted for his management of that important department, to which he was appointed in 1839.

The hammer A (fig. 35), which is of wrought iron of good quality, is fixed into a block or joint of gun-metal by means of a pin a, and a *slot* is made in the hammer to admit of a back motion. The hammer falls by its own gravity upon the vent on the lanyard B C being pulled, when, being liberated from the larger part of the slot in which it is made to turn, it instantly shifts or passes away by the continued action of the lanyard, which pulls it from its first position directly over the vent, as shown by the black lines at A' in the figure. The weight of the hammer is $3\frac{1}{16}$ lbs.

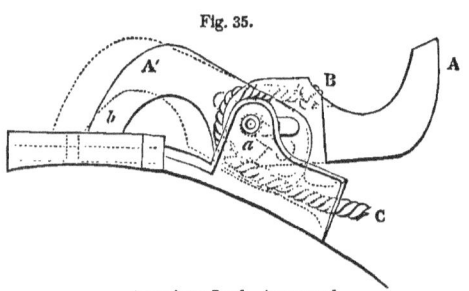

American Lock, improved.

374. This lock, though extremely ingenious and advantageous from the simplicity of the contrivance by which the hammer, having performed its office, is drawn back from the top of the vent by the same force which causes it to strike the tube, is not so quick in its action as it might be if the motion were produced by a spring. The hammer, as will be perceived in the figure, has a considerable space to pass through ere its face strike the tube at b. The motion has to be generated by the pull of a line, which changes its direction by passing round a small axis; and, having a very small leverage at the beginning of the motion, it requires a strong and continued pull to produce a smart blow and instantly draw the hammer back. This lock requires a much stronger jerk than the common spring flint-locks. The hammer has to beat down the head of the tube, and if the pull is not strong and continued the tube is not exploded. It

has been found in exercise that, in nine times out of ten, tubes supposed, from not having exploded, to be defective, will act, on repeating the trial, with a stronger jerk, which shows the failure to have arisen from the captain of the gun not having pulled with sufficient force. It appears, therefore, that a lock constructed upon this principle, with respect to the slot, in which the hammer may be made to act by means of a spring instead of a pull, would be very advantageous, the strength of a spring fully sufficient to produce the desired effect being constant. The author is aware of the objection that may be made to this suggestion, that springs are liable to break, and of the serious consequences of such a failure in close critical action, nor is he prepared to show in what mode precisely the substitution of a spring for a jerk or pull given by hand may be effected; but having considered it fully, he throws out this observation, believing that many of his readers will be attracted to the subject, and that their ingenuity will enable them to devise a lock in which the above method of giving it an immediate escape from the vortex of the vent, may be combined with the action of a spring. The construction ought to be such that, if the spring should fail, the hammer thus detached from the spring may become subject to the direct action of the trigger-line.

On the rapidity of the action of the tubes the promptitude of the delivery of the shot greatly depends. Hence the smallest reduction of the time elapsed between the striking of the lock and the actual exit of the shot, is of vast moment in naval gunnery; and every means which may contribute in the slightest degree to an approximation towards the contemporaneous action of the lock and the tube, cannot be too much attended to.

375. In the French locks (fig. 36), the hammer A falls over the lock, but there is no mechanical arrangement for removing or drawing it aside; and the fluid which escapes by the vent throws the hammer back with violence upon the support or crutch *e*, which has a leathern cushion to break the blow. The whole is of

gun-metal, with the exception of the centre-pin, the ring-bolt *d*, through which the lanyard passes, and the sliding-bar *a*. The last is intended to cover the head of the tube from rain or the spray of the sea, until just before the hammer can strike it: the cover is drawn away by the action of the *dog b* on the rack *c*. The total weight of the lock is 7½ lbs.

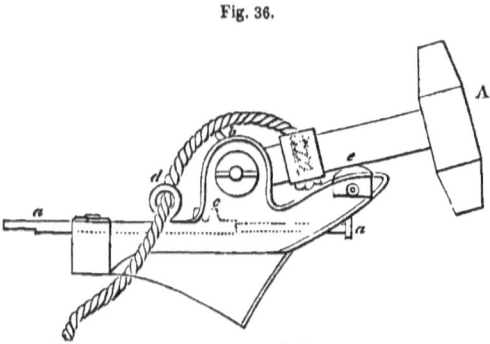

Fig. 36.

French Gun-lock.

376. In the first percussion-lock which was used in the United States' service, the *cock* struck directly over the vent, and being instantly thrown upwards, it was retained in a position clear of the blast from the vent. It struck on a percussion-patch, and there was nothing between the cartridge and the fulminating powder. It is said that powder had been ignited by the stroke when enveloped in seven folds of flannel, and even through seven cartridges inclosing the charge, in a 32-pounder gun. The action of this lock was instantaneous; and if it had stood the work well, it would have been considered, perhaps, the best percussion spring-lock that has been produced; but it was found that the mode of getting the cock out of the way, by causing it to be thrown violently upwards, did not stand repeated firings, and was otherwise liable to considerable objections. In a subsequent contrivance the cock, instantly on striking, was thrown aside by the expansion of a spring acting horizontally, which in the act of cocking had been compressed. The author has witnessed numerous exercises in which the activity and efficiency of this lock were proved, and it is believed to be still in use in the navy of the United States.

377. The cannon lock, lately used in the naval service of Holland, consists of a hammer acted upon by the pull

of a trigger-line. From the translation of the author's 'Naval Gunnery,' by Captain Gobius of the Dutch Navy (now Deputy-Minister of Marine), which, by the addition of foot-notes and other matter, has been adapted to the purpose, it appears that the cannon lock now used in that navy (see fig. 1, plate III., Gobius's translation) is exactly of the same pattern as the French lock. A lock of a different construction has since been introduced for trial in Holland. It is placed on one side, and somewhat above the vent percussion-patch, and has a brass tube directed obliquely to the vent, so that the explosion of the percussion tube is thrown obliquely into the vent, from which it is, however, so far distant that the rush of fire from the vent passes clear of the tube. These locks act with great quickness and force; and the captain of the frigate in which the effects were exhibited at Portsmouth in 1850, said they very rarely missed.

378. The Dutch lock resembles very nearly that which has been in general use for some years in the Austrian naval service; but the action of the hammer is produced, in the Austrian lock, by a spring. The lock is placed on one side just clear of the vent, but in a plane inclined from the vertical. Thus the cock acts in that plane upon a tube which is placed in a hole in the bottom of the pan of the lock, and enters with equal obliquity into the vent, which is enlarged a little to the right. The lock acts, as the author has repeatedly witnessed, with activity and certainty.

379. In the Danish navy the action of the hammer is produced by hand (see fig. 31, pl. XII., '*Danish Naval Gunnery*,' by Captain Michelson, an instructive and important work); but the mode in which the trigger-line is made to act permits the action to be quicker than in the English lock.

380. Some very interesting experiments were made at Bombay in 1844 and 1845, on a proposition made by Major Jacob of the Bombay Artillery, to apply nipples and percussion-caps to land and sea service ordnance in the same manner as they are applied to other fire-arms.

Want of space, which compels the author to omit or abridge much other interesting matter, prevents him from giving to this proposition that attention which it merits; but the delivery of the shot from the gun following almost instantaneously upon the striking of the lock, is a matter of such extreme importance in the service of naval gunnery, and the reports of the military and naval committees on the trials made at Bombay of Major Jacob's proposition certify so strongly the advantage of the nipple and the percussion-cap in respect to quickness of firing, that the author deems it right briefly to notice those trials.

The nipples used in the experiments were of the flat kind, raised about $\frac{1}{4}$ of an inch above the line of sight of the gun, with a vent only large enough to admit a 6-pounder gun-bit and priming-wire, and were made with a screw end about $\frac{1}{2}$ an inch long and $\frac{1}{2}$ an inch in diameter, to screw into the vent of the gun. Caps containing various quantities of fulminate were used, viz.:

1.35 grains of fulminate, or	3	times	
1.8	,,	,,	4	,,
2.25	,,	,,	5	,,
2.7	,,	,,	6	,,
4	,,	,,	8.888	,,
5	,,	,,	11.111	,,

} The charge of an ordinary Musket Cap.

All the rounds from both pieces were fired without the cartridge being pricked: we should not, however, recommend the pricking being discontinued, for even with perfectly efficient caps, we think pricking desirable as a *proof of the cartridge being home*; but this process was dispensed with purposely to test the power of the caps in firing the charge through the serge bag.

No instance of a cap *flying* occurred: this result was, no doubt, attributable to the superior quality of the copper used; and the fact is stated in order to remove any apprehension of danger from the large charge of fulminate used in each cap.

The Committee expressed a decided opinion that Major Jacob's proposition had been verified by experiment, and that the introduction of percussion caps to heavy and light ordnance (for the land-service at least) will supersede the use of matches, portfires, tubes, priming powder, &c., from which will result economy, both as to expense and space, while there will be gain as to rapidity and certainty of fire, and, at the same time, much of the danger attending the ordinary mode of firing guns by priming powder will be avoided.

REPORT OF THE SPECIAL NAVAL COMMITTEE ON PERCUSSION GUN CAPS.

"*To the Superintendent of the Indian Navy.*

" SIR,—We have the honour to report, in compliance with your orders, dated 20th February, 1845, that we assembled in committee (at various periods), and have tried the experiments referred to us for report on the nipple and percussion cap, filled with 3, 4, 5, and 6 grains of detonating composition.

" 2. Our 1st experiment was tried with a 32-pounder 9.6 iron gun of 56 cwt., charge of powder 10 lbs., fitted on a traversing carriage and slide, on board the Honourable Company's ship 'Hastings,' with a steel nipple fitted in the vent and an American hammer lock. Three hundred rounds were then fired, one half with powder only, the remainder with shot, with the view of proving whether the nipple would be injured in quick firing in a hard contested action; the firing was in quick time, and the gun became so heated as to render it requisite to cease firing; the nipple and vent were not in the least degree injured.

" 3. The 2nd experiment was with a 24-pounder brass gun-howitzer, fitted on a land-carriage, with the steel nipple fitted in the vent and the American hammer lock. One hundred rounds, with the full service charge, were then fired with round shot in quick time, and we found we could fire five rounds a minute; the experiment proved that though the gun became heated so as to render it requisite to cease firing, the nipple, on being taken out and examined, was found uninjured.

" 4. The 3rd experiment was with the 32-pounder 9.6 iron gun of 56 cwt., with charges of $\frac{1}{3}$, $\frac{1}{4}$, and $\frac{1}{7}$ the shot's weight, and with percussion caps, filled with 3, 4, 5, and 6 grains of detonating composition, for the purpose of proving under various circumstances their value in the naval service, when we found that copper caps, with 5 grains of fulminating composition, fired the charge in each case when the vent was even wet from having water poured like the spray of the sea over it.

" 5. The 4th experiment has been with percussion caps that have been kept on board the 'Hastings' during the wet season, under the influence of damp air, for the purpose of discovering whether the sea air will, under ordinary circumstances, injure, from damp, the composition, and render the caps unserviceable.

" 6. We therefore beg, after a careful series of experiments, to report that we deem the invention by Major Jacob of great service, and beg leave to recommend the adoption of it for all guns on the upper decks of vessels of war, for all guns fitted for boat service, and for field pieces that are fitted for the Indian navy."

381. The experiments made at Bombay on Major Jacob's propositions are extremely interesting, and are worthy of prosecution in this country. No doubt many objections and difficulties stand in the way, such as interfering with the sight, reducing the diameter of the vent, and counter-sinking the nipple into the gun. No 32-pounder gun fired in the ordinary way is found serviceable after 250 rounds have been fired from it; whereas it appears, by the reports of experiments with steel nipples and percussion-caps, that, after 300 rounds were fired from an iron 32-pounder gun of 56 cwt., with 10 lbs. charges, the vent of the gun remained uninjured, though the nipple was counter-sunk only $\tfrac{1}{8}$ of an inch, and consequently occupied a very small portion of the channel of the vent.

SECTION IV.—ON THE PRACTICE OF FIRING AT SEA.

382. In the practice of naval gunnery it is most particularly important that the actual delivery of the charge from the cylinder of the piece should follow as instantaneously as possible the action of the lock; for whilst the object aimed at is continually changing its relative position, the direction of the gun is varying so rapidly that if the medium which is to convey ignition to the charge act not very smartly, the elevation of the shot's departure may be two or three degrees above or below that at which the gun was pointed when the trigger was pulled. The nature, quality, and care of tubes and priming, therefore, are considerations which may justly be reckoned as most particularly affecting the efficiency of practice; and the most minute differences that can be detected, by the nicest means, in the progress of explosion, should be allowed decisive weight in judging and selecting the medium to be used. For suppose a vessel, in action, be rolling eight degrees, performing each roll in about four seconds of time: when the nature or condition of the tube and priming is defective or bad,

it will very frequently happen that an interval of one second of time, and sometimes considerably more, will take place between the pulling of the trigger-line and the discharge of the piece; and in that time the elevation of the gun would alter two degrees! With any uncertain or sluggish action of this nature, therefore, it is useless to expect much accuracy of effect, even with the best trained men, and with all other means perfect. (Arts. 372, 373.)

383. In every case, when there is much motion (and there will be a great deal more in steam-propelled, than in sailing ships), the shot will not be delivered from the cylinder till its direction is altered, more or less, from that in which the piece was pointed when the trigger was pulled. It is therefore not only vastly important to use those means that are best calculated to produce the most instantaneous discharge possible, but also to consider which direction, and what particular part of a vessel's motions, are most favourable for firing the ordnance with the greatest prospect of effect—whether to fire on the weather or lee-roll,—and at what particular stage or crisis of the motion. These are very important, but very difficult questions, on which the author would rather invite discussion than pronounce any positive doctrine. He will, however, state his opinion, noticing at the same time what may be advanced against it.

A steam-propelled vessel, whether with paddles or a screw, being agitated by rolling and pitching motions, and often, as in a cross sea, by both combined, accompanied also by sudden and violent jerks, will, much more than a sailing-vessel, try the skill and tact of the gunner; in whom, both for shot and shell firing, is required the greatest promptitude of perception, while the utmost intensity in the action of the lock, and vivacity in the action of the tube, are no less necessary. A swift steamer may run across a line of fire, and, in passing, use her broadside guns with effect, herself unharmed. Going at the rate of 11 or 12 knots an hour, a steamer from 150 to 200 feet long will run her own length in from $9\frac{1}{2}$ to 12 seconds. The time of flight of an 8-inch shell or shot, fired at a range of 2725 yards, being 12 seconds, a shot correctly aimed at the steamer's bow would, by the time it reaches the point aimed at, strike astern of her. A steamer acting end-on is a more steady mark to fire at, but a much smaller target to hit, and therefore the gunnery should be still more accurate. This important subject will be resumed hereafter in the work on Steam Warfare.

384. In close action, in smooth water, it is not perhaps material whether the ordnance be fired with or against the roll, provided the captains of guns judge correctly how much their pieces should be pointed above or below the part intended to be hit; but when there is much swell, it is by no means indifferent which of these motions should be preferred, nor what modifications should be admitted for particular cases, in any general maxim that may be established on this subject; and this we shall now endeavour to show. The rule generally laid down for observance in action is to fire when the vessel is nearest on an even keel—that is, upright; and always to prefer a falling side.

385. To deal with considerations respecting the rolling motions only, we shall suppose the vessel to have the wind on the beam; for if hauled upon a wind, the motion would be compounded of rolling and pitching, by the vessel laying across the swell. Now a vessel under sail, with the wind as described, is nearest upright at or near the end of the roll to windward. Were it not for the action of the wind on the sails, she would be upright when she comes to the top of a wave; but this is not the case in a smart breeze, because it requires some degree of counteracting power from the swell, as the vessel sinks upon a wave, to compensate for the heel occasioned by the wind. In a heavy swell, however, a vessel will roll to windward considerably beyond the upright position; but, in stating a case proper for action, we should not suppose the sea to be so rough as to make the vessel incline much to windward. Now a vessel brought to that momentary pause which takes place on the termination of the weather-roll, just before she

begins to feel the rising influence of the next coming wave, must be in the hollow or trough of a sea; and in such a position will have a less commanding view, and *prise* of her enemy, than if he were seen from the top of a wave. This preliminary observation may, perhaps, be considered sufficient to show that the maxim of firing when the vessel is on an even keel should not be too generally or absolutely enforced: and having submitted this, we may proceed to consider the important question that results naturally from it—whether it is most advantageous to fire with a rising or with a falling side.

386. A vessel engaging to leeward—that is, fighting her weather side, must be in the trough of a sea when the side engaged begins to rise; and whilst it is rising she must be performing a lee-roll. The disadvantage of firing from the hollow between two waves having already been shown in the preceding article, the inexpediency of firing at the *beginning* of the rising motion is also proved, for the one ensues immediately from the other; and a very material objection to the practice of firing *during* any part of the rising motion comes from this—that the lee-slope of a wave being always more abrupt or steep than the weather side, the change which takes place in a vessel's position in making a lee-roll, accelerated and increased by the action of the wind, is much more rapid than in rolling to windward; and consequently the direction, or elevation, of the ordnance, will in this case be much more quickly and considerably disturbed in firing with the rising than the falling motion. It appears, therefore, that in fighting the weather-side, we should prefer to fire at the pause immediately before the commencement of the declining motion or weather-roll (unless the vessel heel so considerably as to incur danger from the increased action of the recoil), because the ship being then on the top of a wave, will command a better view of the enemy, and the declining motion will be operating to lessen the slope in the direction of the recoil.

387. In fighting to windward, some of these arguments are reversed. The declining motion of the side

engaged is then a lee-lurch, and at the commencement of that motion the vessel must be in the trough of the sea. We should therefore so far modify the maxim already suggested as to fire at the end of the falling motion of the fighting or lee-side, when the vessel comes to the top of a wave, so that the actual discharge may not take place after the pause which attends the change of motion.

388. But modifications, governed by various circumstances, should be made in all such maxims.—If, in the first case (fighting the weather-side), a ship be heeling under the influence of a strong breeze, her guns, fired at the commencement of the declining motion, or at the pause which precedes it, will rush in with such violence, from the inclination of the deck being in the direction of the recoil, that the breechings and ring-bolts will frequently be incapable of resisting so severe a shock, particularly when the guns are loaded with two shot; and in such cases, consequently, the ordnance should not be fired till the declining motion be partly performed: thus observing in principle, the maxim to prefer firing with a falling side. The rule is so far modified in practice, that in fighting the weather guns they should be laid so that they should bear upon the opponent when the ship comes up to within 1° of the extreme of the weather-roll; and, in fighting the lee guns, to lay them so that they shall bear when the ship has made a portion of about 1° of the lee-roll from its commencement. In both cases this insures that the roll of the ship is sufficient to bring the guns to bear at a time when the rapidity of the rolling motion has been reduced by the action of the wind; and also that the guns are laid more nearly parallel to the plane of the deck—an important point which has not been sufficiently considered. When guns are much depressed, relatively with the plane of the deck, it requires very great experience and tact on the part of the captain of the gun to fire it accurately at the proper moment; for unless he be a very tall man he cannot look over the gun, at the full extent of the trigger-line, when the breech is much elevated; and if

the gun is fired at this great depression, with respect to the plane of the deck, he is endangered by the rapid and heavy recoil, and great strains are moreover occasioned to the breechings and bolts.

Again, when the lee-guns have much elevation, with respect to the plane of the deck, a somewhat similar inconvenience is experienced, though in reverse order, by the captain of the gun having to bring his eye down almost to the level of the deck to look over the gun; and when it is fixed in that position, the recoil being up the very great inclination of the deck, may not be sufficient of itself to bring the gun sufficiently inboard for loading.

For these reasons the modification suggested at p. 239, in the second edition of this work, in fighting the lee-guns, is made general, namely, to fire at the end instead of the commencement of the declining motion, although the vessel must then be nearly in the trough or hollow between two waves; for it very seldom happens to a ship, and particularly a large ship carrying her metal high, to be in action in so heavy a sea as not to have a fair view of her opponent under such circumstances.

389. It may not be improper here to remark that the breechings of naval ordnance, in frigates particularly, are in general considered too weak;[*] and as those on the gun-deck are soon damaged by being continually wet, it is very essential always to have spare breechings ready fitted, and to exercise the people frequently to shift them quickly. Carronades should always be fitted for action with second breechings. (See Art. 121.) In the "Shannon" the preventer-breechings were reaved through holes in the timbers, and toggled on the outside; and to relieve the ring-bolts and breechings, chocks of timber were placed underneath the hinder part of the slides, when fighting the weather-side, so as to lift them nearly to a horizontal plane when the vessel

---

[*] This having been recently represented, it is now established that all 8-inch guns and carronades shall have 9-inch rope for breechings, and all 32-pounder guns and carronades 8-inch breechings. The breeching-loops have all been enlarged accordingly.

was heeling 7° or 8°. Carronades thus fitted, run in with less violence, and are more easily run out; and with this species of ordnance, such precautions should invariably be adopted. If a *gun* break loose, it may be rendered serviceable again in a few minutes; but should a *carronade* break its breeching, or draw a ring-bolt, it is very apt to turn over, or split its slide to pieces. In firing carronades with two shot, these precautions are absolutely indispensable; for although this species of ordnance is not calculated to discharge two projectiles, in common, and double *shotting* is forbidden, yet on very special occasions, when within a few yards of an enemy, double charges may be used with great effect, either composed of round and grape or case shot, or of two round shot, according to the circumstances of the action. Thus, in fighting to leeward of an enemy, the inclination of his ship will expose its deck so much to the effects of grape or case shot, that the double charge should consist of one or the other. In fighting to windward, on the contrary, the weather-side of an enemy heeling off, will be so much exposed below the ordinary water-line as to invite the use of two round shot, whilst the declination of her deck covers or defilades her people from grape or case. Thus modifications in any maxims, as to the most favourable moment for firing, should also be governed by the motion and position of the enemy's vessel.

390. It appears to follow, from what has been advanced, that balls intended to take effect upon the hull of an enemy, should rather be discharged with a falling than with a rising side; but that such pieces as may be appointed specially to act against the masts and rigging, should, on the contrary, be fired with the rising motion, the aim being taken low.

391. In close critical action, the great object should undoubtedly be to hit the enemy's hull. For this purpose, it is better that the pointing of the guns should rather be calculated to take effect low, on the body, than to aim high, and missing the body, to hit the rigging. This is particularly desirable in actions between car-

ronade-armed vessels, or ships armed with shell-guns, because the magnitudes of their shot are so great in proportion to the scantling of the ships, that few body blows are required to drive the crews of the vessels struck to their pumps. One or two large shot, moreover, taking effect below the water-line, and perhaps perforating both sides of a small vessel, will, in general, either force her to surrender or send her to the bottom, though she may not have sustained any other material damage or any loss of men. Such an injury is much more likely to be occasioned by firing with a falling than with a rising side. On the other hand, a ball taking effect high, can only injure the rigging or a mast, and, if the latter, not with much effect; for a mast, wounded aloft, will be more likely to stand, than if the like wound were inflicted by a shot which had previously perforated the ship's side at the upper deck. It is a great waste of means, therefore, to apply round-shot of large calibre, or shells, to the remote chances of destroying rigging, which, branching out from one trunk or stem, may be more effectually felled by a blow otherwise destructive at the same time. The intention of horizontal firing should, indeed, never be lost sight of. When vessels are once fairly engaged in close action, which can only terminate in defeat or victory, the main object should be to sweep the opponent's deck as effectually as possible. Before the contending ships are in such proximity, is the proper time to try what skill and precision can effect upon the enemy's masts and rigging; and vessels fitted with long guns, and manned with expert gunners, will always do well to make previous trials of their arms and qualities; then, with good, circumspect management, close action will not fail to begin with advantage; and the author, from his knowledge of the effects of cannon shot, considers himself justified in concluding that a vessel, equipped in a proper manner, and possessing gunners trained in the way he proposes, cannot fail in tearing to pieces, in ten minutes, any opponent who, not possessing superior advantages in skill or force, is unable to prevent it.

2 E

392. Some of the actions between British and American sloops afford some very instructive illustrations on this important question. In the action between the "Hornet" and the "Peacock," the decisive importance of a few body wounds was unhappily too strongly displayed. The American ship was a good deal injured in her rigging, though comparatively little damaged in the hull; but the British sloop was forced to surrender entirely in consequence of having been hulled so low, that the shot-holes could not be got at; and she sank a few minutes after, having been obliged to yield to this fatal circumstance only!

The "Avon" was brought to the painful necessity of striking to the "Wasp," from being reduced to a sinking state by body wounds; and went down immediately after the last of her brave crew were removed. In this affair the American first crippled the "Avon's" rigging, with dismantling shot from long guns, and then aimed at her hull with fatal success. The "Wasp" does not appear to have been materially injured; for she escaped from a vessel (the "Castilian"), in a short chase that took place, before she was recalled by the "Avon's" signal of distress.

In these two actions it is clear that the fire of the British vessels was thrown too high, and that the ordnance of their opponents were expressly and carefully aimed at, and took effect chiefly in, the hull. The inferior effect of our fire may partly have arisen from such errors in carronade practice as have already been noticed; but it may be suspected to have arisen, chiefly, from not having chosen the most advantageous moment for firing; and this we shall be better able to show, by reviewing the action between the "Frolic" and the "Wasp."

393. This affair will be found to support very strongly what has been said against the measure of firing with the rising motion. The contending vessels were pretty nearly matched in armament; but the "Frolic" went into action under the serious disadvantage of having her mainyard sprung, and useless. The

"Wasp" having the wind, came down and engaged the "Frolic" to windward, on the port side, and consequently fought her lee side against the weather side of the British sloop. The American was considerably injured in her rigging early in the action, and also received a few shot in the hull; but much more serious damage and severe loss were sustained by the "Frolic." This difference of effect may fairly be ascribed, in a great degree, to the crippled state in which our sloop commenced the action; but we cannot hesitate to allow, that it may also have arisen from the circumstance of "her motion being much more rapid and violent than that of the 'Wasp,'" as has been remarked by a very intelligent writer.[a] But the "rapid motion" which so much disturbed the direction of her fire, appears to have been occasioned by the quick dips of lee-lurches; for she fired with a rising side, and, as there was a heavy swell, this motion must have very rapidly disturbed the pointing of her ordnance, whatever was her trim. That the Americans did not fire with the rising motion, we know from the parties themselves; that they could not fire in the hollow of the sea, in such a swell, is evident; and that they did not fire in the lee-lurch is clear from the admitted fact, that the ship rolled her carronade muzzles to the water's edge: we may therefore infer, with certainty, that she fired, in general, from the top of the sea towards the termination of the falling motion. That the British sloop fired with the rising motion is also certain—it is so stated in accounts which, however exaggerated in regard to strength and comparative loss, are unhappily true in the main feature; and the explanation of this affair, from authority, so far as it relates to rapidity of motion, states a cause of error which every seaman knows must be greater in a lee-lurch than in a weather-roll.

If there be any truth or accuracy in these remarks, this case illustrates, very forcibly, what has been offered upon the important question—whether it is most

---

[a] James's 'Naval Occurrences,' p. 146.

advantageous to fire with a rising, or with a falling side; and, together with the losses of the "Avon" and "Peacock," show, that this reasoning applies particularly to cases of fight between small vessels.

A close review of these operations shows clearly, that the Americans prefer to fire with the falling motion; and from what has been said above (Arts. 389, 390) we may fairly conclude that this circumstance had considerable influence in producing the results which we have been considering.

394. When it is expedient to aim partially at the rigging, one or more guns, conveniently placed, and fitted expressly for such purposes, should be named for this service specially, and loaded accordingly. The main-deck guns cannot be elevated sufficiently to effect this, when the enemy's ship is close; and, since case or grape-shot from carronades scatter so much as to be very inefficient, ships should always be provided with at least one piece of ordnance on each side of the quarter-deck and forecastle, fitted for this important purpose. Brass field-howitzers, mounted on carriages made to allow of great elevation, are now provided; and they are also fitted for boat-service, as well as to be used for field-service. Dismantling guns should be capable of being elevated to 30° at least; so that the enemy's main-top may be under the command of a powerful fire of case-shot at close quarters.[a] Guns thus mounted may be fired, en barbette, over the barricades or gangways, and easily brought to bear upon an enemy alongside, or laying across either stern or bow. In the position of the "Cleopatra," when she suffered so much from the "Milan," and of the "Phœbe" when she was so annoyed by the "Didon," before Captain Baker gallantly captured this last, the quarter-deck 9-pounders of

---

[a] "Il y a aussi le tir à *démâter*; mais ce tir, fort incertain, qui fait perdre presque tous les boulets et qui est une des plus graves erreurs de nos dernières guerres, est aujourd'hui presque complètement abandonné. Il ne doit être employé que sur l'ordre exprès du commandant du vaisseau. Ce tir consiste à viser de manière à frapper au trélingage, principalement celui du mât de misaine."—*Instruction sur le Canonnage à Bord.*—PRÉAUX.

those ships might have borne with ease upon the enemy's tops, when no other guns but the stern-chasers could be used.

As, in such positions as these, all broadside guns become useless, ships' crews should be exercised to form themselves, rapidly, upon the deck, boats, and booms, when called for small-arm duties. If, quickly ranging themselves thus, they be instantly supplied with arms, the enemy may soon be driven from their tops; and subduing the fire from thence is a favourable preliminary to the assault by boarding.

395. The best method of opposing the enemy's topmen is to have a few expert marksmen similarly posted; and for this service the quickly loaded rifle-muskets, with cylindro-conical shot (Appendix A), will be highly advantageous; but, at close quarters, ordinary case shot, or large charges of musket balls, may be used with great effect from the elevating guns; and for this purpose some rounds of this nature of charge should always be kept ready for any piece that may have the best opportunity of using them with effect.

396. Dismantling rigging, and carrying away spars, are more likely to be effected when it blows fresh than in light airs. Carrying away a stay, or a few shrouds, or wounding a mast or spar, in a strong breeze, may occasion a serious crash, which in a light wind would not ensue. With respect to sails, in moderate breezes the perforations of shot leave only small holes; but in strong winds a sail frequently splits upwards, as far as the reef-bands at least, as soon as it is perforated.

397. Whether pursuing or pursued, the only chance of stopping an enemy is by bringing down some of his rigging; it is therefore most important to consider the best mode of effecting this. The random aim of a whole broadside battery will be much less likely to accomplish it than the cool and careful use of one well-served gun. Hauling-up, or bearing away, to rake a flying, or a pursuing enemy, always produces a very random volley; for as the change of course must occasion much loss of distance, it is necessary to perform it

so quickly that the effect is seldom good, the distance, or range, altering very much before the vessel comes to a position proper for opening her broadside fire. This alteration of position brings with it a great and unknown alteration in the ship's inclination; consequently a considerable change in the elevation at which the ordnance may have been laid, and which there is not time to correct. It is almost incredible, indeed, how little effect is produced by this sort of raking fire; and the observation requires therefore to be supported by facts. In a certain action, a 74-gun ship bore up across the stern of an 84, to rake her, at a cable's length distance, in moderate weather and smooth water. The 74 had been upon a wind, and not having, perhaps, allowed for the alteration of elevation that would take place after bearing up, not one shot took effect! A proof of what may be effected against the *personnel* of a ship by yawing, and giving a close raking fire of well-directed grape, was gallantly shown by the "Inconstant," commanded by the late Admiral (then Captain) Freemantle, who, keeping in the wake of a line-of-battle ship, gave a raking fire of grape with tremendous effect upon her people, who were very much exposed in striving to clear the wreck of her topmasts, which had been carried away by an overpress of sail.

398. The attack of the American squadron under Commodore Rogers, on the "Belvidera," Captain Byron, furnishes a strong proof of the inefficacy of volleys of raking fire with round shot. Captain Byron, seeing the squadron bearing down upon him in a suspicious manner, and having reason to expect that war had been declared, very prudently kept away also: he gradually made sail, and a chase ensued, during which the "President," outsailing her consorts, came up with the British frigate.

The "President" commenced the attack by firing a heavy bow-gun, by which nine men in the "Belvidera" were killed or wounded; and by continuing to fire from single guns, deliberately aimed, without altering the ship's course, she did much further damage to the chase.

But when, gaining further on the British frigate, the "President" yawed and gave her broadside volley (which was several times repeated), she did the "Belvidera" no further damage beyond cutting a brace or two, and wounding a few spars! Our frigate answered these attacks deliberately, with her stern and quarter guns, with such effect that the "President" (having had a gun burst) suffered more than her expected victim; and the steady and determined manner in which Captain Byron conducted both his defence and his retreat, reflects on him immortal honour, and on his crew a full share of credit. These facts serve to show how much more may be executed by cool, deliberate aim, with single guns, in the hands of well-trained gunners, than by repeated random volleys of whole broadside batteries! Shells, having time-fuzes, and spherical caseshot, may be used with great advantage for dismantling purposes, at very considerable distances, in chasing any vessel; and the moral effect produced by the bursting of a shell is so scaring to a non-combatant vessel, a merchant-ship, pirate, or slaver, that it will rarely fail in bringing-*to* such a chase.

In the chase of the "President," and the action which ensued between her and the "Endymion," some of the best gun-practice ever effected by British seamen was displayed. This operation also affords matter for remark as to the effect of dismantling fire in chasing.— The "Endymion's" sails were completely torn to pieces, and her spars and rigging much cut, by the American dismantling-shot. One of these shot cut away twelve of fourteen cloths of the "Endymion's" foresail, and stripped it almost entirely from the yard.[a]

399. So great a proportion of an enemy's side is opened by the large ports of a carronade battery (4 feet

---

[a] The American dismantling-shot which tore the "Endymion's" sails, was composed of four or five iron bars, each about two feet long, fastened by ringheads to a strong ring. These dismantling-shot were likewise used in other actions (as between the "Java" and "Constitution"), but with little or no effect. The flight is so irregular that they cannot be depended on; and therefore round shot, which may fell a mast, are greatly preferable.

in width and more in height), that grape, or case-shot, seldom fail to commit great execution from entering these large ports, as well as from cutting the rigging. Case-shot for 32-pounders, being composed only of 8 oz. balls, have not power to do much mischief against the *matériel* of an enemy, and should therefore only be used, in good opportunities, specially against the *personnel*; but grape-shot may be used in certain proportions from heavy guns, in any close action, because they are capable of committing infinite ravages against both. Three-pound iron balls, of which there are nine in 32-pounder grape, will penetrate the enemy's barricade defences on the upper deck, and though they cannot penetrate a mast, or by any direct wound bring it down, yet they can break chain-plates, cut shrouds or stays, however thick, and from the number of such chances, will be very likely, in a strong breeze, to dismast the enemy. But it appears, from a close analysis of the reports of the damages sustained by H.M.S. " Shannon" and the United States' frigate " Chesapeake" in action with each other (Art. 254), that very few grape-shot passed through the side of either ship below the barricades of the quarter-deck and forecastle; whereas there were no less than 80 penetrations of grape-shot into the " Chesapeake" through the bulwarks of the forecastle and quarter-decks, which must have produced very fatal effects upon the people stationed there. Nor can the small penetration of grape-shot into the more solid parts of the " Chesapeake"—only from 3 to 5 inches—be easily accounted for. It may therefore be surmised that the upper-deck guns of the " Shannon," chiefly carronades, were overloaded by putting grape-shot over the round shot, which, with a charge of only 2 lbs. 10 oz., may have caused the trifling penetration. However this may be, the facts recorded in these instructive reports may create a doubt whether, in general, a double-loading of round-shot and grape should be preferred to double-shotting all the main and lower deck guns of a ship, in close action, also whether the practice of loading alternate guns with round-shot and grape should not be discontinued, and all such guns loaded with two round shots.

400. The firing of Shrapnel shells, like that of other shells with time fuzes, may be used with great advantage from ships against troops on shore (Art. 257), or against vessels crowded with men on their upper decks, at distances far beyond the effective ranges of grape or caseshot, and in general, wherever the *personnel* of any ship or battery may be seen, and reached. But the service charges for spherical case-shot being necessarily smaller than for other shells, on account of their inability, being thinner, to stand large charges (for which reason those for heavy guns have recently been considerably reduced), the penetrating power, whether of the bullets or the splinters of the shells, is so inconsiderable that this description of projectile is not calculated to produce any effect on the more solid parts of a ship. If the fuze be too long, the shell of the spherical case would most likely break on striking; and, should it explode in a ship, would be far less destructive, from the smallness of the bursting charge, than the common shell; for these reasons, spherical case-shot is of little use in naval actions, and is supplied chiefly for boat-service and the ship's field-guns.

401. Shrapnel shells, to be effective, require to be fired from guns as direct as possible at the body against which they are used; that is, with full service charges, and the least elevation which the case may admit; but this projectile cannot be employed with any useful effect from howitzers at very great elevations, or from mortars; for the bullets, released from a shell which bursts at a considerable altitude in the declining portion of a very elevated trajectory, will strike the plane below with very little horizontal force, and fall to the ground with very limited vertical velocity, approaching, more or less, to the "terminal velocity" due to bullets of that size and density. The circumstances are similar to those in M. Carnot's project of vertical fire with musket bullets, the effects of which that celebrated engineer very much over estimated (see Art. 78, Note) in his system of

defence, by disregarding the effects of the resistance of the air.

402. The first employment of shells fired direct from long guns at bodies of troops at considerable distances, was at the memorable defence of Gibraltar in 1781.—Drinkwater's 'Siege of Gibraltar,' 4th ed. p. 167.

In firing from batteries placed high on the rock, from which the whole interior of the besiegers' trenches and batteries could be seen, round shot fired from heavy guns would evidently have been a wasteful and ineffectual practice against the workmen and troops thus exposed to the direct, though depressed, fire of the fortress. The distance being too great for grape or case-shot, howitzer shells were tried; but shells fired directly from howitzers had neither accuracy nor force sufficient to take full advantage of the command which the batteries possessed over the works on the isthmus below. The charges being small, the projectile velocity of the shells, at the moment of bursting, was not sufficient to impel the fragments forward with the force required to produce the desired effect, whilst the great bursting charges which the shells contained occasioned very great dispersions of their splinters. Guns were therefore substituted for howitzers, and $5\frac{1}{2}$-inch shells fired from long 24-pounders, with as large charges as the shells could resist. The effects were prodigious—the fragments of the shells were driven forward with far greater force—the dispersion was less, on account of the great preponderance of the projectile velocity—and the effects upon the troops and working parties in the enemy's trenches and batteries were extremely destructive.

This remarkable instance of the efficacy of direct shell-firing attracted the notice of all artillerists. The conditions of this description of practice are materially different, as has been already explained (Art. 263), from that of shells whose effects depend wholly upon their explosive force; but the few fragments into which common shells are broken—usually 15 or 16—forming what may be termed a charge of langrage, consisting of a few irregular lumps of iron—are neither suited in form, nor capable in number, of producing any very extensive effects upon large bodies of troops. The late Major-General Shrapnel had the ability and sagacity to perceive that, under such circumstances, the effects of direct shell-firing might be prodigiously increased by filling the shells with musket or carbine bullets, enlarging the charges in proportion, and reducing the bursting charge to a quantity just sufficient to break the shell with as little scattering effect as possible upon the bullets; and to that able and distinguished officer, therefore, is due the credit of the invention which has rendered his name so justly celebrated.

403. Numerous reports having been recently received, from all out-stations where practice is carried on with spherical case-shot from heavy guns, to the effect that the shells, in nearly all cases, burst in the guns when fired with the service charges, and this having been attributed to defects in the fuzes, a series of experiments were made in 1850 to ascertain the real cause of this premature explosion.

The experiments were made from an 8-inch shell-gun, a 32-pounder gun, and a 24-pounder gun, with the full service charges of 10 lbs. for the two former and 8 lbs. for the latter. The result was, that the shells were invariably destroyed within the guns. The same results took place when the shells were filled in the ordinary way with musket balls and bursting powder, *but without fuzes*, the fuze-holes being plugged. The charges were then reduced to 8 lbs. for the 8-inch gun and 32-pounder, and to 6 lbs. for the 24-pounder, when a large proportion of the shells fired burst in the guns as before. The charges were then further reduced to 6 lbs. and 5 lbs. with fuzes fixed, when it was found that the shells resisted the concussion of discharge and burst at the proper time; also, that the penetrations of the bullets were efficient at the ranges tried, viz., 950 yards from the 32-pounder and 24-pounder guns, and 1100 yards from the 8-inch shell-gun; and these charges have been established accordingly for spherical case-shot from all ordnance of those natures and descriptions.

The failures proved that the metal of the shell, in spherical case-shot, is too thin to withstand the necessary charges, not that these were too great for the purpose; for it is essential to the efficacy of spherical case that the shell should be moving with great horizontal velocity at the moment of bursting: it may therefore be doubted whether the evil shown in these experiments might not be better remedied by increasing the thickness of the shell, maintaining the original service charges, than by depriving this important and useful projectile of so much of its power.

In an extensive course of experiments carried on in 1852 with Shrapnel shells, it was ascertained that the frequent failures of the practice by the premature bursting of the shells, are not occasioned by the concussion of discharge, but that the ignition of the bursting charge also arises from the friction of the balls on one another and against the interior surface of the shell, the powder being mixed with the bullets. In order to avoid this evil, it was proposed by Captain Siemen, of the Hanoverian Artillery, to cement the balls in one mass by pouring among them liquid sulphur or plaster of Paris, a cylindrical space being left near the fuze for the bursting powder; and the method appears to have answered the purpose well. But, in 1852, Captain Boxer proposed that the balls in the shell should be completely separated from the bursting powder by means of a wrought-iron plate, or diaphragm, in the shell. This construction was found to succeed admirably; for, in an experimental trial at Shoebury Ness, 119 8-inch shells and 79 24-pounder shells were fired, with 5½-inch fuzes and with the service charge of powder, without a single premature explosion; and, in a trial at Shoebury Ness, with shells which had been long in store at Woolwich (the balls being taken out, the shells cleaned and refilled with balls), of 656 shells projected, only 37 burst prematurely.

It results from this important discovery, and the results of the experiments made with the Boxer fuzes, that the reduction of charges recommended by the committee of 1850, and then adopted, should be forthwith abandoned, in order to restore to Shrapnel shells the power of which, by that reduction, they were deprived; and thus enable field artillery, if properly used, to maintain the relative superiority of that arm to the improved muskets now so generally adopted in European armies. This will not, however, be by Shrapnel shells fired from 6-pounders; these have neither capacity nor power for an efficient use of that destructive projectile; but 9 or 12-pounder guns, making good use of Shrapnel shells, will fully maintain their superiority over any muskets.

404. Range tables, with spherical case-shot from guns, not having been formed for the naval service, the author inserts a table, XXIV., Appendix D, abstracted from practice made on board the "Excellent," of the ranges with 7.9-inch, 6.76-inch, and 6.177-inch spherical case-shells fired from 8-inch (68-pounder) 42 and 32-pounder carronades, and with 5½-inch spherical shells fired from a 24-pounder field howitzer, and the 5½-inch new gun for boats. Tables XXV. and XXVI., Appendix D, are compiled from Lieut.-Colonel Burns' practice-cards, published by authority, which contain the most exact and recent directions for conducting this and all other land-service practice.

405. In close action, rapidity of fire is of the most decisive importance, provided accuracy be not sacrificed to it; for, in proportion as we increase the quantity of fire with equal precision, we in fact increase our force. In close battle, when it is scarcely possible to miss, the vessel that can soonest reload her ordnance, and give her second broadside, supposing both ships to have opened their fire nearly at the same time, must have a prodigious advantage over her opponent. The power of doing this with efficacy, as well as rapidity, can only be acquired by the constant practice of every minute detail relating to the manual exercise and to the pointing of the ordnance.

406. Quick firing depends greatly upon the manual strength of the gun's crew to perform the necessary operations, and particularly that of running the gun

out in the least possible time. But it appears that, in some classes of ships, the number of tackle-men afforded by the total strength of the ship's company, has not been increased commensurately with the great weight of ordnance introduced into the naval service. There is no great difference between the number of men forming a ship's crew in our service and the number forming a crew in the French navy on the peace establishment; but, on the war establishment, a French crew is superior in number to an English crew; and the sea-going ships, the squadrons of evolution, &c., of the French, are, with respect to the number of men, on the war establishment. There is no difficulty in running guns out, however heavy they may be, in a ship which is perfectly or nearly upright, with the regulated gun-crews, aided by the improved mechanical contrivances which have been adopted for this purpose. But these means do not altogether suffice in all the varieties and contingencies of service. In fighting on the weather side, in a strong breeze, even with full complements of guns'-crews, great difficulty is experienced, and much time lost, in running the gun up the inclined plane which the deck then forms; and, if this be so under ordinary circumstances, how much greater must be the difficulty and loss of time in battle, when casualties happen, from men being killed or disabled, or when firemen, sail-trimmers, boarders, &c., are called away! The deficiency of men is very much felt in heavy-armed frigates and in other vessels; and it has accordingly been found necessary to increase the power of the gun-tackles by a double block at the tail as well as at the head; but this is attended with a proportional loss of time in running the gun up; and, though the time thus lost may be little, yet so great is the importance of quick firing, that every effort possible should be made to expedite the operations of loading and running up the gun. The French have only a single block at the tail of the tackle (see Art. 411).

407. The great attention paid to every expedient that can contribute in the slightest degree to quick firing is very apparent in all recent French publications on naval

gunnery, and in the regulations for the manual exercise of their naval artillery (see Art. 424, *et seq.*). "In all artillery battles," writes De la Gravière, vol. ii. p. 236, "nothing can dispense, or be allowed to interfere, with the attainment of the utmost precision and rapidity of fire;" and this maxim is enforced with the greatest weight in the Avertissement (p. 8) to the '*Exercices des Bouches à feu en usage dans la Marine.*'

408. In the first edition of this work the author recommended that, in reloading guns on any occasion in which rapidity of firing is of the utmost importance, the cartridge, shot, and wad should be set home at one operation.[a] It does not appear that there is any other objection to this practice than that which arises from the ball being apt to roll on the tie of the cartridge and thus become jammed in the gun. Such an accident, however, may easily be prevented by cutting the tie short off, or by fastening it round the body of the cartridge. (For the French method see p. 459.) The reduction in the time of loading by such means may appear trifling, but, in close action, the issue of a battle depends so much upon the rapidity with which the first few rounds are fired, that it is justifiable and even necessary to resort to any expedient, the effect of which cannot be detrimental, while its success may ensure victory.

409. Soon after the French translation of the first edition of this work appeared, experiments were made in France to ascertain whether *la charge simultanée*, as it was called, could be adopted with perfect safety, and with the advantage, which the author of this work proposed, of loading and consequently firing with greater rapidity than in the usual way. In a subsequent publication (1845), M. Charpentier, the translator of the 'Naval Gunnery,' gives the results of the extensive trials of that method, made in the French fleet under

---

[a] This method of loading appears to have been practised on one or two occasions during the late war.

the directions of Admiral Lalande, in 1840, extracts of which are given in the subjoined note.*

410. The French, like us, have had much experience of the inconveniences and difficulties in loading chambered guns, and have found it necessary to make a material alteration in the dimensions of the chambers before they could succeed in executing la charge simultanée with the canon-obusier of 80 No. 1, of 1842, which is their principal shell-gun.

When the canons-obusiers were first introduced, each was furnished with two rammers, having heads of different diameters, one to set home the cartridge into the chamber, the other with a hollow in the rammer-head to cover the fuze in pushing home the shell, a wad having previously, by a distinct operation, been put over the cartridge; and, to avoid the necessity of using the worm after every two or three rounds, a small worm was lodged in the end of the sponge, imbedded in the wool, by which means the sponging and worming might be performed by the same operation after every discharge. Next, to obviate the necessity of putting a wad over the cartridge, in order to keep it in its place before the shot was put in, trials were made of wooden

---

* " On a essayé de faire usage de gargousses allongées; et après avoir constaté que la diminution de leur diamètre n'exerçait aucune influence défavorable sur les portées, on a poursuivi les essais, afin de s'assurer qu'elles se prêtaient également bien à l'exécution de la charge simultanée.

" On a tiré 40 coups avec le canon-obusier de 30, en pratiquant les différents genres de tir qui peuvent être exécutés dans un combat; la gargousse s'est constamment rendue au fond de la chambre, sans qu'il ait été nécessaire de refouler.

" En expérimentant avec un canon-obusier de 22 centimètres, No. 1, modèle 1842 (Art. 220), on a également tiré 40 coups, dont les vingt premiers à la charge de 3 k. 50 (6 lbs. 12 oz.), sans éprouver aucune difficulté; pour les vingt autres coups, les gargousses étaient de la contenance de 2 kilogrammes (4 lbs. 6¾ oz.).

" Dans cette série, la charge s'est arrêtée dès le second coup, au raccordement de l'âme avec la chambre, et cette circonstance ayant été attribuée à ce que la gargousse était encore trop courte, on y a remédié en formant un tampon avec l'excédant de la ligature, au lieu de rabattre cet excédant, ainsi que cela se pratique. Cet expédient a réussi.

" Enfin, la charge simultanée s'est parfaitement bien exécutée avec le canon-obusier de 0 m. 22, No. 2."—(Art. 220.)—(See p. 78, *Section on La Charge simultanée des Bouches à feu*, in the Essai of M. Charpentier.)

wads, and hollow tompions made of tin of different lengths, which were attached to the reduced cartridges, to fill the space in the chamber that would otherwise be left void. But neither of these expedients having been found to answer, the wads were suppressed, and the cartridges elongated [a] sufficiently to occupy the whole length of the chamber, as explained in the Note on Art. 217. See also Charpentier, '*Essai sur le Matériel de l'Artillerie,*' p. 79, and the '*Aide Mémoire Navale.*'[b] By this contrivance, simultaneous loading has succeeded perfectly with canons-obusiers without the aid of the small-headed rammer (see Charpentier, p. 346; also, in the present work, Art. 428, p. 454 (*Load*), and Art. 432). Cartridges of this formation were likewise adopted for carronades, and all other chambered ordnance, with this difference only, that, as the chambers of carronades are hemispherical at bottom, so their cartridges terminate in that form.

411. The method of simultaneous loading proposed by the author (see the first edition, Art. 192), and adopted in the French navy in 1840, having lately attracted the notice of naval officers in this country, orders were issued in 1851 to make trials of that system on board H. M. S. "Excellent," and to report the results.

Much prejudice existed at first against this mode of loading, and many evils were anticipated. With 8-inch chambered guns it was found necessary to resort to some

---

[a] "La forme des gargousses ordinaires ne permettait pas la charge simultanée pour les pièces chambrées, mais une forme nouvelle donnée à ces gargousses a rendu cette charge possible, et la même méthode a dû s'appliquer à la charge des pièces chambrées."—*Exercices des Bouches à feu de la Marine, Avertissement*, p. 8.

[b] Canon-obusier de 0.22 (8.67 inches):—

	Charge. lbs. oz.	Diameter of Plug. Inches.	Length of Cartridge. Inches.
No. 1 . .	7 11.5	5.5	10.24
	4 6.6	4.69	9.06
No. 2 . .	6 9.9	5.48	9.81
Canon-obusier de 30:—			
	4 6.6	4.73	8.27
	3 5.9	4.29	7.88

such expedient as that described in the notes p. 98, to ensure the reduced cartridge being properly set home in the chamber simultaneously with the shot. This has been effectually accomplished by placing a cork wad *a* (fig. 37) in the top of every reduced cartridge, of a height sufficient, together with about half the shot, to occupy the whole length of the chamber, upon the principle stated in Art. 123, Note b, instead of putting a grummet wad over the reduced cartridge to keep it in its place, as heretofore practised on board the "Excellent" (see Note a, in the article just quoted), so that the shot on receiving the blow of the rammer-head may cause the cartridge, with the cork wad, to spring into its place in the chamber.

Fig. 37.

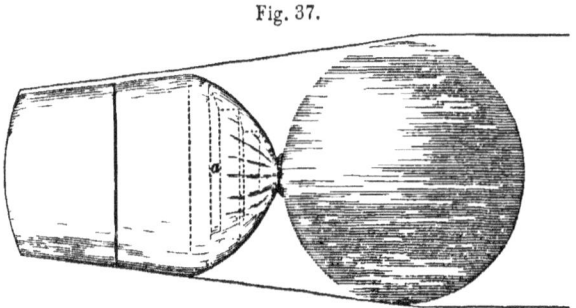

412. After repeated trials with guns of all calibres, from the 10-inch gun to the 6-pounder, with single and double shot, and with shells, no objection whatever was experienced; whilst the advantages to be gained are important by obtaining greater rapidity of fire, and a considerable saving in labour, by ramming the whole charge home at one operation.

These results were forwarded to the squadron of evolution for further trials under all the circumstances of service, and particularly when there was much motion. Such objections or inconveniences, as were found to exist with respect to the chambered guns, were readily obviated; and the Lords of the Admiralty having witnessed the success of the experiments made on board the "Excellent," directed "simultaneous loading" to be established throughout the fleet.

The only disadvantage arising from the adoption of this expedient is that the bulk of the reduced cartridges, and consequently the space required for stowage, is as great as for those which contain the full service charge; but this is so trifling an inconvenience compared with the magnitude of the evil overcome, that it is not worthy of consideration in ships armed with a limited number of shell-guns, according to regulation, and does not amount, we believe, to more than the bulk of two or three ammunition cases in a line-of-battle ship.

413. In this manner simultaneous loading is as easily practised with chambered, as with other guns, and the importance of this method of loading may be stated thus—that in whatever proportion the operation of loading may be quickened, in the same proportion is the actual force of the broadside battery practically increased, as if it consisted of a greater number of guns.

414. The 10 and 8-inch guns were introduced into the British Navy confessedly as analogous guns to the French canons-obusiers, and being like them designed principally for shell-firing and provided accordingly, were then denominated "shell-guns." But all natures and descriptions of naval ordnance being now provided for shell-firing, this nominal distinction has ceased to exist, and those guns should now be contradistinguished from others, by their inherent distinctive characteristic "chambered guns."

415. Seeing that great inconveniences have been experienced both in the French and British services— and it may be added in all others—from the difficulties which the contraction of the chamber, of whatever form, presents to the operations of loading; and that guns having no chambers may be loaded with perfect facility, simplicity, and rapidity, without the aid of any adventitious expedient, it may well be asked how far it is necessary, convenient, and advantageous, or the reverse, that the guns from which shells are to be fired should be chambered; and whether the truncated cone, which we alone have adopted, is the best form of

chambers for guns designed for horizontal firing—as no doubt it is for mortars? (Art. 123.)

416. Much has been written, many elaborate investigations made, and experiments tried, with chambers of all forms and shapes, conical, cylindrical, spherical, and with some of these forms reversed; to ascertain in which description of chamber the impulsive force of the powder acts with the greatest effect against the projectile, and consequently upon the range.[a] The author has carefully and fully consulted all the more important works and documents which treat of this subject, and referred to the experiments which were made at Woolwich (in 1787-9) and elsewhere; and has no hesitation in coming to the conclusion at which Dr. Hutton arrived, "that, however mathematical speculations may show a preference to one form over another, it is found that form is very immaterial, and that in practice the chief point to be observed is to have a chamber of a size just sufficient to contain the charge of powder, and no more, so that the projectile may always be in contact with the powder."[b]

417. To fulfil, completely, this important condition, it is obvious that part of the chamber—the whole, as appears in Fig. 37, p. 434, being, notwithstanding the introduction of the cork wad, too long—might with advantage be removed. Chambers cannot be wholly dispensed with in guns of such high calibre as those of 8 inches and upwards, without augmenting considerably the weight of the gun, and increasing the guns' crews on already crowded decks, and without incurring fresh difficulties in using reduced charges in bores of that diameter; but it does not follow that the existing

---

[a] Though it be agreed that chambers of a spherical form, or in the form of a pear, cause the greatest effects to be produced with a given charge, yet the difficulties which they give rise to, both with respect to the formation of the fire-arm itself and that of the charge (les munitions), have induced artillerists to make the chambers cylindrical to all howitzers and to most mortars. Chambers in the form of truncated zones are disadvantageous on account of the form of the cartridges.—*Traité Élémentaire d'Artillerie*, by Decker.

[b] Dr. Hutton. See, in particular, the article *Chambers*, in his 'Philosophical Dictionary.'

chambers should not be reduced in capacity and length. Although, by the employment of cartridges with wads, the stowage-room of filled ammunition, in boxes, is in no ship increased by more than that which is necessary for six cases, yet every ship must take a certain proportion of spare wads of cork or light wood, for filling and completing the regulated number of reduced cartridges to make up for those that have been expended; and it will likewise be necessary to provide supplies of these wads in all arsenals and stations at home or abroad. Is it not, therefore, desirable to ascertain, by actual experiment, whether the chamber of the 8-inch gun of 65 cwt. might not be sufficiently shortened to fulfil the conditions specified in Art. 416, and so render the cork or any other wad unnecessary? This would reduce the chamber to a form and capacity which, on the one hand, would satisfy the objections to the suppression of the chamber, and effectually remove the evil arising from the chamber being, as at present, too long.

418. It appears by Fig. 37, p. 434, that a space of about 3 inches would be left void between the projectile and the reduced cartridge, were not the cork wad, as represented by the dotted lines, inserted in the head of the cartridge. If, therefore, the bore were continued cylindrically 3 inches further than at present (Fig. 38,

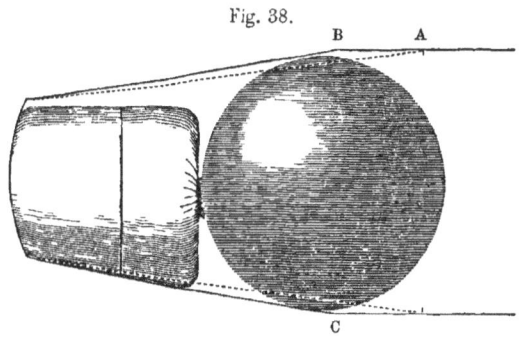

Fig. 38.

see also Plate II. Fig. 14), the conical part to commence where the cylinder produced from A to B ends, the depths cut away at B and C would only be 0.225

of an inch each, the quantity diminishing in proportion to the convergence of the sides of the old and new cone, the total quantity of metal to be removed would be $35\frac{1}{2}$ cubic inches, and the weight only 9 lbs. 6 ozs., and this small quantity would admit of the projectile being in contact with the powder contained in a reduced cartridge.

419. The inconvenience which principally accompanies chambered ordnance is, that the capacity of the chamber, which is calculated for exactly receiving a full charge, is too great to be filled when a reduced charge is introduced. Now the proportion of the length to the mean diameter of a chamber having the form of a conic frustum has not yet been determined; and it appears to the author that no established principle in gunnery would be infringed, but rather benefit would be gained, by shortening the chambers of our 8-inch guns of 65 cwt., and even those of 50 cwt., to the extent requisite for adapting them to the reduced charges. The expedient used by the French artillerists to overcome the inconvenience of the cylindrical chambers in their shell-guns, when reduced charges are used, or when simultaneous loading is practised, is to make all their cartridges of the length of the chamber nearly (Art. 410, Note b), and the areas of the transverse sections proportional to the weights of the charges. Now, although it is stated by M. Charpentier, in his *Essai* above quoted (Art. 409, Note), from some experiments, that the diminution of the diameters of cartridges, the charges being equal, had no influence on the ranges; yet from other experiments which have been made in France, as well as elsewhere, it has been found that, in projecting both solid and hollow shot, with equal charges, the cartridge which has the greatest diameter gives to the shot the greatest initial velocity [a] (Art.

---

[a] It appears from various publications, and from the facts stated in the several parts of this work, that chambered ordnance are losing favour as well in the French land service as in that of the navy; it is proposed[*] to abolish the field service howitzer, and substitute 12-pounder guns without chambers,

[*] Nouveau Système d'Artillerie de Campagne de Louis-Napoléon Bonaparte, Président de la République (now Emperor of the French), par M. Favé, 1851. (See Appendix B.)

154). It would be preferable, therefore, to shorten the chambers of our 8-inch guns sufficiently (about 3 inches) to permit the projectile to be in contact with the cartridge when a reduced charge is employed, and thus enable the expansive force of the powder to exert the greatest effect; while the gentle contraction of the chamber at the place where it unites with the cylindrical bore would permit a cartridge containing a full charge to lie without derangement in the chamber and the adjoining part of the bore.

The shortening of the chamber would be advantageous in another respect: in firing single shot or shells with reduced charges, and particularly in firing double shot, when greater charges than 5 lbs. cannot be used, the unoccupied part of the chase being about half a calibre longer with the diminished than with the usual chamber, the projectile would be acted upon by the charge during a longer time under the guidance of the bore, and thus both the initial velocity would be increased and the direction of the fire more correct.

420. The 8-inch gun of 65 cwt. possesses, as has been shown in Art. 246, prodigious strength in resisting the extraordinary charges with which it was tried à outrance, and which far surpassed anything that could have been expected or designed; it may, therefore, well admit of the abstraction of so small a quantity of metal as that of 9 lbs. If it should appear on trial, however, that the removal of this small quantity of metal would so far weaken the gun as to be inadmissible in existing guns, that alteration might at least be made prospectively in any new gun that may be required, by adding to the thickness of metal in that part of the gun as much as is taken from within. How far this partial removal of the evil arising from the chamber being too long can be carried in the lighter 8-inch guns is a question which actual experiment only can answer; but, however this may be, if the shortening

---

as equally adapted to fire either shot or shells, and to admit of far easier loading and quicker firing, to which the French attach, most justly, as much importance in military as in naval actions.

of the chamber of the 8-inch gun of 65 cwt. can be effected with safety and advantage, considering that this gun enters so much more largely than the lighter 8-inch guns into the armament of our ships and vessels,[a] there can be no reason why that advantage should not be taken, although it may not be procurable in the lighter guns of that class.

SECTION V.—ON THE MANUAL EXERCISE OF NAVAL ARTILLERY.

421. Although, for the reasons stated hereafter (Art. 423), it is not the author's intention to insert in this work the regulated manual for the exercise and service of naval ordnance in the British service in general, as taught on board the "Excellent," he cannot altogether omit to notice a most valuable addition which that important Gunnery School contributes to the security of the country and the efficiency of the navy, in instructing the men of the Coast Guard and of the Preventive Service, in the exercise and practice of Naval Gunnery, whether employed in manning our coast batteries or our ships.

The Coast Guard in Great Britain and Ireland consists of about 3300 men, the Preventive Service of about 560. These men are all trained and practised to the exercise and use of naval artillery in the following manner. One intelligent man of the Coast Guard from each district, and a certain number from the vessels employed in the Preventive Service, are sent to Portsmouth, where a hulk is appropriated specially for their accommodation and instruction, distinct from the establishment of H.M.S. the "Excellent;" and these men are carefully instructed in the gun exercise and practice, according to the instructions taught in the "Excellent," and established in the Royal Navy.

When complete in their course of instruction, and well trained to the practice, these men return to their respective districts, and there instruct and train by

---

[a] See Parliamentary Paper, No. 228.

degrees the whole of the men belonging to the districts; batteries being established in convenient localities, and the ammunition, projectiles, and stores being liberally provided by the Government. Thus all the men of the Coast Guard are drilled to precisely the same exercise as the men in the "Excellent," and fire a certain number of shot every year. Many of these men having served in ships and vessels of war, are moreover possessed of some knowledge of naval gunnery as practised at sea.

The men of the Boat Brigade are drilled to gun-boat service by instructors from the "Excellent;" they are also made to practise target-firing from the fixed land batteries established in the principal arsenals. All the men thus trained are inspected periodically by an officer highly capable of judging of their proficiency.[a]

A reserve of seamen or sea-faring men trained to the guns has thus been created at small expense; and these men, though belonging to a civil department, would, no doubt, enter the naval service in case of any emergency. Their places would immediately be filled by other men, who would be trained in like manner; and these, like their predecessors, would be available in time of need for the naval service.

422. Orders for Manual Exercise for the Boat Brigade and Coast Guard.

On coming to the gun, 1 and 2 see the locks fixed and fit for use, vents clear, sights adjusted to the distance required; they worm, sponge, load, and run out without further orders.

At the word "*Prime*,"

No. 1 opens the tube-box with his left hand, takes out a tube with his right, and primes—"Prime."

At the word "*Point*,"

No. 1 retires to the full extent of the trigger-line, leaning well over on his right knee, keeping his left foot clear of the recoil. The handspike-men pick up their handspikes, keeping them clear of the brackets; the rest of the Nos., with the exception of 2, double in

---

[a] By Commander Jerningham, R.N. (who was appointed to that ship for the special purpose), assisted also by gunners sent from the "Excellent" as instructors.

to the side-tackle falls—"Point." With heavy guns 2 and 11 assist handspike-men.

At the word "*Elevate*,"

the handspike-men place their handspikes on the steps of the carriage, under the breeching, and raise the gun off the coin; 2 steps in with his left foot in a line with the gun, keeping the right clear of the recoil, and withdraws the coin to the full extent; handspike-men lower the gun slowly and steadily. At the word "Well," 2 forces in the coin, and when he feels the weight of the gun, gives the word "Down" to the handspike-men, springing up to the safety position on the right—"Elevate."

At the word "*Ready*,"

No. 2 steps up to the right of the gun, clear of the rear axletree, cocks the lock with his left hand, and retires; the right rear man attends the train tackle.

At the word "*Fire*,"

No. 1 fires with a suitable jerk, springing up to the safety position on the left. Handspike-men ground their handspikes. "Fire."

At the word "*Stop the Vent*,"

No. 1 makes up the trigger-line hand over hand, and lays it across the neck-ring, forces in a vent-plug with his left hand, keeping his thumb on it, and his fingers extended along the vent-field; he half cocks the lock with his right hand. "Stop the vent."

At the word "*Run in*,"

all the Nos. man the train tackle, except 1, and 3, and 4, who overhaul the side-tackles. When the gun is in, 1 gives the word "Well." "Run in."

At the word "*Sponge*,"

Nos. 3 and 4 step inside the breeching together, 3 with his right leg, 4 with his left; 6 faces outwards and takes up the sponge, with his right hand over and left hand under, and gives it to 4, who receives it in the same manner, *forces it hard home to the bottom of the bore in two motions*, gives it two round turns, withdraws it hand over hand, gives it two smart taps under the muzzle, and lays it quietly across the breeching; while the sponge is withdrawing, 6 takes up the rammer. "Sponge."

At the word "*Load*,"

No. 3 receives the cartridge from the powder-man (facing inboard) and enters it seam sideways and bottom first, to the full extent of his arm; 5 gives shot and wad to 3, who enters them; 4 receives the rammer from 6, with his right hand under and left over, and, assisted by 3, forces all home together, hand over hand, giving them two smart blows; they then quit the rammer while 1 pricks the cartridge, to ascertain that it is home, and gives notice to 4.

No. 3 faces the ship's side; 4 springs the rammer, and lays it quietly across the breeching; 6 returns it, 3 and 4 step out together, 5 and 6 throw the side-tackle-falls to the rear. "Load."

At the word "*Run out*,"

all the Nos. man the side tackles except the right rear man, who attends the train-tackle, and 1, who keeps the gun bearing on the object. When the gun is out, 5 and 6 coil down the side-tackle-falls. "Run out."

At the word "*Extreme Train, Muzzle to the Right or Left*,"

Nos. 3 and 5, or 4 and 6, shift the side-tackle to the bolt of the next port; 6 keeps the sponge, rammer, and worm in a line with the gun; right rear man the train tackle; handspike-men assist with their handspikes.

At the word "*Square the Gun*,"

Nos. 4 and 6, or 3 and 5, shift the side-tackle to the quarter-bolt, 6 keeps the sponge, rammer, and worm in a line with the gun; right rear man the train-tackle; handspike-men assist with their handspikes; 1 looks out till the dispart is on with the centre of the port, and gives the word "Well! Muzzle to the right or left."

The detail for fighting both Sides.

Nos. 1, 3, and 4 never leave their guns. 1 primes as the gun goes out, taking care not to cock the lock till the muzzle is clear of the port sill, and not to fire till the side-tackle-falls are coiled down. After the order "Fire," he gives the word "Run in," when all the rest of the Nos. fall in to the train-tackles of the guns on their *left;* and at the order "Sponge and Load," he steps up to the right of the gun clear of the rear axletree, and gives sponge and rammer to 4, at the same time serving the vent. 3 receives the cartridge from the powder-man, provides and enters shot and wad. The rest of the Nos. fall into the train-tackles of the guns on their *right.* After the first round, they go from the train-tackles of the guns they have run in, to the side-tackles of the guns to be run out, taking up their places as at their own guns, and remain at those guns until they are again run in.

Words of Command for fighting both Sides.

"Man both sides"—"Prime"—"Point"—"Muzzle to the right" —"Muzzle to the left"—"Elevate"—"Ready"—"Left guns"— "Fire"—"Sponge and load"—"Fire"—"Sponge and load"—"Run out"—"Fire"—"Sponge and load"—"Run out"—"Fire"—"Cease firing."

At the word "*Secure*" (if out-board),

No. 1 gives the word "Elevate," takes out the coin, and places the parts of the train-tackle under the cascable; 2 unhooks and overhauls the train-tackle; 3 and 4 put in the tompion, haul up the half-port, secure the side-tackles, and hook the train-tackle to the side-tackle-bolts; 5 and 6 assist 3 and 4 in hauling-taut and securing side-tackles; rear men seize the breeching, stationary powder-man returns the powder, and the extra one swabs the deck.

At the word "*Secure*" (if in-board),

No. 1 gives the word "Elevate," takes out the coin, places the bed, sees the gun laid square between the housing-bolts, and that all the numbers perform their duties correctly; 2 prepares the train-tackle and superintends bousing taut the frappings under 1; 3 and 4 put in the tompion and pass the muzzle-lashing; 5 and 6 render the breeching through the clinch, clap on the quarter-seizings, pass the frapping-turns ready for bousing, and prepare the side-tackles for frapping and securing; stationary powder-man returns the powder; extra powder-man swabs the deck.

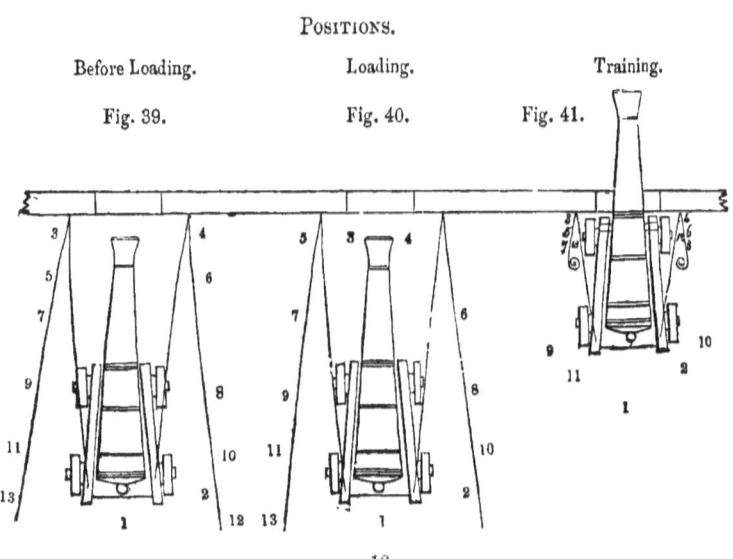

POSITIONS.

Before Loading.   Loading.   Training.
Fig. 39.   Fig. 40.   Fig. 41.

Stationary Powder-man.   Stationary Powder-man.   Stationary Powder-man.
Extra Powder-man.

MAN BOTH SIDES.

Left Guns: 3 remains 3; 5 becomes 4; 7, 6; 9, 5; 11, 2; 13, 7; 1 remains 1.
Right Guns: 4 remains 4; 6 becomes 3; 8, 6; 10, 5; 12, 2; 2, 1.

NINE WORDS OF COMMAND.

Prime.	Ready.	Sponge.
Point.	Fire.	Load.
Elevate.	Stop the Vent.	Run out.

PART IV.   BOAT BRIGADE AND COAST GUARD.   445

GREAT GUN EXERCISE.

Gun Nos.	Station.	To Provide.	Duties.
1	Captain	Provides 6 vent plugs, priming wire, tube box, spare trigger line, a vent bit, sees the lock fixed and fit for use, and places handspikes.	No. 1. The Captain commands, attends the breeching, bed and coin, primes, points, fires, and stops the vent.
2	Second Captain	6 vent plugs, priming wire, tube box, spare trigger line, sees the lock fixed and fit for use, and places handspikes.	No. 2. The Second Captain assists 1, casts loose, hauls up the port, runs out, attends the apron, and port-tackle fall, and cocks the lock.
3	Loader	Shot grummet, spare breeching, wet swab and wads.	No. 3. Takes out the tompion, bears out the port, loads, rams home, runs out and trains.
4	Sponger	Shot grummet, lanterns, fire buckets, hammer or match, wet swab and wads.	No. 4. Takes out the tompion, bears out the port, worms, sponges, rams home, runs out and trains.
5	Assistant Loader	Sponge, rammer, worm and shot.	No. 5. Casts loose, hauls up the port, gives shot and wad to 3, runs out, trains, and spans the breeching.
6	Assistant Sponger	Sponge, rammer, worm and shot.	No. 6. Casts loose, hauls up the port, gives sponge, rammer, and worm to 4, runs out, trains, and spans the breeching.
7 8 9 10 11 12 13 14 15		These numbers, depending upon the size of the Guns, and other circumstances, are to man the tackles, for running out and training, and perform the contingent duties of firemen and others, in preference to Gun numbers. The Auxiliaries assemble on their respective sides, and at once unbar, take off the half ports, cast loose their Guns, and clear their quarters. NOTE.—The last men from the ship's side are always rear men; the handspikemen are always the fourth from the ship's side, with 15, 14, 13, 12, and 11 men; the third with 10, 9, or 8 men; and the second with 6 men. A stationary powder-man is to be stationed at every two guns. An extra powder-man is to be allotted to every four guns, to fetch powder from the magazines to supply the stationary powder-man. With heavy guns, Nos. 2 and 11 assist the handspikemen.	

423. All British naval officers are either in possession of, or may readily procure, the code of instructions for the manual exercise of naval ordnance in their own service. It would therefore be superfluous, and might be objectionable, to publish instructions which are so well taught and generally circulated; but comparatively few naval officers are in possession of the French system of exercise, and it is a sage maxim that officers should be well acquainted with the manual and practical systems of other naval powers: the author therefore inserts the French Exercise, translated into English, that British officers may exactly know the present state of French naval gunnery, and, comparing it with the "Exercise" published in the first edition of this work, trace the progress which has been made in that important branch of practical gunnery since the termination of the late war. Thus they may observe the objects to which the more recent improvements in the French exercise have been directed; and they will not fail to remark, that the principal of these is rapidity in loading and firing.

424. Exercise of the great guns in use in the French navy.[a]

The great attention given of late years to Naval Gunnery has led to useful and important modifications, both in the fittings and also in the manner of working and loading the guns.

These modifications have rendered the last exercise, published in 1834, antiquated, although of recent date, and it has become necessary to re-arrange it.

The principal improvements in the fittings are the elevation sights

---

[a] Translated from 'Les Exercices des Bouches à Feu en usage dans la Marine.' Published by authority of the Ministre de la Marine et des Colonies.

Want of space for the insertion of the new matter which recent improvements in gunnery have rendered necessary, without increasing too much the bulk of the book, has induced the author to make some retrenchments in the Paper on the French Gun Exercise which was given in the third edition. He has done so by leaving out merely the details of the motions executed by the men in the performance of their several duties. All the more important matters, which constitute the course of general instruction—loading, pointing, and firing, with the stowage and the supply of ammunition, have been retained.

(*Hausses*), percussion locks, detonating tubes, stop coins, and split grummet wads.

The elevation sights have given to the pointing an accuracy, a facility, and a quickness, which it never had before. It is true that in close action, when the vision is obscured by smoke, the benefit of this sight is diminished, although it still gives a good indication; but the advantage is restored when, from the distance of the object, the fire is slow, and the smoke is blown away by the breeze.

The percussion hammer, the construction of which is so solid and the use so simple, and the detonating tubes, have replaced with great advantage flint locks and the ordinary tubes.

With percussion locks and tubes has disappeared the use of the powder-horn, which caused accidents and delayed the priming. The fire is instantaneous, and accidents to the lock are more rare and easier to be repaired. The match is only used as a stand-by.

The stop-coins which are used in coast batteries are also an important improvement: they are easily worked. The iron crow (*Pince*), which was so cumbrous, which caused wounds, and wore out the decks, is almost laid aside.

The split grummet wad, from which a piece of about 0.8 of an inch is cut out, has the great advantage of rendering easy the simultaneous loading of the whole charge, and also takes up little room on board. However, the use of junk wads is not altogether exploded; but they are used only in exceptional cases, which will be indicated.

These alterations in the fittings of ordnance have, of course, brought out others in the loading, pointing, and firing.

In working the guns there is a great tendency to too much rapidity. That tendency is blameable, since precipitancy leads to inaccuracy. Now it must be remarked that, in working guns, two distinct features exist, viz., loading and pointing.

One of these, "the pointing," cannot be done with too much care and method: the captain of the gun should be all calmness and attention.

The other, "the loading," on the contrary, cannot be done too quickly, if it do not lead to accidents to the gun or crew.

It is clear that, other things being equal, the advantage in action would be to the quickest loader; and, should it happen that, from the proximity of the ships, pointing is only an accessary, it is evident that the day would be gained by those that were trained to the quickest loading.

These considerations appear decisive; and some trials have lately been undertaken, from the results of which their correctness has been established.

The "pointing," then, should be slow, careful, and methodical.
The "loading" should be done as quickly as possible.
It is with this view that we employ *two loaders*, and the introduction of the *whole charge* at once.

The assistance of two loaders in sponging, loading, and ramming home, makes it quick and sure.

The form of the ordinary cartridge did not allow of the simultaneous loading in chambered guns; but that difficulty has been obviated by a new form, which allows of the same method of loading in chambered and other guns.

From the preceding we see what ought to be the view proposed in this publication.

To prove all the modifications introduced into the exercise, and to give clear forms to the teaching of the pointing; also all the knowledge which so important a branch of the exercise merits.

Also to adopt and teach all methods useful to quicken the fire, which experience has sanctioned. In the detail, it has been tried to keep as much as possible to the old form, the words of which are already in men's memories. In new matter it has been endeavoured to assimilate it as much as possible to the old.

It has been also an object to expunge a great deal of useless verbiage, which has only been hitherto preserved by force of habit.

Everything, not being exactly in the exercise, has been put in the notes, as was done in the former editions.

Lastly, to complete as much as possible what ought to be the course of instruction of a seaman gunner, all the scattered fragments have been reunited in a manual, and submitted to the Minister of Marine.

A TABLE of the NUMBER of MEN for the different GUNS in the FRENCH NAVY.

Designation of the Men.	GUNS.				Carronade.
	36-Pr. and Shell Gun of 80.	30-Pr. and 24-Pr. Long and Short.	18-Pr., 12-Pr. and Shell Gun of 30.	8-Pr.	
Captain of the Gun	1	1	1	1	1
Crew	12	10	8	6	2
Powder Man	1	1	1	1	1
Total	14	12	10	8	4

EXERCISE.

Clearing for Action.

425. The guns on each deck are divided into two divisions, each commanded by an officer; and each of these divisions is subdivided into two sections, commanded by officers or midshipmen.

The clearing for action is denoted, according to circumstances, by—

The ordinary beat of drum;
The quick beat; and
The "générale."

The ordinary beat denotes that there will be only a simple exercise, on one side only, and that it will be on that deck and on that side where the drum is beat.

The quick beat, on the contrary, denotes that the exercise will be general, when hatchways are covered, magazines opened, wings lighted, magazine and passage men at their stations, pumps rigged, stores provided, and guns cast loose on both sides.

The "générale" denotes that every preparation whatsoever for battle should be taken.

At the Ordinary Beat.

The crews repair to their guns on the side where the drum is beat, and place themselves according to the quarter bill. (The guns are supposed to be loaded and secured outboard, as in harbour; if they should be housed or secured outboard for sea, they will be cast loose immediately.)

426. The accompanying plate (Fig. 42) illustrates very distinctly the mode practised in the French navy of passing up by hand, and through scuttles opened in the decks, the cartridge-cases for the supply of the ordnance in the several fighting decks, and of returning the empty cases to the magazine. The frames for receiving the cartridges are not placed in the scuttles until the crew are called to quarters, but are kept on the fighting decks: the scuttle is, till then, closed with a tompion. The man below, placed on the orlop deck, receives the cartridge-case from the magazine, and keeps it in readiness to place it in the frame above, as soon as the preceding cartridge is taken out. The man on the middle or main deck, according as the ship may be a three or a two-decker, is, in like manner, always prepared to slip a case into the frame over his head for the supply of the main deck, or upper deck. The man above is represented as putting an empty case into the canvas hose, down which they slide through all the decks to the orlop. The process shown by the drawing is for the supply of the upper deck and for the return of empty cartridge-cases from it. Each deck is supplied in like manner by a separate arrangement, and the same method is adopted in ships of all classes.

2 G

Fig. 42.

Illustration of the French mode of handing up Cartridges for the supply of the Batteries on the Decks of a first-rate Ship, and of returning the empty Cartridge Cases to the Crlop.

PART IV.   FRENCH EXERCISE.   451

AT THE ORDER "*Provide the Stores.*"

427. The 1st and 1st*ᵃ⁾ men from the ship's side place the sponge
2nd and 2nd   and rammer on the deck-heads inboard, the
sponge inside next the gun. Sponge cap off. Let down half ports,
provide match tub, swab, fire-bucket, and lantern. The 2nd man
from the ship's side, on the left of the gun, provides shot and wads.

The 3rd and 3rdᵃ⁾ take the handspikes, place them on the steps
4th and 4th   of the carriage; raise the breech, so that the
captain of the gun can draw out the wads and train tackle (if they
are kept inside the carriage), and place bed and coin. The handspikes are then placed on the deck, the thick end towards the ship's
side. The same Nos. attend to the side tackles.

The 5th Man on the Left hooks the train tackle, the single block to
the ring-bolt in the deck, the double block to the train-loop.

The 5th Man on the Right puts on the apron, and provides seizings,
rags, &c., which are kept in a cartridge-case.

The last Man on the right of the 2nd gun of each section provides
(and lights if firing is ordered) the match placed in the tub of the
section. At guns of 36-pounders the 6th man assists the 5th. A
sufficient number of powder-horns filled are kept in the magazines
ready for handing up if required.

The Powder-man provides his case, and, if firing is ordered, goes
to the screens for a full case.

The Captain of the Gun provides tube-box; thumb-stall, or vent-
plug, and priming-wire attached to his tube-box; he places his gun
in the best position for firing, and overlooks his gun's crew.

Nos. at a Gun manning one side.

1	1
2	2
3	3
4	4
5	5
6	6
Chief	
	Provider.

At the quick beat of drum.

As soon as the crews are assembled on the side ordered to be
manned, the officer commanding the battery gives the order "Provide stores for both sides," when the men on the right of the gun
turn to the left, and go over to the opposite gun. The sponger
becomes the captain.

The men to the left of the gun and the powder-man remain with
the captain. The 1st and 2nd men from the ship's side pass to the
right of the gun; the 3rd, 4th, and 5th remain on the left.

ᵃ These are men on opposite sides of the gun.

Then at each Gun—

The 1st *Men* do what is prescribed, in manning one side only, by the 1st and 2nd men from the ship's side.

The 2nd *Men* do what is prescribed, in manning one side only, by the 3rd and 4th men from the ship's side, and hook also the train-tackle.

The Captain provides his tube-box, and all the articles ordered in the manning one side only.

As soon as the provisional captain (or that of the opposite gun) has finished everything at his gun, he rejoins his proper gun, and every one resumes his proper No.; the 5th man on the right puts on his apron.

The 5th men on the left, remaining with their proper captains, provide as follow:—

5th *Man on the left of the* 1st *Gun of each Section.*—The match, which is lighted if requisite.

5th *Man on the left of the* 2nd *Gun.*—Vent-bits and worm, which are placed in the rear of the centre gun of each section, if not already there.

NOTE.—The dismounting-gear, spare carriages, and 1 spare breeching for each section, ought to be ready on the lower deck, or orlop.

Boarding weapons are provided and inspected by their owners.

5th *Man on the left of the* 3rd *Gun.*—Boxes containing musket and pistol cartridges, which are placed ready for distribution to boarders.

2nd-*masters of each Battery* take their bags, which are prepared beforehand; passage-men take their screens and funnels, and place them in the hatchways which have been shipped by carpenters and caulkers.

NOTE.—If clearing for action by night, the ports will all be opened together, and only by the express order of the officer of the battery; the half ports must never be lowered in this case without orders.

The " Générale."

The "Générale" comprises all arrangements for battle, both as to manœuvring and working the guns. As soon as it is beat the crews repair to their guns, on the side indicated, cast them loose, provide their stores, according to the orders, and as laid down in the general exercise.

If clearing for action at night, the hammocks are stowed in the nettings by the men stationed in tops; sail-trimmers make all their arrangements for action; the masters of the different departments see that the stores for replacing wounded spars, ropes, &c., are in their proper places.

Mustering.

Mustering is done at each gun at the same time. The captains report to the officers of sections, who see everything correct, and report to the officer of the battery.

Exercise on one Side.

"*Point.*"

428. The captain of the gun raises his sight to the notch indicated by the officer, and according to the charge which he knows to be in the gun. Then he places himself to the right of the train tackle, left foot flat and advanced, knee bent, right leg stretched out, left hand upon the base ring, and the right hand grasping the coin.

The 3rd men from the ship's side, aided by the 4th for heavy guns (24 and upwards), take the handspikes, place them on the steps of the carriage, and raise or lower the breech, at the order of the captain, until the gun is in the proper line—that is to say, when the line of sight is directed as much as possible upon the point which one ought to see, when the ship is in her mean of rolling, or the middle of her roll. (*When, during the roll, the point of sight passes as much above as below the object.*)

If the pointing is very oblique, he commences by throwing the gun forward or aft until nearly in the right direction. Then gets his correct elevation, and waits for the second motion to correct his pointing.

"*Fire.*"—In two motions.

1*st Motion.*—The captain of the gun waits until the motion of the ship brings the object on with the points of his sight; when it is nearly on he gives a signal, then fires by pulling strongly the trigger-line. At the signal from the captain, the Nos. in charge of the side tackle-falls throw them down clear of the trucks; handspike-men ground their handspikes; all the other Nos., except those next the ship's side, go quickly to the train tackle, and draw the gun in to a taut breeching.

The Nos. next the ship's side take the stop-coins and chock the trucks as soon as the carriage is clear of the port; the captain makes up his trigger-line and half-cocks his lock.

The left rear-man chokes the luff of the train tackle; the other Nos. close up.

Note.—In firing, when there is no danger of the gun running out again, the stop-coins may be attended by the 2nd men from the ship's side; the 1st Nos. can then jump in quickly to sponge, and the 2nd man on the right is sooner ready to pass the sponge.

2*nd Motion.*—The 3rd men on the right and left, aided by the 4th for heavy guns, take the handspikes, raise or lower the breech for the captain to place the gun in the best position for loading.

The other Nos. coil down side and train tackle-falls, the handspikes are grounded, and every one resumes his former position.

"*Stop the vent, sponge, and give the rammer.*"

1*st Motion.*—The captain of the gun takes the priming-wire in his right hand, and forces it into the vent to see if it is clear. He then serves the vent with his left thumb until the gun is loaded, not taking it off to see if the charge is home until the loaders have retired clear of the muzzle.

The loader and sponger step to the muzzle, passing over the breeching and side tackle; the 2nd man from the right gives the sponge to the sponger, which he forces into the piece, and then the rammer is placed between the loader and sponger, the staff resting on the port and the head on the deck, touching the fore-truck.

2nd Motion.—The sponger, aided by the loader, turns the sponge several times at the bottom of the bore, so as to make use also of the worm, which is let into the centre of the sponge head; he then withdraws it, turning it round continually, and gives it several blows to throw off any dirt, &c., and then gives it to the 2nd man on the right; he then takes the rammer with his left hand.

The captain introduces the priming-wire into the vent to see if it is clear; if not, he gives the word "sponge," and the 2nd man on the right gives the sponge again to the sponger. The captain serves the vent again.

The right rear-man wipes the hammer and inspects the sponge and rammer-head; the side tackle-falls are passed to the two rear-men, who lay them on the deck.

"*Load.*"

The loader makes a half-face to the left, receives the cartridge from the powder-man, which he enters bottom first; he then enters successively shot and wad, which are given to him by the 2nd man on the left, who resumes his place.

The sponger during this prevents, if necessary, the charge from falling, by placing his right hand before the muzzle.

As soon as the charge is introduced, the sponger and loader force it home to the bottom of the bore by successive motions with the rammer at the full extent of the arm, pressing strongly against it at the last; the sponger makes sure that it is home by the length of the rammer, and notifies it to the captain by striking against the face of the piece. He stretches his right arm out, his left hand upon the muzzle, ready to ram home; the loader places himself in a similar position.

As soon as the powder-man has given his cartridge he goes for another, carrying his case under the left arm, having his right hand on the lid.

"*Ram Home.*"

The sponger and loader give two blows, the loader then returns to his place; the sponger withdraws the rammer, gives it to the next man, who places it on the deck: the sponger then resumes his place.

As soon as this is done the captain pricks the cartridge with one thrust; if he sees that it is not home, he gives the order "ram home;" and the sponger and loader ram home again.

At the same time the left rear-man places the handspike in the square ring, ready for the captain to keep the gun square in the port whilst running out, goes to the train tackle, unchokes the luff, takes the fall in his hand ready to ease away in running out.

The right rear-man and the next No. to the left rear-man give the side tackle-fall to the other men.

The sponger and loader see that the stop-coins are easy to disengage.

If the exercise continues, it is resumed from the second order after "prick the cartridge."

"*Run Out.*"

1st Motion.—The captain of the gun takes the handspike in his left hand, ready to direct the gun square in the port. The sponger and loader take away the stop-coins and place them in their rear, then they lift the breeching to keep it clear.

2nd Motion.—The captain gives a signal with his right hand to run out square in the port. As soon as it is done, the 4th man on the left unships the handspike and places it on the deck. The captain primes and keeps the gun out by placing the side tackle-falls over the button; they are held by the 2nd men on each side.

NOTE.—It would save time to place the tube in the vent as soon as the cartridge is pricked; but in action, it must be left to the intelligence of the captain of the gun to prime at any time before the gun is out.

429. Notes on the different Orders at the Gun Exercise.

On "*Prime.*"

It is essential that the cup of the tube should be pressed close down on the vent patch, so that the blow of the hammer should be efficacious, and that it should not miss fire.

If percussion tubes are not at hand, or if the lock is broken, the priming is done by a powder-horn, which is slung over the shoulder from left to right.

The captain takes the powder-horn in his right hand, his little finger on the trigger, and introduces powder into the vent with his priming-wire; the vent is filled, and a small train laid before the vent, and well bruised to ensure its being ignited by the match.

On "*Point.*"

The distance of the object is given by the captain of the ship. If the captain of the gun has not received any orders, or if there is no officer near him, his own intelligence must supply the want.

To give the proper elevation, he should know what charge, both of powder and shot, is in the gun.

When the captain of the gun wants the breech raised, he places his left hand on the base ring; and so long as his fingers are raised, the Nos. raise the breech. When he lowers his fingers, the breech is lowered also, but slowly and steadily, so that the captain may follow easily the line of sight, and thrust home the coin when the elevation is correct.

If the pointing should be so oblique as to render it necessary to give the direction before the elevation, the captain places his left hand upon the button, and indicates with his right how he wishes the carriage moved. The quickness and fulness of his motions indicate more or less training. When the direction is good, he brings his right hand to the button and holds it horizontal.

As soon as the captain takes the trigger-line the Nos. press the handspikes against the brackets, and train to the right or left, according to orders given with the left hand, holding the arm out at full length; when the object is *on*, he reverses his hand, the palm towards the deck.

The signal "ready" is made by raising the left arm: this—which is preparatory to firing, the signal being to let fall quickly the left arm—ought only to be done the very moment before pulling the trigger-line, so as not to require again the handspikes, which are raised perpendicularly to be clear of the carriage.

Rules for pointing.

The detail of the exercise gives a general idea of the rules of pointing; but in action many things may occur to modify them.

There are different modes of pointing, which are designated by the following expressions:—

En belle.—Horizontal fire, with the gun in the centre of the port.

Direct.—Gun in the centre of the port, but the axis out of the horizontal line according to the distance.

Oblique.—When in pointing it is necessary to throw the breech towards the bow or stern, the order is, "Point for chasing," or "Point for retreating," to express that the gun should be extreme trained.

En plein bois.—When you point so as to hit the hull of your enemy near the mainmast.

For the Water-line.—To point so as to hit the water-line. It is of advantage, in this mode of firing, to avail yourself of the enemy's rolling from you.

There is also firing to dismast, but this very uncertain practice, which causes nearly every shot to be thrown away, and which was one of the gravest errors of our last war, is now almost entirely exploded.

It ought only to be done at the express order of the captain of the ship. In this firing you should point at the fore catharpins principally.

Firing is called direct when the shot strikes the object without any previous graze.

It is called "ricochet" when it strikes the water before the object. This last, which is sometimes very advantageous, should be principally used in smooth water.

Raking fire is when the shot ranges from stem to stern, or *vice versâ*.

Diagonal firing is when the shot takes the direction from cat-head to quarter.

The captain of the gun must bear in mind that in direct firing he must first get his elevation, and then correct his line of direction.

In oblique firing he must get the line of his object nearly on, then his correct elevation, and, lastly, correct his line of direction. If this were not done in oblique firing, one part would derange the other.

When the ship is rolling much, the gun is laid, as near as possible, parallel to the deck, and then you wait until the motion brings the object on with the points of your sight.

If, on the contrary, the rolling is moderate, and the vessel heeling, the captain places his gun for the main heel occasioned by the action of the wind on the sails, *to* which the ship would *steadily* heel if it was not thrown out by the combined action of sea and wind.

When double-shotted, you must point higher as the range is decreased.

If the tangents do not give the elevation for double shot, the intelligence of the captain of the gun will supply it.

The captain must also look out for yawing, but, as this is slow and uncertain, he must not wait as when rolling; and he must not hesitate to point his gun forward or aft, to get his object on at the earliest moment.

If the sea is smooth, you fire when you roll towards your enemy, rather than from him, so as to preserve the chance of ricochets.

In exercise, the captains of guns ought always to point at some object, and follow it through rolling and yawing.

"*Fire.*"

The captain of the gun ought never to fire if his object is not on, or if the side-tackle falls are not clear. To fire at random is to waste stores, heat the gun, fatigue the crew, make smoke, and lose time uselessly. He ought never to fire if not sure of his aim.

To ignite the tube, the hammer ought to strike full upon the cup: to make sure of this, the captain must hold the trigger-line extended, his hand in a line with the hammer, and pull strongly and without jerking.

If the tube miss fire, it is changed; but if it burns priming without communicating with the cartridge, he must not prime again until the smoke has ceased. To do so he comes up to the left of the gun, and then rectifies his pointing before he fires. If any particles of fire have fallen on the deck they are extinguished with a swab.

If the ship be dismasted, and a portion of sails or rigging hanging over the guns, they should not be fired, for fear of setting fire to the ship.

In using "Billette's" tube, which is fixed in the vent with a hook, the captain, after having attached to it the trigger-line, goes to the rear, clear of the recoil; he holds the trigger-line slack, and fires with a strong jerk, as if he wished to break a line fast at one of its extremities.

At the moment when the captain lowers his left arm which he held out, the Nos. let go quickly the side-tackle-falls: if the water is smooth, and lee guns manned, they may be let go at the signal "ready."

It is impossible to haul through the slack of the train-tackle sufficiently quick on the recoil of the gun: therefore, at the word, "ready," the left rear-man takes all parts in his hands, and, when the gun is fired, all the Nos. go smartly to run it in. The sponger and loader hold the stop-coins ready to catch the fore trucks.

If the ship is rolling so that the gun might run out again, the right rearman ought to nipper the train-tackle with his hands whilst the left rear-man chokes the luff.

To enable the stop-coins to be easily withdrawn, they should only be put under the trucks by a small part of their breadth and obliquely.

"Stop the Vent, Sponge."

If the captain of the gun cannot clear the vent, he gives notice to the nearest officer, who causes it to be cleared by the men carrying drills.

It is very important to serve the vent well, to stifle any air and extinguish any fire which might remain in the bore and ignite the newly-introduced cartridge. This is of the greatest consequence; and no matter how quick the firing may be, the captain must never cease to stop the vent as long as the loaders are before the gun.

Most accidents may be traced to this source.

NOTE.—On board some ships a vent-plug, which leaves the captain at liberty, has been used with advantage; but it ought to be made with the greatest exactness, for if badly made it would be dangerous to use it, as the compression of air in the gun, on introducing the charge, would blow it out.

The sponger must pay attention not to strike the sponge against the bottom of the bore, so as to break the worm and render it useless.

The sponging should be done with great care, so as to withdraw anything superfluous, and clean the bore. Whenever the sponge cannot draw out the bottoms of the old cartridges, a worm must be used.

Tapping the sponge on the muzzle ought never to last longer than whilst the captain is ascertaining that the vent is clear. The cleanliness of the sponge ought never to stop the loading of the gun.

As the exercise directs, it is the duty of the right rear-man to examine and clean sponge and rammer.

It is during this motion that the gun should be refreshed if necessary: with this view the sponge is sprinkled by hand before it is put into the gun.

When there is so much sea as to cause the ports to be lowered as soon as the gun is fired, the muzzle is laid fair for the scuttle, so as to allow the sponge to be passed through it.

If two ships are very close, and there is not room to use staff sponges, rope ones must be used.

It so rarely happens that it is necessary to sponge or even to ram home a second time with the common gun, that in these exceptional cases the order may be given *vivâ voce*.

"Load."

In giving the cartridge to the loader the powder-man holds his box open the shortest time possible, to be safe from the fire of the neighbouring guns.

When a cartridge is found burst in the case it must not be taken out for fear of spilling the powder. In this case it is returned to the magazine, and notice given of the accident.

If the cartridge bursts in the bore, so as to leave a train of powder in the gun, a junk wad is used to scrape up the powder.

The bottom of the cartridge must be entered first, so that that part of it which projects beyond the ligature may not hinder the cartridge from going home in the bore, or leave pieces on fire.

Since the adoption of the charges of $\frac{1}{4}$ and $\frac{1}{8}$, it is useless to bleed the cartridges: the object of doing this, which was not to strain the gun, and to make more splinters, has been carried out by the new charges.

To facilitate the simultaneous loading, that part of the cartridge beyond the ligature should be cut to two inches, and made up in the form of a cockade, so that the ball in going home should not jam against the end of the cartridge.

No more than one shot should be used, without express orders from the officer of the battery. If round and grape are used, the latter should be entered last: in this case a junk wad should be used.

The wad is always the last, and only one is used, no matter how many projectiles. A wad is never placed on the cartridge, as it diminishes the range, augments the recoil, and makes the loading longer.

In the Navy two kinds of wads are used. The split grummet wad, which has a section of about ·8 of an inch cut out to render its introduction easier, makes it easy to load simultaneously, and keeps home in the bore very well, acting as a wedge between the bore and the shot. It is also very handy to stow away.

The junk wad is very easily put out of form on board, and it often requires beating and rolling before it can be entered. It renders the simultaneous loading difficult, if not impossible. It ought only to be used when, the cartridge being broken, it is necessary to scrape up the powder along the bore: also when the gun is likely to remain loaded for some time, as with this wad it would be easy to draw the cartridge out without bursting it; but in this case the wad should be rammed home upon the cartridge, to keep off the shot if rolling about in the gun.

If grummet wads be exhausted, they are easily replaced by a simple piece of rope, of a size equal to a grummet wad.

If the shot will not go into the gun, it is put on one side to be cleaned. If it jams in the gun, it must not be forced, but drawn out: this is done by raising the breech and knocking the muzzle against the portsill, or by using the scoop.

The marks on the rammer should be known by night as by day. They are made for the middle charge at $\frac{1}{4}$, and the sponger must judge of the difference according to the charge used.

To avoid bursting the gun, no space should remain between the cartridge and shot; so that the sponger should look carefully to the mark.

In going for another cartridge the powder-man should pass along the side which is not engaged, keeping his case always carefully shut.

"*Ram home.*"

You should not ram home too strongly on the charge, as powder reduced to dust ignites more slowly than in grains, and gives less range.

There is also more danger of driving the bottom of the cartridge into the chamber, which might be dangerous to the succeeding cartridge.

"*Run out.*"

The captain of the gun must run his gun out square in the port, to facilitate the pointing, and so that the half-ports may be shipped if necessary; but if the pointing is likely to be oblique for some time, he must run out the gun on that bearing, taking care to leave room for sponging, &c.

The numbers work together, running out hand over hand, and not by jerks. The left rear-man attends the train-tackle, which he keeps in his hand, easing the gun out, so as not to go too strongly against the ship's side, which would start the charge.

There is another mode of easing the gun out, no matter how heavy the rolling, which is by taking a turn with the fall round all parts, which causes more friction. It is in that case attended by both rear-men—the left rear-man to ease away, and the right to see the fall clear.

430. To leave one side and man the other.

The third man on the right gives the apron to the captain, who places it on the breech, and causes each side-tackle fall to be bracketed. He leaves whatever is necessary on the button.

Numbers on the right of the gun face to the left. Those on the left face to the right. Captain and powder-man make a half-turn.

431. One side being manned, to man the other also.

"*To man both sides.*"

The captains of even guns, if the starboard side is manned—of odd guns, if the port side is manned—leave their tube-box on the button: the right rearman leaves also his apron with small gear.

NOTE.—The guns are numbered from forward.

Men face to the right and left, ready to go to the corresponding gun on the other side.

The three first numbers on the right of the guns which remain face to the left, and the first man on the left faces to the right.

432. Exercise of the Shell-Guns of 22 Centimètres (80), and of 16 Centimètres (30).

In loading with hollow shot, the shell-gun of 30 is worked in the same manner as that of 22 centimètres; and it is only with solid shot that simultaneous loading would be used. In every case hemispherical cartridges should be supplied.

To obviate the inconvenience of these cartridges, and to seek for means to execute the simultaneous loading with the shell-gun of 30, whether with solid shot or shell, it has been ordered to make trials of ordinary cartridges with flat bottoms and elongated with diminished diameters.

These trials have shown that, with these cartridges so modified, simultaneous loading may be done either with shot or shell. In consequence, that mode of loading and the ordinary exercise may be conformed to, whether loading with shot or shell.

If grape is used, a cylindrical wad should be placed between the cartridge and grape.

433. Observations on Fighting both Sides.

If, during independent firing, the order is given to man both sides, the captains who remain at their guns send immediately their men to the gun on the right; the man on the left passes the sponge rammer to the right, and the captain continues his fire with this man as sponger.

The auxiliaries go to their stations, at the guns at which the sponger is become captain.

The captains and crews of the guns which are left go as soon as relieved to the opposite guns; they do this whether the guns left are loaded or not.

Firing on both sides being generally independent, it seldom happens that the guns are being loaded at the same time, so that the powderman is able to supply both.

434. Exercise of the Shell-gun of 22 Centimètres (8.66 in.), and that of 16 Centimètres (6.3 in., or 30 liv.) mounted on a Naval Carriage—using Shell.

"*Point.*"

The captain raises his sight to the degree ordered by the officer, and according to the charge which he knows to be in the gun; then he places himself to the right of the train-tackle, left foot advanced and flat, knee bent, right leg stretched out, the left hand on the base ring, and the right grasping the coin.

The third men, aided by the fourth for guns of heavy calibre, take the handspikes, place them on the steps of the carriage, and raise or lower the breech, according to the orders of the captain, until he has got his proper elevation; that is, when the line of sight is directed as much as possible upon that point, which one ought to see when the ship is in the mean position of her rolling.

If the pointing should be very oblique, he commences by training to the right or left, until he has nearly got his right direction; he then gets his correct elevation, and waits to rectify his pointing.

Fourth Order.—"*Fire.*"

The captain waits until the motion brings the object on with the points of his sight. When nearly on he gives notice, then fires with a strong pull and without jerks. At this signal the men let go the side-tackle falls, handspike-men ground their handspikes. All the Nos. except the loader and sponger go to the train tackle, and run it in to a taut breeching.

The loader and sponger take the stop coins, and chock the trucks as soon as the carriage is clear of the port; the captain makes up his trigger line.

The left rear-man chokes the luff of the train tackle; all the Nos. close up, facing the gun.

"*Stop the vent—sponge—prepare to load with shell.*"

The captain examines the vent with his priming-wire to see if it is clear. He then stops the vent with his left thumb until the gun is loaded, not taking it off even to prick the cartridge, before the loaders are clear of the muzzle.

The sponger and loader step to the muzzle, passing over breeching and side tackles; the 2nd man on the right gives the sponge to the sponger, who forces it into the bore; the 2nd man then takes the rammer with a flat head, and lays it between the sponger and loader, the rammer head touching the fore axle-tree, and the staff resting on the port sill.

The sponger, aided by the loader, turns the sponge several times in the bottom of the bore, so as to use the worm; he then withdraws it, still turning it round, and gives it a few taps on the muzzle, to knock off any particles of dirt; he then gives it back to the 2nd man, and takes the rammer.

The captain introduces the priming-wire into the vent, to see if it is clear; if not, the sponger sponges again, the captain stopping the vent all the time.

The right rear-man cleans the hammer, and examines sponge and rammer; the side tackle-falls are passed along.

"*Load.*"

The 1st man on the left faces half-round, and takes the cartridge from the powder-man, and enters it bottom first; the loader and sponger then press it home to the bottom of the bore by successive motions at the full length of the arm; the sponger makes sure that it is home by the length of the rammer, and gives notice to the captain by striking on the gun; he then stretches out his arm, and, aided by the loader in the same position, remains ready for ramming home.

The powder-man goes for another cartridge, keeping the case under his left arm.

"*Ram home.*"

The sponger and loader give the cartridge two blows, and let go the rammer; the captain ascertains that the cartridge is home; if it is not, he makes the negative sign to ram home again; if it is home, he makes the affirmative sign, and the rammer, with the flat head, is passed to the 2nd man; this man takes the hollow-headed rammer in his left hand, ready to pass it to the sponger. During this time the 1st man on the left stoops down, takes off the lid of the shell-box, and gives it to the 2nd man, who places it behind him.

"*Load with shell and wad.*"

The sponger and loader lift the shell from the box, and place it in the bore, wooden bottom first: the 2nd man takes away the box, and places it behind him; he takes up a wad.

The sponger uncaps the fuze by tearing off a strap upon which the cover is pasted.

The loader places the wad on the shell, and puts his left hand before the muzzle.

NOTE.—The loaders should be most careful not to knock the fuze against the muzzle, so as to expose the shell to burst.

As soon as the shell and wad are introduced, the 2nd man on the right gives the hollow-headed rammer to the sponger, who, assisted by the loader, presses it home strongly to the bottom of the bore.

The sponger ascertains that it is home by the length of the rammer, and gives notice to the captain by striking on the gun; the loader and sponger then remain ready, with arms stretched out, to ram home.

"*Ram home.*"

The sponger and loader ram home with two blows; the loader then returns to his place. The sponger withdraws the rammer, and gives it to the 2nd man on the right, who lays it on the deck.

As soon as this is done the captain pricks the cartridge with a thrust of the priming-wire. At the same time the left rear-man places the handspike in the iron ring, goes to the train-tackle, unchokes the luff, takes the fall in his hands, ready to ease the gun out.

The right rear-man, and the last man but one on the left, stretch along the side-tackle falls. The loader and sponger ascertain that the stop-coins can be easily removed.

435. Observations on the Exercise of the Shell-gun of 22, and of 30 Centimètres.

The shell-gun of 22 centim., mounted on the modified naval carriage, is provided with two handspikes and a roller lever director. To work it, the lever is shipped in the train-loop, and the rear of the carriage is lifted up by weighing down upon it; the two handspikes are also used against the carriage to train the gun.

The shell ought to be entered two or three inches into the bore before it is uncapped.

If by accident the fuze should ignite before it is entered, there would be time enough to throw the shell overboard, as the fuze burns nearly a minute before it communicates with the charge.

If the fuze should ignite when the shell is in the bore, the loader and sponger step outside, and the captain places himself in a line with the men on the right of the gun.[a]

If the shell will not enter, it must be returned to the shell-room.

It is not proper to keep guns loaded with shell, as there are no means of drawing them out again.

It will not do to fire double shell, as one, and sometimes both, burst at the muzzle.

When loading with case-shot, the 2nd and 3rd men on the left place it, like the shell, under the muzzle, so as to be handy for loading.

"*Ram home.*"

It happens sometimes in shell-guns of large calibre that the cartridge is not home, because of the difference between the diameter of the chamber and the diameter of the bore; the flat-headed rammer should not be withdrawn until the captain is quite sure that the cartridge is home.

The negative sign is made by holding the priming-wire perpendicular over the vent-field.

The affirmative sign is made by moving the priming-wire smartly to the right.

The shell-gun of 30, mounted on the naval carriage, and the ordinary gun of 30, loaded with shell, are worked as the shell-gun of 80, with this exception, that the right rear-man has nothing to do with placing the shell: the left rear-man only goes for the shell, and takes back the empty box.

Also when firing shot from the shell-gun of 30, the common exercise and simultaneous loading will be used. It is assumed that cartridges with spherical bottoms are supplied, in which case the hollow-headed rammer only would be used.

If the shell-gun of 30 were mounted on the carriage called "Romme," the pointing being done by screw, would be done as mentioned underneath.

The other orders remain the same nearly.

---

[a] No one who reads this and the previous passage, and the various cautions given in the regulations for conducting shell firing in the British Navy (Arts. 272, 274, 275, &c.), can fail to perceive at a glance that a strong *primâ facie* case of danger to the users stands confessed; and it is quite clear that those dangers must, as General Paixhans most justly observes in his letter (Art. 305), be multiplied in proportion as those treacherous weapons are accumulated on the decks of a ship.

"*Point.*"

The captain raises his tangent to the height ordered by the officer, and according to the charge which he knows to be in the gun; then he places himself to the right of the train-tackle, left foot advanced and flat, left hand on the base-ring, and the right hand on the handle of the screw.

The 3rd man on the right also seizes the handle of the screw to assist in pointing, which is done by getting the line of sight in the same manner as with common guns.

If the pointing is very oblique, the captain proceeds as with common guns.

The captain takes his trigger-line, and acts as with common guns. The 4th man on the left ships the lever director; then, aided by the 4th man on the right, moves the gun to the right or left until the captain has got his line of sight.

As soon as the pointing is done the captain gives the word "ready," when the 4th man on the right unships the lever director and keeps it clear of the trucks.

436. Exercise on both Sides, with the Shell-guns of 80 and 30.

Two auxiliaries must be added for the gun of 80, and one for that of 30. They are necessary to bring up shells.

"*Man both sides.*"

At this order, the captains who leave their guns deposit their gear on the button, as in the ordinary exercise.

The Nos. go over to their guns; the auxiliaries go to the gun of the provisional captain, as 2nd and 3rd men on the left.

437. Shell-Rooms.

The shell-rooms are placed below the orlop decks, before the wine-room.

They are secured against fire and damp, with the same care and in the same manner as the powder magazine.

The bulk-heads which separate them from the main hold, which are not flanked by tanks, should be cased with iron plates. This mode of casing will be applied to all small ships.

The shell-room should be capable of holding its complement of shells. This is easily determined by the number of shells and the cubical contents of the boxes.

NOTE.—A new mode of stowing shells is being tried at the present time in a frigate at Rochefort. It is that they should no longer be stowed in boxes. There are compartments, as in the limbers of howitzers, which unship, and in which the shells are stowed so as to take as little room as possible. This is very advantageous, as the boxes are suppressed and the number of shells doubled without increasing the shell-room. This mode is now indispensable after the increase in the number of shell-guns in ships. Exercising shells may be kept in the wings or in the orlop.

438. System of M. le Capitaine de Vaisseau Lugeol, of stowing Powder, Shells, and Shot.

M. Lugeol having published an excellent pamphlet on the stowage of powder, shells, and shot, this system has been followed in all new ships, and in many of the old ones.

Shot Lockers.—Shot are kept on board in too great numbers to be all placed in the lockers. In frigates, as in line-of-battle ships, the first thing is dryness, and next position. In consequence, they are so constructed that when the vessel is rolling no water can get into them from the hold.

Before the shot are placed in the lockers they should be rubbed lightly with grease, which should be renewed every three months. Every kind of shot should have its separate locker, also one expressly for hollow shot or shells loaded with sand. In some ships this may be placed between the chain-pump wells and the pump well. This depends upon how far the main-hatchway is before the mainmast, but the flooring should be raised so as to be clear of the water in the hold. It should be well caulked and payed.

First Shell-room.—From the pump-well, in the space between it and the old wine-room hatchway, commodious lockers have been built to contain a great part of the shells in their boxes. The doors of these lockers open in the wings, and the passing of the shells is easily arranged either up the main-hatchway or that of the wine-room.

Second Shell-room.—Against the "*sac à terre*," before the powder magazine, lockers 70 centimètres deep are made, which do not interfere with the bins in which the wine is kept, as the tap is always clear.

If these lockers are not sufficient, they may be lengthened on each side before the biscuit lockers.

All these lockers or magazines open upon that part of the wing or platform which passes before them from one side to the other, so as to facilitate the passing up of the shells.

439. Powder Magazines.

The batteries are divided into two parts—sections or divisions, each supplied from the scuttle on its own side.

All powder being made up into good cartridges, shut up in metal cases of uniform dimensions, there is nothing to prevent the complete uniformity of magazines.

The lower planking is well supported as usual. The stanchions shore up the orlop beams so strongly that the stanchions in midships may be removed: this is advantageous, as it increases the room and regularity of stowage.

Part of the large stanchions in midships on the orlop and other decks are not touched.

PART IV. FRENCH EXERCISE. 467

The magazines are well caulked and payed; but as dryness and the working of the ship open the seams, great care should be taken to solder the sheets of copper which strengthen the lower half of the magazines. This operation renders it possible to drown the magazine if necessary.

It being supposed that the vapour arising from the perspiration of the men may be injurious to the cartridges, openings are made in the "*sacs à terre*," in the form of the letter V, so that the exterior should not correspond with the interior.

Each opening is covered with a sheet of copper pierced with holes, and protected by a trap to intercept draughts in case of a storm.

The magazines should be placed as low as possible, to increase the chance of security when fighting to leeward in a strong breeze, or with a land wind against a fort firing red-hot shot. It is thought that the crown of the magazine should be placed from 80 centimètres to 1 mètre below the orlop deck. At this distance, under the planking of the orlop, there is in each magazine a solid piece of timber extending over the whole surface. These timbers are covered with iron plates instead of lead, because it is better for the vegetables which are stowed in bulk upon them.

In frigates, vegetables are not stowed there, but the gunner's spare stores.

Between the crown and the deck the space is measured so as to stow in frigates 5 tier of cases, and in line-of-battle ships 6.

A plank is placed for a man to stand on, so that both in frigates and line-of-battle ships there shall only be 4 cases above the plank, which enables a man to reach the cases.

In line-of-battle ships the upper cases should be used up first. As soon as they are empty the magazine becomes like that of a frigate; and it is not necessary to displace the plank, because underneath it is stowed powder of an inferior quality and boat's ammunition.

It has been thought necessary to separate the handing-rooms, under the magazine passage, from the entrance to the magazines: they have been placed in the angles of the magazine, taking up the space of 4 cases. These "guérites" are entered by scuttles cut in the orlop deck.

There ought to be as many passages as there are batteries to supply: these passages are placed, in frigates, at the foremost angles, and in line-of-battle ships at the four angles.

It remains to see how the passing up powder can be done without error or confusion.

When there are in one battery two different calibres, viz. guns of 30 and shell guns, these last are always the fewest; consequently if their cases passed through the same scuttle as those of the guns, they would interfere with the latter, because they would remain too long on the stands. There must be, then, in each deck two scuttles, and a third for the sleeve or wind-sail for the empty cases. (Fig. 42, p. 450.)

2 H 2

This arrangement is made on each deck, forward and aft. Of the two scuttles the foremost one ought to be appropriated to the shell-guns.

Thus it is very simple—the powder-man finds always at his proper scuttle the full case that he is in want of.

This distinction, so important in the passage of the powder, is not necessary in the magazines: the cartridge-cases of guns and shell-guns enter and pass out by the passage appropriated to each battery. All chance of confusion is avoided by giving to the cases of shell-guns a particular mark, which is easily recognised by the men in the magazine, and also those in the orlop who place them on their stands.

The marks are made in white paint; and, to be easily known in the dark, they have a line with knots according to the battery for which they are destined.

As every battery has its special passage, the cartridges for each battery should be stowed near its own passage :—

The passage at the after starboard angle for the lower deck.
    Ditto   after port angle   middle deck.
    Ditto   foremost starboard angle   main-deck.
    Ditto   foremost port angle   upper deck.

The lighting of the magazines leaves nothing to be required. With a lamp corresponding to the direction of each inner " coursive " (half deck), the lenticular glasses being placed in the exterior and interior of the bulk-heads, there is no danger in placing the lamps of the foremost magazines in the " cambuse " (store-room), and those of the after magazine in the handing-rooms.

SECTION VI.—ON GUNPOWDER AND GUN COTTON.

440. Great errors in the practice of gunnery arise from those diminutions in the strength of powder, which invariably result from any absorption of moisture. Naval ammunition, being particularly liable to be thus injured, should be protected with special care and precaution from the influence of those pernicious damps to which it is continually exposed.

The only sure way of discovering whether a ship's magazine be damp or not is to observe the indications of an hygrometer. In default of a proper instrument of this kind, soak a sponge in a solution of salt of tartar, or common salt and water, and then let the sponge be well dried and weighed; on placing it in the magazine, if the latter be damp, the sponge will become heavier.

Part IV. PROTECTION OF GUNPOWDER FROM DAMP.

441. The most effectual method of protecting gunpowder from damp is to keep it in close vessels with air-tight covers; copper cases have accordingly been provided for this important purpose; and all gunpowder, whether in bulk or made up in flannel cartridges, is stowed in metal-lined cases or barrels, each case containing 120 lbs., and each barrel 90 lbs. The top is removed by means of a key, and the bung-hole underneath is made air-tight, with a tubing composed of bees-wax and tallow. No expense should be spared in providing a supply of these vessels sufficient for a state of war; for so long as gunpowder is exposed to suffer great and unknown losses of strength, all attempts to attain much accuracy in practice, will, in cases of protracted service, be defeated.

442. No degree of care, however, can altogether preserve naval ammunition from receiving some degree of injury by exposure to marine damps; because a quantity sufficient for immediate use must always be kept ready for serving out, and consequently unpacked from the close vessels; and in the exigencies of service in remote regions, ammunition of doubtful quality may be procured from stations which may not have been recently supplied; or it may be necessary to use powder obtained from the capture of an enemy's magazines or vessels. It would appear, therefore, to be desirable that vessels, sent on distant and protracted service, should be provided with means by which to ascertain the condition and strength of their ammunition; otherwise ships stored abroad, or that may have been long at sea without having their powder examined and tried by competent persons at ordnance stations, may be exposed to go into action with ammunition, apparently in good condition, but so deteriorated in strength as to produce very disastrous consequences. This, it may fairly be suspected, has very frequently occurred; perhaps much more frequently and seriously than we are aware of. It must, indeed, have been from such circumstances, that, at different times, an inferior degree of strength has, without reason, been ascribed to English powder.

443. If the condition and strength of naval ammunition were to be ascertained by actual experiment, from time to time, by the proper authorities in the ship to which it belongs, commanders would no longer be exposed to make the mortifying discovery of the weakness of their ammunition in the important moment of service-practice. Provided with the means of doing this previously, officers would either be enabled to report the damaged or unserviceable state of their gunpowder; or, knowing its condition, endeavour to procure a supply of better quality, if their own could not be restored to serviceable efficiency by drying. It is requisite, therefore, that naval officers, and particularly master-gunners of ships, should be qualified to judge correctly of the condition of their gunpowder by inspection, and be taught to conduct those experiments by which its strength may be ascertained. For these important reasons, instruction on this subject is considered indispensable in the course of training established at the depôts of naval ammunition.

444. The only infallible test of the goodness of gunpowder is experiment; but there are certain appearances and indications by which an estimate of its quality may be formed.

Gunpowder should be of uniform colour, approaching to that of slate. The particles should be perfectly granulated, free from cohesion, and should admit of being readily poured from one vessel into another; if otherwise, it may be concluded, either that the powder has been imperfectly glazed, or that it is damp. Gunpowder should be devoid of smell: if it have a disagreeable odour, it may arise from a practice which the author believes is not uncommon—that of heating the nitre excessively, for the purpose of drying it the more effectually. By this process a portion of the nitre may be decomposed; if so, potash,[a] instead of nitrate of

---

[a] Nitrate of potash may be formed by the heat being carelessly applied, but it is doubtful whether pure or free potash is; it requires at least a dull red heat, if not more, for that purpose.

potash, will form a part of the powder. Nitrate of potash, it is true, must be very highly heated, in order to produce potash ; but when over-heated in the process of fusion, too large a portion of nitrate is converted into potash, and this last, being very deliquescent, is not so efficient as undecomposed nitre. To ascertain whether the nitre has been decomposed to the degree of evolving potash, dissolve some of the powder in pure, or rain water, and add a solution of silver to that of the powder. If a black precipitate be formed, it may be concluded that sulphuret of potash exists in the powder.

It is possible, however, that the nitre may have been partly decomposed by having been over-heated, and consequently that the powder will be liable to become damp, from the deliquescent nature of the potash, although no sulphuret of potash may have been formed.

445. Gunpowder will very readily attract moisture from the air, if manufactured with nitre containing deliquescent salts, such as impure common salt : if this be the case, a solution of silver, added to one of the powder, in pure water, will give a white curdly precipitate. If an impurity of this nature be found to exist, or if powder, though originally composed of well-purified ingredients, should once become so damp from the influence of sea-water as to increase its weight beyond the quantity allowed in the proof, no dependence can be placed on the quality of the powder ; for sea-water contains so large a quantity of deliquescent salt, that, as in the preceding case, although the powder may frequently be dried, and sometimes appear not to be damp, yet it will re-attract moisture from a moist atmosphere as often as exposed to it. *Pure* common salt is not deliquescent ; when salt deliquesces it is a proof that it contains muriate of magnesia. The deliquescence of different samples of nitre depends upon the rough nitre having contained *impure* common salt, and that in the process of refining, the whole of the deliquescent salts have not been separated. It is by no means uncommon to adulterate crude nitre with common—that is, with impure salt. Nitrate and muriate of lime, and nitrate of mag-

nesia, are also deliquescent salts; but well-made gunpowder does not get damp by exposure to air, at least not so much as to deteriorate it—the glazing and the other ingredients prevent, or correct the hygrometric property of pure charcoal.

In powder which, however, *has* become damp, large lumps are formed. If the injury be not very considerable, these concretions may be reduced by drying, rubbing, and loosening; but gunpowder thus affected, never altogether regains its lost force.

446. Dampness in powder of good manufacture does not, in general, arise, as is commonly supposed, from the nitre attracting moisture. Pure nitre is not in the slightest degree deliquescent, it is not even hygrometric, whereas charcoal, particularly when newly made, imbibes aqueous vapour with such avidity, that a piece of perfectly dry and well-made charcoal, exposed to the action of the air for a week, will increase in weight about 14 or 15 per cent., and the matter absorbed consists principally of aqueous vapour. Thus powder of the best quality is liable to become damp from a circumstance which cannot be prevented by any degree of care in preparing the ingredients, and which can only be avoided by effectually excluding the atmospheric air. It has been thought right to state these circumstances, for the purpose of showing to the members of the profession the vast importance of a general adoption of air-tight vessels for stowing naval ammunition; and it may be added, that they should be introduced by degrees into all powder magazines.

447. It will not, in general—perhaps never—fall to the share of naval officers to restore damaged gunpowder, by drying it in large quantities by artificial heat: it may, however, become necessary for them to do so occasionally in small quantities. But whatever be the improbabilities of naval officers or gunners being called upon to superintend or conduct such a process, it is proper they should know the precautions that ought to be observed.

In drying gunpowder that may have become damp,

great care should be taken to regulate properly the degree of heat applied to the process; for there are several temperatures, considerably less than that required to explode the mixture, which are nevertheless capable of injuring it extremely. If the heat to which it is exposed be above 140° of Fahrenheit, the sulphur will begin to rise in vapour. At about 240° of Fahrenheit the sulphur will melt, without igniting the nitre,— the uniformity of the granulation will then be destroyed, and a number of small knotty lumps formed. These effects upon the sulphur may easily be shown by scattering a few grains of gunpowder upon a plate of metal heated unequally. The grains that fall upon parts much heated will instantly explode. In other parts the small, blue, lambent flame of the sulphur will be seen to rise and subside without exploding the mixture. A still less degree of heat will cause the sulphur to melt and sublime, and a certain inferior degree of temperature will also cause it to volatilize.

The degree of heat used in the stoves for drying gunpowder, should not, therefore, be above 140° of Fahrenheit.

When gunpowder has become utterly unfit for service, the nitre may be separated by putting it into vessels containing water, by which the nitre will be readily dissolved, and may then be crystallized by evaporation.

448. The strength of gunpowder has been so much increased of late years, that tables of ranges, formed before the commencement of the war (1793-1815), are no longer considered correct rules for practice. This improvement in the strength of ammunition is principally owing to the process of charring wood for the manufacture of gunpowder, in iron cylinders,—hence the term *cylinder powder*. The wood, properly seasoned and prepared, is put into cast-iron cylinders, placed horizontally over stoves, and the front openings closely stopped. Heat is then applied; when the pyroligneous acid passes over, and inflammable gases are evolved through tubes inserted in the back parts of the vessels. The gas is generally suffered to escape—the acid liquor col-

lected in casks—and the carbon left pure in the iron retorts.

449. Lay a drachm or two of powder on a piece of clean writing-paper, and fire the heap by means of a red-hot iron wire: if the flame ascend quickly, with a good report, leaving the paper free from white specks, and do not burn it into holes, the goodness of the ingredients, and proper manufacture of the powder, may be safely inferred.[*]

When good gunpowder is blasted upon a clean plate of copper, no tracks of foulness should be left.

Gunpowder exposed for seventeen or eighteen days to the influence of the atmosphere should not increase materially in weight. One hundred pounds of powder should not absorb more than 12 oz. If it increase in weight more than 1 per cent., it is a proof that deliquescent salt abounds in a degree which should warrant the condemnation of the powder.

450. The modes of examining merchant's powder are, by eye, by hand, and by ladle. The colour should be dark purple; that of a brown colour is of bad quality. If the powder be soft or tender in grain to the touch, or dusty, or if a cubic foot weigh less than 55 lbs., the powder is rejected: if the grain be hard, clean, and good, and the powder pass the trial for specific gravity, it is then proved by three rounds being fired in the Eprouvette mortar, and if it pass with respect to range, it is then put into a box perforated with holes for not less than twenty-one days, having been previously weighed, and it is again weighed on being taken out. Three rounds are afterwards fired with 8-inch solid balls from a mortar, and a mean of the ranges is compared with that of three rounds fired with the original sample, 5 per cent. less than Waltham Abbey Government powder being allowed. If it pass in range on this occasion, "flashing" on copper plates is the last proof.

---

[*] Observations on the Manufacture and Proofs of Gunpowders, by R. Coleman, of the Royal Powder Mills.

451. To prove the strength of large grain or common powder, 2 oz. are fired from 8-inch Gomer mortars, at an angle of 45°, placed on stone beds, and so fixed as not to recoil. These mortars are loaded with shot weighing 68 lbs.; and the average of such ranges, with the Government powder of Waltham Abbey, is 250 feet.

Powder made of common pit charcoal will only project such a ball, under the same circumstances, about 220 feet. Powder that has been re-stoved will only produce a range of from 107 to 117 feet under like circumstances. These facts show the vast reductions in force, and the great loss of accuracy, which invariably and irremediably attend any deterioration in that full-proof condition of gunpowder upon which all rules of practice are formed.

A musket charged with 2 drachms of fine-grained, or musket-powder, should drive a steel bullet through 15 or 16 half-inch elm boards, placed ¾ of an inch from each other, the first board being set at 40 inches from the muzzle of the musket; but with re-stoved gunpowder the bullet will only perforate from 9 to 12 of the boards.

The quality of large-grain powder is ascertained like that of the merchant's powder, by its general appearance, its firmness, glazing, uniformity of grain, and density. The weight of a cubic foot of Government gunpowder must be 58 lbs. The process of "flashing" is also used to ascertain the quality of the gunpowder. About 3 drachms are placed on a copper-plate, and fired with a red-hot iron to see that no residue or foulness be left.

452. The quality of fine-grained powder for muskets, &c. is tested in the same manner as the large-grain cannon-powder; and its strength proved by firing 2-oz. charges from the half-pounder gun Eprouvette.

But the proofs of gunpowder by range, in the manner stated, require means which can only be found at the principal Ordnance establishments at home: they cannot be resorted to by the navy on most foreign stations, and are not always to be had on home service.

453. To practise with the Eprouvette, charge it with a small quantity of loose powder, by means of a ladle. Raise the muzzle by pushing the piece forward, and press the powder lightly together with a rammer-head; then set the index, by pressing the limb in contact with the head of the frame, and note the division pointed out. The piece should be primed with a thread of quick-match, and fired by a port-fire. After the Eprouvette has made freely its first recoil, the vibrations should be gradually stopped, taking care not to touch the index or its limb. The division pointed out on the quadrant will then show the effect of the charge; and the strength of the powder may be estimated by comparing the extent of the vibration with that which is known to be due to powder of proof strength.

The vibrating Eprouvette being fired with 2 oz. of powder, the index should describe an arc of recoil of 26° with proof powder; and with re-stoved and all other descriptions of powder, not less than 24°.

ON GUN-COTTON.

454. In 1846 some experiments were made in Paris with the recently-invented gun-cotton, in a musket suspended from a horizontal axis similarly to the gun-pendulum or eprouvette; and the following is a table in English denominations of the initial velocities obtained with different charges of the cotton, the charges being so pressed that their lengths might be nearly proportional to their weights:—

Charge in Troy Grains.	Velocity in Feet.
15.44	497
30.89	1035
46.33	1348
61.78	1565
77.22	1700

On comparing these velocities with those which were observed to result from equal charges of gunpowder, it was found that 77.22 grains of gun-cotton produced an effect on a musket-ball equal to that of about half an ounce of common gunpowder; and from some experiments made by M. Arago, the force of the cotton appeared to be about three times as strong as that of gunpowder.

The report is said to have been strong, but less fatiguing to the ear than that of gunpowder: there was no appearance of smoke, a short flame only being seen at the mouth of the piece. After each fire there was left in the bore a quantity of condensed vapour of water, with a little carbon; and unless the bore was cleaned after each firing, the velocity was sensibly diminished.

It may be observed here that the successive increments of velocity arising from the successive increases of 1 gramme (.035 oz.) in the charge, form a regular decreasing series. This law is peculiar to the gun-cotton, and seems to result from the regularity of its chemical transformation : common gunpowder, being an imperfect compound, does not exhibit such regularity.

In firing against a plate of cast-iron at the distance of 15 yards, with a charge equal to 4.6 grains troy of gunpowder, the leaden ball was compressed as much as half its diameter; while an equal effect was produced by 0.76 grains only of gun-cotton.

Experiments with gun-cotton have been carried on at Washington; and from these it was found that the initial velocity of a ball whose diameter was 5.69 inches, and weight 24.33 lbs., with a windage equal to 0.135 inch, and a charge of cotton equal to 1 lb., was 1080 feet; with a charge of 2 lbs. the initial velocity was 1413 feet. A musket-barrel loaded with 120 grains of cotton, and carrying one ball, with a wad, burst at the breech.

455. In order to ascertain the relative force of gunpowder and gun-cotton for mining operations, the French experimenters caused a hole to be bored in a piece of rock having the form of a prism, and placed in it 4·6 ounces of gunpowder; the hole was tamped, and the charge being fired, the rock was split by the explosion in two lines at right angles to one another: thus dividing the mass into four equal fragments. · A

similar piece of rock, of the same size nearly, being bored, and the hole charged with 1.66 ounces only of gun-cotton, the explosion split the rock in one line, dividing it into two nearly equal parts, thus producing the effect of nearly three times the quantity of gunpowder.

456. The following circumstances, from the second report of Major Mordecai on his experiments at Washington in the years 1845, 1847, and 1848, for the purpose of determining the fitness of gun-cotton as a substitute for gunpowder in the military service, will be read with interest:—

1. Explosive cotton burns at 380° Fahr., therefore it will not set fire to gunpowder when burnt in a loose state over it.

2. The projectile force of explosive cotton, with moderate charges, in a musket or cannon, is equal to that of about twice its weight of the best gunpowder.

3. When compressed by hard ramming, as in filling a fuze, it burns slowly.

4. By the absorption of moisture its force is rapidly diminished, but the force is restored by drying.

5. Its bursting effect is much greater than that of gunpowder, on which account it is well adapted for mining operations.

6. The principal residua of its combustion are water and nitrous acid : therefore the barrel of a gun would be soon corroded if not cleaned after firing.

7. In consequence of the quickness and intensity of its action when ignited, it cannot be used with safety in the present fire-arms.

8. An accident on service, such as the insertion of two charges before firing, would cause the bursting of the barrel ; and it is probable that the like effect would take place with the regular service charges if several times repeated.

PART V.

ON THE TACTICS OF SINGLE ACTIONS AT SEA, WITH OBSERVATIONS ON SOME NAVAL OPERATIONS BETWEEN THE SHIPS OF GREAT BRITAIN AND THE UNITED STATES.

457. ALTHOUGH the alliance between steam and wind in the propulsion of ships of war, whether the two motive powers act in combination to produce the tactical evolutions of a fleet, or whether steam-power is merely an auxiliary to that of wind in the movement of a single ship (called by the French in this case *bâtiment mixte*), will no doubt make a great revolution in the modes of maritime war—a subject which will be treated by the author in a forthcoming work on Steam Warfare; yet, as has been well observed in a recent publication,[a] sailing-ships, particularly frigates, in seas remote from Europe, will probably for many years have to contend in the ordinary manner, and according to the principles of the existing tactics for sailing-vessels. On this account the author purposes to repeat, with some corrections, what he had stated in the former editions of this work concerning the tactics of single ships, with the observations he had made on some naval actions which had taken place during the late war between Great Britain and the United States.

458. A review of the tactics of some of our naval actions with the United States will be found to yield many useful deductions in support of much that has been advanced in the course of this work.

The reasons which have induced the author to attempt a review of these important occurrences, are to show that the United States' commanders so circumspectly

---

[a] Note par le Prince de Joinville.

and cautiously adapted their tactics to the superior powers of their armament, that even when opposed to very inferior numbers and quality of ordnance, they would neither approach, nor permit us to join in close battle, until they had gained some decisive advantage from the superior faculties of their long guns in distant cannonade, and from the intrepid, uncircumspect, and often very exposed approach of assailants who had long been accustomed to contemn all manœuvring, and who only considered how to rush soonest into yard-arm action. Such, unquestionably, was the character of these proceedings. The uncircumspect gallantry of our commanders led our ships, unguardedly, into snares which wary caution had spread; and in point of fact our vessels were, in almost every instance, so crippled in distant cannonade, from encountering rashly the serious disadvantage of making direct attacks under the powerful fire of whole broadside batteries, that all those close actions which terminated unfavourably to us may fairly be considered to have been fought under very disadvantageous tactical circumstances, even had the force of the contending ships been equally matched.

459. In the action between the "Macedonian" and the "United States," the American frigate avoided close action for a full hour after fire commenced. Captain Carden states, "that from the enemy keeping two points off the wind, the British frigate was not enabled to get so close to her as was desired; and that it was not till after an hour's cannonade, when the enemy backed and came to the wind, that close battle commenced." This shows that Commodore Decauter's plan of operation was to keep at long-shot distance for some time, to try the effect of relative precision of fire—to avail himself of the superiority of his long 24-pounders over the 18-pounders of the "Macedonian"—and by edging away from the British frigate, which gallantly attempted to close directly from the windward, to prolong a preliminary operation so much in his own favour. How far he succeeded is shown by the opinion of the court-martial, "that the 'Macedonian' was very materially damaged before close

PART V. "MACEDONIAN" AND "UNITED STATES." 481

action commenced." When the British frigate was completely crippled, the American came to the wind: the event is well known. As a display of courage, the character of the service and of the country was nobly upheld; but it would be deceiving ourselves were we to admit that the comparative expertness of the crews in gunnery was equally satisfactory. The author's object being to press home the absolute necessity of training to expert practice master-gunners, their crews, and captains of guns, he supports his opinion of the vast national importance of such a measure by strong, impartial, and unreserved appeals to facts. Now taking the difference of effect, as stated by Captain Carden, we must draw this conclusion — that the comparative loss in killed and wounded (104 to 12), together with the dreadful account he gives of the condition of his own ship, whilst he admits that "the enemy's vessel was comparatively in good order," must have arisen from inferiority in gunnery, as well as inferiority in force. That our frigate should be captured was not at all surprising, considering the great odds against her; and the comparative ravages in the two vessels indicate the disadvantages under which this gallant officer was compelled to engage.

460. Let us apply to these and other actions what has been said (Arts. 131, 133, and 147) on the comparative powers of long and short guns: the qualities of long 24-pounders, and their decisive superiority in such distant preliminary operations as those with which the Americans introduced, and will continue to commence, close action. Do not such facts exact, *absolutely*, the timely preparation of means and qualities necessary to attain the most minute precision for distant cannonade, in which superior training is, in fact, superior power; but which cannot be sufficiently acquired in our present system? Do not these and the facts mentioned in Arts. 145, 146, incontrovertibly support the reasoning (Arts. 391, 392) as to the vast advantages that may be reaped in distant cannonade with powerful guns, directed with every resource of refined, minute expedient, to gain accuracy? The action between the "Java" and "Con-

2 I

stitution" was also a very gallant display of that undaunted resolution which carried the British frigate forward to dare her antagonist to close action; but this was not a very favourable method of joining in battle with a cautious enemy, who knew well the advantages he might reap by opposing circumspect caution to the open audacity of his too bold assailant.

461. So much depends upon the way in which a vessel is approached, or brought to action, that the author finds it impossible to avoid making a few observations upon the tactics of single actions, more particularly as it appears, from a close and attentive study of the manœuvres of the American vessels, in action with ours, that the tactics of those operations were not matters of chance, nor of individual determination, but a general predetermined plan of operation, expressly calculated to procure those advantages which our resolute, straightforward, but not very prudent methods of attack were expected to present.

In entering on this part of his subject, the author is not without apprehension that he may be considered as touching unnecessarily upon a professional question rather foreign to the subject he has undertaken, and quite beyond his powers; but, to bespeak indulgence for this obtrusion, he must submit that an essay on the service-practice of Naval Gunnery, according to the view the author has taken, and to the plan he has followed, is so inseparably connected with tactics that he cannot properly avoid touching upon them, notwithstanding his inability to treat the important subject as it deserves.

462. The form of action which, in tactical circumstances, has been most unfavourable to us, is the attack from the windward. Of this character were the actions in which the "Macedonian" and "Java" were captured. The serious disadvantage of running down directly upon an enemy to leeward, to engage him in close battle, is so obvious that we can only consider such a plan of operation to have been adopted from the unguarded confidence inspired by the intoxication of glory in our

naval warfare, which had taught us to despise all manœuvring for position, and to consider circumspection as unnecessary. Guided by such sentiments, it appears to have become an established maxim of the profession, that whenever an enemy can be attacked, the only method worthy of our flag is to come at once to the point; and, in the emphatic terms of the seamen, to seize the bull by the horns : but this metaphor implies that the excited animal rushes also to meet the foe ; whereas, in the actions above referred to, we have seen that the opposite party was perfectly aware that such a mode of attack would be adopted, and were so coolly and cautiously prepared to receive and prolong it, that this plan of approach, bold and dignified though it be, may fairly be considered to have contributed very materially to the great loss of British blood, if not to that of the material trophies now so proudly exhibited. If this be so, there is abundant reason for endeavouring to bring back that respect for manœuvring for position which was formerly entertained in the profession, and which is particularly necessary in attacking from the windward ; and so to regulate audacity by scientific circumspection, as to know when it is necessary to *blind the bull*, before it is prudent to attempt to seize him. Borrowing an established maxim from the tactics of land operations to support this admonition, let it be well remembered that he is considered the best officer who effects most by manœuvre.

463. There is this fundamental difference, however, between naval and military tactics, that great advantages belong to the offensive operations of well-constituted and well-commanded armies, whereas in naval operations the point of attack is so undisguised that, if the defensive operation be well managed, the attacking force must be exposed to great and often decisive effect previously to close action. That generous intrepidity, therefore, which belongs to the commonest person of our nation, and which disdains anything that can be called foul play, should, when opposed to an enemy, intelligent, cautious, and, in force, generally superior, be

restrained by some circumspection—Circumspection!!
the very word startles the author as he writes it down;
and it may convey a similar shock to the feelings of
those who may read this, and who may also have
contemned circumspection in all such cases; but the
observation is introduced with such support from facts,
to show that our resolution, courage, and want of circumspection, have materially contributed to the successes of a wary enemy, and therefore the author
ventures to repeat the admonition. The stigma which
rash men might attach to a prudent observance of circumspection, in the mode of bringing on close action
with a vessel of superior or equal force, should not deter
our officers from acting with that discretion which will
lead, upon even terms, to a close terminal struggle, in
which their native courage will be free from the obligations of caution. But, to reconcile this to the noble
spirits of our glorious navy, without wounding them by
any implication, the author will endeavour to show how
this contempt of manœuvring for position has been
promoted, and why this error ought to have been corrected in some late affairs. In doing this he avails
himself, with infinite satisfaction, of the supporting
observations of Sir Philip Broke, the distinguished captain of the "Shannon" (to whom the author is greatly
indebted for information upon many professional points),
some of which are given nearly in his own words.

464. In a great many of the single actions of the
late French war (1794–1814), the enemy thought of
nothing but escape; and, in bringing on a running
fight, he deprived himself of those advantages which a
ship may reap by coolly awaiting her enemy's approach,
with appropriate manœuvre of sail and helm. By such
a measure the enemy gave us all that British valour
desired—an opportunity of coming at once to close
action without previous loss, as soon as our vessel could
come up with her antagonist. Thus all scientific manœuvre became superfluous, and under such circumstances was properly disregarded.

That the intrepidity of our ancient naval commanders

had given them likewise a dislike of refined manœuvre, is very evident in the pages of our naval history; but this disregard of all caution in approach to action is only a character of the time, arising out of peculiar circumstances, and should now yield to considerations of a very different nature.

Our most accomplished officers, who commanded ships at the beginning of the war arising from the French Revolution of 1793, not only displayed more caution in preparing their ships for action than was subsequently used in a more triumphant period of the naval war, but also held in high respect all scientific manœuvring, with a view to gain the most favourable position for action. That this *was* the professional feeling is well known— that it *is not* so now is the consequence of that deterioration in the European navies already noticed (Art. 3), which first led us to relax our warlike circumspection, and then taught us to contemn it entirely.

465. Our modern ships of war, particularly frigates, are not perhaps so well calculated for manœuvring as the old-fashioned ships were, being so much slower in turning, on account of their flat futtock and great length; but any *system* of manœuvring was hardly ever thought of in latter times, and indeed would seldom have been appropriate, sure as we were that if we could only outsail the enemy, we should be able to bring him to close action on very advantageous terms. Under such circumstances it would have been absurd, and justly injurious to a commander's character, to have attempted any manœuvring for advantageous position, even in the most scientific way, when it was in his power to lay his enemy alongside without difficulty. For there is always some danger in bracing and trimming sails, and particularly in backing them, after ships have exchanged a few well-directed rounds, and are under each other's fire. Engaged in such manœuvres, should a chance shot have crippled the assailant, the use the enemy would have made of the accident would have been to escape, in which case a British commander would justly have been severely reprimanded for losing

by his theories a victory which the courage and ability of his officers and seamen would have ensured had they been led at once to an abrupt, impetuous attack. Thus, in our warfare with European navies, the escape of an enemy after we had brought him to action was felt by us as a defeat, and indeed the enemy boasted of it as a victory. These circumstances and considerations seem to justify, and even to demand, a bold and uncircumspect approach to that critical position in which, though we might be disabled ourselves, we were generally certain of preventing the enemy from leaving us, having him so much under our fire that any refit would be impossible. Thus stood the case, as it regarded single actions in general, during the French war; but when we come to meet an enemy so much nearer our own stamp of character, and whose warlike navy (so long as it is inferior in proportion to his mercantile navy) must be well manned, and who will try, in most classes of his ships, to keep some advantage of size and armament,—if such an enemy seek to join science to his other advantages, we must be prepared to answer him in his way when we cannot have our own.

466. In the tactic of the action between the "Guerrière" and "Constitution," there was a good deal of manœuvring, yet the general courses of the two ships gradually converged towards each other in a degree which admitted of at least an hour's occasional cannonade before close action commenced. The wind was fresh from the north. When the vessels first distinguished each other, the "Guerrière" was to leeward, close hauled upon the starboard tack, the American on the weather-beam, standing S.S.W. The "Guerrière" opened her fire first (which it is said fell short), and soon afterwards the American opened his battery, and continued to fire occasionally as he came down. When he began to draw near the "Guerrière," the British frigate several times wore to avoid being raked. This prudent counter-manœuvre, to avoid a serious disadvantage, operated against closing, which accordingly did not take place till about an hour after the action had

commenced. Thus, from the time the fire opened till close battle began, the vessels, steering free, kept up a sort of running fight upon courses gradually converging towards each other. Various very untoward circumstances conspired to terminate this affair in an unfortunate manner. Our frigate was very short of hands; her powder, it may fairly be asserted, from the testimony of the American commander, who says " that her shot fell short," was deteriorated by long keeping and damp (Art. 442). Several of her guns and carronades broke loose, owing to the perished condition of their breechings (Art. 389), and the decayed state of the timbers through which the long bolts passed. The armament, in guns (28 long 18-pounders), was very inferior to the 30 long 24-pounders opposed to her. The loss of the mizen-mast by a chance carronade shot, and its unlucky fall to windward, threw the ship up in the wind in a singular manner; and her other masts, which fell soon afterwards, had been crippled previously by stress of sail and decay. These untoward circumstances are quite sufficient to account for the capture of the frigate, and to show, indeed, the impossibility of preserving her, notwithstanding the gallantry with which she was defended; but they are not sufficient to account for the great disparity of loss in killed and wounded, viz., 78 to 14. If the author is at all correct in what he has stated in Arts. 397, 398, respecting the incredibly trifling effect usually produced by random broadside volleys, given in sudden changes of position, he would say that, in wearing several times to avoid being raked, and in exchanging broadsides in such rapid and continued alterations of position, and consequent elevation of her ordnance, the " Guerrière's" fire was much more harmless than it would have been had she given it in a more steady position.

467. It appears that the most advantageous way in which a vessel to leeward can receive a direct attack, and bring on close action, with an enemy coming down the wind for this purpose, is to come to the wind herself, and there wait, making as little way as possible,

whilst the offensive movement is in progress. This opinion requires some introductory explanations.

If two ships, A and B, Fig. 43, move at an equal rate, upon courses equally inclined to each other, they will approach gradually upon equal terms, and come close to each other, when arrived at C. If the two

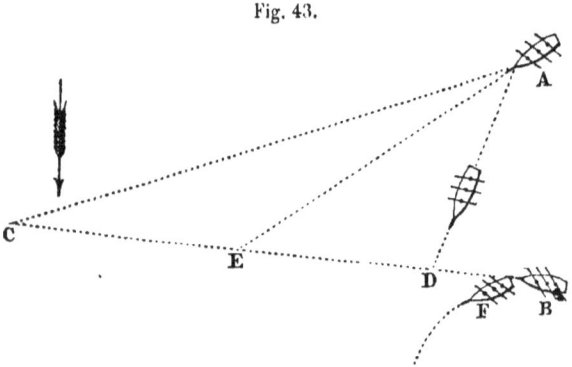

Fig. 43.

vessels be equal in force and quality of crew, an action brought on in this way would be alike favourable to both; but if either should possess any superior power of guns, such as might induce her commander to prefer commencing with distant cannonade, and approximate gradually to close battle, then it is evident that the other vessel should vary its plan of operation, to defeat the purpose which the enemy has in view, and which is soon perceptible in his actions.

According to the well-known principle of chasing to leeward, the course should be so regulated that the chase always bear on the same point of the compass. If she be found to draw ahead, the chaser must haul more up; if the chase draw aft, the pursuer must keep more away. Applying this to Fig. 43, the more B draws aft, that is, the slower he goes, the more direct must be A's approach, according to this principle of chasing. Now suppose B, instead of standing-on rapidly in the line B C converging gradually to A's course, were to remain as stationary as possible, keeping her broadside turned towards A, it is evident that A (whom we suppose desirous of coming to close action) cannot approach under the fire of B without obvious disadvantage; consequently the slower B moves on the line B C, the more inclined to that line must be the course of A's advance, as A D, and the more he will be exposed to a raking fire in coming down. If the ship A be so circumspect as to come down on a line A E, out of range, B should not, upon any account, stand-on to meet him, if the relative force of the ships demand circumspection on the part of B; for doing this would be acting exactly in the manner A wishes, as is evident by such a movement falling in with his plan of attack. But if A come down in any line A D within range, then B should follow him with his broadside steadily bearing, as F, and in this way should not object to come to close action, the previous advantage having been his.

468. If, having moved upon the line A E, Fig. 44,

Fig. 44.

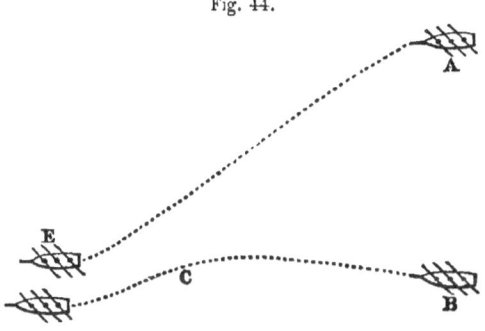

out of range, the ship A should come to the wind at E on the same tack as B, and there wait, B should then run up to close action, and raking A's stern, as at C, engage him to leeward, if he will permit. If A should decline this, and bear up, as F, Fig. 45, to avoid being raked, B may either do so too, and engage him going

Fig. 45.

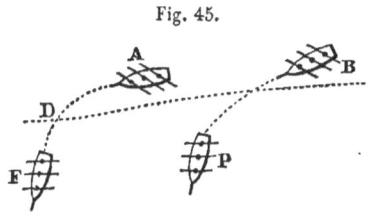

free, as at P, on even terms, or stand-on, and, crossing his stern at D, keep his wind, and manœuvre afresh.

If A, having come down in the line A E, out of range, should haul up at A 2, Fig. 46, on the contrary tack,

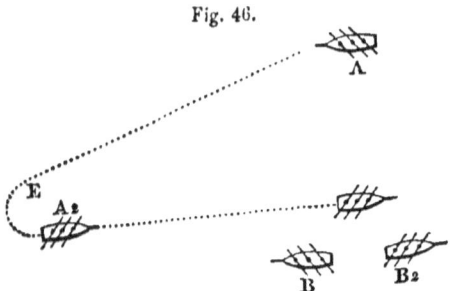

Fig. 46.

B, if he is desirous of close action, should immediately wear, or tack, and stand-on slowly, as B 2. If A keep his wind, B should wait for him, and thus engage to leeward upon equal terms; but if A attempt to bear up, across B's stern, to rake him in passing to leeward, then B may either bear up to avoid being raked, and thus, as in Fig. 45, engage A close, going free; or waiving this, accept close battle upon A's terms, as the "Shannon" did the "Chesapeake." To execute this, B should not reduce his sail, but by keeping the maintopsail to the mast, and the others shaking, or, when off the wind, by *bracing-by* as flat as possible, just keep the vessel under the influence of her helm, and no more.

469. But it may be said, a cautious, intelligent enemy, attacking from the windward, will come down abaft B's line of fire, Fig. 47, as the "Chesapeake" did upon the

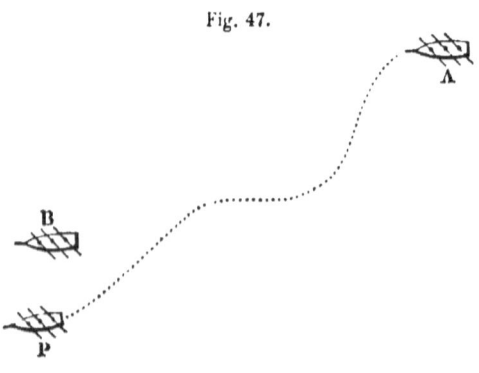

Fig. 47.

"Shannon," and, when nearly in his wake, either run up to windward, or pass to leeward, as he may choose, if B will wait for him, or if A outsail B. But whether the action is to be thus fought or not, will neither depend upon B's sailing nor upon A's pleasure, if B manœuvre properly; for if he have any reason for not desiring such a plan of action, and should not think proper to give A an opportunity of raking his stern, in passing to engage him to leeward, he should tack or wear at a convenient time, and stand-on slowly the other way. Thus if A 1, Fig. 48, perceiving B 1, laying

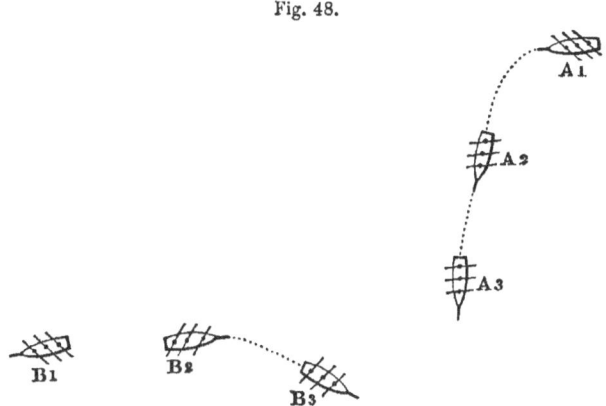

Fig. 48.

to leeward, shape a course to run down into his wake, B 1 should tack or wear in time, and stand on as B 2, towards A 2; and this manœuvre will bring the case exactly to that which has been considered in Figs. 43, 44, 45. If B 1, neglecting or waiving this, stand-on, and let A 1 get close in his wake, then A 1 may bear up, and, raking B 1's stern, engage him to leeward. This is an obvious advantage which the "Chesapeake" might have availed himself of (as Sir Philip Broke admits) instead of ranging up to windward of the "Shannon;" and it is one which, had it been taken, would most probably have gained some previous advantage. There is no way in which B 1, having permitted A 1 to come close in his wake, can now avoid sustaining some previous disadvantage, if A 1 should try to rake his stern. For if B 1 tack to avoid it, he will first expose his stern, B, Fig. 49, to be raked;—he will be

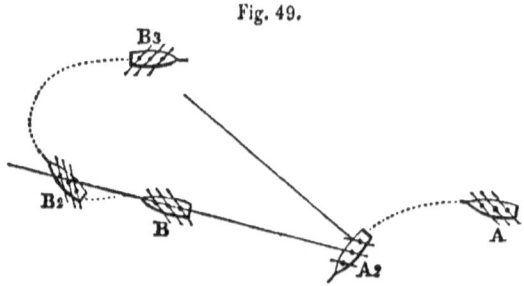

Fig. 49.

severely punished whilst in stays by a fire in great part diagonal,—if he hang in stays he will be utterly destroyed; and in coming round upon the other tack he may fall off, nearly end-on towards A 2, as at B 3. No good officer, indeed, would attempt such a method of avoiding being raked; and if, on the contrary, B bear up, as B 2, Fig. 50, to prevent this, her opponent A

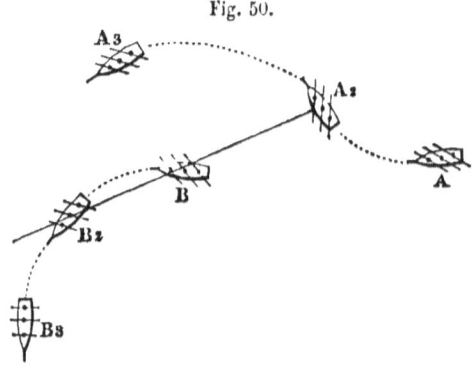

Fig. 50.

may luff-to, and rake him before B can get away, and then manœuvre for fresh advantage.

Now if, on the contrary, B should have tacked, as suggested in Art. 468, and stand-on towards A, as B 2, Figs. 48 and 51, then, if the offensive movement be continued within range, B should deaden his way as much as possible, and open his fire upon A coming down, keeping his broadside, as at B 3, B 4, steadily bearing, and thus follow the movement of A 2, A 3, gradually, till both ships come close; and thus again the commander of B could have no objection to close action, the previous advantage having been his.

Fig. 51.

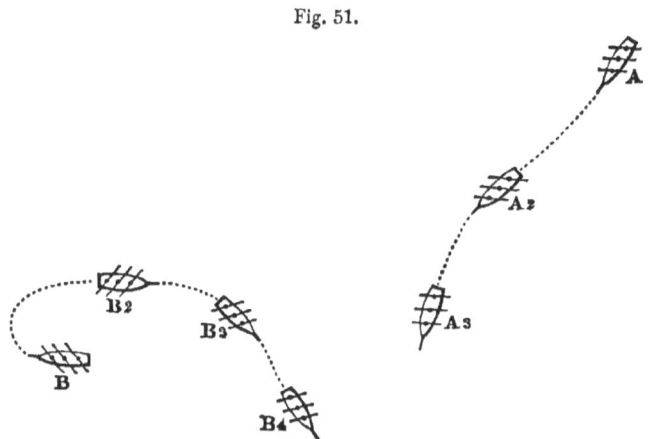

470. If this reasoning be correct, the best way for a vessel B, Fig. 52, to leeward, to receive an attack with circumspection, from a vessel A, to windward, is, never to let A come down into his *wake*; but having tacked in time, as B 2, stand-on slowly till A approach within B's fire, from which time B should keep as stationary as possible. Supposing the vessels to be of nearly equal force, it may be assumed that A has no intention of avoiding action; but after he is once brought to the position A 4, it is evident he cannot approach nearer to B manœuvring thus without receiving a mass of fire which he cannot return. If he shape his course to cross B's bows, the counter-manœuvre which B should apply is not velocity, but gradual change of position, in steady broadside bearing, with as little way as possible, following A's bow with the broadside, so long as he tries to cross B's bow,—an attempt which can only be continued till A come close to the wind on the larboard tack; and here again there would be no objection to bring on close action in this way, the previous advantage having been to B. If, in thus rounding-to, the vessels should

get foul of each other, it will be in a position favourable to B, as E, F, Fig. 52, if the manœuvre have been properly and steadily executed; and this will bring on a

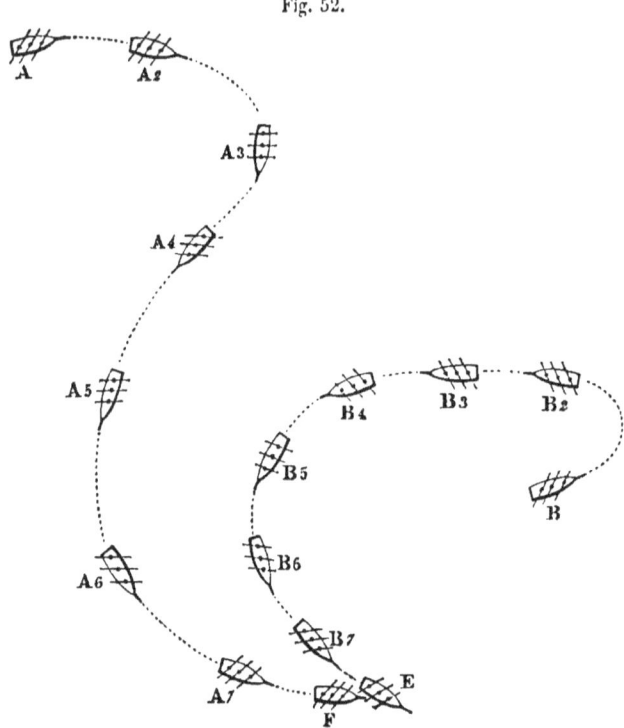

Fig. 52.

character of combat (boarding) which we always desire. These manœuvres will, at all events, refuse to A the opportunities of which we have supposed him to be desirous, viz., previous distant cannonade on his own terms; and therefore it appears that this method of manœuvring, in receiving an attack from the windward, is favourable for ships which are not at liberty to receive battle under any disadvantageous tactical circumstances.

471. The author is quite aware that there may be difficulties as well as facilities in executing these manœuvres that will require modifications, and in some cases perhaps abandonments, in the theories he has endeavoured to suggest. But many of those cases which

demand modifications, or which present facilities, have already been noticed in the several articles (391 to 398 inclusive) which relate to distant practice, close battle, raking fire, dismantling shot, &c. There are, however, some general considerations which it will always be necessary to keep in mind, with strict application to the immediate case, by which the experienced commander will not fail to be correctly guided. From the improved construction of our ships in quality of sailing, they are not so well calculated to manœuvre quickly as our old ships were. In a considerable breeze it is extremely difficult to moderate their velocity, when nearly before the wind, by *bracing-by*. This renders it very difficult to keep a free course for any particular bearing of the guns, without either going too fast to preserve that bearing, or yawing frequently. In luffing-to, also, our long ships are very slow; and in executing it run over a great space of water; so that in the smoke of a broadside discharged in this act, there is great risk of a ship getting a-back, and being obliged to *box-off*; or of losing her head-way, and remaining for a considerable time in this position, which is forcibly termed being *in irons*. To get her out of this awkward position the ship must be *paid off*, by backing the head-sails. This exposes her to the chance of making a strong stern-board (particularly with the fore-sail set), and will, at all events, throw her long *out-of-hand* before she recovers head-way; and then heavy yards are to be braced round, perhaps under the guns of a closing enemy. Before any manœuvre is attempted, therefore, the actual position of the contending ships, the state of the sea and weather, should be considered, and experience consulted, as to the time the ship may take to tack or wear, compared with the distance the enemy may run in that time, or the time required for the enemy to make whatever countermanœuvre may seem to suit the case, and the consequent positions to which the two ships will thus be brought.

472. The action between the "Macedonian" and the "United States" was, in tactical circumstances, of a nature

different from those cases which have been considered in Arts. 467 et seq. The British frigate was to windward, and ran gallantly directly down upon the American; but in doing this was so severely damaged that the upper deck was almost entirely disabled by the raking fire of the "United States," then laying steadily to leeward.

In the action between the "Java" and the "Constitution," the British frigate was to windward. The American vessel tacked and stood away free soon after she was discovered. At 11 A.M. the "Java" hauled up, bringing the wind on the port quarter. At 1 50 P.M. the "Java" shortened sail, bore down upon the "Constitution's" quarter, and received her first two broadsides, which did little or no damage to the "Java." The British frigate then luffed-to on the American's weather-beam, almost touching, gave her first broadside, which killed and wounded between forty and fifty of her people; but from the total inexperience in gunnery of the crew of the "Java," she appears to have seldom hit her opponent afterwards. In manœuvring, the British frigate was well handled, and obtained positions which, had they been as effectually taken advantage of by good gunnery as they were conceived and executed with great nautical skill, must have produced a very different result from that which unhappily ensued. The "Java" passed at one time under the stern of the "Constitution" so close, that her ensign almost touched, when the latter put her helm up and made sail, upon which the "Java" instantly put her helm down, passed up close under the stern of the "Constitution," and got again on her weather-quarter. Soon after this the "Java" lost the head of her bowsprit, and became much damaged from the fire of her opponent; when it was determined, as a last hope, to board the "Constitution," and the "Java" bore up for that purpose. Whilst doing so, the head of her bowsprit actually rubbed along the quarter of the "Constitution," which brought the "Java" up in the wind; and in this position she remained for upwards of an hour, a mark for cool target practice. The "Constitution" at length came under the broadside of the "Java," when, finding her fire far from being silenced, the American frigate made sail, and went out of action for an hour, leaving the "Java" with her mainmast only standing, with the weather-half of her main-yard broken aloft, which she soon afterwards rolled away. A topgallant-mast was rigged as a jury-foremast, and a stay-sail set, forward, in the vain hope to get before the wind. The "Constitution," seeing this attempt, returned to the fight, and took a position right a-head of the "Java." Further resistance would have been unavailing, and the "Java" accordingly struck her colours. It is impossible to pass from the relation of this action, in which so much nautical skill and gallantry were unavailingly displayed by Captain Henry Lambert, who fell mortally wounded by a musket-ball fired from one of the tops,[a] and by her first lieutenant, H. D. Chads, without noticing, as a remarkable and interesting incident in the career of Rear-Admiral Chads, that he was, by his great merits, placed in a situation which—if sufficient extension and permanency be given to the corps of seamen-gunners to meet the wants and exigencies of the service (Arts. 27, 32)—will effectually prevent the recurrence of disasters arising from such a cause, by the zeal, experience, and practical skill with which this distinguished officer directed and superintended the important institution over which he presided, for teaching the science and improving the practice of gunnery in the British navy. But it is painful to reflect that, after twenty years of labour devoted to this most important duty by Rear-Admiral Chads and his distinguished predecessor, Sir Thomas Hastings, the establishment of the "Excellent" is still so limited as only to have produced about 2000 men qualified to serve as captains of guns, with no reserve of well-trained gunners ready in cases of emergency (Arts. 32, 36).

473. In this action there was, on the side of the British frigate, excess of gallantry, but very little cir-

---

[a] This fatal case may well be associated with the fate of Nelson in urging what is stated in Appendix A, on the introduction of the improved rifle and musket into the Naval service.

cumspection; and it is melancholy to reflect that such gallantry should have terminated so disastrously. If the British frigate had manœuvred, the "Constitution" would not have run away. Its commander might have hesitated to meet the manœuvre, and join fairly in close battle: he would probably have waited to see whether or not the British commander would be so incautious as to approach him, knowing well that such a step would operate in his favour. If our frigate had run down astern, the American would either have tacked or *waited*; if not, he had fled, and we had triumphed: but no—if we had been *more* cautious, he would have been *less* so. It appears, therefore, that battle should first have been offered upon equal terms, by commencing the ordinary manœuvre of running down in the wake. If the British frigate, represented by A, Fig. 53, declining

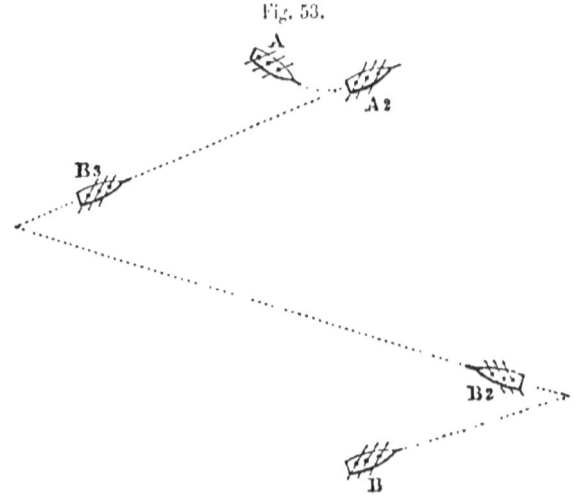

Fig. 53.

this, had brought-to, as at A 2, the American, B, fancying her rather shy, would certainly, after some time, have approached. This he probably would have done by tacking, as at B 2, and standing close upon the starboard tack into A's wake, and thence tacking towards her, as at B 3. Now, if B tack in A's wake, B cannot go to windward of A, nor rake him, except partially, by luffing-up in the wind, or by keeping

PART V.   THE "JAVA AND "CONSTITUTION."   499

away, both of which would be random and very inefficient volleys (Arts. 397, 398). But if B should stand-on, as at B 1, Fig. 54, and tack, as at B 2, to windward of A's wake, then it would be advisable for A to tack also, as at A 2, because B, by acting thus, may

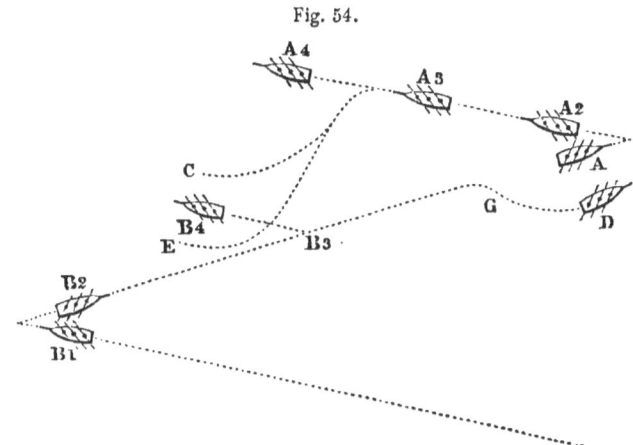

Fig. 54.

be suspected of an intention of crossing A's stern, to rake him before he engage him close to leeward, as at D. Now if A tack, it is evident that upon this course also he will go to windward of B; and if B proceed to B 3, A 3 may lay across his bows and rake him. This B will not, of course, suffer; and to prevent it, must either wear or tack again. If he tack and there wait, as at B 4, A may run up alongside, and engage him to windward at C, in close action; or, crossing B's stern, fight him to leeward, as at E; in either of which cases A will accomplish, by previous manœuvre, the very nature of action which he had offered at first, but which B then declined. If, from the position B 2, Fig. 54, B keep away, as B 2, Fig. 55, when he sees that A has tacked, as at A 2, Figs. 54 and 55, then A (being still either out of range or at very random distance) should wear round to engage B, going free, if he goes away as B 4; but if B try to cross A's bows, he should wear round gradually, deadening his way as much as may be

2 K 2

Fig. 55.

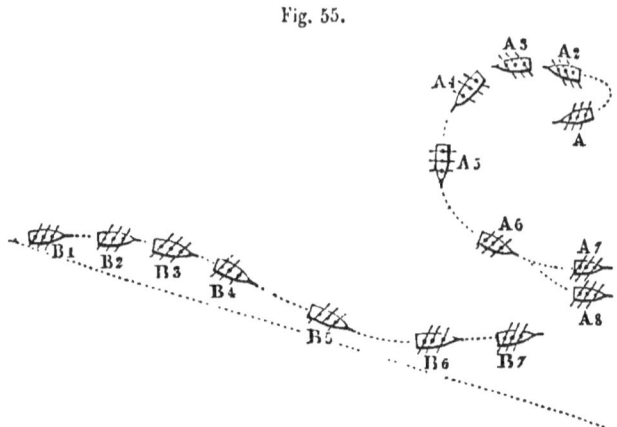

necessary to keep his broadside bearing upon B, as A 5, A 6, A 7; and thus have no objection to close action, either by hauling up, as A 7, and waiting for B, or by standing-on, A 8, and crossing B's bows to rake him.

474. The action between the "*Shannon*" and the "*Chesapeake*" reflects upon the victors immortal honour, and no discredit on the vanquished. Its characteristics are, that though the enemy did not, as usual, commence with distant cannonade, yet he was so far circumspect in his approach, as not to have been pre-exposed to the "Shannon's" fire, having come down astern, and only received the fire of the British frigate's after-main-deck gun and quarter-deck carronade before he opened his own fire. The rapidity and precision of the "Shannon's" fire were irresistible;—the enemy was beaten in eleven minutes! [a]

---

[a] It is not easy to account for the facts stated in the report given in the notes to Art. 254, respecting the deficient powers of penetration of the 18-pounder solid shot fired from H. M. frigate "Shannon" into the "Chesapeake." It appears by that Report, which was made from an actual survey, that sixteen 18-pound shot did not perforate, but only penetrated, some 4, 5, and 6, and others 8 and 10 inches; and that, of the shot which passed through, the greater number were 32-pounder carronade shot. Some unexpected circumstances with respect to relative position may perhaps account for this. It is stated in James's 'Naval History,' vol. vi. p. 201, that the Captain of the "Chesapeake," finding he was ranging too far a-head, brought his ship so sharp to the wind in attempting to deaden her way, that she lay, in consequence, for some time with her stern and quarter exposed to her opponent's broadside, the "Shannon's" aftermost guns firing diagonally into the "Chesa-

PART V.   THE "SHANNON" AND "CHESAPEAKE."   501

In this remarkable combat, however, it must be observed that an error in principle was committed on both sides. The "Shannon" gave the "Chesapeake" a great advantage in permitting her to approach as she did. The "Chesapeake" committed an error in not taking advantage of it as she might have done; for the American frigate might, when near, having considerable way, suddenly have put her helm up, and raked the "Shannon" by her stern, and then either hauled off, or, coming up again, engaged the "Shannon" to leeward. To frustrate the operation of being raked by the stern, which the "Shannon" had every tactical authority to believe was the American's intention, the "Shannon" should have kept away or tacked as soon as she saw such a course shaping towards her; but these manœuvres would have brought on a running, and perhaps a less decisive fight. The captain of the "Shannon," under the peculiar circumstances of the case, abiding by the terms of the defiance he had sent to the captain of the "Chesapeake,[a]" resolved to offer the advantage which his position presented, and not to appear to shrink from the contest he had long sought.

" 'Twas vain to seek retreat, and vain to fear;
Himself had challenged, and the foe drew near."

The gallant foe disdained to avail himself of the

---

peake;" and that (p. 202) the "Chesapeake" making, it would appear, a sternboard, fell with her quarter upon the "Shannon's" side, just before her starboard main-chains, and hooked with her quarter port the fluke of the "Shannon's" anchor stowed over the chest-tree. These changes of position may be considered as illustrations of what has been said—(Art. 105)—that naval actions are subject to sudden and unforeseen mutations in the positions of the contending vessels, in consequence of which one of them may suddenly be exposed to oblique, diagonal, or raking fire, and thus require that the loads of metal, and charges of powder, should be varied so as to insure penetration into and through the enemy's ship, however she may be struck. The penetrating power adequate to this end, in every position of the opponent, should never be sacrificed to the practice of using only the load and charge, which may be barely sufficient for penetrating in direct broadside action. May not inattention to this important rule have been omitted in the practice on board the "Shannon?" It appears moreover that the shot fired from the "Chesapeake" were likewise deficient in penetrating power.

[a] James's 'Naval History,' vol. vi. p. 199.

tactical advantage which his bold opponent offered, and for each it may be said,—

> "But open be our fight, and bold each blow;
> I steal no conquest from a noble foe."

The captured frigate was taken to Halifax, and there her gallant captain, killed in the action, was buried with military honours, his funeral attended by the civil authorities, the troops in garrison, and the surviving officers of the "Shannon," with the exception of her victorious chief, who was severely wounded in the fight.

> "The stone shall tell the vanquish'd hero's name,
> And distant ages learn the victor's fame."

APPENDIX.

(A.)

ON RIFLE MUSKETS.

(1.) THE common, and the rifle musket are weapons of which sufficient use has not hitherto been made in the British navy; but the latter, in its present improved state, is become of such importance, that in all probability it will henceforth be very generally used in the naval as well as the land service, and will become a matter of great public interest. An account of the new muskets for projecting elongated rifle-shot of various forms (see Arts. 175, 176, 177) will therefore be deserving of special notice in the present work.

(2.) Naval history furnishes many melancholy proofs of the danger which is incurred of setting fire to the sails and rigging of a ship in action by firing muskets from the *tops;* and Nelson was so well convinced of the reality of such danger, that he forbade the practice on board the "Victory" in the action of Trafalgar. The destruction of the "Alcide" and "L'Orient" on former occasions, and of the "Achille" and "Redoubtable"[a] in that engagement, prove that his apprehension of such a catastrophe was well founded; but it is remarkable that a shot fired from the rigging of the "Redoubtable" inflicted on this country the severest loss it ever sustained in battle—the death of the gallant Admiral himself.

We refer to these facts, and might quote numerous instances on record, of ships in action being burnt or disabled by an indiscreet and reckless firing of musketry from aloft, amongst the sails; and we would strongly urge the necessity of prescribing by regulation the circumstances under which, only, musketry should be used on board ship, the extent to which it should be adopted, the positions of the marksmen with respect to the sails, and the direction of the wind; we would also urge that every precaution possible should be taken to avoid this danger, the most formidable that can happen to a ship in action.

(3.) The new rifle muskets are wholly indebted to the adoption of the elongated projectile for their efficiency and celebrity; nor have the instruments from which they are projected any other

---

[a] De la Gravière. Translated by the Hon. Captain Plunket. See Art. 300, Note.

merit than that of facility and quickness in loading. Elongated shot possess the properties of being less resisted by the air, having longer ranges and greater penetrating power than spherical projectiles of the same diameter. In accordance with these views, it is proposed to give a short account of the circumstances which led to the introduction of the elongated shot in the French army, and to explain the several modes by which it has been proposed to force the shot into the rifle state in the several muskets and carbines that have since been constructed; also to describe more particularly the several improvements that have been made up to the present time in that description of rifle musket which has been adopted in the British service.

(4.) It is a remarkable fact, that the use of the rifle as a military arm was abandoned by the French in the early campaigns of the Revolutionary War;[a] and it was not revived in the service till after the "Restoration," when it was brought forward by M. Delvigne in the novel form which bears his name.

To obviate the difficulty and loss of time in loading ordinary rifles, by forcing the ball into the barrel by repeated blows of the ramrod or a mallet, on account of which the use of that arm had so long been suspended, M. Delvigne proposed that the bullet should have sufficient windage to enter freely into the barrel, in order that, when stopped by the contraction of the chamber with which this arm was furnished, it might be forced to expand and enter into the grooves, on receiving a few smart blows; thus, the piece being fired, the bullet would come out a forced, or rifle ball, without having been forced in.

But this ingenious contrivance was not found to answer. The edge of the chamber on which the ball lodged not being opposite to the direction of the blow, did not form a sufficient support upon which to flatten the ball when struck by the ramrod, and thus cause the bullet to expand; whilst portions of the charge of powder, previously poured in, having lodged on the contraction, cushioned and still further impeded the expansion of the shot; and as, obviously, no patch could be used, the grooves were liable to get foul, and to become leaded, to an extent which by no means could be effectually obviated.

(5.) To remedy this defect, Colonel Thouvenin proposed in 1828 to suppress the chamber, and substitute a cylindrical tige

---

[a] Favé, 'Des Nouvelles Carabines, et de leur Emploi,' p. 3. Paris, 1847. In 1793, a very small number of French light infantry were armed with carabines rayées, loaded in the ordinary way by forcing the ball with a mallet; but, as the French armies in the campaigns of that year were little familiarized with arms of this complicated description, and could not be sufficiently trained to their uses, the rifle was given up; and during the whole of the wars of the Republic and of the Empire was not thought of.

or pillar of steel (C, Fig. 56, p. 506), screwed into the breech in the centre of the barrel, so that the bullet, when stopped by, and resting upon the flat end of the pillar, directly opposite to the side struck, might more easily be flattened and forced to enter the grooves.

(6.) But here another defect appeared. The pillar occupying a large portion of the centre of the barrel, and the charge being placed in the annular space which surrounds it, the main force of the powder, instead of taking effect in the axis of the piece, and on the centre of the projectile, acted only on the spherical portion of the bullet which lies over this annular chamber, and thus the ball receiving obliquely the impulse of the charge, was propelled with diminished force. (*De la Création et de l'Emploi de la Force Armée*, pp. 44, 45, Paris, 1848.) The next improvement, which was proposed by M. Delvigne, was to make the bottom of the projectile a flat surface; the body cylindrical, and to terminate it in front with a conical point (A, Fig. 56, p. 506), thus diminishing the resistance of the air comparatively with that experienced by a solid of the same diameter having a hemispherical end. The form of the projectile was, therefore, an approximation to that of Newton's solid of least resistance. (Art. 175.)

(7.) While engaged in the conquest of Algiers, a French army of one hundred thousand men was long kept in check by the nomadic inhabitants of that country, a people ill armed and quite destitute of military organization. Favoured by their power of rapid movement, the Arab horsemen, keeping themselves at a distance, directed against their opponents, who were deficient in cavalry, a destructive fire of matchlocks, and immediately retired beyond the range of the muskets carried by the European infantry, whose solid columns, encumbered with artillery and baggage, were unable to follow with sufficient rapidity. The necessity of arming the French infantry with weapons capable of affording, with considerable precision of fire, a more extensive range than could be obtained from common muskets, was immediately felt; and ten battalions of chasseurs (infantry), which were organized in 1840, were armed with the pillar-breech rifle musket. (Delvigne, *De la Création et de l'Emploi de la Force Armée*, pp. 14, 15, 16, 45.) It is probable that the circumstances above mentioned drew the attention of military men in general to improvements in the musket and rifle. In 1841 a patent was obtained by Captain Tamisier for his method of giving steadiness to the flight of cylindro-conical shot, by cutting three sharp circular grooves, each 0.28 inch deep, on the cylindrical part of the shot, by which the resistance of the air behind the centre of gravity of the projectile being increased, the axis of rotation was kept more steadily in the direction of the .

trajectory; the grooves being to this projectile what the feathers are to the arrow, and the stick to the rocket.*

(8.) The following is a brief account of the French Pillar-Breech Musket, with its latest improvements:—

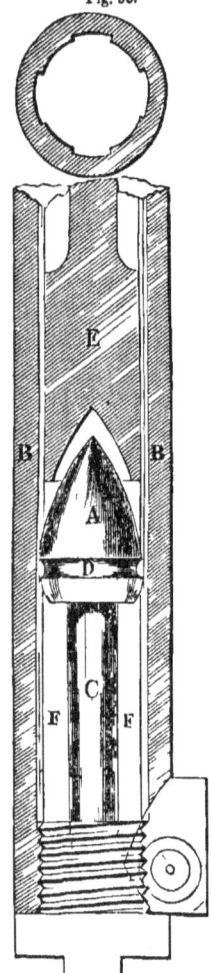

Fig. 56.

"The pillar-breech musket is loaded at the muzzle with a leaden projectile of the form shown at A, in Fig. 56 annexed; this projectile is 0.657 in. diameter, and weighs 728 grains. The barrel, B B, of the musket is 34 inches long, rifled with four grooves, and has an elevating scale or sight 3¼ inches high; the 'tige' or pillar, C, is screwed into the face of the breechpin. The cartridge, containing 2½ drachms of powder, is made of strong paper, which is tied round the projectile at the groove D, near its base.

"In loading, the soldier breaks the cartridge, when the powder falls into the space, F F, round the pillar, and he throws away the paper of the upper part of the cartridge; the ball, which is nearly of the same diameter as the bore of the piece, is then made to rest with its flat end on the head of the pillar; the end, E, of the rammer being countersunk of the same form as the point of the ball, the soldier gives three or four smart blows upon the latter, which, being supported by the pillar, is shortened in length and widened in diameter, so as to force the lead and paper round it into the grooves of the rifle. The point of the ball is held in the axis of the barrel by the head of the rammer, which is so nearly of the same diameter as the bore, that no sensible variation in the position of the ball can take place: when fired, the projectile is constrained to follow the grooves of the rifle, and the paper protects the barrel from being leaded."

(9.) This method of forcing the shot into the rifled state has been adopted by the Austrians and some of the smaller German states; and the French Chasseurs d'Orléans are still armed with the carabine-à-tige. The method of forcing the shot into the grooves of the rifle has also been adopted in this country by Mr. Lancaster, in his pillar-breech rifle, which is much used by sportsmen, and was proposed

---

* Some very interesting experiments have recently been made with rockets without sticks, a rotatory motion on their axes being given by making the orifices out of which the inflamed and compressed composition rushes, oblique to the axis instead of direct. By this contrivance the rotatory and propelling powers are combined; and so long as the former continues, the rocket will proceed in the direction of the trajectory, at right angles nearly to the plane of rotation. (Art. 324.)

by him for adoption in the military service; the length of his musket is 2 feet 8 inches, the charge of powder 2½ drachms, and the weight of his cylindro-conoidal shot 710 grains (troy). The grooves, of which there are four, are straight to the extent of 18 inches from the breech; they, then, take a spiral form, gradually increasing to the muzzle (*gaining twist*), making a quarter of a turn from its commencement: this, it was presumed, would give to the projectile great accuracy of flight, with the least recoil; but it appears that, whatever may be the advantages of the *gaining twist* in firing a spherical bullet or a conical *picket* (an American practice), it is totally inapplicable to a cylindro-conoidal shot for reasons ably stated by Lieut.-Col. the Hon. Alexander Gordon, in his pamphlet entitled '*Remarks on National Defence*,' Appendix, p. 32.

(10.) But the pillar-breech musket having been found inconvenient in cleaning, the chamber round the stem becoming soon fouled, the pillar liable to be broken, and, after firing some rounds, the operation of ramming down so fatiguing to the men as to make them unsteady in taking aim, M. Minié, previously distinguished as a zealous and able advocate for restoring the rifle to the service in an improved form, proposed to suppress the tige, and substitute for it an iron cup, *b* (Fig. 57), put into the wider end of a conical hollow, *a*, made in the shot: this cup being forced further in by the explosion of the charge, causes the hollow cylindrical portion of the shot to expand and fix itself in the grooves, so that the shot becomes forced at the moment of discharge.[*] A slip of cartridge-paper is wound twice round the cylindrical part of the projectile, so that, as the latter does not become forced or rifled till the charge is fired, it fits so tightly to the barrel as to be free from any motion which would be caused by the carriage of the rifle on a march, or by its being handled before the shot is fired.

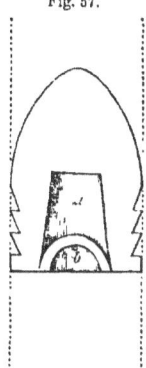

Fig. 57.

The reader will perceive that, unless the cup *b* (Fig. 57) be driven, by the first action of the explosion of the charge, so far into the conical space in which it is placed, as to cause the lead to enter into the grooves of the rifle before the shot moves, there will be no rotation—the paper wrapt round the shot not sufficing for this purpose.

In the experiments of 1850 it was found that the hollow part of the Minié cylindro-conical shot was very frequently separated

---

[*] This is the expedient for forcing the shot, so well described in an article on rifle shooting in Ceylon, which appeared in the 'Times' of the 29th of March, 1851, and which the reader of that article will perceive is the Minié method.

entirely from the conical part by the force with which the cup was driven into the hollow part of the shot, and sometimes remained so firmly fixed in the barrel that it could not be extracted; but in the more recent trials with shot made by compression and with better lead no such failure occurred.

(11.) The principle of placing and igniting the charge in front of the projectile by means of a needle, was patented in England by Abraham Mosar on the 15th of December, 1831; his musket was submitted to the Board of Ordnance for trial in 1834, but the method of loading, namely, at the muzzle, was very complicated; and the inventor not having pecuniary means sufficient to improve and carry out his invention, no trials were made. While efforts were being made in France to augment the power and accuracy of small arms, loaded at the muzzle as already described, M. Dreyse, of Sommerda, in Thuringia, was led to try whether the inconvenience of ramming down and flattening the shot might not be got rid of by loading the barrel at the breech—an old project (see Art. 227); and he suggested a plan for this purpose, which has been adopted to a great extent in the Prussian army.

(12.) The Prussian rifled musket for firing cylindro-conical shot is designated "zundnadelgewehr," from the ignition of the charge being produced by passing a needle through the cartridge, to strike the percussion powder placed in the wooden bottom, or spiegel, as shown in Fig. 58, p. 509. The following is a description of the musket, which is loaded at the breech:—

" The barrel, A A, which is 34 inches long, is rifled with four grooves, and has a 'hausse' or sight adapted to distances of 600 metres; it is screwed into the end of a strong open guider or channel, B B: the chamber, properly so called, is bored out from behind, conically in a slight degree, so that when the cartridge is placed in it, the shoulder, C D, of the ball shall meet and be stopped by the projections or ribs of the rifling, the body of the shot being of sufficient diameter to fill the full depth of the grooves. Inside of the guider slides an iron tube, E E, with a strong 'hebel' or handle, F, attached, and having at the front end a space, G G, of about 1¼ inch in length. In the middle of this space is a 'tige' or pillar, H, which, instead of being solid, like that of the pillar-breech musket, is pierced with a small hole in its entire length, and through this passes the needle, N, which is to ignite the charge: the steel tige is screwed from behind into a solid plate of iron, J J, left in the tube E E; and this plate it is which (like the breech-pin of the ordinary musket) receives the whole reactionary force of the charge. Behind the plate, J J, there is a second tube of iron (which could not be shown in the drawing) having a double catch-spring attached, and carrying within it a small inner tube of iron, K K, (Fig. 59,) having two projecting rings, L L, (Figs. 58, 59,) on one moiety of its length, and a spiral spring, M M, sliding on the other half. Through the tube K K, passes the needle, N N N N, which is a thin steel wire about ·03 inch diameter, bluntly pointed at the end, which is to ignite the charge; at the other end it is screwed into a brass head, O, and this again screws into the inner tube, which carries the spiral spring. The trigger, which is of a peculiar form, and has a bolt movement in firing, could not be shown intelligibly in the diagram. It has a knuckle-catch, so that, when

App. A.   ON RIFLE MUSKETS.   509

pressed down, it admits of the whole mechanism of the tube, E E, being drawn out from behind, when the parts can be taken to pieces, cleaned, and put together again by the soldier in a few minutes, there being no pins whatever and no screw, except that by which the needle is connected with the inner tube.

"The shot for the 'needle-prime' musket is of the form shown by the dotted lines in the upper part of Fig. 60, and weighs 437½ grains, or exactly one ounce avoirdupois; its diameter at the shoulder is .632 in. Underneath the shot is the 'spiegel' or bottom, of equal diameter, formed of wood, covered with brown paper and hard rolled, with a hollow at the upper end to receive the lower end of the ball; beneath it is a small cup (P, Fig. 58), to contain the igniting composition, which is pressed hard by mechanical means. The cartridge is made of paper somewhat thicker than that used in our service; a small square piece is first pressed by the hand against one end of the wooden *former*, and this constitutes the end or bottom; another oblong piece is pasted on the edge of one side and one end, and rolled once round the *former*, the pasted end being pressed close and flat round the bottom; when dry, the powder (62 grains, or about 2¼ drachms) is first put in; after which the spiegel with the priming composition is placed downwards upon the powder, and then the shot. The paper is tied round the point of the shot, and its end is cut off smooth. This end of the cartridge is dipped into melted tallow as far as the shoulder (C D) of the projectile.

"At the lower part of the barrel is the sliding tube through which passes the needle for igniting the charge: this tube is capable of being moved backwards or forwards in the barrel, near the breech, by means of a pin or handle in its side, which passes through a perforation similar to a bayonet notch in the side of the barrel. When the tube is drawn as far as the perforation will allow towards the stock, there is formed an open chamber between its extremity and the nearest extremity

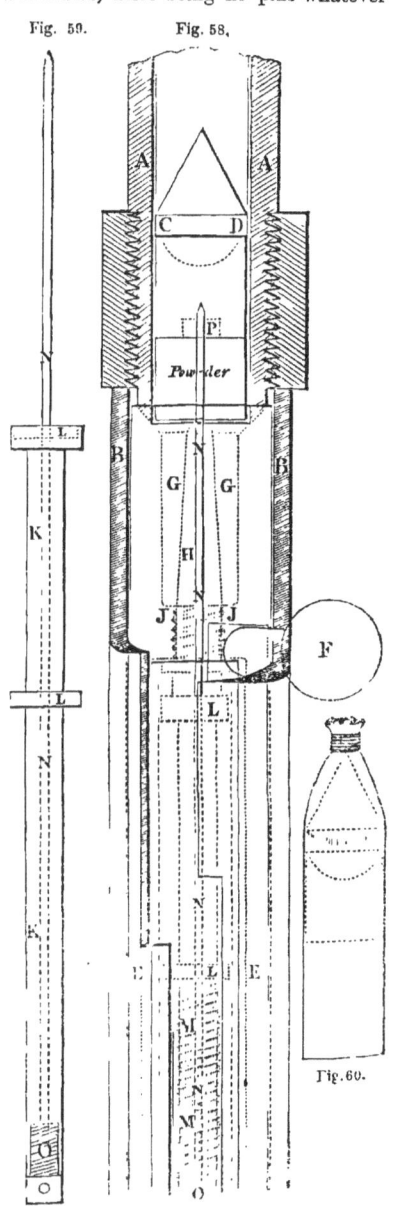

Fig. 59.   Fig. 58.

Fig. 60.

of the barrel; and, by means of this chamber, the charge is introduced into the barrel. The tube is then pressed forward till its extremity, which is in the form of a frustum of a cone, is in contact with the rear extremity of the barrel; this extremity having the form of a hollow cone to receive the end of the tube: the pin, or handle, being then turned round in the notch, the tube is, as it were, locked in close contact with the barrel. In this state the needle in the tube is in connection with the trigger of the lock, and the musket is ready for being fired."[a]

(13.) The escape of gas at the junction of the chamber and barrel is considered by all, as a great objection to the needle-prime musket: it is stated that the point of the igniting needle soon becomes furred, so that it is difficult, and, after a time, impossible to draw it back by the thumb. The Prussians, however, appear to be quite confident of the superiority of the latter over other rifle muskets; their Government is said to have caused 60,000 stand of these arms to be executed, and, at least, half as many more are ordered. Their fusiliers, who are armed with the needle-prime musket, have also a short sword, with a cross hilt: this they plant in the ground; and, lying down, they use the hilt as a rest for the purpose of taking a steady aim.

In 1850 some experiments were made at Woolwich, under the direction of the Committee on Small Arms, when it was found that in the operation of opening and closing the breech, by withdrawing to load, and replacing when loaded, the bolt (which acts in a manner resembling that of fixing the bayonet) required a great deal of manual strength, particularly when the piece got heated and foul. Also, while attending the experiments of 1850, at Woolwich, the author was very much struck with an important objection in the very apparent escape of gas from the breech. This defect increased to such a degree in continued firing, even with a new piece, that the flash indicating a copious escape of gas became very apparent, and was at length sensibly felt in the face of the man who fired, and the soldier on his left. The escape of gas will evidently be much greater in long protracted firing, from the effects of friction both upon the bolt and the barrel. The escape of gas takes place chiefly from the left side of the breech; which indicates an imperfection in the contact between the chamber and the barrel on that side, arising from the want of central or direct support for the tube which carries the charge, when it is pressed up to the top of the chamber, in order to close the breech, previously to the musket being fired. This defect in the contact,

---

[a] The use of carbines loaded at the breech is much recommended by General Rémond. The General proposes, with them, to employ infantry formed three, and even five deep—the front rank men being not exposed to the fire of those in the rear, which is the case when the muskets are loaded at the muzzle, by the men raising the right hand to the head in using the ramrod. (P. 6.)

apparently goes on increasing during the repetitions of the firing. It is possible that the evil may be remedied by giving to the lower part of the barrel the form of a hollow frustum of a cone, and to the upper part of the tube carrying the charge a corresponding form, so that, in closing up the tube to the barrel, the contact of the conical surfaces may be sufficiently close to prevent the escape of gas. By this construction, even in the event of these surfaces being in part worn away by friction, it will be possible, by pressing the tube still closer to the barrel, to preserve the accuracy of the contact.

(14.) The advantage of placing the composition above the gunpowder is supposed to be, that it secures the ignition of the whole charge. The mode of producing ignition by placing the percussion powder in the body of the cartridge is, however, very objectionable. If the cap put on the nipple of an ordinary lock should fail, another may be instantly put on; if ignition fail to be produced in the Prussian cartridge, it must be withdrawn bodily, and it becomes useless till made up afresh.

Another very serious objection to cartridges so made up is, that the fulminating composition being packed with the cartridges in barrels, tumbrils, waggons, or other receptacles for musket ammunition, would, if these should be entered by shot, infallibly explode, and cause the whole to be blown up. The ordinary caps are capable of being stowed in small bulk by themselves.

(15.) It has been reported that many of the Prussian needle-prime rifles taken by the Danes in the late engagements (1850) were found to have become defective; but the author learns from an authentic source, his informant being a Danish officer of distinction, that no arms of this description were captured by the Danes, the small arms taken being either common muskets or rifles of the ordinary kind, loaded at the muzzle; but they fired conical shot with remarkable effect.* It requires only an inspec-

---

* In the Danish official report of the battle of Idstedt, on the 25th of July, 1850, it is stated that "The enemy's skirmishers, under cover of a hedge, fired with pointed balls (*spitzkugeln*), at a distance of 100 and 150 yards. It was in vain that a couple of guns threw grenades at a short range among the skirmishers; it was in vain that a body of the cavalry made three several attacks; in vain it was endeavoured to bring up the infantry from Oberstolk, which was now in flames, while a fierce engagement was going on in it from the house windows and in the streets. In less than an hour we suffered a great loss. The brave General Schleppegrell fell, mortally wounded, during one of the attacks; the *chef* of his staff, Lieut.-Col. Bulow, was severely wounded; the commander of the battery, Capt. Baggesen, was made prisoner, and two of his guns taken by the enemy. Several other officers were also killed, and among them Lieutenant Carstensen, while endeavouring to rescue Capt. Baggesen, with about 70 subalterns and privates: at least 90 horses were killed or taken.—(In what manner the pointed balls were forced, is not stated,

tion of the figure, page 509, to be convinced that, whatever be its merits in other respects, the needle-prime musket is too complicated and delicate an arm for general service.

(16.) A very serious disadvantage arising from the use of cylindro-conical shot is the excess of their weight above that of a common musket-ball, on which account the already overloaded soldier must carry a greater weight of ammunition in an equal number of rounds, or the number of rounds must be diminished, and either of these results would be very prejudicial. So quickly may the new rifle-muskets be loaded, that a soldier may fire away all his ammunition in a very few minutes, which soldiers are prone to do as quickly as they can; and unless additional and extraordinary measures be adopted for replenishing in the field, and in action, the speedily exhausted cartridge boxes, the rifleman may, in a protracted action, be left long in a state in which he can neither annoy the enemy, nor defend himself.

(17.) It is, no doubt, in some respects, an important advantage in the Prussian rifles, that they may be loaded more quickly than the ordinary musket or rifle; but here, too, we agree with M. Favé (*Des Nouvelles Carabines*, etc., p. 40), that rifle actions are generally decided, not by mere rapidity of fire, but by each soldier taking time to use his arm in the most efficient manner possible. Although, as has been already stated, the use of the rifle was suspended in the French armies throughout the whole of the general war (1794-1815), yet the French infantry, armed with the common musket, were well trained to act *en tirailleur*, and showed great aptitude for that kind of service, in which a mere militia or new levied troops, in many circumstances, render as much service as veteran soldiers.

(18.) The infantry of the line, and light infantry in the French service, are identical, excepting a slight difference in their uniform; they carry the same nature of arm, and are trained nearly in the same way, though the regiments of light infantry are more particularly employed in the service of the advanced posts. The companies of voltigeurs of the infantry are armed with a musket shorter by 0ᵐ .054 (2 inches) than that of the other companies. The only troops armed with the carabine *à tige* are the ten battalions of Chasseurs d'Orléans. We observe, too, that French writers, in discussing the comparative merits of the musket and the rifle, appeal, as well they may, to the well-directed and well-sustained rolling fire of British musketry of the line, which, as M. Favé states (p. 44), gave us the victory in battle: therefore,

---

nor their exact form given, in the report; but the author has ascertained from authority at Copenhagen, May 3, 1851, that they were cylindro-conical, fired from carabines rayées, à tige.

we should be very careful not to compromise that established efficiency by any very general or extensive adoption of the new arms, according to a theory, the effects of which have been tried only in circumstances which cannot exist on service.[a] A French writer admits the efficiency of our rifle firing, and states that, at the battle of Waterloo, almost all the officers of the First Regiment of Infantry of the Line, including the colonel himself, were wounded by rifle balls, or, as the colonel called them, officers' balls, because the English troops, as he supposed, directed their fire at the officers without regarding the private soldiers.[b]

(19.) Public attention having been excited by M. Favé's interesting work, '*Les Nouvelles Carabines,*' as well as by the efficient use made of the *carabine à tige* by the Chasseurs d'Orléans in Algiers, and by the adoption of its rival, the needle-prime musket, by the Prussian government, experiments with muskets of the latter construction, which had been made in England from foreign patterns, were, in 1850, carried on in this country, in order to compare them with the British regulation-musket and the British rifle carrying the belted ball. It was proposed to ascertain the relative merits of those arms, and particularly to discover whether the method of loading at the breech, as in the needle-prime musket, considered under all aspects, could be adopted with advantage in the British service. The results of these experiments showed that the Prussian needle-prime musket was by far the quickest in loading and firing. The English regulation-musket came next, Lancaster's pillar-breech the third, and the

---

[a] The following is one of the numerous special occasions, related in Napier's History, on which the British musket and bullet, in good and gallant hands, proved their efficacy at considerable distances, without other artillery.

Combat of the 10th of December, 1813.—" The practice of the guns, well directed at first, would have been murderous if the musketry from the church-yard had not made the gunners withdraw their pieces a little behind the ridge, which caused their fire to be too high. The General (Kempt), thinking the distance too great (it was between 350 and 400 yards), was inclined to stop this fire; but the moment it lulled, the French gunners pushed their pieces forwards again, and knocked down eight men in an instant. The small arms then recommenced, and the fire of the French guns was again too high. The French were in like manner kept at bay by the riflemen in the village and mansion-house, and by the 52nd on the ridge adjoining."—*Napier*, vol. vi. p. 382.

[b] " Je crois ne pas me tromper en disant que si l'on voulait se donner la peine de fouiller dans les cartons du Ministère de la Guerre, on y trouverait un rapport important de M. le Colonel Lebeau, du premier régiment d'infanterie de ligne. On y lirait qu'à la bataille de Waterloo presque tous les officiers de ce régiment et le colonel lui-même furent blessés par des balles de fusils rayés, par des balles que M. Lebeau appelait des balles d'officiers, car les riflemen Anglais qui tiraient sur son régiment, dédaignant le commun des soldats, avaient visé les officiers, et, comme vous voyez, ne les avaient pas manqués."
—*Discours de M. Arago à la Chambre des Députés*, Juin, 1839.

English regulation-rifle the slowest; that the hits, in 60 rounds fired at a target 6 feet square at the distance of 150 yards, were, respectively, in the order in which the arms have been already named, 40, 29, 50, and 30; and lastly, that the average percentage of hits at different ranges up to 600 yards, were, in the same order, 33, 25, 35, and 37; but, for the reasons already stated, muskets loaded at the breech were condemned as arms for general service in the field, however useful they might be in the hands of a few expert men for special purposes.

(20.) Good patterns having been obtained of the Delvigne carabine à tige, the French and the Belgian Minié rifles, experiments were made at Woolwich in 1851 with these three arms and with Lancaster's pillar-breech rifle, in order to test their relative merits in firing at a target 6 feet square, at 400 yards distance. The results of these experiments were considered to have so fully established the peculiar advantages of M. Minié's method of quick loading, and forcing the shot into the rifled state, that the manufacture in this country of a large supply of what has been called the regulation Minié musket was ordered.

The form of its projectile, which is simply conoidal, is given in Fig. 61 annexed.

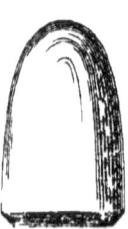

Fig. 61.

(21.) His Grace the late Duke of Wellington was strongly in favour of the leaden bullet for line firing, and he objected to a reduction of the calibre of the musket; consequently, on account of the weight of the elongated shot proposed being greater than that of the usual spherical bullet, there remained no alternative but either to overload the soldier by the weight of 60 rounds of ammunition, or to reduce the number of shot carried by him. In order therefore to conciliate the Duke, it was made a regulation by the Board of Ordnance that the Minié muskets should be so constructed as to be as easily loaded with a round bullet of one ounce weight as with its own ammunition, so that either description of shot might be used, as circumstances might require, for line and rifle firing. It soon appeared, however, that the fulfilment of this condition was impracticable, and that the attempt would have the usual fate of endeavours to accomplish two purposes essentially different by one and the same means—attempts which generally result in the failure of both.

(22.) The only mode in which elongated shot can be employed without either increasing the weight of the ammunition carried by the soldier or reducing the number (60) of rounds in his pouch, is to diminish the calibre of the fire-arm, so that the elongated shot may not be heavier than the regulation bullet. With a view to effect this object most conveniently, the Master-General of the Ordnance, Viscount Hardinge, invited some of the principal gun-makers of England to submit pattern muskets, in the hopes of obtaining a lighter and more efficient arm with a smaller bore. The following makers accordingly prepared, and sent in, muskets for trial—Mr. Purday, Mr. Westley Richards, Mr. Lancaster, Mr. Wilkinson, and Mr. Greener: Mr. Lovell, the inspector of small arms, also prepared a new musket at the government manufactory.

(23.) The following table of details (see p. 516) respecting the regulation Minié rifle, and those of more recent pattern which have been made the subjects of experimental trials, is both important and interesting. (*Remarks on National Defence*, by Lieut.-Col. the Hon. A. Gordon, and *A Handbook of Field Service*, by Capt. Lefroy, R.A., F.R.S.)

(24.) The projectiles for all the muskets in the table are conoidal or cylindro-conoidal. Mr. Wilkinson's projectile is cast solid, and has two deep grooves round the lower part. It is intended to be used without paper or patch of any kind, the two grooves being merely filled with grease; and the method of loading is the same as that which is now in use with the rifle regiments of the service, the powder being made up in a small cartridge by itself, and put into the barrel before taking the ball out of the ball-pouch. The muzzle of the barrel has been countersunk, so as to receive the ball before it is rammed down, which is a great convenience. The grooves are of a spiral form, having one turn in 6 ft. 6 in. Mr. Purday's projectile is cylindro-conoidal, with a belt round the lower part: one of the two kinds of projectile has the simple Minié cavity, and the other has a cavity with a plug, which is driven into the cavity by the force of the fired powder. The grooves have the character of the *increasing* spiral; they commence with one turn in 6 ft., and end with one turn in 4 ft. 9 in. Mr. Lovell's heavy shot (No. 1) made excellent shooting at all distances, but the difficulty of loading, and the great weight of the shot, renders it unfit for troops of the line. Even with a strong wooden rammer it was sometimes difficult to drive the shot home. The lighter shot (No. 2) made very good shooting up to 400 yards; but sometimes, after firing a dozen

	Weight of Musket with Bayonet.	Barrel.					Projectile.				Charge.
		Length.	Weight.	Diameter		Number of Grooves.	Diameter.	Length.	Weight.	No. of Shot to the Pound.	
				At Muzzle.	At Breech.						
	lbs. oz.	Ft. in.	lbs. oz.	In.	In.		In.	In.	Grs.		Drs.
Common Regulation Musket	11 11	.. ..	.. ..	..	..	..	..	..	634.4	14	..
Regulation Minié Rifle	10 8¾	3 3	4 10	.702	.702	4	.690	1.03	680	..	2½ F. G.
Wilkinson's	9 5	3 3	4 1	.530	.531	5	.527	1.075	500	31	2½ F. G.
Purday's Minié Plug	9 1½	3 3	3 15¼	..	..	4	.643 .643	.910 1.050	487 610	17	2½ F. G.
Lovell's— No. 1 No. 2	9 1½	3 3	4 1½	..	..	4	.630 .628	1.145 .948	686 562	18	2½ F. G.
The Brunswick Rifle	11 5½	2 6	3 14	.704	.704	2	.696	..	557	..	2½ R. A.
The New Enfield Rifle	9 3	3 3	4 2	.577	.577	3	.568	.960	520	..	2½ F. G.
Lancaster's, with elliptical bore	9 9	3 3	4 14	Major axis. .550 Minor axis. .540	.557 .543	..	.532	1.125	542	30	2½ F. G.

shots, it became uncertain, and the difficulty of loading was as great as with the heavier ball. The grooves are regular spirals with one turn in 6 ft. 6 in. The grooves of the Brunswick rifle have one turn in 2 ft. 6 in.

Mr. Lancaster, who invented the ordnance with an elliptical bore, spirally formed, as described in Art. 190, and the pillar-breech rifle (Art. (8)), proposed also a nature of musket having a bore of a similar kind. No grooves are cut in the interior surface of the barrel; but, in a transverse section, the bore has the form of an ellipse of small excentricity and it is *freed*[a] at the breech: the projectile is cylindro-conoidal, with a circular base, and, when heated by the fired gunpowder, it expands so far as to take a form corresponding to the elliptical section of the bore. The bore, being a continuous spiral, fulfils the object of grooves, and causes the shot, in passing along it, to acquire a rotatory motion on its axis. The spiral is not uniform in its whole length, but has what is called by the Americans a *gaining twist* or *an increasing spiral*. The advantages of this rifle are supposed to be,—greater accuracy of practice, less recoil than other muskets have, and no tendency to cause the rifle to turn over sideways.

In November of last year (1853) experiments were made at the school of musketry at Hythe, under the direction of Colonel Hay, for the purpose of trying the accuracy of Mr. Lancaster's musket; and the result was, that, in firing twenty rounds at an iron target 6 feet square, at the distance of 600 yards, all the 20 shot hit the target; of these, 10 struck above, and 10 below the centre; 11 struck on the right hand and 9 on the left.

Five of these muskets were tried at Hythe against the Enfield musket with a smooth, elliptical bore; and, after a prolonged trial, the Enfield rifle was found decidedly inferior to that of Mr. Lancaster as regards accuracy of shooting, and its shot *stripped*[b] oftener in the barrel. Further trials were made at Woolwich, at Enfield, and at Hythe, in December, 1853, with Lancaster's musket having an elliptical bore, an increasing spiral, and being freed at the breech; this was in order to compare its shooting with that of an Enfield rifle with three grooves, having a uniform spiral with one

---

[a] To be freed at the breech signifies that the bore is made larger there than in the anterior part of the barrel: the term is also used to denote an enlargement of the grooves at the same place. The Americans enlarge in this manner the whole of the barrel up to within 1¼ inch of the muzzle; English gunmakers only as far as 2 or 3 inches from the breech. The object is to facilitate the operation of ramming home. Not much value is attached to the construction in this country: it may be advantageous for the American rifles, which carry smaller shot than ours; and the like may be said of the American practice of increasing the twist of the spiral grooves towards the muzzle.

[b] To *strip*, speaking of shot, signifies to pass out of the barrel without becoming rifled.

turn in 6 feet 6 inches, and an equal bore throughout, 0·577 inch diameter: the object being, apparently, to simplify rifles, if possible, by dispensing with the *increasing* spiral and the *freed* breech. The report of these trials was in favour of the Enfield rifle, Lancaster's musket having evinced a strong tendency to *strip*: and at the longer ranges this defect was very marked. The cause of it has not been satisfactorily ascertained, but it is presumed to be in part owing to the wax which entered into the lubricating material; this may have caused some of the paper of the cartridge to adhere to the bore about the greater diameter, and thus render the bore nearly circular. It was decided then to adopt the three-grooved rifles; and it is said that a great number of them have been, or are now being made.

The gaining twist is considered objectionable, in offering to the shot's progress a resistance continually increasing, while the shot itself is subjected to a continually increasing force of impulsion. A projectile is set in motion gradually, and should attain its greatest velocity just before quitting the muzzle; there is consequently a tendency of the shot to yield to the increasing force of the powder, and pass directly out of the barrel without following the spiral. Should, however, the projectile, which is in part cylindrical, follow a twist of the *gaining* kind, it will still be subject to a continual change of figure during its passage along the bore; and it may even be said that the rear end of the shot is being twisted while the front is passing out of the muzzle. An advantage derived from the absence of grooves is that there are no edges to be worn away by working the ramrod, and a smooth bore is less likely to become foul than one with grooves. The plug-bullet used by Mr. Lancaster does not appear suitable for military purposes; generally the plug is driven too far into the lead during the making up of the cartridge; and it is desirable that bullets for military service should be of solid lead, without cup or plug.

An objection to Mr. Lancaster's musket, and this applies also to the Enfield rifle of 1853, is, that it requires a slender cartridge ($3\frac{1}{4}$ inches long and $\frac{9}{12}$ inch diameter), which, when made up with the heavy projectile, is liable to become torn at the neck, so that the powder, in part at least, escapes: this objection exists in a greater degree, in the Minié rifle, the projectile of the latter being heavier than that of either Lancaster's or the Enfield rifle. The paper, moreover, being stiff, does not always expand after it has been compressed in the act of biting off the end; in consequence of which the powder is not readily poured into the barrel, and much of it is often thrown away with the paper.

The argument in favour of an uneven number of grooves for a rifle, as in Mr. Wilkinson's and the Enfield rifle, over an even number of grooves is apparently very sound: the *lands* in the former kind being opposite the grooves in the barrel, instead of the grooves being opposite to each other, as in the latter kind of rifle; the expanded lead is thereby more effectually pressed into the grooves, and the projectile consequently acquires a more steady rotation on its axis.

(25.) The chief objection to the new Minié rifle-musket consists in its great liability to get fouled, and, as it would appear, the grooves partially leaded, after firing a number of rounds, more or less, so as to render it sometimes impossible to force the shot into and down the barrel. This difficulty appears to have been foreseen by Colonel Gordon, since, in his pamphlet (Appendix, p. 18), he expresses regret that the bore of the new rifle had not been made a few thousandths of an inch larger; he states that, after firing a few rounds, the shot jams in the barrel, and cannot be forced down to its place; and as all the Minié rifles and their ammunition, with which our forces in the East are armed, are exactly of the dimensions assigned to the Regulation Minié rifle in the table page 516, it may be apprehended that this difficulty in loading will be experienced on service, as it unquestionably is to a great extent in practice at home.

(26.) To remedy this objection, Lieut.-Colonel Hay, who now ably superintends the establishment for training soldiers in the use of the Minié musket, proposes to diminish the diameter of the shot by .005 inch (Fig. 62): this small quantity, Colonel Hay has, after

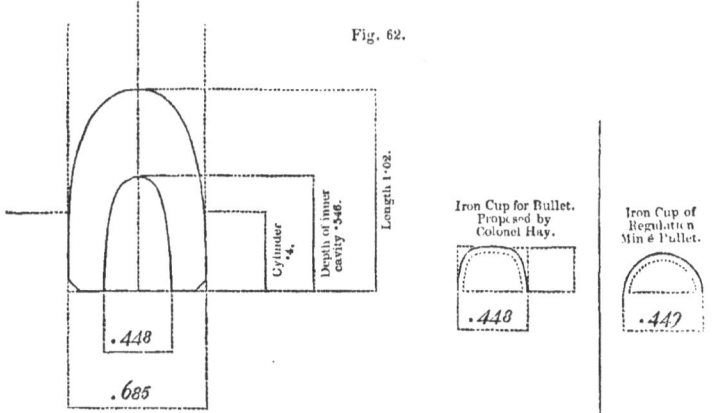

Fig. 62.

many trials, found sufficient to prevent jamming, and permit the shot to be in contact with the surface of the bore when the arms are in any position whatever; it also allows the musket to be loaded quickly when repeated firing becomes necessary. Colonel Hay's proposition having been approved, ammunition of that size was supplied for further trial in the school. That which was first sent succeeded admirably; but, subsequently, the cartridges loaded too easily. On examination, the paper was found to have been reduced both in thickness and quantity compared with that with which the former cartridges were made. This evil, however, was easily remedied. Colonel Hay has likewise introduced an important improvement in the shape of the *culot* or cup, and in the figure of the cavity into which it is forced on the firing of the charge. The reader will perceive, on referring to Fig. 57 (page 507), that the cavity a in the Minié shot has the form of a frustum of a cone, while that of the cup b is a hemisphere : now all who have examined the shot picked up after having struck an iron target or penetrated into earth, find that the hemispherical cup is very liable to be canted or turned instead of being forced directly into the hollow space : the lead of the shot is not driven equably into the grooves of the rifle. For this evil Colonel Hay has proposed a remedy, in giving both to the culot and the cavity in the shot conoidal forms (see Fig. 62), by which means the former must, by the force of the powder, proceed directly forward in the hollow space, and thus uniformly expand the lower part of the shot, in the bore.

(27.) The fouling of the Minié rifle-musket arises, probably, in part from an adhesion of the lead rubbed off from the shot during its passage along the bore, and in part from the paper between the powder and the shot being charred on the explosion of the gun-powder, and leaving a deposit adhering to the barrel: whatever be the cause, unless the cartridge be well greased, it becomes impossible, after firing a few rounds, to get the shot into and down the barrel.

It has happened that soldiers, finding that the shot has jammed in the barrel, have endeavoured to force it down by striking the ramrod with heavy stones. The inconvenience and loss which must be sustained in such a case may be easily conceived, when troops armed with the Minié rifle only, having fired away each man ten or fifteen rounds of his ammunition, are compelled to stand inactive while an enemy is advancing against them for the purpose of bringing on a close and decisive action. It is remarked that the Minié rifle does not so soon become foul when loaded in a vertical, as when loaded in a horizontal position, the rifleman lying down. The Minié rifle is, however, too long to be conveniently loaded in this last position. The bullet proposed by Colonel Hay has sufficient windage, before it is forced, to allow the loading to be effected even though the grease may have been rubbed off from the cartridge.

The exercise for loading with blank cartridge is performed with the Minié rifle as with the ordinary musket,—the soldier bites off the end and puts the cartridge into the musket, as in the usual method ; but in shot-firing the end must be bitten off and the cartridge reversed in loading, otherwise the cartridge will stick in the barrel, and the musket will be rendered useless during the rest of an action. To prevent this from occurring on the battle field, through the nervous excitement which during an action must be felt even by veteran soldiers, but in a higher degree by such as have never before been under fire, the loading motions with blank cartridges have now been assimilated to those with shot cartridges; so that the soldier will be habituated to perform, in both cases, the reversion of the cartridge after having poured in the powder.

(28.) A great inconvenience attending the use of the Minié rifle is, that every cartridge employed must be carefully greased, in order, by lubricating the barrel at every discharge, to prevent it from becoming so foul as to be unserviceable after a few rounds have been fired. It is well remarked by Lieut.-Col. Gordon, in his useful pamphlet above quoted, (Appendix, p. 8), that it is doubtful whether or not a soldier would grease even the rounds which he carries in his pouch, and that it would be scarcely possible for him to grease those which he might receive, on a second or third supply, during a protracted action, or on a sudden demand for ammunition in case of a disturbance of the peace, or of troops landing in an enemy's country. Greased cartridges could not, certainly, be kept in store without being deteriorated by vermin or by the effects of climate; unless, indeed, some method could be found of preparing paper with an unctuous matter, which should be capable of resisting the effects of heat and moisture and the attacks of vermin, and which should remain for months in a state fit for use. The greased cartridges which were sent to India for trial with the Minié rifle were, on their arrival, found to be totally unserviceable.

(29.) The following table exhibits the ranges, to the first graze, of the different kinds of new muskets, in comparison with the range of the regulation Minié musket. The experiments on which the table is formed were carried on at Enfield during the summer of 1852.

The muskets were fired horizontally at a height of 4 ft. 7½ in. above the ground.

	Ranges.		
	Yds.	ft.	in.
The Regulation Minié musket	177	1	7
Mr. Wilkinson's musket (naked shot)	185	0	3.3
——————— (cartridge)	172	1	5½
Mr. Purday's musket, with the plug ball	180	2	4
Mr. Lovell's musket, No. 1, or heavy shot	190	0	5
——————, No. 2, or light shot	176	1	6½
The Brunswick musket	173	0	2½
Mr. Lancaster's musket	194	0	11

The angles of elevation which should be given to the several arms, in order that the projectiles may attain different ranges, from 100 yards to 1000 yards, are as follow:—

Ranges.	Regulation Minié.	Wilkinson.	Lancaster.		Purday.	The Brunswick.
			Heavy Projectile.	Light Projectile.		
Yards.	° ′	° ′	° ′	° ′	° ′	° ′
100	0 14	0 14	0 11	0 14	0 15	0 8
200	0 39	0 28	0 26	0 26	0 40	0 34
300	0 53	0 47	0 49	0 52	0 57	0 54
400	1 22	1 5	1 9	1 18	1 17	1 26
500	1 51	1 25	1 34	1 47	2 21	Irregular.
600	2 23	2 0	2 2	Irregular.	2 58	..
700	3 5	2 29	2 32	..	3 41	..
800	3 25	2 44	2 50	..	..	..
1000	..	4 31	4 16	..	..	..

(30.) The following table, kindly communicated by Colonel Hay, on the subject of the rifle practice at Hythe, shows the comparative accuracy of shooting at different distances, with the common percussion musket of 1842 and the rifle musket of 1851; the former carrying a spherical bullet, and the latter the regulation Minié shot (a cylindro-conoidal projectile). Twenty men fired ten rounds each, 5 in file firing and 5 in volley firing, against a target 6 feet high and 20 feet broad, equal to a front of 11 file of infantry, or 22 men. The muskets were 4 ft. 6 in. from the ground.

Distances.	Percussion Musket, 1842.				
	Number of Hits in the			Hits.	
	Bull's Eye.	Centre.	Outer.	Total.	Per Cent.
Yards. 100	7	48	94	149	74.5
200	3	20	62	85	42.5
300	4	9	17	32	16.
400	2	..	7	9	4.5

Distances.	Minié Rifle Musket, 1851.				
	Number of Hits in the			Hits.	
	Bull's Eye.	Centre.	Outer.	Total.	Per Cent.
Yards. 100	10	68	111	189	94.5
200	9	47	104	160	80.
300	6	32	72	110	55.
400	5	29	71	105	52.5

Colonel Hay remarks that the shot from the common musket, which missed the target, fell from 20 to 50 feet wide of it, whereas the Minié shot which missed fell within 2 or 3 feet only of the target. The same officer, in a letter to the author, observes that, with careful training, soldiers of the line may be made to put half their shot into such a target at 400 yards, and two-thirds at 300 yards.

Sept. 17th, 1853, at Hythe, practice was made with the regulation Minié musket by four men, each of whom fired ten rounds, at distances from the target varying from 700 to 200 yards. The target was 16 feet long and 6 feet high. The men advanced in skirmishing order, according to regulation; fired kneeling, and exercised their own judgment respecting their distances from the object. Of the 40 rounds fired, 8 hits took place in the bull's-eye, 16 in the central part (a circle 6 feet diameter) and four beyond, but very near the central part; in all 28 hits, and 12 misses. Of the hits, 18 were below the centre and 10 above, 16 were on the right and 12 on the left. The wind was blowing strongly, and the atmosphere was cloudy. The same day similar practice was made by four men with the musket having an elliptical bore, at the same distances from the target, the men advancing in skirmishing order, and kneeling. Of 40 rounds fired 6 hits took place in the bull's-eye, 22 in the centre circle, and 6 beyond it; six shots only missed the target. Of the hits, 25 were below the centre, and 9 above; 15 on the right, and 19 on the left.

(31.) In August, 1852, two muskets were made at the Royal Manufactory at Enfield, in which were embodied the improvements and alterations suggested by the experience obtained during the course of the trials with the experimental arms, and which, it was hoped, would possess the necessary requirements of a military weapon on the Minié system.

The musket, including the bayonet, to weigh about 9 lbs. 3 oz.; and the bore decided upon was .577 inch; the barrel to be, in length, 3 feet 3 inches; and its weight 4 lbs. 6 oz. It was to have three grooves, each a constant spiral of 1 turn in 6 feet 6 inches. The ramrod to have a swell near the head; the bayonet to be fixed by means of the locking-ring, and the lock to be made with a swivel.

One of the muskets was to be made with a block-sight for 100 yards, and two leaves, for 100 and 200 yards, in addition to the block-sight. This was proposed for troops of the line, militia, &c.

The other was to be made with a block-sight (as before), and a modification of Mr. W. Richards's sight, calculated for use up to 800 yards. This was proposed for the Rifles and the picked men of other corps.

Mr. Pritchett, a gunmaker of London, was requested to adapt to the bore of this musket a projectile without cup or plug, of a pattern (Fig. 63), which he had recently been trying with great success.

Fig. 63.

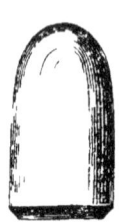
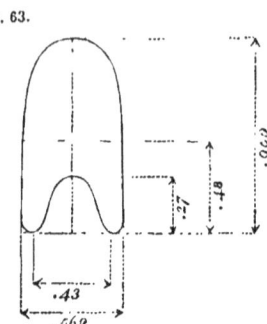

The shooting of these muskets, with this bullet, at distances up to 800 yards, was found to be superior to any that had yet been tried, when loaded according to the usual Minié style, with greased cartridges, reversing the bullet. They were also tried with the same bullet as before, made up into cartridges, loading without reversing, like the old spherical ball. The shooting was tolerably good, but the barrels fouled immediately above the powder for about 2 inches, probably from the ungreased paper, between the powder and the greased bullet, being charred on the explosion of the gunpowder and leaving a slight deposit adhering to that part of the barrel. In the Minié system the greased paper around the bullet lies close above the powder and lubricates the entire barrel at every shot.

The experiments alluded to in the foregoing remarks were ordered by Lord Hardinge for the purpose of obtaining, if possible, a lighter and a better musket than that in use by the army when he became Master-General of the Ordnance in 1851; his predecessors at the Horse Guards and the Ordnance Office having previously sanctioned the change of system from that of a smooth bore with spherical balls to that of Captain Minié, viz., a rifled barrel with greased expanding bullets and the cartridge reversed; making however, with this change, no abatement of the weight to be carried by the soldier, but, on the contrary, a slight addition; for although the weight of the musket itself was slightly reduced, the increased weight of the Minié bullet caused an increase of about half a pound in the total weight of the musket, with 60 rounds of ammunition.

The object of the Master-General of the Ordnance, in the experiments above alluded to, was to obtain the best rifled musket which could be constructed, on the system which had already received the sanction of the military authorities; and such a musket has now, it is presumed, been obtained. The new rifled musket possesses the following advantages:—

1. A saving in weight of about 3 lbs. for every soldier has been effected, although the new bullet itself is 30 grains heavier than the old spherical ball.

2. The 60 rounds for each man have been retained.

3. The strength of the musket has been very much increased.

4. The accuracy of shooting, of a musket which only costs (without bayonet) about 2*l*. 10*s*., has been improved, so that at a distance of 300 yards a good marksman can generally hit a bull's-eye with a six-inch radius.

5. The manufacture of the projectile has been very much simplified, the Minié bullet originally adopted with the Minié muskets having been altered from an inconvenient form, and a compound of lead and iron, requiring great care in the preparation, to a simple form of lead only.

6. An indirect advantage of the new rifled musket is, that any of the improvements which are constantly being made in the form and composition of elongated projectiles will be more easily adapted to a barrel of this diameter than to one of the former size.

This being the state of the case, it now rests with the military authorities to decide how far the Minié system can with safety be adopted by the army.

The improvements which have been suggested in the pattern of the musket, viz., the swivel lock, locking-ring for the bayonet, bands instead of loops and pins, &c., are, if found to be advantageous, equally applicable to the old smooth-bore musket as to those made for the new Minié system.

(32.) In 1852 Colonel Le Couteur, commandant of the militia in Guernsey, instituted a very important course of experiments to determine the relative penetrating powers of musket-shot of different natures. The arms employed were: the Regulation percussion musket, the Brunswick rifle, and the Regulation Minié rifle. The projectiles were: the spherical ounce-ball, the belted bullet, the Minié shot, with and without the cup, and a sharp conoidal shot proposed by Colonel Le Couteur. The objects fired at were a box of 3-inch deal filled with fine sand saturated with water, and a hogshead also filled with sand saturated with water.

The penetrations of spherical shot, both from percussion muskets and the Brunswick rifle, were very small; the shot, after passing through the front of the box, entering to the depths of 2 and 4 inches only in the sand; while the Minié shot and Colonel Le Couteur's sharper projectile, after piercing the front of the box, entered to depths, the mean of which was about 10 inches, into the sand; the advantage being rather on the side of the last-mentioned shot.

In the same year some experiments were made at Woolwich for a like purpose with the belted Regulation shot and Colonel Le Couteur's conoidal shot, both kinds being fired from the Brunswick rifle, also with Lancaster's ellipsoidal shot fired from the barrel with elliptical-spiral bore. The target, which was at 30 yards from the muzzle of the musket, consisted of several half-inch elm planks saturated with water and placed one inch apart. The following Table exhibits the results of the experiments:—

	Number of planks penetrated.
Colonel Le Couteur's sharp conoidal ball . . .	11
The belted regulation round ball	9
Colonel Le Coute 's ball	13 (stuck in the 14th).
The belted regu t' on ball	9
Colonel Le Couter's ball	12
The belted ball	9
Ditto	9
Lancaster's ellip ti al ball and rifle	9 (hit a knot).
Ditto litto	10
Colonel Le Conte r's smooth conoidal ball . .	13
Lancaster's ellipt' al ball and rifle	10
Colonel Le Couteur's smooth ball	11

Colonel Le Couteur observes that sharply-pointed balls have the greatest power of penetration; but the points being destructive of the paper, when made into cartridges, they have been laid aside. If a wad c uld be applied to cover the point, he conceives that they would exceed all other shot in penetrating power; and would be capable of sinking a boat, on its approach for the purpose of a hostile landing. The following experiments show that this opinion is not destitute of probability:—

At 500 yards' distance, three shot fired from the long four-grooved rifle musket buried themselves in a 3-inch oak target; at 700 yards, one ball passed through a 3-inch deal target, and was flattened on the iron at the back; at 800 yards the shot buried itself in a 3-inch elm target; and at 900 yards one ball buried itself in a 3-inch deal target. It may, hence, be presumed that a volley of such shot fired by fifty men against a boat, at the distance of 500 yards, would sink her.

(33.) The elevations requisite to attain given ranges vary much in different arms: frequently the sight-mark for 800 yards will be correct for one musket, and afford a range of 900 yards for another.

A musket which projects a heavier ball requires a little more elevation to attain the same range; but the five natures of Government muskets require less than 6 degrees, to enable them to send their shot to the distance of 1000 yards.

In order to ascertain the number and extent of the ricochets made by shots fired from the Regulation rifle just mentioned, the axis of the fire-arm was placed in a horizontal position, 5 feet above the plane of the sea-shore. The grazes were as follow:—

Number of Grazes.	1	2	3	4	5	6	7	8	9	10
Extent of the Intervals between the Grazes, in Yards . . .	Yds. ft. in. 150 0 11 from the *Rest.*	..	..	..	151	39	25	17	7	13

Distance of the last Graze from the *Rest* = 402 yds. 0 ft. 11 in.

The conviction of this officer is that men who have been taught to fire conical shot from rifle-muskets with elevating sights ought to drive any troops not similarly armed out of the field (in any open country) at distances between 200 and 800 yards; that is, before men using the old round ball with common muskets could render their arms available.

(34.) In 1853 a committee appointed by Major-General Love, and of which Colonel Le Couteur was the president, caused some experiments to be made in order to try the relative merits of the following natures of small-arms:—
A sea-service musket, 30-inch barrel, rifled with three grooves;
A musket, 39-inch barrel, rifled also with three grooves;
A sea-service musket, 30-inch barrel, rifled with four grooves;
A musket, 39-inch barrel, rifled also with four grooves;
Also the new Regulation Minié rifle.

From these experiments, in which 100 rounds were fired from each of the four first kind of muskets, the men firing from a rest, it resulted that the number of hits and the number of misses respectively, were very nearly equal: the means, in 100 rounds, were $74\frac{3}{4}$ hits, and $25\frac{1}{4}$ misses (nearly as 3 to 1). In 50 rounds of the new regulation Minié rifle, there were 26 hits and 24 misses: from which it would seem that this arm is inferior in accuracy of fire to either of the others, except, perhaps, the sea-service musket. The long four-grooved musket appeared to be the most advantageous in this respect.

The wind, which was high at the time the experimental trials were made, appeared to affect the balls in proportion to its strength, and to the distance of the shot from the musket; beyond 400 yards the lateral deflection was from 10 to 50 feet to leeward of the line of aim.

Similar experiments were carried on with the four first kinds of musket, 300 rounds being fired from each, and the men firing from the shoulder; the means, in 100 rounds, were $54\frac{3}{7}$ hits, and $45\frac{7}{8}$ misses. From these it would appear that the advantage was in favour of the longer musket, the number of hits being, to those made by the shorter musket, in the ratio of 9 to 7, nearly: other experiments have, however, shown that the practice from a barrel 2 feet 6 inches long was quite as good as that from one of 3 feet 3 inches. On comparing the shooting from a rest with that from the shoulder, it was found, 100 rounds being fired in each of the two ways, that a column of men consisting of 13 files in front, would receive 79 hits from the rests, and 54 from the shoulder.

Colonel Le Couteur observes that some Minié shot which had ranged from 920 to 1051 yards, and had fallen into the sea, which was then from 20 to 30 feet deep, were picked up on the sands at low water, scarcely altered in form by their impact on the sea.

(35.) Till within the last twenty years no *sight* was considered necessary for a common musket, the stud at the muzzle being sufficient for the purpose of taking aim. When percussion arms were first introduced, a fixed block-sight for 120 yards was adopted; and when the Rifle Brigade was supplied with two-grooved rifles, a block-sight for 200 yards and a leaf for 300 yards were affixed to the fire-arm. Whilst the order, in 1851, for constructing 28,000 stand of Minié muskets was being executed, a very serious defect in the sighting was discovered.[a] The

---

[a] The Committee appointed to report on the effects of conical ammunition fired from common service-muskets rifled, recommended that the scale of sights should commence at 100 yards, and not at 200 yards, as in the new Minié musket, as they considered that, otherwise, the firing would be too high in close action: they proposed, also, to have folding sights up to 300 yards, as these would be more readily shifted, for different distances, than the slide in the long sight.

In a letter accompanying the report, the Committee stated the relative number of hits per cent. in shot fired from a common service-musket rifled, and a common musket, as in the following table:—

Distance.	Size of Target.	Position of the Musket.	Number of Hits per Cent.		Penetration in Inches.
			With the Rifle.	With the Common Musket.	
Yards.					
200	6 ft. square	{ At the shoulder, without a Rest. }	59	29	..
300	..	With a Rest . .	78	8 or 9	..
500	12 ft. square	..	59	0	3 to $4\frac{1}{2}$
1000	9 ft. square	..	52	..	$2\frac{1}{2}$ to $3\frac{1}{2}$
1300	39 ft. by 21 ft.	..	46	..	$2\frac{1}{4}$

lowest, or fixed sight, was so placed as to give a range of 200 yards, considered to be the point-blank range; instead of which it was found to be a line of metal range, with an elevation of about 1°; hence the height of the trajectory above the line of sight was such that, at 100 yards distance, shot aimed directly at a mark by the fixed sight struck, on an average of many rounds, 2 ft. 6 in. above the point aimed at. The only remedy for this defect in sighting was to explain to the soldier the error in the sighting of his weapon, and direct him, whenever his enemy advanced within 200 yards, to aim lower and lower, up to a certain point, when the deviation was greatest, and then reverse the process. As soon as this defect became known to Lord Hardinge, then Master-General of the Ordnance, the subject was referred to the select committee, and the error was rectified by adapting the fixed sight to 100 yards, and carrying on, from that point, the system of sights as now established.

At present it is proposed to furnish every musket with a complicated and delicate sight; and those sights which have been submitted for trial are, chiefly, in accordance with the pattern adopted for the 28,000 muskets which were ordered for the army in 1852: they are subject to some serious defects. The parallel sides become, in use, slightly collapsed, and thus lose their power of supporting the transverse slider containing the notch; the slider consequently falls down by the concussion produced at every shot; and, before firing again, the soldier has to re-adjust it: besides this, the under part of the retaining spring, which soon gets foul, cannot be cleaned. The sight proposed by Mr. Richards, appears, however, as far as can be judged at present, to be free from these defects. (*Remarks on National Defence*, Appendix, p. 10.)

(36.) The force required to draw the trigger of an ordinary musket has been estimated at between 16 and 28 lbs.; and it is evident that, in the exercise of this force, the attention of the soldier must be diverted from the object aimed at; consequently the precision of his fire must be greatly diminished, or entirely destroyed. In order to remedy this evil, Lieutenant Harris, of the Royal Marines, invented what is called a circular trigger, by which the force required to throw down the hammer is greatly reduced. This is accomplished by making the acting surface of the trigger in the form of an excentric circular arc, which, on pulling the lever or cock, causes that lever to rise gradually, as if it moved on a circular inclined plane. The mechanical power thus obtained is that which not only permits the exercise of a smaller effort in drawing the trigger, but allows the force to be employed without a *jerk*: the more or less gradual development of the moving power depends on the less or greater excentricity of the curvilinear inclined plane.

(37.) From the previous statement it appears that the armament of the British military force is now in a very unsettled and complicated condition, there being now in the hands of the British soldiers the old regulation musket with the round bullet; the old regulation rifle musket with the belted ball; the new regulation Minié musket, and the new Enfield small-bore musket; but in what proportion these arms are borne by our troops in the East, the author cannot say. The question of armament being undecided when the troops embarked, some regiments were provided with two sets of arms, viz. the regulation smooth-bore musket and the Minié rifle musket; other corps were armed, some wholly with the old musket, and some with the Minié musket; in other corps, with both the new and the old arms in certain proportions, requiring vast quantities of their appropriate ammunition in the artillery stores, and great complications in the arrangements for the supply in the field. To avoid as much as possible these complications, great care has been taken of late years to introduce uniformity of calibre and simplification in ammunition, both in the navy and the army. Uniformity of calibre, and simplification in the supply of ammunition, is also the ruling principle of the Emperor Louis Napoleon's new system of field artillery (see Appendix B). In the British service, uniformity and simplification are objects in the accomplishment of which great difficulty exists. Are we prepared to sacrifice the value of the vast quantities of small-arm ammunition in the numerous military stations which our troops occupy in all parts of the world, and send thither supplies of the new ammunition to replace the old? Or must we supply all our military stations with both the new and old ammunition? How a regiment, a brigade, or an army, having a mixed armament, is to be supplied with different sets of ammunition in the field, is a question of vast importance and difficulty, which the present state of the armament of our land forces necessarily raises. But the question of a mixed armament of muskets is still more important, when we consider the effects it may produce upon the efficiency of any corps so armed; for if rifles be required in the position which the corps armed with both is required to act, they would require to be wholly armed with rifles; if not there should be no rifles. The use of the long-ranged rifles commences where that of the regulation musket ceases; and the use of the musket begins where rifles are not required, and may even be detrimental, as in line firing, which may best be accomplished with a spherical bullet and a smooth barrel.

GENERAL OBSERVATIONS ON RIFLED MUSKETS.

(38.) Whilst we fully admit the vast importance of the rifle-musket as a special arm, we must be permitted to doubt the correctness of the opinion, that it will prevent artillery from keeping the field. Shrapnel shells will, undoubtedly, still prove an overpowering antagonist of infantry acting in swarms, *en tirailleur*, in the manner in which it is proposed to employ infantry armed with long-range rifle-muskets. One of the first occasions in which the author observed the effects of Shrapnel shells on service, was that in which they were fired from a light 6-pounder at a gun, which, at Elvina, in 1809, had been brought up, by the French, at a distance of 1400 yards, to support their skirmishers, when warmly opposed by our advanced posts. The first shell knocked down more than one-half of the men about the gun.[a] The 8-pounder gun (French) cannot stand against the carabiniers, who, beyond 650 or 700 yards, struck the gunners without a single ball of grapeshot reaching them.—*Rémond*, p. 192.

Field artillery, 9 and 12-pounder guns in particular, placed far beyond the reach of even the most random range of these rifles, may, by means of Shrapnel shells, pour upon swarms of skirmishers, musket bullets which, after having described in the shell a trajectory of 800 or 900 yards, and then being dispersed by the bursting of the shell, will produce an effect as destructive as a gun charged with common case-shot at a distance of 300 or 400 yards; and an important improvement in a short-range fuze, well adapted to the service of spherical case-shot, has lately been made by an artillery officer of great talent and promise.

Under the powerful effect of Shrapnel shells, together with the menaces and charges of cavalry, clouds of infantry, acting *en tirailleur*, will either be compelled to rally into masses, or to retire upon their supporting bodies, columns or lines, when round shot will exercise its wonted power, and thus the battle will become general in the ordinary way. The three great arms, artillery, cavalry, and infantry combined, will act according to their distinctive faculties; and the general who, according to the proposed scheme, had hoped, by infantry armed with rifle-muskets, to drive artillery out of the field, and overpower infantry and cavalry in a general skirmish, will only commit the serious error of bringing on a general action under circumstances highly

---

[a] " Si, comme nous l'avons dit, les combattants s'écartent les uns des autres, et si les troupes sont plus clairsemées, on a moins besoin de lancer un mobile qui ait une grande force, que d'en lancer un grand nombre avec une force moindre. C'est pour cela que nous pensons que les Shrapnels (spherical case-shot) acquièrent, dans l'état actuel des choses, un intérêt particulier, et que l'artillerie est naturellement amenée à tourner ses études de ce côté."— Favé, *Des Nouvelles Carabines*, &c., p. 47.

disadvantageous to himself; since a commander, forced to fight in a manner different from that which he had intended, and for which he had prepared, is always, as has been well said, more than half beaten. The opponent following up with all his arms the advantages which well combined movements must produce, the army which should rely upon the random range of the new rifles would be penetrated, thrown into confusion, and even driven off the field.

(39.) Now, whatever be the power of range, to the first graze, which the French and Prussian rifles are said to possess, it need scarcely be repeated that uncertainty increases in proportion as great elevations are used, that is, as the range increases. And admitting that the trajectories of these projectiles are more true than those of others, yet the chance of hitting objects must necessarily diminish in proportion as the descending branch of a trajectory approaches the vertical direction. In all cases of gunnery the great object is to have the path of the shot as nearly horizontal as possible.

The shot employed with the French musket being heavier than that which has been generally used hitherto, will necessarily have a greater angle of descent, particularly at long ranges; and it is rightly observed by M. Favé, that the shot will meet an object of determined magnitude, as a man, only in a very small portion of its path.[a] This circumstance is highly disadvantageous in war practice, which requires that the shot should meet as many objects as possible; and, obviously, this can only be the case when the path of the shot is nearly horizontal.

(40.) In practice, with the Minié rifle as at present made, it is found that the elevation necessary to obtain a range of 1000 yards is not above 5 degrees: theory indicates that it should be 8 degrees; and the length of the *sight* for that range has hitherto been adapted to that indication. The trajectory, in this case (see Art. 268, Note, and Fig. 19, Pl. II.), acquires its greatest elevation at about 600 yards, and it is there about 150 feet above

---

[a] " Il résulte de là un fait important pour la pratique ; car, la balle tombant à terre sous un angle plus grand, ne peut rencontrer un but d'une hauteur déterminée, un homme par exemple, que dans une moindre étendue de son parcours. Ainsi à une grande distance, beaucoup plus difficile deja à apprécier qu'une petite, la même erreur d'appréciation aura une influence beaucoup plus nuisible dans la pratique de la guerre."—Favé, *Des Nouvelles Carabines*, &c., p. 35.

" La justesse de tir dépend beaucoup sur la trajectoire. L'effet des bouches à feu sur un champ de bataille dépendrait beaucoup de leur chance de toucher le but directement. L'angle d'incidence influe beaucoup sur la chance d'atteindre : si cet angle est petit, le projectile pourra toucher, dans une grande partie de son parcours, un but elevé; s'il est grand, le contraire aura lieu."—*Nouveau Système d'Artillerie*, 1851, edited by Favé, pp. 28, 29.

App. A. GENERAL OBSERVATIONS ON RIFLED MUSKETS. 533

the horizontal plane passing through the axis of the piece, at the muzzle. But however this may be, whether 5 degrees or 8 degrees, the angle of descent—upon which more than on the angle of elevation the effect upon the battle plane depends (Art. 342)—is so great, that the shot would pass over the head of a man placed a few yards in front of the farthest extremity of the range, and the secondary effect, by ricochet, would be small and uncertain. With an elevation of 5 degrees the shot is said to make good ricochets; but the grazes, instead of being in the direction of the previous flight, would, in consequence of the elongated form of the shot, be (Art. 183) greatly and irregularly deflected.

(41.) On a plane surface the graze of a musket or rifle bullet of a spherical form is remarkably true, and as straightforward as if traced with a ruler, as the author has often witnessed; and a very large portion of the men struck in action in the field are so by bullets which had previously grazed. On this account the well-known caution to "level low" rather than high, is enjoined; but when ranges of the extent above mentioned are required, the caution must be reversed. It is justly observed by M. Favé that the ricochet holds an important place in general actions, particularly in a plane country, since it serves to obviate the errors which may be committed in the appreciation of distances; but any shot, when fired with a considerable elevation, and cylindro-conical projectiles even when fired at low elevations, the first on account of the abruptness of the angle of descent, and the other, either from not rebounding or from the deflection or irregularity of the graze occasioned by the figure and rotation of the shot on striking the ground, must be nearly ineffectual when directed against troops in line.

The trajectory of a shot from a common musket or Minié rifle, in a range of 400 yards, is not above 11 or 12 feet high at the highest point; it has therefore so much horizontality as to command a large portion of the surface of the plane or field over which it flies, and, if correctly aimed, can scarcely fail to come in contact with some of the masses or persons acting upon that plane.

(42.) It has been said by a French writer (Favé, *Des Nouvelles Carabines*, &c., p. 45) that the increase in the range of small-arms tends to diminish the influence and to weaken the action of cavalry; and that writer suggests the re-examination of a question often controverted, whether it would not be possible to train cavalry to act dismounted, at least as tirailleurs, and, in that case, to arm them with the new muskets; it is proposed, in fact, to restore the long-disused system of mounted riflemen.[a]

---

[a] Mounted riflemen are universally condemned, excepting for desultory service, such as that in the late war against the Kaffirs at the Cape of Good Hope, and against hordes of Arabs in Algiers: to return to this organization

(43.) The opinion that the Minié rifle is capable, with unerring certainty, of picking off the artillerymen, and thus of silencing any field-battery, has led to the extraordinary and ill-judged expedient of arming artillerymen with the Minié carbine, in order that they may protect themselves against infantry armed with that rifle. No doubt riflemen will, as heretofore, be pushed forward, whenever this can be done, for the purpose of taking off the officers directing and men serving the guns. Captain Geary was, by a rifleman who had crept forward unobserved, shot through the forehead at the battle of Vimiero, while in the act of laying a gun; and a battery of artillery may be attacked by a sudden rush of cavalry whenever it may appear to be not properly supported and protected by troops. If rifles are required to protect artillery against riflemen, they should be in the hands of riflemen; but to allow artillerymen, under any circumstances whatever, to lay down the sponge and take up the rifle is subversive of the object for which they were organized; and the attention of artillerymen should never be diverted from their proper arm. When so posted as to see and be able to reach the enemy's lines or masses of troops, the artillery should fire upon those troops, but it should never turn its fire upon guns brought up to attack it: artillery should never fire upon artillery except when the enemy's troops are screened from its fire, and its own troops are exposed to that of the enemy's artillery. Should artillery be attacked by the enemy's riflemen, it should have recourse to spherical-case shot; but, if the circumstances of the case should be such that spherical-case shot, improved as it now is, does not accomplish the end, then bodies of riflemen should be employed to do for the artillery what the latter cannot do for itself—repel those of the enemy. In all cases, when artillery is attacked by riflemen, it is with a view of drawing the fire of the artillery from the object for which it has been placed in position; and this attempt should be counteracted either by the fire of the protecting troops, or by that of some battery brought up from the reserve.

In a letter lately received from the seat of war in the East, it is stated that the rifles used by the Russians at the battle of the Alma were of good construction; they are said to have been formed with two grooves, and to have carried solid conoidal shots, each weighing 767 grains, equivalent in weight to a spherical bullet of 9 to the pound; consequently much heavier than the English regulation Minié shot. The annexed cut, fig. 64, is an exact representation of one of these projectiles: they are flat at

---

in England, by transforming some of our yeomanry corps into mounted riflemen, should not be sanctioned; for such a combination must result in making bad cavalry and worse riflemen.

the base, and have projections at the sides to correspond to the grooves of the muskets; they are not hollowed, and, of course, have neither cup nor plug. The great weight of these missiles is very objectionable; the soldiers who carried them must have been much distressed by the loads in their pouches, or these must have contained a smaller number of shot than are carried in the field by English or French soldiers.

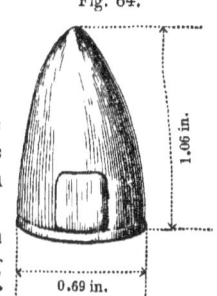

Fig. 64.

1.06 in.

0.69 in.

The Russian missile is more pointed than the English Minié shot; and, no part being cylindrical, it must be liable to irregular movements in the barrel; and, consequently, to unsteadiness in its flight. It has the designation of a Minié shot, a term now generally, but improperly, applied to all elongated shot for musketry, since they differ from one another both in weight and form. The issue of the recent actions on the Danube, as well as that which has just now taken place on the Alma, may be considered as a strong confirmation of the argument used by the author in Art. (38), p. 531, in opposition to the opinion that the new rifle muskets will become the means of compelling artillery to quit the field, of paralyzing the action of cavalry, and of reducing general actions to combats of skirmishers. The battle of the Alma has proved that the three arms, artillery, cavalry (though of these there were lamentably few), and infantry, if duly combined, will still prevail in action over the most extensive employment of rifle musketry.

If the country had been at this time in a state of peace, the author might have added to the above historical notice of the new rifled muskets for projecting elongated shot a more particular examination of these arms with respect to their relative merits, with an inquiry into the extent to which they might be adopted for the armament of the British infantry: but being at war with a gigantic power, and in circumstances which may bring forward an extensive employment of these arms by the enemy as well as by the British troops and their allies, the relative efficiency of the several arms will, probably, be determined by the most sure of all possible tests—experience in actual warfare. On this account the author has thought proper to postpone the practical inferences which he has deduced from the preceding statements; and with which he, at one time, intended to have terminated this Appendix.

(B.)

THE EMPEROR LOUIS NAPOLEON'S NEW SYSTEM OF FIELD ARTILLERY.[a]

Simplicity, mobility, and efficiency are the great objects sought to be obtained in the various changes which have been introduced in the organisation and service of field artillery during the last four hundred years.

Until the time of Gribeauval, armies in the field were embarrassed in their movements by prodigious trains of ordnance of all natures and descriptions, from the 48 to the 4-pounder gun, besides a variety of howitzers, for siege operations as well as for field service, for both of which services the like natures of artillery were employed.

Gribeauval separated field from siege artillery, and appropriated to the former 12, 8, and 4-pounder guns and 6-inch howitzers.

In 1792, and throughout the war, the field artillery of France consisted of 4, 8, and 12-pounder guns, 6-inch howitzers, and in some campaigns (particularly in Italy) of 8-inch mortars.

In 1827 the 4-pounder having been abolished, the field equipment was reduced to the 8 and 12-pounder guns, to which were added howitzers of 15 and 16 centimetres calibre (5.9 and 6.3 inches).

To simplify the appropriation still further, the new system provides for the ultimate abolition of the 8-pounder gun (very nearly equivalent to our 9-pounder), and the retention of only one piece of ordnance for field service, viz., a 12-pounder unchambered gun, which may at pleasure be used either to fire shot or shells, and which M. Favé therefore designates a canon-obusier.[b]   Exclusive of the batteries composed of those guns, there are reserve batteries of howitzers of 15 and 16 centimetres attached to the parks or reserves of artillery, ready at hand to be brought up when

---

[a] The Emperor Napoleon's New System of Field Artillery. Edited by Captain Favé. Translated into English by Captain William Hamilton Cox of the Royal Artillery. Parker, Furnivall, and Co., Military Library, Whitehall

[b] The gun described in M. Favé's pamphlet is considered in England to be a howitzer. It is however, in fact, an unchambered 12-pounder gun of reduced weight, from which, in addition to round and case shot, shells are to be fired, and is, therefore, called a canon-obusier. The reader will keep this in mind, wherever the new gun is so mentioned.—AUTHOR.

needful, and especially for the attack of posts or other field defences.

The simplification is to be effected—first, by reducing the charge of the new 12-pounder gun for solid shot from a third to a fourth of the weight of the shot; secondly, by reducing the weight of the gun in the same proportion as the charge; and thirdly, by mounting the gun on the present 8-pounder carriage; and, whilst new 12-pounders are preparing, the existing 8-pounders are ordered to be bored up to the calibre of 12-pounders, and thus immediately made available as such. Shells are to be fired with a reduced charge, to be determined by experiment, but which shall be greater than that with which shells are fired from howitzers.

Field artillery, at the present period, consists of two descriptions of gun carriages, four sorts of ordnance, and eight kinds of projectile.[a]

The new system of field artillery consists of only one gun carriage, one description of ordnance, and three projectiles. Therefore it abolishes one gun carriage, three sorts of ordnance, and five projectiles (the 12-pounder shell being already employed for Mountain Artillery). They are about to adopt, in France, spherical case shot, in imitation of the artillery of several European countries, which would introduce four new projectiles into their former system, but would add only one to the proposed system; so that they will really suppress, by their project, nine projectiles of different kinds in the service of their field artillery.

The advantage of range, of accuracy, and of penetration, which the 12-pounder has over guns of smaller calibre, will give to the French batteries of six of those guns a decided superiority over those of the former system, consisting of one 12-pounder and five 8-pounders.

The ruling principle upon which this simplification proceeds is so much in point with observations made in several parts of this work, on the comparative merits and demerits of unchambered and chambered ordnance, of guns and howitzers for sea and land service (Art. 339 *et seq.*), that the author cites with much satisfaction the following passage from the Emperor Napoleon's very able and interesting work:—

"There is an essential advantage in the proposed system to which we call the attention of all scientific men, viz., the power of firing at will, either round shot or shells, from all the pieces, for it has the effect of tripling the actual power of a battery as regards hollow projectiles, or of increasing by one-third the effect as regards solid shot. At present a battery composed of four guns

---

[a] Round shot and canister for guns, and shells and canister for howitzers. —*Note by the Translator.*

and two howitzers does one of two things: it either makes use alternately of the four first and the two second, or it fires these pieces together. In the first case, the effect of its fire is considerably reduced; in the second, it is acknowledged that one of the two is sacrificed to the other.

"In fact where the round shot is advantageous, the shell is less so, and *vice versâ*; for this reason foreign powers have batteries entirely composed of howitzers.

"We shall employ, it is true, hollow projectiles of a smaller size than are now used from howitzers, but we shall fire them from guns with greater charges, and we hope to obtain superior accuracy. We are, besides, certain to have for the same weight of metal a greater number of splinters."[a]

The weight of the new 12-pounder gun is determined, conformably to the Principle No. 1, by reducing the weight (880 kilo., or 17½ cwt.) of the old gun, in proportion to the diminution of the charge from one-third to one-fourth;[b] the weight of the new gun is therefore fixed at 660 kilo., or 13 cwt. very nearly. The lengths of the reinforces are the same as in the 8-pounder, but the length of the bore is less by 254 millim. (10 inches) than that of the old 12-pounder; that length is still 14.6 calibres, therefore this reduction of the charge will not cause any very material diminution in the initial velocity of the shot.

Extensive experiments, of which the programme is given in the work above quoted (p. 35), were made in 1850 at Metz, Strasbourg, Toulouse, and Vincennes, with two new and one old 12-pounder gun, one 8-pounder gun, a howitzer of 16 cent., and a howitzer of 15 cent., to determine the comparative precision of practice, the effects of the projectiles, and the strength of the carriages.

The several Reports of the trials made at the four principal artillery schools or stations having been digested and compared, the following results were announced:—

I. *Accuracy of Fire.*

"The shell-gun, firing 12-pounder round shot with a charge of a fourth of its weight, has put a greater number of shots per cent. into the target at all ranges than the present 12-pounder gun, except at that of 600 metres, where they are equal.

"The shell-gun has put, taking the average of the five distances, 48 per cent. into the target, the 12-pounder gun has put in 46,

---

[a] The Emperor Napoleon's New System of Field Artillery. Translated by Capt. W. H. Cox, R.A., pp. 27, 28.

[b] The charge of the former gun was 2 kilo., or 4.4 lbs.; and that of the new gun is 1.5 kilo., or 3.3 lbs.

that is an advantage of 1 in 23 for the shell-gun; at the distance of 900 metres, the shell-gun had 33 and the 12-pounder gun 32 per cent.

"The shell-gun has put, at all distances, more shots in the target than the 8-pounder gun.

"The average on all the ranges is 48 per cent. with the shell-gun, and 40 per cent. with the 8-pounder gun; that is to say, an advantage of one-fifth for the shell-gun. At the distance of 900 metres, the shell-gun has put in 33 per cent. and the 8-pounder only 28; the difference is 5 in 28; that is an advantage amounting to between one-fifth and one-sixth, for the shell-gun."

II. *With respect to shell practice, the averages of the results are stated as follow :—*

"The shells of 12 cent. (4.73 in.) fired with a charge of 1.225 kilo. (2.69 lbs.) struck the target in a much greater proportion than the shells of 16 cent. and 15 cent. (6.3 in. and 5.9 in.), with the small charge (0.75 kilo. or 1.6 lbs.). The proportion of shells of 16 cent. and 15 cent. that struck the target, when fired with the small charge, is not, at any distance, equal to half that of those of 12 cent., excepting the shells of 16 cent. at 500 metres (550 yards nearly).

"It is also to be remarked that at 900 metres (980 yards) the howitzer of 15 cent., out of 60 rounds with the small charge, has not put one into the target, while at this distance the shell-gun put in 33 per cent. The shell-gun, taking an average of the effect at all the distances, has put 49 per cent. of shells into the target, which is a greater proportion than for any other piece fired during the experiments.

"The proportion per cent. that struck the target from the howitzer of 15 cent., with the large charge, at ranges of 800 and 900 metres, is much less than from the shell-gun at the same distance.

"The howitzer of 16 cent., with the large charge, at the same ranges, is a little superior to the shell-gun; at 800 metres it gives 38 rounds per cent. through the target, the shell-gun has 35 rounds at the same distance; at 900 metres the proportion is 36 per cent. for the first, and 35 per cent. for the shell-gun. It will be seen by the preceding table that the slight superiority of accuracy for the howitzer of 16 cent. is only applicable to a very small number of rounds, as a battery of reserve carries only 26 shells of 16 cent. to be fired with the large charge."

III. *Penetration of the Projectiles.*

The average penetrations of solid shot in well-rammed earth, from experiments at the four stations, were as follow. The charges are given in pounds avoirdupois, the distances in yards, and the penetrations in feet.

Nature of Ordnance.	Charge.	Distances.	Penetration.
	lbs.	Yards.	Feet.
12-pounder gun	4.3	32	5.2
8-pounder gun	2.7	32	4.46
New 12-pr. shell-gun	3.3	32	4.69

The average penetrations of shells into well-rammed earth, from experiments at the four stations, were as follow :—

Nature of Ordnance.	Distances.	Charges.	Penetrations.	Charges.	Penetrations.
	Yards.	lbs.	Feet.	lbs.	Feet.
12-pounder gun	54	3.3	3.38		
	87	3.3	3.67	1.65	3.3
	108	3.3	5.66	1.65	4.62
8-pounder gun	54	2.2	3.08	1.1	
	87	2.2	3.11	1.1	2.9
	108	2.2	4.3	1.1	3.57
New 12-pounder shell-gun	87	2.7	2.75		
	108	2.7	4.26		
	216	2.7	2.1		

No shells were broken by the explosion on leaving the gun, or from ricocheting on the ground, but some of the shells of 12 cent. broke in the earth at the short ranges.

IV. *Common case, or canister shot.*

"The practice with case shot took place at ranges of 440, 550, and 660 yards: it was satisfactory as far as regards accuracy, and the effect was good in proportion to the number of balls. The howitzer of 15 cent. throws, with a charge of 2.2 lbs., case-shot containing 70 balls,—weighing nearly 26.5 lbs. The 12-pounder case-shot weighs 22 lbs., and it should be adopted. There would, perhaps, be some advantage in filling the 12-pounder case with the balls belonging to the 8-pounder, of which it would contain a greater number. We are led to believe that, in giving the proposed system three case shot per box, each weighing from 10 to 11 kilogrammes (22 to 24 lbs.), it would have a superiority over the present system, as regards the firing of case shot."

V. *Recoil.*

"The recoils varied much; but those of the new shell-gun, when fired with solid shot, shells, and case shot, were within the limits of the recoils of the ordnance at present in use."

VI. *Resistance of the Carriages.*

"The trials made to determine this point were made by firing the pieces at angles of 5° and 10° with the plane on which the carriage was placed. The carriages have all resisted well the practice which was carried on to determine the accuracy of firing. At three of the stations the carriages suffered no damage except what was considered slight and accidental. At the fourth station, however, the howitzer of 15 cent. caused considerable damage to its carriage, apparently from a want of harmony in the construction of the piece and carriage. On the whole it is concluded that, with respect to this point, the proposed system has the advantage over the present system."

VII. *Comparative battering powers of the proposed and present systems of field ordnance.*

"In making this comparison, with regard to the effect produced by the ordnance in firing solid shot against any obstacles, let it be supposed that they fire at a range of 900 metres (980 yards) against a wall 3 metres (3¼ yards) high and 30 metres (33 yards nearly) long; then, by calculation, it is found that, if the *moving force*[a] of a solid shot from the old 12-pounder be represented by 1, that from the new 12-pounder shell-gun will be 0.95, and that from the 8-pounder 0.53. The former 12-pounder battery will exercise a *moving force* represented by the number of shots that strike, that is, by 276. The new shell-gun battery will exert a force represented by the number of shots that strike, multiplied by 0.95; that is to say, by 233.

"The 8-pounder battery of division will exert a force represented by the number of shots that strike, multiplied by 0.53; that is to say, 123.

"The shell-gun battery of division will exert a moving force represented by the number of rounds striking, multiplied by 0.95; that is to say, 165.

"If we then calculate the *moving force* for one battery of reserve and five batteries of division of the two systems, we find the *moving force* of the six batteries of the present system represented by 891, and that of the six batteries of the proposed system by 1058.

"The difference is 167, that is, nearly one-fifth in favour of the proposed system.

"If we make a similar comparison as to the effects that would be

---

[a] Moving force (*force vive, vis viva*) of a body in motion is the whole mechanical effect which it will produce in being *brought to a state of rest*, no regard being had to the time in which the effect is produced, and it varies with the weight of the body multiplied by the *square of the velocity.—Note by the Translator.*

produced by shells at 900 metres, we find that the object would have been struck by 19 shells of 16 cent., thrown with the large charge, and 31 shells of the same calibre with the small charge; such is the effect of a 12-pounder battery of reserve. But the shell-gun battery of reserve would have struck the object with 245 shells of 12 cent., that is, nearly five times as many rounds.

" Whatever may be the proportion between the *moving force* of these projectiles, the shells of 12 cent. have such a superiority in point of accuracy, that the advantage must generally remain with them.

" The 8-pounder battery of division would strike the object with 10 shells of 15 cent. with the large charge, and 4 with the small.

" The shell-gun battery of division would have struck with 174 shells, that is, 12 times more.

" Six batteries of the present system would strike the object with 120 shells, that is, 19 shells of 16 cent. fired with the large charge, 31 shells of 16 cent. with the small charge, 50 shells of 15 cent. with the large charge, and 20 shells of 15 cent. with the small charge.

" The six corresponding batteries of the proposed system would strike the object with 1115 shells of 12 cent., that is, 9 times more.

" The advantages of the proposed system would not always be exactly the same at all distances, nor under all circumstances, but what we have just said will give some idea of its superiority at 900 metres. We may add that at greater ranges the *moving force* of the shot from the shell-gun will approach nearer to that of the shot from the 12-pounder, and will increase more and more above that from the 8-pounder gun; therefore the superiority of the fire with round shot from the proposed system as regards *moving force* would go on increasing. At 500 metres the shot from the shell-gun and the shot from the 8-pounder have the same velocity; therefore at this distance the moving forces are in proportion to their weights, that is to say, of 1.5 to 1. From thence the advantage of the first over the second increases as the distance.

" The 276 shells of 16 cent. of the present 12-pounder battery of reserve give 5796 splinters, of which 4692 weigh more than 0.1 kilogrammes (3½ ounces)."

The 744 shells of 12 cent. of the shell-gun battery of reserve give 12,648 splinters, of which 10,416 weigh more than 0.1 kilogrammes.

The 300 shells of 15 cent. of the present 8-pounder battery of division give 6600 splinters, of which 5700 weigh more than 0.1 kilogrammes.

The 528 shells of the shell-gun battery of division give 8976 splinters, of which 7392 weigh more than 0.1 kilogrammes.

Now that experience has proved all the superiority, as to accuracy, of the shell of 12 cent., it is established that many more of the splinters will be effective at the point of explosion, as it must generally be nearer the object.

The new 12-pounder gun having been adopted in the French service, an experiment was made in the spring of the present year, 1854, in France, to determine its powers in breaching the revetments of fortresses. For this purpose, four of these pieces of ordnance were placed in battery in the covered-way of Fort Valerien at the distance of about 30 yards from the escarp, which is 32 feet high. The artillery was pointed so as, by its fire, to make, with solid shot, a horizontal groove 8 feet deep, without entirely piercing the revetment, the shots fired being at length arrested on striking those which had previously lodged in the groove: two vertical grooves were afterwards in a similar manner cut in the wall, and the mass between these was then brought down, by salvos of shot. After firing 850 rounds, a breach 18 yards broad was made, but not being practicable for a column of troops, 250 rounds of shells were fired into the breach, by which it was rendered completely practicable. It was remarked that the masonry, which had been standing ten years, was not perfectly dry in the interior of the wall.

"In the French service the different descriptions of ordnance still retain their old designations; that is to say, guns are distinguished by the weight of the shot in lbs.; but ordnance for throwing hollow projectiles are designated by the diameter of the projectile, expressed in centimetres. The only exception to this is the mountain howitzer, which is generally called the 12-pounder howitzer, from being of the same calibre as the 12-pounder gun; but its shell is called a shell of 12 cent.

"The new measures have been adopted for artillery since 1839, the millimetre being taken for unity.

"The French field guns are somewhat longer than the British; the 12-pounder, weighing 17½ cwt., is about 7 feet 2 inches in length; the 8-pounder, of 11¼ cwt., is a little more than 6 feet; the howitzer of 16 cent. is 6 feet 2 inches; that of 15 cent. is 5 feet 7 inches.

"The lengths of the corresponding guns in the British service are, 12-pounder, 6 feet 6 inches; 9-pounder, 6 feet; 32-pounder howitzer, 5 feet 3 inches; 24 pounder howitzer, 4 feet 8½ inches.

"The French 12-pounder mountain howitzer weighs about 2 cwt., and is 2 feet 10 inches long.

"The gun carriages and waggons now in the French service have all been introduced since 1827, and are copied from the English; but they consist of only two varieties. For instance, there are only two sorts of field-gun carriage, one for the 12-pounder gun and howitzer of 16 cent., the other for the 8-pounder and howitzer of 15 cent.; and only one diameter of wheel, one pattern limber and two axle-trees, for all kinds of field-gun carriages and waggons.

"The ammunition waggon (*caisson à munition*) is the same for all descriptions of field artillery; a spare wheel is carried behind it on a sort of axletree arm, where our rear foot-board is placed.

"The store waggon (*chariot de batterie*) conveys tools, spare articles, and materials necessary for making repairs.

"The '*chariot de Parc*' is destined to carry reserve ammunition.

"The pole is retained in the draught in preference to the British system of shaft for all carriages.

"Two sorts of projectile, viz., round shot and common case or canister, are fired from guns; shells and case from howitzers; both shot and shells have wooden bottoms; the round shot is attached to the cartridge (*cartouche à boulet*); the shells and common case are separate.

"Two charges are employed for howitzers, a large charge of $\frac{1}{4}$th the weight of the shell, and a small charge of $\frac{1}{12}$th; a small charge is carried for every shell, but only a few large ones, as they are supposed to injure the carriage.

"The shells are packed in the boxes with the fuzes fixed, and the bursting charges in them; the fuzes are calculated so as to explode the shell at the extremity of its flight.

"The cartridges for howitzers have wooden ends (*tampon*); that for the small charge is sufficiently long to fill up the vacant part of the chamber.

"A 12-pounder battery in the French service has thirty carriages, viz., four guns, two howitzers, three ammunition waggons for each piece, two spare gun carriages, two store waggons, and two forges.

"An 8-pounder battery has also thirty carriages, viz., four guns, two howitzers, two ammunition waggons for each piece, two spare gun carriages, two store waggons, two forges, and six small-arm ammunition waggons.

"The proportion of artillery with an army in the field is from 1 to 3 pieces to 1000 men, according to the nature of the country, two-thirds of which are guns, one-third howitzers; one-sixth of the whole are heavy guns.

"From two-thirds to three-fourths of the whole quantity of artillery is intended to be with the divisions of the army, and from one-third to one-fourth in reserve; the quantities of ammunition for each piece, in round numbers, are 200 rounds with the battery, and 200 more in the reserve.

"Six horses are the allowance for guns and waggons, and four for spare carriages."

App. B. NEW SYSTEM OF FIELD ARTILLERY. 545

"The British Field artillery, as at present organized (1854), consists of the following natures and descriptions of ordnance:—

Nature of Gun.	Length.		Weight.	Calibre.	Charge.		Spherical Case.			Common Case.			
							Filled.		No. of Balls.	Filled.		No. of Balls.	Weight of each Ball.
	Ft.	in.	cwt.		lbs.	oz.	lbs.	oz.		lbs.	oz.		oz.
Light 6-pr.	5		6	3.668	1	8	5	7½	27	9		41	3½
9-pounder.	6		13½	4.2	3		8	1½	41	13	2½	41	5
12-pounder medium	6	.66	18	4.62	4		10	13¼	63	16	14	41	6½
5½ in. howitzer *	4	8.6	12½	5·72	2	8	21	4	128	19	8	140	2
4⅖ in. howitzer	3	9.2	6½	4.58	1	4	10	13½	63	11	9	84	2

RANGES of FIELD GUNS.
LIGHT 6-POUNDER.

Solid Shot.		Spherical Case.				Common Case.	
Elevation.	Range.	Letter on Fuze and Length.	Elevation.	Range.		Elevation.	Range.
				From	To		
Degrees.	Yards.	In Tenths.	Degrees.	Yards.	Yards.	Degrees.	Yards.
P B	200		1	380	640	P B	100
¼	300	B .2	1¼	570	800	¼	125
½	400	C .3	1⅝	720	930	½	150
¾	500	D .4	2¼	845	1045	¾	175
1	600	E .5	2⅞	955	1145	1	200
1¼	650	.6	3¾	1060	1240	¼	225
1½	700	.7	4	1160	1330	½	350
1¾	750	.8	4⅝	1255	1415	¾	275
2	800	.9	5¼	1345	1500	2	300
2¼	850	1.0	5⅞	1430	1580		
2½	900	1.1	6⅝	1510	1655		
2¾	950	1.2	7¼	1585	1725		
3	1000	1.3	7⅞	1655	1785		
3¼	1050	1.4	8¼	1720	1840		
3½	1100	1.5	8⅞	1780	1890		
3¾	1150	1.6	9½	1835	1940		
4	1200	1.7	10¼	1885	1980		
4¼	1250	1.8	11	1953	2020		
4½	1300	1.9	11⅝	1980	2055		
4¾	1350	2.0	12⅜	2025	2090		
5	1400						

* In conformity with well established usage and technical exactitude, as stated above (p. 543) with respect to the French service, and in p. 374, in the British service, as well as in all others, the author has called 24 and 12 pounder howitzers—5½ and 4⅖ inch howitzers.

Ranges of Field Guns—*continued.*

9-Pounder.

Solid Shot.		Spherical Case.				Common Case.		Ricochet.		
Elevation.	Range.	Letter and Length of Fuze.	Elevation.	Range. From	Range. To	Elevation.	Range.	Charge.	Elevation.	Range.
Deg.	Yards.	In tenths.	Deg.	Yards.	Yards.	Deg.	Yards.	Oz.	Deg.	Yards.
P B	300	B .2	1⅜	640	920	P B	150	7	5	500
¼	400	C .3	1⅞	800	1060	¼	175	6	6¼	500
½	500	D .4	2¼	930	1180	½	200	5	6¾	500
¾	600	E .5	2¾	1050	1290	¾	225	7	6½	600
1	700	.6	3¼	1160	1390	1	250	6	7½	600
1¼	775	.7	3¾	1260	1480	1¼	275	5	9½	600
1½	850	.8	4⅜	1360	1570	1½	300			
1⅝	925	.9	5⅛	1455	1655					
2	1000	1.0	5¾	1550	1740					
2¼	1050	1.1	6⅜	1640	1820					
2½	1100	1.2	7	1725	1895					
2¾	1150	1.3	7⅝	1805	1965					
3	1200	1.4	8¼	1885	2035					
		1.5	8⅜	1960	2100					
		1.6	9⅜	2030	2160					
		1.7	10	2095	2215					
		1.8	10¾	2165	2275					

Medium 12-Pounder.

Solid Shot.		Spherical Case.				Common Case.		Ricochet.		
Elevation.	Range.	Letter and Length of Fuze.	Elevation.	Range. From	Range. To	Elevation.	Range.	Charge.	Elevation.	Range.
Deg.	Yards.	In.	Deg.	Yards.	Yards.	Deg.	Yards.	Oz.	Deg.	Yards.
P B	300	B .2	1⅜	660	960	P B	150			400
¼	400	C .3	1⅞	820	1110	¼	175			400
½	500	D .4	2¼	960	1230	½	200	6	6½	500
¾	600	E .5	2¾	1080	1340	¾	220	5	7	500
1	700	.6	3¼	1195	1445	1	250			600
1¼	775	.7	3¾	1305	1545	1¼	275			600
1½	850	.8	4⅜	1415	1645	1½	300			
1¾	925	.9	5⅛	1520	1740					
2	1000	1.0	5¾	1620	1830					
2¼	1050	1.1	6⅜	1720	1920					
2½	1100	1.2	7	1815	2005					
2¾	1150	1.3	7⅝	1905	2085					
3	1200	1.4	8¼	1990	2160					
		1.5	8⅞	2070	2230					
		1.6	9⅜	2140	2290					
		1.7		2200	2340					

Ranges of Field Howitzers.

5½ Inch Howitzer.

Common Shells.			Spherical Case.			Common Case.		Ricochet.		
Fuze.	Elevation.	Range.	Fuze.	Elevation.	Range.	Elevation.	Range.	Charge.	Elevation.	Range.
In.	Deg.	Yards.	In.	Deg.	Yards.	Deg.	Yards.	Oz.	Deg.	Yards.
	P B	250	.1	1	450	P B	150	6	7½	400
	¼	300	.1½	1¼	500	¼	175	9	4¾	
	½	350	.2	1½	550	½	200	8	9	500
	¾	400	.2½	1¾	600	¾	225	10	7½	
.1	1	450	.3	2	650	1	250	11	6	
.1½	1¼	500	.3½	2¼	700	1¼	275	11½	5½	
.2	1½	550	.4	2½	750	1½	300	12	5¼	
.2½	1¾	600	.4½	2¾	800	1¾	325	14	5	
.3	2	650	.5	3	850	2	350	9	7¾	600
.3½	2¼	700	.5½	3¼	900	2¼	375	12	6½	
.4	2½	750	.6	3½	950	2½	400	1 lb.	4¾	
.4½	2¾	800	.6½	3¾	1000					
.5	3	850	.7	4	1050					
.5½	3¼	900	.7½	4¼	1100					
.6	3½	950	.8	4½	1125					
.6½	3¾	1000	.8½	4¾	1150					
.7	4	1025	.9	5	1175					
			.9½	5¼	1200					

4⅗ Inch Howitzer.

Common Shells.			Spherical Case.			Common Case.		Ricochet.		
Fuze.	Elevation.	Range.	Fuze.	Elevation.	Range.	Elevation.	Range.	Charge.	Elevation.	Range.
In.	Deg.	Yards.	In.	Deg.	Yards.	Deg.	Yards.	Oz.	Deg.	Yards.
.1	1	400	.1	1¼	450	P B	100			
.1½	1¼	450	.15	1½	500	¼	125			
.2	1½	500	B .2	1¾	550	½	150			
.2½	1¾	550	.25	2	600	¾	175			
.3	2	600	C .3	2¼	650	1	200			
.3½	2¼	650	.35	2½	700	1¼	225	6	7	600
.4	2½	700	D .4	2¾	750	1½	250	8	6	600
.4½	2¾	750	.45	3	800	1¾	275	10	5	600
.5	3	800	E .5	3¼	850	2	300			
.5½	3¼	850	.55	3½	900					
.6	3½	900	.6	3¾	950					
.6½	3¾	950	.65	4	1000					
.7	4	1000	.7	4¼	1025					
.7½	4¼	1025	.75	4½	1050					
.7¾	4½	1050	.775	4¾	1075					
.8	4¾	1075	.8	5	1100					
.8¼	5	1100	.825	5¼	1125					
.8½	5¼	1125	.85	5½	1150					
.8¾	5½	1150	.875	5¾	1175					
.9	5¾	1175	.9	6	1200					

A 6.3 inch howitzer has been adopted in the service, and is to be associated with the medium 12-pounder in batteries of position; but is not included in the above category on account of the great discrepancies in the ranges resulting from the peculiar services as at Shoeburyness, for three successive years, for which reason the printed tables of the Howitzer have not yet been completed.

VIII. On the new Organization of the Personnel of the French Field Artillery.

The new system for the simplification of the French Field Artillery, as described in the previous pages, required corresponding modifications in the ordonnances of 1829, 1833, with respect to the *personnel*, in order to simplify it, and adapt it to the new organization of the matériel. This important proposition was first referred to the consideration of a select committee of Artillery officers; and having been recommended by them, and approved by the minister of war, was carried into effect by decree of the Emperor. That new organization we are now to explain.

During the wars of the Revolution and of the Empire, the *personnel* of the Field Artillery was divided into Foot Artillery, Horse Artillery and Train-d'Artillerie, from which latter were taken the horses and drivers appropriated to the draught of the gun and other carriages; the Drivers being a distinct corps, clad in a different uniform, and paid separately from the gunners, are without any training as gunners, which might qualify them to supply the place of artillery-men killed or wounded in action.

In 1829 the Train-d'Artillerie was re-formed into a new corps composed of Canonniers-conducteurs (Gunner-drivers), who, besides their special duties in the management and driving of horses, were trained sufficiently as Canonnier-servans, to enable them, in case of need, to replace gunners *hors de combat*.

Detachments or squadrons of the corps of Canonniers-conducteurs, required to horse the gun, and other carriages of batteries of Horse or Foot Artillery, were permanently incorporated in those bodies; and the train-d'artillerie, reduced by so much, appropriated to the general services of the park of artillery—to the batteries which move with it, to the transport of stores and provisions, to horse the reserves of every description, and likewise the carriages of the Pontoon Train.

Great advantages resulted from these arrangements, particularly with respect to the Horse Artillery; and, by a subsequent improvement, to the batteries of Foot Artillery, by adapting the carriages to carry, when needful, the artillerymen attached to the service of the guns, by which that large portion of Field Artillery, called

APP. B.   NEW SYSTEM OF FIELD ARTILLERY.   549

batteries montées, has been created; thus placing at the disposal of a general officer possessing a competent knowledge of the powers and general uses of artillery—a qualification which it is essential, and even indispensable, that all general and staff officers should have acquired,[a]—a new branch of that arm, capable of moving, without fatigue to the Gunners, with as much celerity as Cavalry or any Infantry, and capable of being used, according to circumstances, in combination with troops of the other arms, with promptitude and effect, whether in position or in the performance of defensive or offensive movements.

But the organization, and composition of the Regiments of Artillery, still remained in a very defective and complicated state. The whole artillery force was divided into regiments, each consisting of three batteries of Horse Artillery; six batteries montées, and seven batteries of foot; besides a sufficient number of squadrons of the Train-d'Artillerie to *horse* the various carriages belonging to the Horse Artillery, and mounted batteries of each regiment.

Thus the organization of 1833, whilst it remedied some of the inconveniences of the previous system, created others scarcely less serious. All the batteries on a peace establishment to which horses were attached were either Horse, or mounted Artillery; while none were attached to the Foot Artillery—the one being very frequently

---

[a] This is well provided for in the French service. A large proportion of the officers of the permanent staff—the corps d'état major—is taken from the artillery; and no officer is appointed to that corps without having obtained a competent knowledge of that branch of the military art, by having gone through a regular course of instruction at l'École Spéciale d'Artillerie at Metz. There is no exclusion of artillery officers from the command of armies, divisions, or brigades, and we cannot adduce better evidence of the importance attached to this qualification in France, than by quoting the following emphatic passage from the well-known work of a very distinguished artillery officer, himself a proof, amongst many others, of the vast advantages which the French have so largely experienced in employing officers of the artillery in the command of armies:

"Non que je prétende qu'un général d'armée doive s'appesantir sur les détails de l'artillerie; mais il faut, au moins, qu'il en connaisse en grand la théorie et les principes, et surtout qu'il ne soit pas exposé, faute de pouvoir juger des ressources ou des besoins de cette arme, à lui donner des ordres ou inutiles ou inexécutables."

"Qu'on ne dise pas pour dispenser un général d'armée de savoir au moins manier l'artillerie en grand, le jour de bataille, comme je l'exige ici, qu'il suffit que dans toute affaire de quelqu'importance il se concerte avec le général de l'artillerie. Se concerter avec le chef d'une arme, suppose qu'on connaisse la tactique de cette arme; sans quoi c'est le concert de l'aveugle que l'homme éclairé conduit: or l'homme éclairé ose-t-il toujours conduire celui qui ne l'est pas, et qui ne peut ni le juger, ni le disculper si le conseil donné ne réussit pas?"—*L'Espinasse,* p. 87.

converted into the other, as at present with us, for a course of exercise and practice in the field, or for other requirements. Thus, for the exigencies of war, it frequently became necessary to convert a considerable number of batteries of foot into mounted batteries, and to resolve into Foot Artillery, those that had been mounted; so that this liability of the one to become the other, rendered it necessary to give to the officers and under-officers, of both, a competent knowledge of the care, training and managing of horses; which knowledge was no longer required when the battery ceased to be mounted. In 1838 six mounted batteries were transformed into as many permanent foot batteries for service in Algiers, and in 1848 eighteen new foot batteries were formed for the service of the interior.

To remedy the great inconveniences arising from those changes and fluctuations, the new organization of the personnel was formed on the principle, that there should be as many distinct regiments of artillery, as there are specialities in the modes of service. These specialities are:

1. Horse Artillery destined to manœuvre and move with cavalry, or wherever, from the celerity and facility of their movements, they may be found useful.

2. Mounted Batteries destined to move and manœuvre with divisions or brigades of infantry on the field of battle, or to be kept at hand as reserves.

3. Foot Artillery for the general service of the parks, and for batteries of reserve which move with it; for the reserve howitzer batteries; and for the service of siege and other trains.

These three distinct services, instead of being, as heretofore, united in the same regiment, are by the new system resolved into three several elements, by a homogeneous organization which admits of the attention of the officers and soldiers of each special branch of the arm being exclusively devoted to the improvement and efficiency of that particular portion to which they are permanently attached.

In conformity with these principles, the *personnel* of the artillery is divided into distinct regiments, viz.—

    5 regiments of foot artillery,
    1   ,,   of artillery pontonniers,
    7   ,,   of mounted artillery,
    4   ,,   of horse artillery,
    1 cadre de dépôt, monté.

Each regiment of Foot Artillery consists of a regimental staff and of artillerymen sufficient to man

    12 batteries of foot artillery,
    6   ,,   of park artillery,
    1 cadre de dépôt, monté.

The regiment of Artillery Pontonniers consists of
> Staff,
> 12 companies of gunner-pontonniers,
> 4 „ of gunner-drivers and draught horses,
> 1 cadre de dépôt, monté.

Each regiment of Mounted Artillery,
> Staff,
> 15 mounted batteries,
> 1 cadre de dépôt, monté.

Each regiment of Horse Artillery,
> Staff,
> 8 batteries,
> 1 cadre de dépôt, monté.

Thus the total number of batteries on the present establishment (March, 1854) is
> 60 foot batteries,
> 12 companies of gunner-pontonniers,
> 34 „ of gunner-drivers,
> 105 mounted batteries.
> 32 horse artillery batteries,
> 17 cadres de dépôt.

Total 260 batteries, companies, and cadres.

The comparative powers of French and British Field Artillery, anterior to the alterations recently introduced into the French system, being now very greatly changed, a few practical observations on the effects which those alterations produce on the relative powers of the new French system and ours as it exists, are offered, with a view to inquire how far it is expedient to imitate in any degree the principles upon which the new French system is established.

The range of the new French 12-pounder, reduced in weight from 17½ cwt. to 13 cwt., which is very nearly the weight of the British 9-pounder, with a maximum charge of one-fourth of the weight of the shot, is inferior to the range of that gun with a service-charge of one-third the weight of its shot; because, though the charges of one-fourth of the former, and one-third of the latter shot, are equal, the weights projected are as 4 to 3. The momentum of the 12-pounder shot will, however, be greater than that of the 9-pounder, fired with the same charge, in the ratio of their respective weights; and the penetrating and battering powers lie likewise in favour of the new French 12-pounder. The 8-pounder, bored up to the calibre of 12, will no longer be a match, in any respect, for the British 9-pounder. If, as has been proposed, the British 9-pounder were bored up to the calibre of 12, in imitation of the enlargement of its equivalent, the French 8-pounder, 75 lbs. of metal must be taken out, and our gun would lose in power of range, a faculty that would not be compensated by what it would gain in momentum and penetrating power for general service in the field, or by its additional capacity for shell firing. The British 9-pounder is a capital and efficient field-service gun; and, relatively with the reduced powers of the new French 12-pounder, and of the bored-up 8-pounder, the British system of Field artillery is comparatively raised—the British 9-pounder, from inferiority, to superiority over the French 12-pounder, and the French 8-pounder, bored up to 12, is no longer a match, in range, for that gun. The French 12-pounder has greater capacity for firing shells; but when shell-firing is required, special batteries of howitzers should be employed. Mixed armaments in the same battery, and consequently complications of ammunition, should be avoided. The author fully concurs with what is so well stated above, page 538—that in a mixed battery of guns and howitzers, the functions, effects, and powers of one of the two are sacrificed to the other.

With respect to the new organization of the *personnel* of the French artillery, and their reasons for having abandoned the former system, there may be found perhaps something worthy of imitation. That some such inconveniences as those which induced that change, must have been experienced in our present practice of appropriating by temporary arrangements portions of one vast universal corps to the various duties of artillery, may be inferred from the various propositions that have been made to divide our establishment of artillery into its main branches—garrison artillery, foot artillery, field artillery, horse artillery, and gunner drivers—as the French have done expressly to avoid the evils of one universal system.

When the British artillery consisted of only four battalions, and the system of field artillery was in a very simple and defective condition (1792), when there were no field batteries, but pairs of battalion guns, drawn by teams of contract horses driven by hired drivers, no horse artillery, and when the general duties of artillerymen were far less multifarious and intricate than they are now, it was perhaps possible to train all artillerymen to a degree of capability which might enable them to perform either of those special duties, to which they might be posted, in the comparatively rude manner of that time. But now, with one vast corps, consisting of 12,000 or 13,000 men, and with the present greatly improved condition of field and other artillery, the requirements of the service appear to have outgrown that system. It would be easy to show, for instance, that the frequent conversions of a battery of foot artillery, unhorsed, into a field battery to which horses are

NEW SYSTEM OF FIELD ARTILLERY.

attached, and the reverse; the constant transformation of the one into the other for exercise and drill, for the purpose of instructing artillerymen to discharge either, or all of their duties, including those of riding and driving, must render the men less expert in those duties than if they remained permanently attached to the special service in which they had been made perfect. This routine system is undesirable, inconvenient, and objectionable in peace; it must evidently be more so in a state of war; and, though this may not appear in the condition of the artillery sent to the East, from the admirable manner in which it was fitted for service, yet it may be suspected that its efficiency has been obtained at some sacrifice of the effective state of the corps from which the draughts have been made; and that, in the prosecution of the war upon a large scale, the inconvenience of one general system will be severely felt.

The author is well aware of the difficulties that beset this question, and that much may be said on both sides; yet it does appear to him that something in the direction of the French system of the *personnel* might be done to avoid at least some of the inconveniences of having but one general, instead of several special branches, of the artillery corps.

The French system of mounting the gunners on the carriages, in every movement, whatever the movement may be, as regularly as horse artillerymen mount their horses, in manœuvring with troops, is attended with great advantages. We have our mounted batteries likewise, but the men are not carried on the waggons, excepting on special occasions, when some great object is to be attained by a movement which would exhaust the artillerymen did they march on foot. The main consideration in the marches and movements of artillery is to spare as much as possible the horses, by not overloading the carriages, and thus increasing the draught, in moving across a country, by the greater penetration of the wheels into soft ground. According to the French system, the waggon that carries the artillerymen must always be in close attendance on the gun, and is brought up with it for action; it is, in consequence, equally exposed to the fire of the enemy.

The system of mounted batteries is not new. The Austrians had their Wurst waggon; we had our car-brigades fifty years ago. In both, the gunners were mounted on the carriages in all movements; but, in our present field batteries, the gunners are never mounted on the bodies of the waggons or on the limbers of both gun and waggon, except on special occasions, when great objects are at stake, in order to get up batteries for action with their gunners fresh and unfatigued, when that can be done without overworking the horses. The waggons are never brought up with the gun into action, but are left, if possible, out of sight, under shelter of any slope or hollow ground, immediately in the rear.

In the French service the bridge department is attached to the artillery; with us it is assigned to the engineers: this, perhaps, is immaterial. Both corps are highly competent to the charge; but it is very material that the equipages of the pontoon train should be permanently and efficiently equipped with horses and drivers, other than artillery soldiers, before they are required for service in the field; and, from this not having been done, serious disadvantages and difficulties must have been experienced, when, as in the camp at Chobham, artillery horses driven by artillery gunners were taken from the field batteries.

The French *Train-d'Artillerie* resembles, in many respects, the corps of Gunner-Drivers which we had during the late war for the general uses of the park, and for the service of batteries of Foot Artillery; the Horse Artillery having, incorporated with each troop, drivers sufficient for the draught horses of all the carriages belonging to it.

At the conclusion of the war, the whole of the Gunner-Driver Corps in the British service was reduced; not even a cadre or nucleus was retained; and the very expensive, and, as we shall now in war find, objectionable and inconvenient expedient was adopted, of making well-trained gunners, of whom we have by no means sufficient, lay down the sponge and the rammer, and take up the rein and the whip.

A very reasonable apology for the adoption of this temporary expedient appeared in the difficulty there would otherwise be in providing drivers for field artillery in the colonies; and the measure was adopted accordingly in the British Artillery, as one of economy and necessity in time of peace. In order to carry out this arrangement effectually it would have been necessary that all gunners should be taught to ride and drive, so that they might have been taken promiscuously from their particular duties for the purpose of acting as drivers; but the practical working of the measure consists in assigning that service to men who had been accustomed from youth to the management of horses; and who, in other respects, were the least qualified for the duty of gunners. Thus, in every battery, there exists a practical distinction between those who perform the important duties of gunners and those who execute the secondary service of drivers, which strongly indicates the necessity of making the distinction of duties systematic.

The full complement of gunners to a 9-pounder field battery, consisting of 6 guns at 8 men for each, is 48: each gun being drawn by three pairs of horses, each of the six ammunition waggons by two pairs of horses, and the other six carriages by the like number, 42 gunners are appropriated to the duties of drivers in every battery, so that there are nearly as many artillerymen acting as drivers as of those serving the guns. Taking into account the number of batteries now horsed for active service in the field, at home and abroad, and for the various services of the park of artillery, as well as for pontoons and bridge equipments, for which no special provision is made, we see that a very large portion of artillerymen are abstracted from their special duties, to act as drivers.

Whenever, therefore, artillerymen are required, as at present, for garrison, field, and coast battery service at home, and for all artillery services in the colonies, and for the war in which we are engaged, a large portion of well-trained gunners may be immediately made available by abandoning this objectionable system, and reverting to the institution of a special corps of Gunner-Drivers. The race of post-boys is not yet extinct, and there would be no difficulty in obtaining in a very short time any number of young active men of light weight, habituated in various ways to the management of horses, and who would soon become expert military drivers; whereas, if the present system remain in force, the whole corps will have to be augmented, not so much on account of want of gunners,—for there are nearly as many of these employed as drivers as in serving the guns,—but to train recruits to serve in both capacities, at the sacrifice of much time and cost.

The French manage this matter much better. They train drivers to act when needful as gunners, not gunners to act as drivers; and assuredly they are right: for besides that more horses are disabled than drivers, in action, and that some of these soon become available, they fill up vacancies in the higher and more important class of soldiers from an inferior class, instead of abstracting from the efficiency of the artillery corps in general, and more especially from the service of the guns in the field, by calling upon artillerymen to act in the inferior capacity.

The comparative advantages of having, or not having, a special corps composed of drivers and draught-horses for the service of field batteries, and of the Park in general; also *équipages militaires* for the transport of provisions, ammunition, baggage, and medical stores; hospital conveyance and ambulances for the wounded, as in the French service, may be perhaps painfully contrasted with the serious consequences of the utter want of such establishments in the British service. In the course of our dear bought experience during the late war, after struggling with many difficulties, we experienced the necessity of creating and well providing for all these important services. We had then an extensive corps of gunner-drivers; we had also our waggon corps for the transport of baggage; our ambulance and hospital conveyance; our staff-corps, and our bridge department. Not a nucleus of these existed when, suddenly, we were caught in a political storm, and forced into a mighty war, very inadequately provided with the means which are absolutely indispensable to enable an army to take the field.

(C.)

ABSTRACT OF THE FRENCH NAVY.

THE following Table exhibits the number of ships and vessels in the sailing and steam navies of France in 1851: it is extracted from the 'Etat Général de la Marine' for that year, and from other authentic sources of information.

	Afloat.	Being built.
SAILING NAVY.		
Ships of the line of the four classes	25	21
Frigates of the three classes	37	19
Corvettes of the two classes	30	5
Brigs of the two classes	44	4
Small vessels	43	—
Transports	32	—
STEAM NAVY.		
Ships of the line, with screws	1	2
Frigates (450 to 650 H. P.)	20	—
Batimens mixtes	4	—
Corvettes	27	5
Steam Packets	57	3

The decrease in the number of ships of the line in the French navy, compared with those of which it has been composed in all times since France became a great naval power, and the large disproportion between the number of line-of-battle ships and other vessels shown in the preceding return, were circumstances not fortuitous, but the result of a deliberate and long-considered change of system in future naval operations. While relaxing so much of their wonted exertions in constructing ships of the line, the French have greatly increased the number of their frigates, corvettes, and smaller vessels, and have created an extensive steam navy.

What that change of system was to meet may be gathered from various works published at that time. It was the adoption of a desultory system of naval warfare, as recommended to Napoleon in 1805 by his Minister of Marine, Admiral Decrès,[*] by sending out numerous small active squadrons, or quick-sailing frigates and other vessels singly, to act against our commerce, and the remote possessions of our widely extended empire, to attract thither as much of our naval force as might be required to protect and defend those possessions; and taking advantage of

---

[*] Thiers' History of the Consulate and Empire, vol. v. pp. 484-486.

this diversion, to act vigorously against our coasts and trade in the narrow seas. This was acted upon, after the defeat off Trafalgar, with a considerable degree of success in the Eastern seas,[a] the Mediterranean,[b] and elsewhere; and France, with a very much increased force of frigates, corvettes, and small vessels, and a steam navy of upwards of 100 ships and vessels, possessed the means of rendering that system more formidable than heretofore: we have seen that a desultory, corsair-like system of naval warfare has recently been urged by an illustrious personage[c] and other professional men[d] in France, in preference to acting in large fleets, and fighting general actions with a navy which experience has long shown they could not defeat. This, in fact, is the origin and essence of the Paixhans' system.

In the Notes on ' *L' Enquête Parlementaire sur la Situation et l'Organisation des Services de la Marine Militaire,*' it is stated

---

[a] James's Naval History, vol. iv. p. 213; vol. v. pp. 19, 146.
[b] James's Naval History, vol. v. pp. 4, 7, et seq.—Gantheaume's Squadron.
[c] " By the plan which I shall develope by and bye, we could act efficiently against her commerce, spread over the whole surface of the seas. . . . . We should possess, at all points of the globe, divisions of frigates ready to follow in the steps of those glorious squadrons which so nobly struggled for the country in the Indian seas: they would sail around our colonies; around those new points seized on the distant ocean by our thoughtful policy, destined to serve as a basis for their operations, as well as to become the asylum of our privateers."

. . . .

" It is an incontestable fact, that although during twenty years the war of squadron against squadron was almost always against us, the exertions of our privateers were almost ever crowned with success. Towards the end of the Empire divisions of frigates sailing from our ports with the mission of skimming the sea, without compromising themselves uselessly against an enemy superior in number, have inflicted considerable losses on the English commerce; harassing their commerce is touching the vital principles of England— 'tis tearing their heartstrings." . . . . " To attain this end France might establish well-organised privateers in all the quarters of the globe. In the Channel and in the Mediterranean this service might be well granted to steam vessels: those which are employed as packets during peace would form, from their extreme swiftness, excellent privateers in time of war!"—*Prince de Joinville:* translation by W. Peake, Esq. Parker's Military Library, Whitehall, 1844.

[d] " J'ai dit plus haut qu'il nous importait de rechercher comment une Marine numériquement inférieure pouvait soutenir une lutte inégale. Je ne connais point d'autres moyens d'atteindre ce but que ceux que je viens d'indiquer. Armer à l'avance, faire peu pour faire bien, ne point nous préoccuper du nombre des navires que nous enverrons à la mer, mais de la manière dont ils s'y présenteront."—De la Gravière, *Guerres Maritimes,* tom. ii. p. 266.

" Les Américains ont, dans cette guerre (1812), montré beaucoup d'habileté et de résolution. . . . . C'est à nous qu'il appartient de juger avec impartialité ces événements maritimes trop exploités peut-être par un orgueil national que l'on est tenté d'excuser."—De la Gravière, *Guerres Maritimes,* tom. ii. p. 273.

that, in the sitting of the 3rd of February, 1851, it was carried unanimously that the number of ships of the line afloat, which by the decree of 1846 was fixed at 24, should be increased to 30, and always in a condition fit for immediate service ; some to form the squadron or squadrons of evolution, a considerable number in *Commission de Port*, and all the rest ready to be immediately armed and manned. It was likewise carried that the squadron or squadrons of evolution should consist of 10 sail of the line, either acting together or in separate squadrons, with a certain number of frigates and war-steamers as circumstances might require.

It was likewise decreed that no more ships of 100 guns should be built, for that their force in number of guns, compared with that of the 90-gun ship on two decks, and with that of three-deck ships, is not proportioned to the difference of expense between the 90 and the 100-gun ship, nor to the difference of cost between the latter and a ship of the first class.

In the sitting of the 7th of February, 1851, it was decided unanimously that steam propulsion, by the screw, should be applied as soon as possible to all line-of-battle ships ; that, of the 27 ships then afloat, such only should undergo the necessary alteration and transformation as were in so good a state as not to require repairs exceeding $1\frac{1}{4}$ *degrees* towards completion; that, of the rest, 13 should remain sailing vessels, their age and condition being such that the expense of transformation would not be compensated by their ulterior services; that these, therefore, should be fitted for, and continued to be used for service in peace, whilst new vessels, adapted to screw propulsion, should be constructed, to attain the number of 30, fixed by the Commission.

It was accordingly resolved, that of the ships of the line afloat, 12 should be immediately adapted to receive *moteurs à vapeur* (screw propellers).

The number of vessels *en chantier* (on the stocks) is proposed to be reduced to 15 ; and, in conformity with the decree of the 7th of July, 1851, "that steam propulsion should be applied to all ships of the line," it was decided that the 15 *en chantier* should undergo the necessary transformation, as has been done with respect to the "Turenne," at Rochfort, laid down in 1827. It was likewise decided that the 15 vessels in course of construction should severally be so far advanced as to be brought forward for launching in succession, at the rate of, at least, two in every year in peace ; but at the same time to bring the greater number to such a state of advancement as to be able to provide much more rapidly for any great exigency.

1. The "Austerlitz" and the "Jean Bart," ships of the third class now being constructed, and which were to be launched in 1851.

2. The "Charlemagne," third class, which has been just launched, and which is provided with a steam engine: lastly, the steamer "Napoléon," now being armed at Toulon.

The fleet would then be, provisionally, composed of 13 sailing ships, 16 auxiliary screw ships, and 1 ship of superior power.

In the sitting of the 12th of February, 1851, it was resolved unanimously that 20 frigates of the first class should be fitted for, and provided with steam power *à grande vitesse;* others transformed into *batiments mixtes* (auxiliary screws), giving sufficient speed " pour escorter les navires rapides qui pourraient transporter des troupes de débarquement."

In the sitting of the 7th of March it was resolved unanimously that the corvette force should consist of 50, and that steam propulsion, *à grande vitesse*, should be applied as soon as possible to vessels of that class; that sailing transports for the conveyance of troops should be abolished; and that twenty steam-propelled transports of large size, each adequate to carry 1000 men, should be provided to get rid of the difficulty, uncertainty, and risk of taking and keeping in tow, under fire, transports containing troops " de débarquement."

The "Napoléon" was lengthened at Toulon, in order to increase her displacement sufficiently for the additional weight she would have to carry, without prejudice to the conditions on which speed depends. She has eight boilers, and her engines work up to 1500 horse power. She carries coal sufficient for ten days' steaming at full speed; and, in her trial voyage from Toulon to Ajaccio, her speed was 12 knots. This vessel has answered admirably, and is the type on which others will be constructed. The great use of these vast ships is defined to be to escort *convois à vapeur* (steam transports) conveying troops, with all their material of war, and protect their debarkation against any force whether on the sea or on the coast.—(Report of M. Daru, vol. ii. p. 141.)

The French navy is at present (1854) composed as follows:—

Sailing Navy.

Ships of the Line.

9 of 120 guns, carrying	1,080	guns.
14 of 100 ,, ,,	1,400	,,
13 of 90 ,, ,,	1,710	,,
11 from 86 to 82 guns, carrying	914	,,
Total 53 ships.	5,096	guns.

Frigates.

42 from 60 to 50 guns, carrying	2,286	guns.
16 from 46 to 40 ,, ,,	670	,,
Total 58 frigates.	3955	guns.

Corvettes.
39 from 30 to 14 guns, carrying . . . . . 868 guns.
Brigs, Schooners, and Cutters.
101 from 20 to 4 guns, carrying . . . . . 1,066 guns.
Transport Corvettes, Lighters, &c.
39, carrying together 788 guns, and measuring 18,500 tons.

STEAM NAVY.

3 Ships of the Line, 20 Frigates, 30 Corvettes, and 64 Avisos, representing a power of 28,750 horses.

In the above return are included several ships still on the stocks.

The following decree, issued in 1849, will fully explain the present armament of the French navy.

In the name of the French People,

THE PRESIDENT OF THE REPUBLIC,

On the Report of the Minister of the Marine and Colonies, taking into consideration the deliberation of the Council on Naval Works, dated April 4, 1849; also the deliberation of the Council on Naval Affairs, dated June 18–21, 1849,

Decreed :—

ART. 1.

The Armament of the ships hereafter indicated shall henceforth be as follows :—

Ships of the First Class.

		Guns.
Lower Deck . .	4 howitzers of 22 centimetres, No. 1; 6 50-pr. guns;[a] 22 30-pr. guns, No. 1	32
Middle Deck . .	6 howitzers of 22 cen., No. 2; 28 30-pr. guns, No. 2 . .	34
Main Deck . .	34 30-pr. guns, No. 3	34
Forecastle and Quarter Deck	12 30-pr. guns, No. 4	12
	Total	112

Ships of the Second Class.

		Guns.
Lower Deck . .	4 howitzers of 22 cen., No. 1; 6 50-pr. guns;[a] 22 30-pr. guns, No. 1	32
Main Deck . .	6 howitzers of 22 cen., No. 2; 28 30-pr. guns, No. 2 . .	34
Forecastle and Quarter Deck	24 30-pr. guns, No. 3	24
	Total	90

---

[a] The 50-pr. guns have since been withdrawn from the broadside batteries of line-of-battle ships and frigates, in conformity with the distribution in the " Enquête Parlementaire," and replaced by 30-pr. guns, No. 1. For the dimensions, weight, &c. of French naval ordnance, and the ranges, compared with those of British naval ordnance, see Tables XVIII. and XX. Appendix D.

Ships of the Third Class (New Model).

		Guns.
Lower Deck	4 howitzers of 22 cen., No. 1; 6 50-pr. guns; 20 30-pr. guns, No. 1	30
Main Deck	6 howitzers of 22 cen., No. 2; 26 30-pr. guns, No. 2	32
Forecastle and Quarter Deck	20 30-pr. guns, No. 3	20
	Total	82

Ships of the Third Class (Old Model, called Model of 86).

Lower Deck	4 howitzers of 22 cen., No. 1; 26 30-pr. guns, No. 1	30
Main Deck	4 howitzers of 22 cen., No. 2; 28 30-pr. guns, No. 2	32
Forecastle and Quarter Deck	18 30-pr. guns, No. 4	18
	Total	80

Ships of the Fourth Class (New Model).

Lower Deck	4 howitzers of 22 cen., No. 1; 4 50-pr. guns; 4 30-pr. guns, No. 1	28
Main Deck	4 howitzers of 22 cen., No. 2; 26 30-pr. guns, No. 2	30
Forecastle and Quarter Deck	16 30-pr. guns, No. 4	16
	Total	74

Ships of the Fourth Class (Old Model, called 74).

Lower Deck	4 howitzers of 22 cen., No. 1; 24 36-pounders	28
Main Deck	30 long 18-pr. guns	30
Forecastle and Quarter Deck	12 30-pr. guns, No. 4	12
	Total	70

Frigates of the First Class.

Main Deck	2 howitzers of 22 cen., No. 1; 2 50-pr. guns; 26 30-pr. guns, No. 1	30
Forecastle and Quarter Deck	2 30-pr. guns, No. 1; 18 30-pr. guns, No. 3	20
	Total	50

Razés.

Main Deck	2 howitzers of 22 cen., No. 1; 2 50-pr. guns; 24 30-pr. guns, No. 1	28
Forecastle and Quarter Deck	2 30-pr. guns, No. 1; 20 30-pr. guns, No. 3	22
	Total	50

Frigates of the Second Class.

Main Deck	2 howitzers of 22 cen., No. 2; 2 50-pr. guns; 24 30-pr. guns, No. 2	28
Forecastle and Quarter Deck	2 30-pr. guns, No. 1; 16 30-pr. guns, No. 4	18
	Total	46

Frigates of the Third Class (New Model).

Main Deck	2 howitzers of 22 cen.; 2 50-pr. guns; 22 30-pr. guns, No. 2	26
Forecastle and Quarter Deck	2 30-pr. guns, No. 1; 12 30-pr. guns, No. 4	14
	Total	40

	Afloat and on the Stocks.	Guns.
Main Deck	2 50-pr. guns, or 2 howitzers of 22 cen., No. 2; and 24 30-pr. guns, No. 2	26
Forecastle and Quarter Deck	2 30-pr. guns, No. 1; 8 30-pr. guns, No. 4	10
	Total	36

Corvettes (New Model).

		Guns.
Main Deck	2 howitzers of 22 cn., No. 2; 2 30-pr. guns, No. 2;[a] 14 30-pr. guns, No. 3	18
Forecastle and Quarter Deck	2 30-pr. guns, No. 1	2
	Total	20

Corvettes à barbette (New Model).

Forecastle and Quarter Deck	2 30-pr. guns, No. 2; 14 30-pr. guns, No. 3 . . . Total	16

Brigs of the First Class (New Model).

Forecastle and Quarter Deck	12 30-pr. guns, No. 4 . . . . Total	12

ART. 2.

The artillery of the other ships of the fleet will remain as was determined by the ordonnances of the 1st Feb. 1837, and the 22nd Nov. 1846.

ART. 3.

The decree of the 20th July, 1848, remains in force.

ART. 4.

The Minister of the Marine and Colonies is charged with the execution of this decree.

Done at Paris, at the Elysée-National, the 27th of July, 1849.

(Signed)   L. N. BONAPARTE.

The Minister of the Marine and Colonies.

(Signed)   V. TRACY.

The first idea of having an inclined plane to turn about a longitudinal axis, for the purpose of giving motion to a ship in the water, is due to Daniel Bernouilli, who investigated the subject in a *Mémoire* presented to the French Academy in 1753; and from the principle there enunciated has arisen, by successive developments, the powerful agent now so much employed in the warlike and commercial navies of Europe.

Down to the year 1845 ships of war drew comparatively little advantage from steam as a moving power, the paddle-wheels till then used preventing the broadsides of the ships from receiving the armament which would place them on an equality, in this respect, with sailing vessels. In the summer of 1850 the French launched at Toulon their first great steam-ship of war, the "Na-

---

[a] These four pieces of ordnance are placed at the middle of the side.

poléon:" it is propelled by a screw with four vanes, whose pitch is 10 yards, and it carries 92 guns or obusiers. The screw, which is wholly under water, can be detached from the machinery, and, without raising it out of the water, the impelling power of the steam may be exchanged for that of the wind in the sails, the resistance opposed by the screw in this state to the motion of the ship being inconsiderable; also, the disturbance produced in the water by the revolutions of the screw is found to have no effect on the action of the rudder. The power of the engine is computed to be equal to that of 900 horses, and the maximum velocity attained by the ship, when impelled by steam, is $13\frac{1}{2}$ knots per hour. The "Napoléon" was found capable of towing three ships at once, one of them a three-decker, at the rate of more than 5 knots per hour. By experiment it was ascertained that, at each revolution of the screw, the ship advanced about $9\frac{1}{2}$ yards (8.6 mètres).—(From the *Rapport* to the Académie des Sciences, by Baron Charles Dupin.)

In England, the screw was first applied to ships of war in 1841; and the "Rattler," which was built at Sheerness, and launched in 1843, was furnished with that means of propulsion.

The following account of the Russian navy at the present time cannot but be interesting.

The Russian Naval force in the Black Sea, in May, 1854, consisted of—

Line-of-Battle ships—6 of 120 guns each, and 8 of 84 guns each.
Frigates—7 carrying from 44 to 60 guns each.
Corvettes—5 carrying from 18 to 20 guns each.
Brigs—12 carrying from 12 to 18 guns each.
Schooners—8 carrying from 8 to 16 guns each.
Cutters—7 carrying from 10 to 12 guns each.
Steamers—28 varying from 40 to 400-horse power.
Besides *Gun-boats, Transports,* &c.

It is reported that 4 sail of the line are now being built and fitted for screws.

In the Baltic the Russians have at present in an efficient state about 20 line-of-battle ships, two of which carry 120 guns, and two 112 guns each; the rest are 84 and 74-gun ships; they have six frigates of 44 guns, with five brigs and corvettes of 20 guns each. All these are sailing ships, and they have, besides, schooners, transports, and luggers, with about fifty gun-boats. The steam navy in the Baltic consists of ten paddle-wheel vessels carrying from 6 to 16 guns each, and varying in power from 200 to 450 horses.

The Russian ships are not only well provided with shells, in the use of which their gunners are highly efficient, but are likewise heavily armed with solid shot; and in the affair at Sinope they fired with great effect both 68 and 42 pounder solid shot, many of which were picked up on that coast. The largest solid shot gun in the British navy is the 32-pounder, which is about equivalent to the 30-pounder, the general gun in the armament of French ships of war (the use of the 36-pounder having been abolished by a regulation, in ships of the new model), while our 8-inch shell-gun is superior to the canon-obusier of 80 : but of what avail are 32-pounder shot and 8-inch shells against the 42-pounder and 68-pounder solid shot of the Russian ships, or against the heavy ordnance of the Russian batteries, whose parapets of stone will cause hollow shot to fly in pieces before them ?

(D.)

TABLE I.

Ranges obtained at Deal, in the Year 1839, with the undermentioned natures of Ordnance.
(Below the corresponding Elevations.)

Nature of Ordnance.	Length.			Weights.			Windage.	Charge.		Height of the Gun above the plane on which the Shot fell.	Point blank.	¼ degree.	½ degree.	¾ degree.	1 degree.	1½ degree.	2 degrees.	3 degrees.	4 degrees.	5 degrees.	7 degrees.	9 degrees.	12 degrees.	15 degrees.
	ft.	in.		Cwt.	qrs.	lbs.	Inch.	lbs.	oz.	Feet.	Yards.	Yards.	Yards.	Yards.	Yards.	Yards.	Yards.	Yards.	Yards.	Yards.	Yards.	Yards.	Yards.	Yards.
32-pounder (A.)	9	6		49	1	18	.198	8	0	22	556	632	701	793	862	1007	1163	1443	1662	1939	2231	2546	3092	3513
,, (B.)	8	6		44	2	0	.175	7	0	22	552	625	638	773	851	1006	1124	1440	1716	1939	2242	2498	3175	3391
,, (C.)	8	6		40	3	22	.175	6	0	22	525	656	677	794	905	1022	1185	1476	1722	1894	2305	2514	3083	3546
,,	8	0		31	1	24	.125	5	0	22	513	586	655	692	774	888	1101	1382	1593	1738	2155	2566	2852	3193
,,	6	0		24	3	14	.125	4	0	22	469	523	607	621	714	812	953	1206	1458	1636	1966	2383	2868	3152
,, *	6	0		63	3	7	.233	12	0	22							1366	1581	1832	1998	2318	2682	3050	3359
,,	6	7						10	11	22							1365	1657	1775	1964	2275	2584	3033	3469
,,	6	7						10	8	22							1252							
10-inch	8	0		49	1	0	.233	8	0	22							1183	1466	1607	1835	2372	2489	2919	3384
8-inch	9	4		85	1	7	.15	12	0	8½	312	391	478	568	637	838	1033	1252	1489	1642	2097	2579	3028	3546
10-inch Howitz.	5	0		65	0	14	.125	10	0	22							1133	1323	1602	1920	2248	2577	3016	3360
8-inch Howitz.	4	0		42	0	14	.16	7	0	15†	323	409	422	496	511	646	709	934	1073	1270	1663	2133	2402	2554
32-pounder	6	6		21	3	0	.16	4	0	15†	330	392	450	479	513	620	706	907	1186	1309	1553	1810	2231	2492
,,				32	0	14	.175	5	6	22							948	1254		1817				
,,	8	6		46	2	11	.195	6	0	22					705		1053	1337		1753				3585
,,	9	0		51	0	14	.215	7	0	22					734		1151	1435		1953				3634
,,	9	6		61	2	12	.175	8	0	22							1233	1410		1877				3499
,,	9	6		61	2	14	.2	10	0	22					942		1252	1564		1978				
,,								10	0	22					926		1175	1437		2017				3444

* With this gun, the elevation being 30°, the range was 4785 yards; the deflection 154¾ yards.
    ,, 31 ,, ,, 4765   ,,   ,,   154¾
    ,, 32 ,, ,, 4860   ,,   ,,   233¾
    ,, 33 ,, ,, 4795   ,,   ,,   3   At this round the gun burst.

† This height was continued till the elevation was 2°, at which elevation, and those which follow, the height was 22 feet.

TABLE II.

Ranges obtained at Deal, in the Year 1839, with the undermentioned Natures of Ordnance.

(Below the corresponding Elevations.)

Nature of Ordnance.	Length.		Weight.			Windage.	Charge.		Height above Plane.	Point Blank.	¼ degree.	½ degree.	¾ degree.	1 degree.	1¼ degree.	Remarks.
	Ft.	in.	Cwt.	qrs.	lbs.	Inches.	lbs.	oz.	Feet.	Yards.	Yards.	Yards.	Yards.	Yards.	Yards.	The ranges here stated are the averages of a certain number of rounds, varying from 3 to 5; the times of flight, which are not given in this Table, were taken throughout the practice by the same officer, Lieut. Morris, with the instrument belonging to the department of the Inspector of Artillery, which measures seconds and fortieths of seconds; one officer, Lt. Mitchell, also laid every gun—he allowed for the wind so as to bring the shot as near into the range as he could; his constant practice rendered his judgment very mature; and one officer, Lieut. Tyler, measured every first graze.
32-pr..	9	0	49	1	18	.198	8	0	22	556	632	701	793	862	1007	
									15	490	564	671	697	772	1047	
									8	395	490	600	673	814	983	
32-pr..	8	6	44	2	0	.175	7	0	22	552	625	638	773	851	1006	
									15	499	552	630	708	815	969	
									8	363	473	559	638	766	930	
32-pr..	8	0	40	3	22	.175	6	0	22	525	656	677	794	905	1022	
									15	430	524	579	667	734	909	
									8	315	427	557	612	713	890	
32-pr..	6	6	32	1	24	.125	5	0	22	513	585	655	692	774	888	
									15	411	491	561	635	732	872	
									8	308	412	476	586	664	777	
32-pr..	6	0	24	3	14	.125	4	0	22	469	523	607	621	714	812	
									15	381	464	550	600	674	776	
									8	273	373	452	516	596	740	
32-pr..	9	6	56	0	0	.233	10	0	15	475	564	671	749	877	996	
									8	380	491	603	627	793	951	
32-pr..	8	0	42	1	0	.233	8	0	15	446	525	629	718	813	1016	
									8	364	445	545	641	780	980	

RANGES OBTAINED AT DEAL.

Table III.
Ranges of different 32-pounder Guns and an 8-inch Gun. From Experiments at Deal, in 1839.
(Below the corresponding Elevations.)

Nature of Gun	Length	Weight	Windage	Charge	Height of the Gun above the Plane	Point Blank	¼ degree	½ degree	¾ degree	1 degree	1½ degree	2 degrees	3 degrees	5 degrees
	Ft. in.	Cwt. qrs. lbs.	Inch.	lbs. oz.	Feet.	Yards.	Yards.	Yards.	Yards.	Yards.	Yards.	Yards.	Yards.	Yards.
32-pounder	9 6	56 0 0	.233	10 0	...	...	...	...	...	877	...	1211	1467	1904
	9 6	49 3 14	.175	8 0	...	...	...	...	...	875	...	1244	1483	1973
	9 6	32 0 14	.175	5 0	...	...	...	...	...	705	...	948	1254	1817
	6 6	32 1 24	.125	5 0	...	...	...	...	...	732	...	977	1303	1840
	9 0	49 3 14	.175	6 0	...	...	...	...	...	853	...	1080	1425	1893
	9 0	49 1 18	.175	6 0	...	...	...	...	...	717	...	1087	1276	1877
	8 6	44 2 0	.175	6 0	...	...	...	...	...	767	...	1139	1374	1899
	8 6	40 3 22	.175	6 0	...	...	...	...	...	734	...	999	1317	1720
	7 6	39 1 2	.175	6 0	...	...	...	...	...	782	...	1130	1461	1903
	6 6	32 2 14	.175	6 0	...	...	...	...	...	734	...	1053	1337	1753
	6 0	49 1 0	.233	8 0	...	...	...	...	...	813	...	1181	1436	1907
	8 0	40 3 22	.175	6 8	...	...	...	...	...	756	...	1068	1327	1900
8-inch	8 0			8 0	15	401	478	548	692	749	818	...	...	...
	8 0		.233	8 0	8	284	391	492	568	665	...	...	...	...
	8 0		.175	10 0	15	474	485	566	717	805	888	...	...	...
	8 0			10 0	8	302	438	531	587	698	...	...	...	...

Table IV.
Comparative Ranges of 42 and 56-pounder Guns. From Experiments at Deal, in 1839.
(Below the corresponding Elevations.)

Nature of Gun	Length	Weight	Windage	Charge	Height of the Gun above the Plane	2 degrees	3 degrees	4 degrees	5 degrees	6 degrees	8 degrees	10 degrees	12 degrees	15 degrees	20 degrees	30 degrees	31 degrees	32 degrees	33 degrees	34 degrees	35 degrees	Remarks.
	Ft. in.	Cwt. qrs. lbs.	Inch.	lbs. oz.	Feet.	Yds.	Yds.	Yds.	Yds.	Yds.	Yds.	Yds.	Yds.	Yds.	Yds.	Yds.	Yds.	Yds.	Yds.	Yds.	Yds.	
42-pr.	10 6	80 3 0	.175	14 0	22	1346	1605	1842	2086	2381	2735	3080	3428	3732	4381	5446	5285	6607	5432	5392	5437	Solid shot.
				14 0	22	1321								3867				5720*			5200	Shell filled with lead.
				14 0	22									4087							5600	Solid shot.
56-pr.	11 0	97 2 26	.175	15 0	22	1394								4001								Ditto.
				17 0	22	1330	1743	1986	2261	2523	2958	3118	3465	4029								Ditto.
				17 0	22	1391																Shell filled with lead.

* The longest round of this average.

Table V.

Ranges with Sea-service Iron Ordnance, Single-shotted, obtained on board H.M.S. "Excellent." Elevations by Spirit-level in 1839.

Nature of Guns.	Weight.	Length.	Diameter of Bore.	Charge.	Height above the Plane.	Ranges in Yards, with corresponding Times of Flight.																	Remarks.
						⅜	1	2	3	4	5	6	7	8°	9°	10	11	12	13	14	15	16	
Prs.	Cwt.	Ft. in.	In.	lbs.	Ft. in.																		
10-inch	84	9 4	10	12	5 4	300 ¾'	600 2"	880 3¼'	1180 4"	1440 4¾'	1670 5¾'	1800 6½'	1960 7¼'	2100 8¼'	2270 9¼'	2450 10¼'	2565 11'	2645 12'	2800	2900	3050	3170	
8-inch	{65 {60	9 0} 8 6}	8.05	10	5 4	330 ¾'	630 2¼"	1000 3¼'	1225 4'	1515 5¼	1700 6'	1855 7"	2050 8'	2190 8¾	2365 9½'	2500 10"	2630 11	2800 11¼	2910 11½	3015 12'	3100 12¼	3200 13	56 lbs. Shot.
						⅝	⅞°	1⅛	1¼	1½	1¾°	2⅛											
8-inch	{65 {60	9 0} 8 6}	8.05	8	5 4	300	400	500	600	700	800	900	..	..	..	..	..	..	..	..	..	..	56 lbs. Shot.
						¼	1°	2	3°	4°	5°	6°	7°	8	9°	10°	11	12	13	14°	15	16	
8-inch	{65 {60	9 0} 8 6}	8.05	10	5 4	400 ¾"	645 2¼"	1030 3 "	1270 4"	1460 5¼"	1650 6"	1845 7"	1960 8"	2110 8¾"	2285 9½"	2340 10"	2490 11"	2530 11¼"	2630 11½'	2725 12"	2800 12¼"	2900 13'	51 lbs. Shell, loaded.
						½°	1¼°	1½°	1¾°	2⅛°													
8-inch	{65 {60	9 0} 8 6}	8.05	8	5 4	300	600	700	800	900	..	..	..	..	..	..	..	..	..	..	..	..	51 lbs. Shell.

RANGES WITH IRON ORDNANCE.

Gun	Charge (lbs)		Length	Wt. shot		½°	¾°	1°	1¼°	2°	3°	4°	5°	6°	7°	8°	9°	10°		Nature
8-inch	{65 / 60}	{9 0 / 8 6}	8.05	5	4	200	300	400	500	..	..	..	..	..	..	..	..	..	..	51 lbs. Shell, loaded.
8-inch	50	6 8	8.05	8	7	320 (⅜°)	550 (1°)	800 (2°)	1040 (3°)			1250 (4°)	1435 (5°)	1600 (6°)	1750 (7°)	1880 (8°)	2000 (9°)	2100 (10°)	..	51 lbs. Shell, loaded.
8-inch	50	6 8	8.05	5	7	200 (1½°)	300 (⅞°)	400 (1°)	500 (1¼°)	..	..	..	..	..	..	..	..	..	..	51 lbs. Shell, loaded.
8-inch	50	6 8	8.05	8	7	300 (⅜°)	530 (1°)	770 (2°)	1010 (3°)			1230 (4°)	1460 (5°)	1670 (6°)	1810 (7°)	1900 (8°)	2080 (9°)	2190 (10°)	by approximation.	56 lbs. Shot.

REMARKS.

The experiments with Shells made on board the "Excellent," in the autumn of 1838, and on which the Tables for Shell-practice have been formed, are corroborated by experiments which were made at Woolwich on the 5th November, 1838. From which it appears that, at 1250 yards, 10 lbs. of powder forced the Shell through one side, and fixed it in the other side of a section of a line-of-battle ship at the lower deck. With 8 lbs. of powder the Shell passed through the first side and rebounded from the second. At 900 yards, with charges of 8 and 10 lbs. of powder, the Shell passed through the first side and lodged in the second. At 600 yards, with a charge of 5 lbs. of powder, the Shell passed through the first and lodged in the second. At 300 yards, with charges of 8 and 7 lbs. of powder, the Shell passed through both sides, and buried itself in the butt. With the 56 lb. Shot, and charge 10 lbs. of powder, at 1250 yards, the Shot passed through the first side and buried itself in the second.

From this practice it would appear that the lowest charge should be used, which will give sufficient velocity to penetrate; especially as it was observed with the 12 lbs. charge (since abandoned) that the Shells frequently burst at the muzzle of the Gun. The "Eprouvette" before trial, showing a recoil of 23 20';
The powder used in obtaining these ranges was from various barrels mixed together. after trial, a recoil of 23 50'.

TABLE VI.

Ranges with Sea-service Iron Ordnance, Single-shotted, obtained on board H.M.S. "Excellent." Elevations by Spirit-level, 1845.

Nature of Guns.	Weight.	Length.		Diameter of Bore	Charge.	Height above the Plane.		Ranges in Yards, with corresponding Times of Flight.														
		Ft.	in.	In.	lbs.	Ft.	in.	¼°	1°	2	3°	4°	5°	6°	7°	8°	9°	10°	11°	12	13°	14°	15°	16
Prs. 68	Cwt. 113	10	10	8.13	20	5	3	340 4¾"	833 2¼'	1247 3¼'	1558 4¾"	1737 5¼"	2035 6¼"	2317 7¼"	2440 8¼"	2640 9¼"	2797 11	3011 11¼"	3233 12¼"	3460 13¾"	3553 14¾"	3673 15¼"	
56	98	11	0	7.7	16	5	4	290 ¾"	753 2'	1267 3"	1663 4¼"	1890 5¾"	2067 6¼"	2260 7¼"	2378 8¼"	2557 9¼"	2785 10¼"	2923 11	3022 12	3208 13¼"	3327 14	3410 14¾"	3547 15¾"
56	87	10	0	7.7	14	5	3	340 ¾"	821 2¼'	1234 3¼'	1516 4¼"	1793 5¼"	2040 6¼"	2193 7¼"	2433 8¼"	2630 9¼"	2818 10¼"	2991 11¼"	3110 12¼"	3243 12¾"	3420 13¾"	3500 14¾"	3642 15¾"
42	84	10	0	7.	14	5	8	317 ¾"	775 2'	1183 3"	1500 4¼"	1792 5¼"	2002 6¼"	2190 7¼"	2467 8¼"	2663 9¼"	2820 10¼"	2953 11¼"	3063 12¼"	3173 13	3287 13¾"	3427 14¾"	358 16
42	75	10	0	7.	12	5	5	257 ¾"	775 2'	1170 3'	1510 4¾"	1713 5¼"	1958 6¼"	2190 8'	2332 8¼"	2603 9¼"	2747 10¼"	2803 11¼"	3033 12¼"	3120 13¾"	3287 14¾"	3317 14¾"	349.3 15¾"
32	56	9	6	6.41	10	5	4	343 ¾"	700 2¼'	1100 3'	1513 4¾"	1760 5¾"	1930 6¼"	2100 7¼"	2300 8¼"	2477 9¼"	2662 10¼"	2800 11					
								3°xx	1°	1¼	1¾	1¼	2¾											
32	56	9	6	6.41	8	5	4	350 ¾"	600 2¼'	700 3"	800 4"	900 5"	1000 6"										
32	50 48	8	0	6.41	8	5	6	320 ¾"	760 2"	1116 3"	1320 4"	1516 5"	1696 6"	1835 7"	2060 7¾"	2246 8¼"	2313 9¼"	2493 10"					
32	50	9	0	6.3	8	5	6	346 ¾"	747 2"	1173 3"	1435 4¼"	1698 5¼"	1900 6¼"	2127 7¼"	2290 8"	2453 8¾"	2600 9¼"	2777 10"					

								¼°	1°	2°	3°	4°	5°	6°	7°	8°	9°	10°	Remarks.
32	45	8	6	6.3	7	5	6	333 ¾"	716 2"	1040 2⅞"	1320 4"	1600 5¼"	1800 6¼"	2026 7¼"	2180 8"	2340 8¾"	2510 9¼"	2697 10"	
32	40	8	0	6.3	6	5	6	326 ¾"	700 2"	1026 3"	1300 4"	1566 5¼"	1710 6¼"	1890 7¼"	2105 8¼"	2250 8¾"	2446 9¼°	2576 10"	
32	46	9	0	6.35	6	5	6	300	730	1026	1303	1580	1760	1940	2120	2300	2440	2570	
32	{42 40	8 7	0 6}	6.35	6	5	4	300	640	985	1280	1480	1660	1820	2060	2250	2370	2440	
								½°	1°	2°	3°	4°	5°	6°	7°				Should it be desired to fire with the reduced charge and a single shot with the 50 cwt. 32-pounder, of 8 feet, the ranges obtained with a 5 *lbs.* charge for the 6 feet 6 in. 32-pounders, may be used as an approximating guide for practice; and the rule may be applied to other guns *differing in length* and *weight, but having* reduced and full charges of the same weight and nature.
32	32	6	6	6.3	5	5	4	320 ¾"	500 1½"	825 2½"	1100 3½	1246 4¼"	1510 5¼"	1660 6¼"	1800 7¼"				
32	25	{6 5	0 8}	6.3	4	5	4	300 ¾"	470 1¼"	736 2"	1000 3¼"	1200 4½"	1418 5½"	1580 6¼"	1730 7¼"				
								¼°	1°	2°	3°	4°	5°	6°	7°	8°	9°	10°	
24	{50 48	9 9	6 0}	5.823	8	5	4	360	763	1123	1486	1646	1850	1960	2090	2240	2406	2630	
								⅜°	1°	2°									
24	{50 48	9 9	6 0}	5.823	6	5	4	340	600	980									
24	40	7	6	5.823	8	5	4	350	735	1153	1413	1593	1828	1980	2070	2220	2400	2600	
								⅜°	1°	2°									
24	40	7	6	5.823	6	5	4	330	600	980									

TABLE VII.

Ranges with Sea-service Iron Ordnance, Single-shotted, obtained on board H.M.S. "Excellent." Elevations by Spirit-level.

Nature of Guns.	Weight.	Length.		Diameter of Bore.	Charge.	Height above the Plane.		Ranges in Yards below the corresponding Elevations.								Remarks.
		Ft.	in.	Inches.	lbs.	Ft.	in.	¼°	1°	2°	3°	4°	5°	6°	7°	8°	
18-pounder	{42	9	0	5.292	6	5	4	350	600	993	1335	1558	1770	1920	2020	2130	Sea practice will fall generally within 5°, though the ranges have been given in some cases with elevations as high as 16°. Where fractional parts are denoted, the ranges have been obtained by pointing at an object close to the water. This applies to all the Tables.
	38	8	0	5.292	6	5	4										
18-pounder	{42	9	0	5.292	4	5	4	260	570	900	1078	1300	1586	1668	1841		
	38	8	0	5.292	4	5	4					1030	1260	1540	1620	1790	
18-pounder	22	7	6	5.18	3	5	4	230	543	850							
18-pounder	22	7	6	5.18	2	5	4	190	500	800							

TABLE VIII.

Ranges with Carronades, Single-shotted, obtained on board H.M.S. "Excellent." Elevations by Spirit-level.

Nature of Carronade.	Weight.	Length.		Diameter of Bore.	Charge.	Height above the Plane.		Ranges in Yards below the corresponding Elevations.					
		Ft.	in.	Inches.	lbs.	Ft.	in.	¼	1°	2°	3°	4°	5°
8-inch	Cwt 36	5	4	8.05	5 0	5	4	270	500	730	940	1100	1260
42-pounder	22	4	6	6.84	3 8	5	4	230	430	700	900	1050	1200
32-pounder	17	4	0	6.25	2 8	5	4	220	380	600	800	975	1170
24-pounder	13	3	9	5.68	2 0	5	4	200	360	580	770	950	1120
18-pounder	10	3	4	6.16	1 8	5	4	180	340	550	745	920	1050
12-pounder	6	2	8	4.52	1 0	5	4	150	310	520	715	890	970

TABLE IX.

Ranges with Sea-service Iron Ordnance, double-shotted, obtained on board H.M.S. "Excellent."
Elevations by Tangent Sight.

Nature of Guns.	Weight.	Length.	Diameter of Bore.	Charge.	Height above the Plane.	Ranges in Yards above the corresponding Elevations.					Remarks.
						100.	200.	300.	400.	500.	
	Cwt.	Ft. in.	Inches.	lbs.	Ft. in.	°	°	°	°	°	
32-pounder	64 56	9 7 9 6	6.41	6	12 6	$\frac{1}{4}$	$\frac{3}{8}$	1	$1\frac{1}{4}$	$1\frac{1}{2}$	NOTE.—The 10 and 8-inch Guns and Carronades are never to be double-shotted.
32-pounder	50 48	8 0	6.41	5	12 6	$\frac{1}{4}$	$\frac{1}{2}$	$\frac{3}{4}$	$1\frac{7}{16}$	$1\frac{3}{4}$	
32-pounder	46 40 39	9 0 7 6	6.35	4	19 6	$\frac{1}{4}$	$\frac{1}{2}$	$\frac{3}{4}$	$1\frac{3}{8}$	2	
32-pounder	32	6 6	6.3	3	19 4	$\frac{1}{2}$	1	$1\frac{1}{2}$	$2\frac{1}{4}$	$2\frac{3}{8}$	
32-pounder	25	6 0 5 4	6.3	2	19 4	$\frac{1}{2}$	$1\frac{3}{4}$	2	$2\frac{3}{8}$	$2\frac{5}{8}$	
24-pounder	50 48	9 6 9 0	5.823	4	19 4	$\frac{1}{3}$	$\frac{3}{4}$	$1\frac{1}{8}$	$1\frac{1}{4}$	$1\frac{1}{2}$	
24-pounder	41	7 6	5.823	4	19 4	$\frac{1}{3}$	$\frac{3}{4}$	$1\frac{1}{4}$	$1\frac{3}{8}$	$1\frac{5}{8}$	
18-pounder	42 38	9 0 8 0	5.292	3	19 4	$\frac{1}{3}$	$\frac{3}{4}$	$1\frac{1}{4}$	$1\frac{1}{2}$	2	
18-pounder	22	7 6	5.18	2	33 6	$\frac{3}{4}$	1	$1\frac{1}{2}$	2	$2\frac{5}{8}$	

TABLE X.

Ranges, by Ricochet, with Sea-service Iron Ordnance, Single-shotted, obtained on board H.M.S. "Excellent." Elevations by Spirit-level.

Nature of Guns.	Weight.	Length.		Diameter of Bore.	Charges.	Height above the Plane.		Elevation.	First Graze.	Extreme Graze before Deflection.	Number of Grazes.	Remarks.
	Cwt.	Ft.	in.	Inches.	lbs.	Ft.	in.	°	Yards.	Yards.		
10-inch	84	9	4	10	12	12	6	0	280	2850	16 ⎱ Hollow	The Ricochet Ranges were generally obtained when the water was perfectly smooth. It has been observed that a slight ripple diminishes the range considerably.
								½	658	2700	14 ⎰ shot,	
								1				
								2	770	2680	18 ⎱ 84 lbs.	
8-inch	65 & 60	9 0		8.05	10	12	6	½	340	2900	32 ⎱ Hollow	Ricochet Ranges are not given for shell, because the present fuzes are always extinguished when so used.
		8 6						1	626	2766	23 ⎰ shot,	
								2	970	2576	14 ⎱ 56 lbs.	
8-inch	50	6 8		8.05	7	5	4	½	331	2150	14 ⎱ Hollow	It has been observed, in Ricochet firing, that the nearer the gun is laid to the water (other circumstances being alike), the farther it ranges the shot. It might therefore, in some cases, be advisable to heel the ship over, by running to the opposite guns.
								1	653	2017	14 ⎰ shot,	
								2	933	2333	9 ⎱ 56 lbs.	
32-pounder	64 & 56	9 7		6.41	10	12	6	½	350	2850	24	
		9 6						1	800	2900	15	
								2	1220	2650	12	
32-pounder	50 & 48	8 0		6.41	8	5	4	½	306	2613	12	
								1	706	2593	14	
								2	1117	1907	8	

APP. D.   RANGES WITH IRON ORDNANCE.   575

Gun									
32-pounder	50	9 0	6.3	8	5 6	0	346	3183	33
						1	747	3033	26
						2	1173	2490	13
32-pounder	45	8 6	6.3	7	5 6	0	333	2450	15
						1	716	2150	8
						2	1040	2260	8
32-pounder	40	8 0	6.3	6	5 6	0	326	2366	19
						1	700	2100	8
						2	1026	2480	16
32-pounder	46	9 0	6.35	6	5 4	0	300	2800	32
						1	730	2510	14
						2	1026	2210	10
32-pounder	42 / 40 / 39	8 0 / 7 6	6.35	6	12 6	0	333	2003	9
						1	637	1867	8
						2	857	2007	18

TABLE X.—continued.

Ranges, by Ricochet, with Sea-service Iron Ordnance, single-shotted, obtained on board H.M.S. "Excellent." Elevations by Spirit-Level.

Nature of Guns.	Weight.	Length.	Diameter of Bore.	Charge.	Height above the Plane.	Elevation.	First Graze.	Extreme Graze before Deflection.	No. of Grazes.	Remarks.
	Cwt.	Ft. in.	Inches.	lbs.	Ft. in.	°	Yards.	Yards.		
32-pounder	32	6 6	6.3	5	12 6	0	300	1980	15	
						1	530	1800	10	
						2	800	1950	15	
32-pounder	25	5 4	6.3	4	5 4	0	270	1980	21	
						1	480	2027	16	
						2	990	1670	6	
24-pounder	50 & 48	9 6 / 9 0	5.823	8	12 6	0	406	2143	14	
						1	880	2060	10	
						2	1186	1637	5	
18-pounder	42 & 38	9 0 / 8 0	5.292	6	12 6	0	410	2943	34	The variation is attributable to the fact of the water being very smooth when the ranges for the 18-pounders were taken, and there being a slight ripple when the ranges for the 24-pounder, and those of the 32-pounders, of 42, 40, and 39 cwt., were taken.
						1	680	2717	24	
						2	1093	2233	9	
18-pounder	22	7 6	5.18	3	5 4	0	250	2660	29	
						1	596	2500	20	
						2	926	2293	9	

RANGES WITH IRON ORDNANCE.

Ranges with Sea-service Iron Ordnance, with Single Round and Grape, obtained on board H.M.S. "Excellent."
Elevations by Tangent Sight.

Nature of Guns.	Weight.	Length.	Charge.	Eleva- tion.	First Graze of Grape.	First Graze of Round.	Extreme Range of Grape.	Spread of Grape, at Yards.				Extreme Spread of Grape.	Remarks.
								500	400	300	200		
	Cwt.	Ft. in.	lbs. oz.	°	Yds.	Yds.	Yds.	Yds.	Yds.	Yds.	Yds.	Yds.	
32-pounder	56	9 6	6 0	0	120	250	1280	..	..	20	..	50	The following remarks were made on practice obtained with Round and Grape, Double Grape, and Double Case, in December, 1833:— The grape and round ranged closer together when the round was put in first, but the range and the velocity of the round were decreased by being so used, and that of the grape increased. Although the gun was six times fired at the target with round and grape, two only of the grape struck. At the last, round the gun had a double charge of grape, and was fired with 1½° of elevation, at the 400 yard target; they fell very close together, at about 350 yards, two of the shot passing through the target, and the extreme range was about 1400 yards. Two 32-pounders, 9 ft. 6 in. in length, were charged with round and grape, the round being entered first at one gun, and the grape at the other, and fired at the section of a 74-gun ship's side, 300 yards distant; they ranged closer together when the round was put in first, but required more elevation: the grape, having less velocity, impedes the round. They were again charged, the one with double grape, the other double case, and the balls grazed the water at about 100 yards' distance, and few of them appeared to reach the object, whereas the grape ranged tolerably well together to 300 yards. The number of grape fired at the ship's side was 45, out of which 18 struck, some of them passing through the side, viz. 5 inches of fir, and the rest penetrating from 2 to 3 inches in oak. The number of case tried was 140, of which only 21 struck, 6 penetrated 1 inch in the fir, one passed through 2¼ inches, the rest penetrated from ¼ to 1 inch in fir. February 26, 1844.—The block fired at was that used before, 12 feet by 12 feet, and two pieces of canvas were stretched out from each end, making it equal to three ports and the intervals between them. The men were directed to fire at the wooden or central port, and 10 of the round shot passed just above, below, or close to the side sill of it. A piece of canvas had been nailed over it. Out of 99 grape shot, 41 struck the side, but 7 only entered the ports. The centre port was aimed at, and the round shot all struck within the breadth of the port.
		9 0	5 0	1	70	380	1100	..	..	..	40	90	
				2	50	600	1170	20	..	..	..	50	
32-pounder	50	9 0	5 0	1½	150	300	1300	..	20	..	..	45	
				1	190	350	1400	..	20	..	..	55	
				2	300	400	1450	..	5	..	..	45	
32-poun	48	8 0	5 0	1½	230	lost	1210	..	10	..	..	80	
				1	210	lost	1170	..	20	..	..	110	
				2	150	550	1020	25	..	..	..	80	
32-pounder	39	7 6	4 0	1½	130	lost	1150	..	10	..	..	80	
				1	200	200	1300	..	15	..	..	40	
				2	200	500	1300	..	15	..	..	80	
32-pounder	25	{6 0, 5 8}	2 8	1½	170	460	820	..	25	..	..	115	
				1	260	500	650	..	25	..	..	60	
				2	290	300	750	..	20	..	..	32	

2 P

TABLE XII.

Ranges with Sea-service Iron Ordnance, with Double Grape, obtained on board H.M.S. "Excellent." Elevations by Tangent Sight.

Nature of Guns.	Weight.	Length.	Charge.	Elevation.	First Graze of Grape.	Extreme Range of Grape.	Spread of Grape, at Yards.										Extreme Spread of Grape.
							500	450	400	350	300	250	220	200	150		
	Cwt.	Ft. in.	lbs. oz.	°	Yards.	Yards.	Yds.	Yds.	Yds.	Yds.	Yds.	Yds.	Yds.	Yds.	Yds.		Yards.
32-pounder	56	9 6	6 0	0	40	1210	.	.	.	.	.	.	.	40	.		90
				1	60	1240	.	.	35	.	.	.	30	.	.		120
				1	230	1250	.	.	.	.	.	.	.	18	.		90
32-pounder	50	9 0	5 0	2	130	1450	.	.	.	.	.	.	.	.	.		65
				½	110	1300	.	.	.	25	.	.	.	.	.		63
				1	130	1250	35	.	.	.	.	.	.	.	.		80
32-pounder	48	8 0	5 0	2	180	1110	.	.	.	.	.	.	.	.	.		100
				½	170	1320	.	.	35	.	30	.	.	.	8		120
				1	50	1150	.	50	.	.	.	.	.	.	.		100
32-pounder	39	7 6	4 0	2	190	1270	.	.	.	.	.	20	.	.	.		120
				½	100	1320	30	.	.	.	.	.	.	.	.		80
				1	250	1320	.	.	.	.	15	.	.	.	.		80
32-pounder	25	{ 6 0 / 5 8 }	2 8	2	140	770	.	.	.	.	.	.	.	.	.		55
				½	150	920	.	.	.	.	.	10	.	.	.		65
				1	200	620	.	.	60	.	.	.	.	.	.		50

TABLE XIII.

Ranges with Sea-service Iron Ordnance, with Single Grape, obtained on board H.M.S. "Excellent." Elevations by Tangent Sight.

Nature of Guns.	Weight.	Length.	Charge.	Elevation.	First Graze of Grape.	Extreme Range of Grape.	Spread of Grape, at Yards.								Extreme Spread of Grape.
							600	550	500	400	380	300	150	100	
	Cwt.	Ft. in.	lbs. oz.		Yards.	Yards.	Yards.	Yards.	Yards.	Yards.	Yards.	Yards.	Yards.	Yards.	Yards.
32-pounder . .	56	9 6	6 0	½°	90	950	..	..	..	12	..	..	..	..	60
				1	100	1250	..	50	..	..	..	..	..	..	80
				2	100	1450	..	..	..	..	..	25	..	..	90
32-pounder . .	50	9 0	5 0	½°	100	1350	..	..	..	..	..	..	..	18	90
				1	120	1400	..	50	..	20	..	..	..	..	35
				2	50	1500	..	..	..	..	..	..	60	..	120
32-pounder . .	48	8 0	5 0	½°	80	1010	..	..	..	15	..	..	..	..	110
				1	250	1120	..	..	..	10	..	..	..	..	85
				2	410	1250	..	20	..	..	..	..	..	..	100
32-pounder . .	39	7 6	4 0	½°	130	1450	..	..	..	15	..	..	..	..	100
				1	160	1400	..	..	..	8	..	..	..	..	80
				2	230	1400	..	10	..	..	..	..	..	..	70
32-pounder . .	25	{ 6 0 / 5 8 }	2 8	½°	290	770	..	..	..	25	..	..	..	..	70
				1	190	905	..	..	..	..	10	..	..	..	50
				2	190	960	..	..	..	40	..	..	..	..	70

TABLE XIV.

Ranges with Sea-service Iron Ordnance, with Single Case Shot, obtained on board H.M.S. "Excellent." Elevations by Tangent Sight.

Nature of Guns.	Weight.	Length.	Charge.	Elevation.	First Graze of Case.	Extreme Range of Case.	Spread of Case, at Yards.						Extreme Spread of Case.
							500	450	400	300	200	80	
	Cwt.	Ft. in.	lbs. oz.		Yards.	Yards.	Yards.	Yards.	Yards.	Yards.	Yards.	Yards.	Yards.
32-pounder	56	9 6	6 0	0	40	1150	..	..	..	..	15	..	140
				½	30	1080	70	..	20	..	..	..	160
				1	50	1090	..	..	..	..	..	..	130
32-pounder	50	9 0	5 0	½	60	1090	..	..	..	..	..	50	120
				1	40	1020	..	80	..	..	..	..	125
				2	120	1200	80	..	..	..	..	..	175
32-pounder	48	8 0	5 0	½	90	1000	..	..	60	..	..	..	120
				1	150	1000	..	..	35	..	..	..	130
				2	110	1000	..	..	30	..	..	..	130
32-pounder	39	7 6	4 0	½	100	1300	..	..	..	..	60	..	170
				1	110	1150	..	..	..	60	50	..	200
				2	100	1000	..	..	..	..	..	..	200
32-pounder	25	6 0 } 5 8 }	2 8	½	160	750	..	..	45	..	..	..	75
				1	60	700	..	..	..	..	40	..	120
				3	100	950	..	..	..	60	..	..	125

Table XV.

A useful and practical Table of Approximate Ranges, with the Elevations and Times of Flight.

Nature of Gun.	Charge of Powder. lbs.	oz.	RANGES IN YARDS. The Upper Horizontal Line corresponding to each Charge consists of the Elevations; the Lower, of the Times of Flight. 200	300	400	500	600	700	800	900	1000	1100	1200	1500	1800	2000
For all 8-inch guns with these charges.	10	0	.. ..	¼° ¾″	½° 1″	¾° 1⅜″	1° 1¾″	1¼° 2″	1¾° 2¼″	2° 2½″	2¼° 2¾″	2½° 3¼″	3° 3¾″	4° 5¼″	6° 7″	7° 8″
	8	0	¼° ½″	½° ¾″	¾° 1″	1° 1½″	1¼° 1⅞″	1¾° 2¼″	2° 2½″	2¼° 2⅞″	2½° 3¼″	3° 3⅝″	3½° 4¼″			
	5	0	½° ½″	¾° ¾″	1° 1¼″	1¼° 1⅝″	1¾° 2″									
For all 32-pounders with these charges.	10	0	.. ½″	¼° ¾″	¼° 1″	½° 1¼″	¾° 1⅜″	1° 1⅝″	1¼° 1⅞″	1½° 2¼″	1¾° 2⅝″	2° 3″	2¼° 3⅜″	3° 4½″	4° 6″	5° 7″
	8	0	.. ½″	¼° ¾″	½° 1″	¾° 1¼″	1° 1½″	1¼° 1¾″	1½° 2″	1¾° 2⅜″	2° 2¾″	2¼° 3¼″	2¾° 3¾″			
	6	0	¼° ½″	½° ¾″	¾° 1⅛″	1° 1⅜″	1¼° 1⅝″	1½° 1⅞″	1¾° 2⅛″	2° 2½″	2¼° 3″	2¾° 3½″	3¼° 4″			
	4	0	¼° ½″	½° ¾″	¾° 1⅛″	1° 1½″	1½° 1⅞″	2° 2⅛″	2½° 2½″	2¾° 2⅞″	3¼° 3¼″	3½° 3⅞″	4° 4½″			
	2	8	½° ¾″	¾° 1″	1° 1½″	1½° 1⅞″	2° 2¼″	2¼° 2⅜″	3° 3″	3½° 3½″	4° 4¼″					

Note.—The best way to commit this Table to memory, is to learn the Elevations for the 8 lbs. charges, and subtract ¼ for the 10 lbs. charges, and add ¼ for the 6 lbs., which will be very nearly correct up to 1000 yards.

TABLE XVI.

Penetrations into Oak. (From the *Aide Mémoire Navale*.)

The Initial Velocities were obtained from Experiments with the Ballistic Pendulum, and the Velocities at the instant of Impact were computed.

Solid Shot. Density, 7.15 (water=1).

Nature of Ordnance.	Charge.	Initial Velocity			Distance of the object from the mouth of the Gun in Yards.						
					109	219	438	656	875	1094	
	lbs. oz.										
Canon de 36	13 3.7	1578	Velocity at Impact	Feet, &c.	1476	1394	1223	1089	974	876	
			Penetration	Yards	4 6.8	4 0.7	3 5.8	2 11.4	2 6.4	2 1.6	
	9 15.8	1482	Velocity at Impact	Feet, &c.	1391	1305	1158	1033	958	836	
			Penetration	Yards	4 1.2	3 9.2	3 2.5	2 9.	2 3.6	2 0.	
	6 9.9	1302	Velocity at Impact	Feet, &c.	1227	1154	1030	761	833	754	
			Penetration	Yards	3 1.9	2 9.	2 9.1	2 4.	2 0.	1 8.4	
	11 0.5	1591	Velocity at Impact	Feet, &c.	1482	1384	1217	1076	959	856	
			Penetration	Yards	4 1.9	3 10.1	3 3.	2 8.6	2 3.6	2 11.2	
Canon de 30 (long)	8 4.3	1492	Velocity at Impact	Feet, &c.	1394	13.5	1151	1020	912	817	
			Penetration	Yards	3 10.4	3 6.5	3 0.2	2 6.2	2 1.4	2 2.4	
	5 8.2	1305	Velocity at Impact	Feet, &c.	1223	1151	1020	912	814	738	
			Penetration	Yards	3 3.4	3 0.2	2 6.2	2 1.5	1 9.6	1 6.5	
	8 13.2	1610	Velocity at Impact	Feet, &c.	1489	1384	1200	1050	925	820	
			Penetration	Yards	3 10.4	3 6.5	2 11.4	2 5.2	2 0.4	1 8.	
Canon de 24 (long)	6 9.9	1505	Velocity at Impact	Feet, &c.	1397	1299	1132	994	879	740	
			Penetration	Yards	3 6.8	3 3.4	2 8.6	2 3.1	2 10.4	1 6.8	
	4 6.4	1305	Velocity at Impact	Feet, &c.	1217	1138	997	8.2	784	699	
			Penetration	Yards	2 11.8	2 8.6	2 3.1	1 10.4	1 6.8	1 3.7	
	6 9.9	1630	Velocity at Impact	Feet, &c.	1496	1374	1174	1013	8.2	774	
			Penetration	Yards	3 6.5	3 2.2	2 7.1	2 1.2	1 8.4	1 4.8	
Canon de 18 (long)	4 15.4	1519	Velocity at Impact	Feet, &c.	1397	12.6	1165	958	840	.35	
			Penetration	Yards	3 3.	2 11.	2 4.7	1 3.4	1 6.8	3.4	
	3 4.9	1309	Velocity at Impact	Feet, &c.	1210	1122	971	946	741	656	
			Penetration	Yards	2 8.6	2 5.2	1 11.6	1 7.3	1 1.2	1 6.	
	4 6.4	1640	Velocity at Impact	Feet, &c.	1479	1341	1118	945	807	675	
			Penetration	Yards	2 10.9	2 8.6	2 4.6	1 8.	1 3.7	1 1.2	
Canon de 12 (short)	3 4.9	1532	Velocity at Impact	Feet, &c.	13.7	1263	1056	925	767	662	
			Penetration	Yards	2 9.8	2 5.9	1 11.6	1 6.4	1 2.5	0 11.5	
	2 3.3	1325	Velocity at Impact	Feet, &c.	1207	1105	935	797	645	597	
			Penetration	Yards	2 4.3	2 1.2	1 7.7	1 3.4	0 11.8	0 9.5	

Hollow Shot, diameter=16 centi.=6.3 inch. Density 4.93 (water=1)

Canon de 30 (long)	8 4.3	1742	Velocity at Impact	Yards	1578	1423	1183	1000	853	735	
			Penetration	Feet, &c.	3 0.96	2 9.8	2 1.9	1 8.4	1 4.	0 6.	
	4.9	1584	Velocity at Impact	Yards	1420	1292	1062	913	787	682	

APP. D. LIST OF BRITISH NAVAL ORDNANCE. 583

TABLE XVII.
List of Naval Ordnance in use for the British Navy.

	Nature.	Weight.	Length.	Diameter of Bore.	Windage.	Charge of Powder.	Whose Pattern.	Remarks.
		Cwt.	Ft. in.	Inches.	Inches.	lbs. oz.		
	IRON:							
1	68-pr.	95	10 0	8.12	0.2	16 0	Col. Dundas', new	These are for *Steamers*.
2		88	9 6			14 0		
3	56 ,,	98	11 0	7.65	0.175	16 0	Mr. Monk's, new	These give place to the above when Ships at sea are recommissioned.
4		87	10 0			14 0		
5	42 ,,	67	9 6	6.97	0.2	10 8	Col. Dundas', new	Ditto Ditto.
6	10-in.	86	9 4	10.0	0.16	12 0	Gen. Millar's modified Shell Gun.	These are to be substituted, in Steamers, for the 68-pr. Guns. See observations. Art. 212, p. 226.
7		84						
8	8 ,,	65	9 0	8.05	0.125	10 0	Gen. Millar's Shell Gun	These are for large heavy vessels.
9		60	8 10			10 0	Gen. Millar's Shell Gun	Frigates.
10		52	8 0			8 0		First Class 6th rates.
11		61¼	9 6		0.2	10 0	Col. Dundas'.	
12	Guns	58ª	9 6		0.175	10 0 & 8 5	Col. Dundas'.	The old Gun, which will probably hereafter give place to one of 58 cwt.
13		56	9 6	6.41	0.233	10 0	Sir Thos. Blomefield's	A Gun of 9 ft. and 50 cwt. will take the place of this.
14	32-pr.	48 to 50	8 0			8 0	Dickson's and Millar's	
15		41	8 0	6.35	0.173	6 0	Bored up from Sir Thos. Blomefield's 24-Pr.	Bored up Guns, which, when unserviceable, will be replaced by the 32-pr. of 42 cwt, and 8 ft. in length.
16		39	7 6			6 0	Do. Sir W. Congreve's do.	
17		40	7 6			6 0		
18		32	6 6			5 0	Do. Blomefield's do.	Bored up Guns used in brigs and some 1st rates.
19		25	6 0	6.3	0.123	4 0	Do. do. 18-Pr.	Bored up, in general use.
20		25	6 0			4 0	Col. Dundas', new	Gun made new.

ª 32-pr. Guns, 9 ft. 6 in. long, windage 0.2 inch, and weighing 58 cwt., are now in general use.

TABLE XVII.—*continued.*
List of Naval Ordnance in use for the British Navy.

	Nature.		Weight.	Length.		Diameter of Bore.	Windage.	Charge of Powder.		Whose Pattern.	Remarks.
			Cwt.	Ft.	in.	Inches.	Inches.	lbs.	oz.		
21	IRON:		50	9	6	6.375	0.198	8	0	New Gun A ⎫	⎧ These, which may be denominated
22		⎧ 32-prs.	45	9	6			7	0	New Gun B ⎬ Monk's new.	⎪ Medium Guns, enter largely into
23	Guns.	⎨	42	8	6	6.35	0.173	6	0	New Gun C ⎭	⎨ the armament of the Navy.
24		⎪	22	8	0			3	0	Bored up from 9-pr.	⎪
25		⎨ 18-prs.	22	7	0	5.17	0.071	3	0	Do. light 12-pr.	⎪ These are made for the smaller class
26		⎩	20	6	6			2	0	Do. light 9-pr.	⎩ of brigs, and other small craft.
27		68-pr.	15	5	0	8.05	0.125	5	8	⎫	
28		42 ,,	63	5	6	6.84	0.078	3	11	⎪	
29	Carro-	32 ,,	22	4	6	6.25	0.073	2	8	⎬ Carron Company.	
30	nades.	24 ,,	17	4	0	5.68	0.068	2	0	⎪	
31		18 ,,	13	3	9	5.16	0.061	1	8	⎭	
32		12 ,,	10	3	4	4.52	0.066	1	0		
33		6 ,,	6	2	8	3.6	0.05	0	10		
34	Mortars	⎧ 13-in.	4¾	2	9	13.0	0.16	20	0	⎫ Sir Thomas Blomefield's.	
35		⎩ 10 ,,	101	4	5	10.0	0.16	9	8	⎭	
			52	3	9¼						
36	BRASS:	⎧ 9 pr.	13¾	6	0	4.2	0.1	3	0	⎫	For Steam-packets.
37	Guns.	⎨ 6 ,,	6	5	0	3.868	0.1	1	8	⎬	One, mounted on a field carriage, on board all large ships for land service.
											For boat-service, and to be landed on field carriages when required. 24-pr. (2) on board each 1st, 2nd, and 3rd rate; 12-pr. (2) 10 cwt., on board 4th and 5th rates, and 12-pr. (2) 6½ cwt. on board 1st, 2nd, 3rd, 4th, 5th, and 6th rates, and 1 on board brigs and smaller vessels.
38	Howit-	⎧ 24 ,,	13	4	8½	5.72	0.125	2	8	Gen. Millar's, new	
39	zers.	⎨	10	4	7	5.58	0.125	2	0	Col. Dundas', new	
40		⎩ 12 ,,	6¼	3	9¼	5.58	0.125	1	4	Gen. Millar's, new	

May 5th, 1848.

W. B. DUNDAS.

TABLE XVII.—*concluded.*

Land Service.—Iron Ordnance.

Nature.		Weight.	Length.		Diameter of Bore.	Windage.	Service Charge.		
		Cwt.	Ft.	in.	Inches.	Inches.	lbs.	oz.	
Guns	56-pr.	98	11	0	7.65	0.175	16	0	
	8-in.	65	9	0	8.05	0.125	10	0	
		50	6	8½			8	0	
	32-pr.	56	9	6	6.41	0.233	10	0	
		32	6	6	6.3	0.122	5	0	Bored up from 24-pr. of 33 cwt.
	24 ,,	50 or 48	9 / 9	6 / 0	5.823	0.211	8	0	
		20	6	0	5.75	0.138	2	8	Bored up from 12-pr. of 21 cwt.
	18 ,,	42	9	0	5.292	0.193	6	0	
		20½	6	0	5.17	0.071	3	0	Ditto ditto.
	12 ,,	34	9	0	4.623	0.1	4	0	
		21	6	0	4.623		3	0	
	9 ,,	17	5	6	4.2	0.1	3	0	
	6 ,,	17	6	0	3.668	0.1	2	0	
Howitzers.	10-in.	41	5	0	10.0	0.16	7	0	
	8 ,,	21	4	0	8.0	0.14	4	0	
	5½ ,,	15	3	4¾	5.62	0.025	2	8	
Mortars	13-in.	36	3	0¾	13.0	0.16	9	0	
	10 ,,	16½	2	7½	10.0	0.16	4	0	
	8 ,,	8¼	2	1¼	8.0	0.14	2	0	

Land Service.—Brass Ordnance.

Nature.		Weight.	Length.		Diameter of Bore.	Windage.	Service Charge.		
Guns	12-pr.	18	6	6	4.623	0.1	4	0	With the new Shot this charge will be reduced probably to 3 lbs. 8 oz.
	9 ,,	13½	6	0	4.2	0.1	2	8	
	6 ,,	6	5	0	3.668	0.1	1	8	
	3 ,,	3	4	0	2.913	0.09	0	12	
		2¼	3	0			0	10	
Howitzers.	32-pr.	17½	5	3	6.3	0.125	3	0	
	24 ,,	13	4	8½	5.72	0.125	2	8	
	12 ,,	6½	3	9¼	4.58	0.122	1	4	
	4⅗ ,,	2½	1	10½	4.52	0.066	0	8	
Mortars	10-in.	12¾	2	3	10.0	0.16	4	0	
	8 ,,	6½	1	9½	8.0	0.14	2	0	
	5½ ,,	1¼	1	3	5.62	0.025	0	8	
	4⅗ ,,	¾	1	0¾	4.52	0.066	0	4	

June, 1847.

TABLE XVIII.

Comparative Ranges of French 30-livre Guns, long and short, with English 32-Pounders. The Ranges of the French Guns are taken from the general Table formed on the Experiments made at Gavre between 1830 and 1840, interpolated for the Charges and Elevations here given. The Weights and Dimensions are in English denominations. The Ranges of the English Guns are taken from the Table of Experiments carried on in the year 1845 on board the "Excellent."

Nature of Ordnance.	Weight of Gun.	Diameter of Bore in Inches.	Charge.	Weight of Shot (solid).	Elevations. Ranges.									
					0° 15′	1°	2°	3°	4°	5°	6°	7°	8°	9°
	Cwts. qrs. lbs.	In.	lbs.	lbs. oz.	Yards.	Yards.	Yards.	Yards.	Yards.	Yards.	Yards.	Yards.	Yards.	Yards.
Fr. long gun, 30 liv.	58 3 17	6.48	10	33 13	230	620	1020	1315	1580	1830	2040	2257	2456	2648
Short ditto . . .	49 0 0	6.48	10	33 13	239	630	975	1317	1572	1810	2016	2225	2426	2620
(Windage 0.2 in.)														
Eng. 32-pr. . . .	56 0 0	6.41	10	..	343	700	1100	1513	1760	1930	2100	2300	2477	2622
(Windage 0.233 in.)	9 ft. 6 in. long.													
Fr. long, 30 liv. .	58 3 17	6.48	8	33 13	231	590	940	1246	1400	1732	1947	2149	2345	2534
Short ditto . . .	49 0 0	6.48	8	33 13	237	580	938	1231	1488	1715	1922	2118	2295	2532
(Windage 0.2 in.														
Eng. 32-pr. . . .	56 0 0	6.41	8	..	350	600	970	..	..	..	..	..	..	..
(Windage 0.233 in.)	9 ft. 6 in. long.													
Ditto	50 } 8 ft. long.	6.41	8	..	320	760	1116	1320	1316	1696	1835	2060	2246	2313
(Windage 0.233 in.)	48 }													
Fr. long, 30 . .	58 3 17	6.48	7	33 13	218	581	955	1248	1492	1715	1939	2134	2329	2526
Short ditto . . .	49 0 0	6.48	7	33 13	225	547	940	1213	1462	1685	1900	2099	2247	2510
(Windage 0.2 in.														
Eng. 32-pr. . . .	45 0 0	6.3	7	..	333	716	1040	1320	1600	1800	2026	2180	2340	2510
(Windage 0.173 in.)	8 ft. 6 in. long.													
Fr. long 30 . .	58 3 17	6.48	6	33 13	210	516	850	1255	1352	1604	1784	1935	2163	2360
Short ditto . . .	49 0 0	6.48	6	33 13	220	512	828	1113	1345	1556	1755	1954	2134	2327
(Windage 0.2 in.)														
Eng. 32-pr. . . .	40 0 0	6.3	6	..	326	700	1026	1300	1566	1710	1890	2105	2250	2446
(Windage 0.173 in.)	8 ft. long.													

APP. D. COMPARATIVE RANGES OF FRENCH AND ENGLISH GUNS.

Ranges, in Yards, of French Canons-obusiers of 80 (liv.), with 50-pounder and 36-pounder French Guns. From the Experiments at Gavre in 1846, 1847.—(Corrected.)

Nature of Ordnance.	Charge.		1°	2°	3°	4°	5°	6°	7°	8°	9°	10°	15°	20°	25°	30°	Remarks.
	lbs.	oz.	Yards.	Yards.	Yards.	Yards.	Yards.	Yards.	Yards.	Yards.	Yards.	Yards.	Yards.	Yards.	Yards.	Yards.	
Canon Obusier 80, No. 1 (1842)	7	11·5	443	731	998	1201	1395	1580	1645	1840	2070	2195	..	..	..	..	Hollow Shot.
Ditto ditto 2	6	10	430	660	898	1060	1230	1390	1455	1625	1780	1900	..	..	..	..	Ditto.
50-pounder (Fr.) Gun	17	10	632	1051	1347	1669	1900	2129	2322	2504	2674	2831	3610	4288	4704	4923	Solid Shot.
36-pounder (Fr.) Gun	13	3·75	687	1083	1391	1647	1866	2072	2270	2458	2636	2811	3605	4245	4510	4543	Ditto.

TABLE XX.

Comparative Ranges of the Original Canon-obusier of 80, and the English 8-inch Shell Gun.

The Ranges of the French Gun at the Elevations below stated are taken from a General Table formed on the Experiments made at Gavre between 1830 and 1840.
The Ranges were reduced to English Yards, and afterwards, by interpolations, reduced to the particular Elevations in the subjoined Table.
The Ranges of the 8-inch Gun (English) are taken from the Table of Experiments carried on in the Year 1839, on board the "Excellent."

Nature of Ordnance.	Weight of Gun.			Length of Gun.		Diameter of Bore.	Weight of Shot.		Charge.	Elevation.												
										22′ 30″	1°	2°	3°	4°	5°	6°	7°	8°	9°	10°	11°	12°	13°
	Cwt.	qrs.	lbs.	Ft.	in.	Inch.	lbs.	oz.	lbs.	Yards.	Yards.	Yards.	Yards.	Yards.	Yards.	Yards.	Yards.	Yards.	Yards.	Yards.	Yards.	Yards.	Yards.
Canon Obusier 80, Windage=0.1318 inch	73	3	14	9	3,8	8.79	58	8	10	460	679	847	1077	1390	1640	1700	1900	2070	2169	2320	2540	2657	2830
8-inch Gun, Windage—0.125	65	0	0	9	0	8.05	56	0	10	330	630	1000	1225	1515	1700	1855	2050	2190	2365	2500	2630	2800	2910
	60	0	0	8	6	8.05	51	0 Shell loaded.	10	400	645	1030	1270	1460	1650	1845	1960	2110	2285	2340	2490	2530	2630

										Elevation.							
										37′ 30″	52′ 50″	1° 7′ 30″	1° 15′	1° 30′	1° 45′	2° 7′ 30″	
										Yards.	Yards.	Yards.	Yards.	Yards.	Yards.	Yards.	
Canon Obusier 80, Windage=0.1378 inch	73	3	14	9	3.8	8.79	58	8	8	350	440	537	565	650	736	835	
8-inch Gun, Windage—0.125	65	0	0	9	0	8.05	56	0	8	300	400	500	600	700	800	900	
	60	0	0	8	6	8.05	51	0	8	..	..	..	600	700	800	900	

588 NAVAL GUNNERY. App. D.

TABLE XXI.

Heights in English Measure, above the Water, of the different parts of French Ships of War and their Masts, according to the following Rates (1850).

	Line-of-Battle Ships.											Frigates.				Corvettes.		Brigs.	
	120.		100.		90.		86.		82.		60.		44.		24.		18.	
	Guns.		Guns.		Guns.		Guns.		Guns.		Guns.		Guns.		Guns.		Guns.	
	Ft.	in.	Ft.	in.	Ft.	in.	Ft.	in.	Ft.	in.	Ft.	in.	Ft.	in.	Ft.	in.	Ft.	in.
Portsills of lower deck	5	3	7	10	6	7	5	8	5	8								
,, ,, middle ,,	13	1																
,, ,, upper ,,	19	4	15	1	14	9	12	9	12	9	6	6¾	6	11	5	10	4	3
,, ,, quarter ,,	25	7	21	3	1	8	19	2	19	2	13	9	13	4	12	0		
After part of poop plank sheer	35	5	29	2	29	6	25	4	25	4								
MAINMAST:—																		
Upper side of mainyard	79	0	77	1	76	5	71	0	66	0	67	0	59	0	41	0	39	0
Under side of maintop	93	6	91	10	90	2	84	0	78	0	79	0	70	0	49	0	47	0
Upper side of cap to mainmast	109	6	109	10	108	3	101	0	94	0	95	0	84	0	59	0	56	0
,, main topsail yard	148	7	140	9	141	0	139	0	130	0	131	0	114	0	79	0	72	0
,, crosstrees to main topmast	158	5	150	11	150	11	148	0	138	0	139	0	121	0	85	0	77	0
,, cap to main topmast	166	4	161	8	155	0	156	0	147	0	148	0	128	0	89	0	81	0
Head of topgallant rigging	190	7	185	4	185	4	173	0	170	0	171	0	151	0	104	0	96	0
Truck	219	9	212	7	201	9	205	0	192	5	187	9	168	6	120	7	112	0
FOREMAST:—																		
Upper side of foreyard	72	10	72	2	71	2	64	0	61	0	67	0	51	0	34	0	34	0
Under side of foret p	86	7	84	11	83	7	77	0	72	0	73	0	63	0	43	0	41	0
Upper side of cap to foremast	103	0	101	8	100	0	92	0	87	0	88	0	76	0	53	0	50	0
,, fore topsail yard	136	5	128	11	129	7	128	0	117	0	118	0	101	0	70	0	66	0
,, crosstrees to fore topmast	145	8	139	1	137	9	136	0	126	0	127	0	109	0	76	0	70	0
,, cap to fore topmast	153	6	148	11	145	0	144	0	135	0	136	0	116	0	81	0	75	0
Head of top gallant rigging	173	10	170	3	169	7	168	0	156	0	157	0	136	0	95	0	89	0
Truck	200	1	194	10	184	8	188	0	176	0	172	1	152	0	109	6	103	6
MIZENMAST:—																		
Upper side of crossjack yard	73	1	71	10	70	2	64	0	61	0	62	0	52	0	32	0		
Under side of mizentop	83	4	83	7	81	4	75	0	71	0	72	0	62	0	39	0		
Upper side of cap to mizenmast	93	6	96	5	93	9	85	0	82	0	83	0	71	0	47	0		
,, mizen topsail yard	121	4	116	9	117	1	110	0	107	0	108	0	94	0	56	0		
,, crosstrees to mizen topmast	127	11	125	4	122	4	117	0	112	0	113	0	99	0	61	0		
Head of topgallant rigging	154	2	150	11	145	8	143	0	136	0	137	0	120	0	74	0		
Truck	177	9	171	7	158	9	160	0	153	0	150	0	133	6	85	0		

APP. D. ANGLES SUBTENDED BY FRENCH MASTS. 589

TABLE XXII.

Showing the Angles subtended by the Mainmasts of French Ships of War, between the Water-line and the Truck, also between the Water-line and the Crosstrees, at distances expressed in English yards and in Cables' lengths. The eye of the observer is supposed to be 20 feet above the level of the Water.

Distances.		Line-of Battle Ships.								Frigates.				Corvettes.		Brigs.	
		120 Guns.		90 Guns.		82 Guns.		60 Guns.		44 Guns.		24 Guns.		18 Guns.	
Yards.	Cables' lengths.	Ft. in. 219 9 Truck of Mainmast.	Ft. in. 158 5 Maintop-mast Crosstrees.	Ft. in. 201 9 Truck of Mainmast.	Ft. in. 150 11 Maintop-mast Crosstrees.	Ft. in. 192 5 Truck of Mainmast.	Ft. in. 138 0 Maintop-mast Crosstrees.	Ft. in. 187 9 Truck of Mainmast.	Ft. in. 139 0 Maintop-mast Crosstrees.	Ft. in. 168 6 Truck of Mainmast.	Ft. in. 121 0 Maintop-mast Crosstrees.	Ft. in. 120 7 Truck of Mainmast.	Ft. in. 85 0 Maintop-mast Crosstrees.	Ft. in. 112 0 Truck of Mainmast.	Ft. in. 77 0 Maintop-mast Crosstrees.
120	½	32 14	24 13	29 59	23 3	28 47	21 11	28 11	21 12	25 37	18 51	18 50	13 25	17 31	12 11
240	1	17 7	12 29	15 46	11 50	15 4	10 54	14 43	10 55	13 15	9 35	9 34	6 45	8 52	6 7
360	1½	11 33	8 22	10 37	8 56	10 8	7 18	9 54	7 19	8 54	6 24	6 24	4 30	5 56	4 5
480	2	8 42	6 17	7 59	5 57	7 38	5 29	7 27	5 30	6 41	4 48	4 48	3 23	4 27	3 4
600	2½	6 59	5 2	6 24	4 46	6 7	4 23	5 58	4 23	5 21	3 51	3 51	2 42	3 34	2 27
720	3	5 49	4 12	5 20	3 59	5 6	3 39	4 58	3 39	4 28	3 13	3 13	2 15	2 58	2 3
840	3½	4 59	3 36	4 35	3 24	4 22	3 8	4 16	3 8	3 50	2 45	2 45	1 56	2 33	1 45
960	4	4 22	3 9	4 1	2 59	3 49	2 44	3 44	2 44	3 21	2 24	2 24	1 41	2 14	1 32
1080	4½	3 53	2 48	3 34	2 42	3 24	2 26	3 19	2 26	2 59	2 9	1 9	1 30	1 59	1 22
1200	5	3 30	2 31	3 21	2 23	3 4	2 12	2 59	2 12	2 41	1 55	1 55	1 21	1 47	1 14
1320	5½	3 11	2 17	2 55	2 10	2 47	2 0	2 43	2 0	2 26	1 45	1 45	1 14	1 37	1 7
1440	6	2 55	2 6	2 40	1 59	2 33	1 50	2 29	1 50	2 14	1 36	1 36	1 8	1 29	1 1

590 NAVAL GUNNERY. APP. I

TABLE XXIII.

Tangent Practice with 8-inch Guns, weighing 65 and 60 cwt., carrying a single shot weighing 56 lbs., or a shell weighing 51 lbs., and charged with 10 lbs. of powder. The line of sight is parallel to the axis of the bore, and the Gun is 5 feet 4 inches above the level of the water.

Elevations.	Distances.	Heights of the parts aimed at above the Water.	The points aimed at.	
			In a Line-of-Battle Ship of 82 Guns.	In a 44-Gun Frigate.
° ′ ″	Yards.	Ft. fm.		
0 22 20	330	11 10	One foot below the port-sills of the main-deck.	One and a half foot below the port-sills of the quarter-deck.
0 35 0	435	18 7	Seven inches below the port-sills of the quarter deck and forecastle.	Five feet above ditto.
0 41 0	535	27 4	Two feet above the after part of poop plank sheer.	Twice the height from the water to ditto.
0 47 30	630	38 4	About midway between the water-line and the under side of the fore-yard.	Three times ditto.
1 0 0	756	58 0	Three feet below the upper side of the fore-yard. [maintop.	Upper side of the main-yard.
1 20 0	879	82 0	Upper side of cap to mizenmast; or four feet above the under side of the maintop.	Two feet below the upper side of the cap of the mainmast.
2 0 0	1000	110 0	Two feet below the crosstrees of the mizenmast.	One foot above the upper side of the crosstrees of the foretopmast.
2 20 0	1078	137 0	One foot below the crosstrees of the maintopmast.	One foot a ove the head of the top-gallant rigging, foremast.
2 40 0	1153	166 0	Four feet below the head of top-gallant rigging.	Two feet below the truck of the mainmast.
3 0 0	1225	198 0	Six feet above the truck of mainmast.	

Tangent Practice with 32-pounder Guns, weighing 56 cwt., carrying one solid shot, charge 10 lbs. The line of sight being parallel to the axis of the bore, and the Gun is 5 feet 4 inches above the water.

0 15 0	343	9 10	Three feet below the port-sills of the upper-deck.	Three feet above the port-sills of the upper-deck.
0 30 0	435	16 3	Four feet above ditto.	Three feet above the port-sills of the quarter-deck.
0 41 0	525	24 5	One foot below the after part of poop plank sheer.	About mi tw av between the water and the under side of fore-yard.
0 52 0	6 3	33 0	Midway between the water and the upper side of main-yard.	One-third of distance from the water to the crosstrees of mizen-
1 0 0	700	42 0	About midway between the water and the upper side of the cap of the mizenmast.	Midway between the water and cap of mainmast. [topmast.
1 20 0	835	63 8	About two feet below the upper side of the main-yard.	The under side of foretop.
1 40 0	968	90 0	F ur feet below the upper side of the cap of the mainmast.	Four feet under the mizen topsail yard.
2 0 0	1100	120 0	Three feet above the upper side of foretopsail-yard.	One foot below the upper side of the crosstrees of maintopmast.
2 20 0	1240	157 0	One foot above th head of the top gallant rigging, foremast.	Five feet above the truck of f remast.
2 40 0	1378	198 0	Five feet above the truck of mainmast.	

Tangent Practice with 32-pounder Guns, weighing 50 and 48 cwt., carrying one solid shot, charge 8 lbs. The line of sight is parallel to the axis of the bore, and the Gun 5 feet 6 inches above the water.

0 15 0	320	9 8	Three feet below the port-sills of upper-deck.	Three feet above the port-sills of upper-deck.
0 26 0	437	15 5	Three feet above ditto.	Two feet above the port-sills of quarter-deck.
0 37 0	547	23 0	Two feet below the upper part of poop plank sheer.	One-third of the distance from the water to the under side of maintop.
0 48 0	655	33 0	Midway between the water and the upper part of poop plank sheer.	One-third of the distance from the water to the upper sill of the crosstrees, mizentopmast.
1 0 0	760	45 0	About midway between the water and the upper side of the cap of mainmast.	One-third of the distance from the water to the head of the top-gallant rigging, foremast.
1 20 0	881	67 0	One foot above the upper side of main-yard.	Three feet below the under side of maintop.
1 40 0	999	93 0	One and a half feet below the upper side of the cap of mainmast.	One foot below the upper side of mizen topsail-yard.
2 0 0	1116	122 0	Four feet below the head of the top gallant rigging, foremast.	One and a half feet above the upper side of the crosstrees, main-
2 30 0	1218	165 0	Five feet below the upper side of the crosstrees of foretopmast.	Three feet below the truck of the mainmast. [topmast.
3 0 0	1320	213 0	Eighteen feet above the truck of the mainmast.	

Table XXIV.

Ranges with "Shrapnel's Shells" for Sea-service Ordnance.

Nature of Gun.				Description of Shell.				
Carronade.	Weight.	Charge.	Calibre.	Diameter.	Weight empty.	Weight filled.	Number of Balls.	Bursting Charge.
	Cwt.	lbs.	In.	In.	lbs.	lbs.	No.	Oz.
8-Inch . .	36	..	8·05	7·9	32	59½	395	15
42-Pounder	22	..	6·84	6·762	19¼	35¾	240	7½
32-Pounder	17	$2\frac{10}{16}$	6·25	6·177	15	27½	178	7

Carronade.	Lengths of Fuze Composition, in inches or Decimals of an Inch.	B ·2	C ·3	D ·4	E ·5	F ·6	G ·7	·8	·9	1·0	1·1	1·2	1·3	1·4	1·5	1·6	1·7	1·8	1·9	2·0
8-Inch	Elevations in Degrees .	2⅛	3⅛	4¼	5⅛	5⅞	6¼	7⅛	7¾	8⅝	9	9¾	10¼	11⅛	11¾	12½	13¼	14	15	
	Range in Yards . . .	500	750	950	1100	1225	1350	1475	1600	1700	1800	1900	2000	2100	2200	2300	2375	2450	2525	
42-Pounder	Elevations in Degrees .	2⅛	2¾	3⅜	4	4⅝	5⅛	5¾	6⅜	6⅞	7½	8⅛	8⅞	9⅝	10⅜	11⅛	12	12¾	13⅛	14½
	Range in Yards . . .	500	700	850	975	1100	1220	1330	1440	1550	1650	1750	1840	1930	2020	2110	2200	2280	2360	2440
32-Pounder	Elevations in Degrees .	2¼	2⅞	3¼	3⅞	4½	5⅛	5¾	6⅜	6⅞	7½	8⅛	8⅞	9⅝	10⅜	11	11⅝	12⅛	13	14¼
	Range in Yards . . .	500	700	850	975	1100	1200	1300	1400	1500	1590	1680	1770	1860	1950	2040	2120	2200	2280	2360

The letters B, C, &c., designate the fuze for the various times of flight, or rather length of fuze. This is necessary in the field, where the distances vary rapidly. After ·7 fuze or G you must cut your fuze accordingly.

TABLE XXV.

Ranges with "Shrapnel's Shells" for Field Ordnance.

Nature of Gun.			Description of Shell.						Lengths of Fuzes (Fuze Composition) with the Corresponding Elevations and Ranges.															
Field Howitzer.	Weight.	Charge.	Calibre.	Diameter.	Weight empty.	Weight filled.	Number of Balls.	Bursting Charge.	Lengths of Fuze Composition, in Inches or Decimals of an Inch.	A ·1	B ·2	C ·3	D ·4	E ·5	F ·6	G ·7	·8	·9	1·0	1·1	1·2	1·3	1·4	1·5
	Cwt.	lbs.	In.	In.	lbs.	lbs.	No.	Oz.																
24-Pounder . . .	12½	2½	5·72	5·595	11 1/16	21½	128	6	Elevations in Degrees . . .	1	1½	2	2½	3	3½	4	4½	5	5½					
									Ranges in Yards	450	550	650	750	850	950	1050	1125	1175	1225					
5½-Inch New Gun for Boats . .	10	2	5·62	5·595	11 1/16	21½	128	6	Elevations in Degrees	1¾	2¼	2¾	3½	4¼	4¾	5½	6¼	6⅞	7½	8⅛	8¾	9⅜	10	10⅝
									Ranges in Yards	350	475	600	700	800	900	1000	1100	1200	1300	1400	1475	1550	1625	1700
12-Pounder . . .	6½	1¼	4·58	4·454	5 1/16	10·82	63	4½	Elevations in Degrees	1¼	1¾	2¼	2¾	3¼	3¾	4¼	5	6						
									Ranges in Yards	450	550	650	750	850	950	1025	1100	1200						

TABLE XXVI.

Ranges with "Shrapnel's Shells" for Field Ordnance.

Nature of Gun.				Description of Shell.						Lengths of Fuses (Fuze Composition) with the Corresponding Elevations and Ranges.																		
Field Gun.	Weight.	Charge.	Calibre.	Diameter.	Weight empty.	Weight filled.	Number of Balls.	Bursting Charge.	Lengths of Fuze Composition, in inches or Decimals of an Inch.	B ·2	C ·3	D ·4	E ·5	F ·6	G ·7	·8	·9	1·0	1·1	1·2	1·3	1·4	1·5	1·6	1·7	1·8	1·9	2·0
	Cwt.	lbs.	In.	In.	lbs.	lbs.	No.	Oz.																				
12-Pounder	18	4	4·62	4·45	$5\frac{10}{16}$	10·82	63	4½	Elevations in Degrees	1⅜	1⅞	2¼	2¾	3¼	3¾	4⅜	5⅛	5⅝	6⅜	7	7⅝	8¼	8⅞	9⅜	10			
									Range in Yards . .	750	900	1050	1175	1300	1400	1500	1600	1700	1800	1900	1975	2050	2125	2200	2250			
9-Pounder	13½	3	4·2	4·033	4·59	8·08	41	3½	Elevations in Degrees	1⅜	1⅞	2¼	2¾	3¼	3¾	4⅜	5⅛	5¾	6⅜	7	7⅝	8¼	8¾	9⅜	10	10¾		
									Range in Yards . .	725	875	1000	1125	1240	1340	1430	1520	1610	1700	1780	1860	1940	2010	2080	2140	2200		
6-Pounder	6	1½	3·660	3·55	$3\frac{2}{16}$	5·44	27	2½	Elevations in Degrees	1¼	1⅝	2¼	2¾	3⅜	4	4⅝	5¼	5⅞	6¼	7⅛	7⅝	8¼	8⅞	9⅝	10¼	11	11⅝	12⅜
									Range in Yards . .	650	800	900	1000	1100	1200	1300	1390	1480	1560	1640	1700	1760	1820	1870	1920	1960	2000	2040

TABLE XXVII.
Ranges of Sea-service Iron Mortars.

At Woolwich in 1798, Elevation 40°.

		1	2	3	4	5	6	7	8	9	10	12	14	16	18	20
A 13-Inch Mortar, weighing 82 cwt. 1 qr., weight of shell 201 lbs, and its diameter 12·85 inches	Charge in lbs.	..	2	..	4	..	6	..	8	..	10	12	14	16	18	20
	Range in Yards	..	690	..	1400	..	1900	..	2575	..	2975	3500	3860	3900	4000	4200
	Time of Flight	..	13″	..	18″	..	21″	..	24″·5	..	26″·5	29″	29″·5	30″	30″·5	31″
A 10-Inch Mortar, weighing 41 cwt. 1 qr., weight of shell 88 lbs, and its diameter 9·85 inches	Charge in lbs.	1	2	3	4	5	6	7	8	9	9¼					
	Range in yards	680	1340	1900	2500	2800	3200	3500	3800	3900	4000					
	Time of Flight	13″	18″	21″	24″·5	26″	27″	29″	30″	30″·25	30″·5					

At Landguard Fort in 1806, Elevation 45°.

		1	2	3	4	5	6	7	8	9	10	12	14	16	18	20
A 13-Inch Mortar, weight of shell loaded 200 lbs.	Charge in lbs.	1	2	3	4	5	6	7	8	9	10	12	14	16	18	20
	Range in Yards	420	620	980	1242	1646	1939	2180	2478	2800	2960	3338	3617	3849	4 04	4200
	Time of Flight	9″	11″·25	14″·12	16″·36	18″·6	20″·56	21″·9	23″·4	24″·25	25″·3	26″·9	28″·1	29″·1	29″·8	30″·4
A 10-Inch Mortar, weight of shell loaded 92 lbs.	Charge in lbs.	1	2	3	4	5	6									
	Range in Yards	1032	1663	2180	2562	2891	3090									

APP. D. LIST OF THE ROYAL NAVY. 595

TABLE XXVIII.

List of the Royal Navy, on Saturday, March 19, 1854, from the Official Navy List.

Rate	Sailing Ships: Com.	Ord.	Build.	Sub.	Screw Propellers: Com.	Ord.	Build.	Sub.	Paddle-wheel Propellers: Com.	Ord.	Build.	Sub.	Total: Com.	Ord.	Build.	Sub.	Guard, Troop, Store, &c., Ships	Tenders
First Rate	7	3	3	13	1	0	3	4	..	..	..	..	8	3	6	17		
Second Rate	8	24	3	35	9	..	6	15	..	..	..	..	17	24	9	50		
Third Rate	5	18	..	23	1	1	..	1	..	..	..	..	6	18	..	24		
Fourth Rate	5	22	7	34	5	2	7	14	..	..	..	..	10	24	14	48		
Fifth Rate	5	34	..	39	3	1	1	5	..	..	..	..	8	35	1	44		
Sixth Rate	12	16	..	28	2	3	1	6	..	..	..	..	14	19	1	34		
Sloops	25	38	2	65	9	..	10	19	..	..	..	..	34	38	12	84		
Frigates (Paddle-wheel)	..	..	..	..	..	..	..	..	16	2	..	18	16	2	..	18		
Sloops (Paddle-wheel)	..	..	..	..	..	..	..	..	24	8	..	32	24	8	..	32		
Guard, Troop, Store, &c., Ships				32				6				2				40	40	
Tenders				30				67								97		97

Grand Totals, exclusive of Tenders, Troop, Store, &c., Ships:—
Sail 237
Screw 64
Paddle-wheel . . 50
                ———
                351

2 Q 2

(E.)

Copy of the Prospectus compiled by Sir S. J. Pechell from the 'Naval Gunnery' for the Establishment on Board the "Excellent," as stated in his Letters to the Author, Note, pp. 9, 10, Part I., and in the Text, p. 10.

THEIR Lordships having had under their consideration the propriety and expediency of establishing a permanent corps of seamen to act as Captains of Guns, as well as a depôt for the instruction of the officers and seamen of His Majesty's Navy in the theory and practice of Naval Gunnery, at which a uniform system shall be observed and communicated throughout the Navy, have directed, with a view to the formation of such Establishment, that a proportion of intelligent, young, and active seamen shall be engaged for five or seven years, renewable at their expiration, with an increase of pay attached to each consecutive re-engagement, from which the important situation of Master Gunners, Gunners' Mates, and Yeomen of the Powder-room shall hereafter be selected, to instruct the officers and seamen on board such ships as they may be appointed to, in the various duties at the guns, in consideration of which they will be allowed 2s. per month, in addition to any other rating they may be deemed qualified to fill, and will be advanced according to merit and the degree of attention paid to their duty, which, if zealously performed, will entitle them to aspire to the important situations before mentioned, as well as that of Boatswain.

Their Lordships have therefore directed the "Excellent," with her present fittings (already placed in a situation where practice may be carried on with shot without risk of injury to any individuals), to be established as a sixth rate, with a complement of 200 men, and appointed Captain . . . . . . to the command of her.

The following instructions are sent for your guidance and that of Captain . . . . . . in the execution of these duties:—

APP. E. PROSPECTUS.

Complement.

Captain	.	1
Lieutenants	.	4
Surgeon	.	1
Purser	.	1
Assistant-Surgeon	.	1
Midshipmen	.	15
Clerk	.	1
Warrant officers	.	8
Ship's cook	.	1
Cook's mate	.	1
Carpenter's crew	.	2
Armourer	.	1
Purser's steward	.	1
Sick-boy	.	1

			Boys of Second Class.
Captain's servants	.	2	1
Gun-room ditto	.	2	4
Midshipmen's berth	.	2	1
Warrant officers' ditto	.		3
Purser's steward	.	.	1
Marines	.	34[a]	
Seamen gunners	.	116	
		190	10

[a] Including 1 officer of Marine Artillery, 3 non-commissioned officers, and 2 privates.

As in the establishment of the officers and crew of the "Excellent," a Lieutenant, three non-commissioned officers, and two privates of the Marine Artillery are included in her complement of Marines, it is intended that the theoretical instruction required for the officers and seamen-gunners should be furnished by them, and you will take care that every facility and assistance be given them to insure the performance of this duty, the most material points of which are the names of the different parts of a gun and carriage, the dispart in terms of lineal magnitude and in degrees, how taken, what constitutes point blank and what line of metal range, windage, the errors and loss of force attending it, the importance of preserving shot from rust, the theory of the most material effects of different charges of powder applied to practice, with a single shot, also with a plurality of balls, showing how these affect accuracy, penetration, and splinters; to qualify them to judge of the condition of gunpowder by inspection; to ascertain its quality by the ordinary tests and trials, as well as by actual proof, these being very indispensable qualifications; to instruct them also in the laboratory works required for the naval service,[b] such as making rockets for signals, filling tubes, new priming them and filling cartridges, precautions in airing and drying powder, care and inspection of locks, choice of flints, correct mode of fitting them, &c. &c.

[b] Laboratory works being dangerous, ought to be taught on shore.

You are to understand that it is the intention of their Lordships that the Gunners from His Majesty's ships in ordinary, and also from the ships in commission when they can be spared, should assemble on board the "Excellent," in divisions of such numbers as the Commander may deem convenient, to carry on (assisted by the Marine Artillery already embarked in her) the fullest experiments as to the power and ranges of the various natures of sea-ordnance from point blank to the highest elevation the ports of the "Excellent" will admit of (or as may be safely tried without danger of the shot reaching the shore beyond the mud-banks), also the ranges at similar elevations with different reduced charges of powder, likewise the difference in the ranges when two shots are introduced instead of one,[c] and in such cases to observe and note down the apparent divergence of both shots from the direct line. Also, if it can be tried with safety, the range and force of grape and canister shot; and, in short, every experiment of such description which will tend to give the gunners and others who may attend such practice the most perfect knowledge of the exact powers of each nature of gun in every manner in which it can be tried.

[c] Two shots are as many as ought to be used for safety and effect.

To facilitate these experiments their Lordships consider that Beacons may be fixed in the mud at different measured distances from the ship, say at every hundred or every two hundred yards, or at such other distances as may be found most convenient.

At the same time that the above-mentioned experiments are going forward it will be the duty of the Captain and of the Lieutenants to assist him, to endeavour to ascertain the comparative value of the several descriptions of sights for cannon which have been submitted by various individuals, some of every kind of which the Board of Ordnance has been desired to cause to be put on board the "Excellent."

It is also their Lordships' intention that the efficiency of the improved tube boxes, powder flasks, and all other implements of every description connected with sea-gunnery practice, should be proved, as far as may be done, on board the "Excellent," and Captain . . . . . . is to consider it an important part of his duty to report impartially his opinion on all the implements in question, and to submit for their Lordships' consideration any alteration of any of them deemed likely to prove advantageous by himself or any of the officers assisting him in conducting the duties hereby ordered.

The Captain is also to make known to their Lordships any improvements he may have been informed of, either in guns themselves, or in the mode of mounting, or fitting, or fighting them, or of the implements for serving them which may not have been furnished by the Ordnance Department to the "Excellent," in order that they may cause them to be also supplied for trial and report.[d]

[d] The improvements or inventions ought to be first submitted to the Admiralty.

Another material branch of the duty of Captain . . . . . . will be to perfect the gunners and all others who may attend on board the "Excellent" for that purpose, in the established exercise or service of the guns, to the end that each of them may fully understand and be able to explain the object of every movement ordered; that they may likewise understand perfectly the principle of the sights, moveable targets, and everything used in gun practice, either for exercise or real service.

All these points, after being fully considered and tried in the various exercises and practice to be daily carried forward, are to be fully reported upon to their Lordships, in order that they may give directions for the general adoption of that system which shall be found and admitted, upon full and fair trial, to answer best in practice.[e]

[e] A register to be kept on board of all experiments, that the Commanding Officer may be able to contrast the results.—A second register to be kept of the exercises which are carried on daily, containing a nominal list of the persons under instruction.

It is further to be understood that any ships in commission, the Captains of which are desirous of sending any portion of their officers, captains of guns, or others of their crews, to attend the practice or exercise on board the "Excellent," to gain instruction on any of the points detailed, are to be at liberty to do so, and it is to be an essential part of the duty of Captain . . . . . . and the Lieutenants to give every useful information to persons so sent for instruction, and to advance them on the points most useful for them to understand, to such extent as the short time they can probably be spared will admit.

Their Lordships have requested the Board of Ordnance to give instructions to their officers to render every assistance in forwarding the objects of these instructions, and to supply such quantities of ammunition or other articles as may from time to time be required by Captain . . . . . . and approved by you for the purposes above detailed.

Captain . . . . . . is to be assisted in conducting these duties by the officer of Marine Artillery to the utmost of his abilities, and the latter officer is to be directed to obey whatever directions he may receive from Captain . . . . . . or the Lieutenants for the objects stated whilst on this service.

Their Lordships desire that you send a copy of this letter to Captain . . . . . ., that he may be fully apprised of the duties he is to execute, directing him to govern himself and those placed under him accordingly. And you are also to give such further directions and assistance as you may deem necessary or advisable for the more perfect accomplishment of the objects explained; and you are to cause Captain . . . . . . to make a weekly return to you of each day's transactions and practice, noting the number and descriptions of persons attending on board each day; to which also is to be added any remarks the Captain may deem it right to offer relative to the occurrences or details; and these weekly returns are to be regularly transmitted by you to me for their Lordships' information.^f

^f Certificates of qualification, founded upon the register before mentioned, should be given to each man, according to his merit and proficiency on completing his instruction; and copies of the same transmitted to the Admiralty.

(F.)

ARMAMENT OF THE BRITISH NAVY.*

(*Dated July*, 1848.)

First Rates. "*Caledonia*" *Class.*ᵃ

Lower Deck { 8 8-inch guns, 65 cwt., 9 feet.
            { 24 32-prs. of 56 cwt., 9¼ feet.
Middle Deck { 4 8-inch guns, 65 cwt., 9 feet.
            { 30 32-prs.   50 ,, 9 ,,
Main Deck . . . 34 32 ,,   42 ,, 8 ,,
Quarter Deck and { 6 32 ,, 45 ,, 8½ ,,
Forecastle . . { 14 32-pr. carronades, 17 cwt.

120 guns. Complement of men, 1000.

First Rates. "*Royal Frederick*" *Class.*

Lower Deck { 8 8-inch guns, 65 cwt., 9 feet.
            { 22 32-prs.   56 ,, 9¼ ,, .
Middle Deck { 4 8-inch guns, 65 ,, 9 ,,
            { 26 32-prs.   56 ,, 9½ ,,
Main Deck . . . 32 32 ,,   42 ,, 8 ,,
Quarter Deck and { 8 32 ,, 45 ,, 8½ ,,
Forecastle . . { 16 32 ,, 25 ,, 6 ,,

116 guns. Complement of men, 950.

ᵃ "ROYAL ALBERT'S" ARMAMENT. (*June*, 1854.)

Lower Deck . . . . . 8 in. of 65 cwt. 9 ft. 0 in. long, 32 guns.
Middle Deck . . . . . 32-prs. 56 ,, 9 6 ,, 32 ,,
Upper Deck . . . . . . 32 ,, 42 ,, 8 0 ,, 32 ,,
Quarter Deck and Forecastle { 32 ,, 42 ,, 8 0 ,, 24 ,,
                            { 68 ,, 95 ,, 10 0 ,, 1 ,,

Total 121 guns.

ARMAMENT FOR BOATS.

Brass Ordnance { Howitzers { 24-pr. 13 cwt. 2 guns.
               {   { 12 ,, 6½ ,, 2 ,,
               { Guns . .   6 ,,   6 ,,   1 ,,

The Number of SHELLS.

8-inch filled . . . . 500 shells.
32-pr   ,, ... 500 ,,
24   ,, ... 70 ,, } Shrapnel.
12   ,, ... 70 ,, }
24   empty   . . . . 30 ,, } Ditto.
12   ,, ... 30 ,, }

---

* This Table is not introduced as containing a correct statement of the present armament of the British Navy, but to prove what is stated in this work (pp. 307, 308) concerning the vast increase in the number of shell-guns now employed, and the great extent to which the shell-system is now carried in our naval service.

Second Rates. "*Princess Charlotte*" *Class.*

Lower Deck	{ 8 8-inch guns, 65 cwt.,	9 feet.
	{ 20 32-prs. 56 ,,	9½ ,,
Middle Deck	{ 4 8-inch guns, 65 ,,	9 ,,
	{ 26 32-prs. 48 ,,	8 ,,
Main Deck . . .	30 32 ,, 32 ,,	6½ ,,
Quarter Deck and	{ 6 32 ,, 45 ,,	8½ ,,
Forecastle . .	{ 10 32 pr. carronades, 17 cwt.	

104 guns. Complement of men, 850.

Second Rates. "*Rodney*" *Class.*

Lower Deck	{ 18 8-inch guns, 65 cwt.,	9 feet.
	{ 14 32-prs. 56 ,,	9½ ,,
Main Deck . . .	{ 6 8-inch guns, 65 ,,	9 ,,
	{ 28 32-prs. 56 ,,	9½ ,,
Quarter Deck and	{ 2 8-inch guns, 52 ,,	8 ,,
Forecastle . .	{ 24 32-prs. 42 ,,	8 ,,

92 guns.

Second Rates. "*Albion*" *Class.*

Lower Deck	{ 18 8-inch guns, 65 cwt.,	9 feet.
	{ 14 32-prs. 56 ,,	9½ ,,
Main Deck . . .	{ 6 8-inch guns, 65 ,,	9 ,,
	{ 26 32-prs. 56 ,,	9½ ,,
Quarter Deck and	{ 2 8-inch guns, 65 ,,	9 ,,
Forecastle . .	{ 24 32-prs. 42 ,,	8 ,,

90 guns.

Second Rates. "*Asia*" *Class.*

Lower Deck	{ 8 8-inch guns, 65 cwt.,	9 feet.
	{ 22 32-prs. 56 ,,	9½ ,,
Main Deck . . .	{ 4 8-inch guns, 65 ,,	9 ,,
	{ 28 32-prs. 48 ,,	8 ,,
Quarter Deck and	{ 6 32 ,, 42 ,,	8 ,,
Forecastle . .	{ 16 32-pr. carronades, 17 cwt.	

84 guns.

Second Rates. "*Superb*" *Class.*

Lower Deck	{ 16 8-inch guns, 65 cwt.,	9 feet.
	{ 12 32-prs. 56 ,,	9¼ ,,
Main Deck . . .	{ 4 8-inch guns, 65 ,,	9 ,,
	{ 24 32-prs. 50 ,,	9 ,,
Quarter Deck and	} 24 32-prs. 42 ,,	8 ,,
Forecastle . .		

80 guns.

ARMAMENT OF THE BRITISH NAVY.

Third Rates. "*Bellerophon*" *Class.*

Lower Deck . . { 6 8-inch guns, 65 cwt., 9 feet.
                 24 32-prs.   56 "   9½ "
Main Deck . . . { 2 8-inch guns, 65 "   9 "
                 30 32-prs.   45 "   8½ "
Quarter Deck and { 6 32 "   42 "   8 "
Forecastle . . { 10 32-pr. carronades, 17 cwt.

78 guns.

Third Rates. "*Agincourt*" *Class.*

Lower Deck . . { 6 8-inch guns, 65 cwt., 9 feet.
                 22 32-prs.   56 "   9½ "
Main Deck . . . 28 32 "   42 "   8 "
Quarter Deck and { 4 32 "   42 "   8 "
Forecastle . . { 12 32-pr. carronades, 17 cwt.

72 guns.

Third Rates. "*Boscawen*" *Class.*

Lower Deck . . { 8 8-inch guns, 65 cwt., 9 feet.
                 18 32-prs.   56 "   9½ "
Main Deck . . . { 2 8-inch guns, 65 "   9 "
                 26 32-prs.   50 "   9 "
Quarter Deck and } 16 32 "   42 "   8 "
Forecastle . . }

70 guns.

Note.—All line-of-battle ships of the 1st, 2nd, and 3rd rates, have 2 24-pounder brass howitzers, 13 cwt., and 2 12-pounder howitzers, light, 6 cwt., for boat service; and 1 6-pounder light brass gun for boat service and short practice.

Fourth Rates. "*Ajax*" *Class, for Block Ships with auxiliary Steam-power Screw.*

Lower Deck . . 26 42-prs.   66 cwt., 9¼ feet.
Main Deck . . . 22 32 "   42 "   8 "
Quarter Deck and { 4 56 "   87 "   10 "
Forecastle . . { 4 10-inch-guns, 86 "   9 ft. 4 in.

56 guns. Complement of men, 500.

Fourth Rates. "*Vernon*" *Class.*

Main Deck . . . { 8 8-inch guns, 65 cwt., 9 feet.
                 20 32-prs.   56 "   9½ "
Quarter Deck and { 4 8-inch guns, 65 "   9 "
Forecastle . . { 18 32-prs.   45 "   8½ "

50 guns. Complement of men, 500.

Fourth Rates. "*Portland*" *Class.*

Main Deck	{ 8 8-inch guns,	60 cwt.,	8 ft. 10 in.
	22 32-prs.	50 "	8 feet.
Quarter Deck and	{ 4 32 "	45 "	8¼ "
Forecastle	16 32 "	25 "	6 "

50 guns. Complement of men, 450.

Fourth Rates. "*Arrogant*," *auxiliary Steam-power Screw.*

Main Deck	{ 12 8-inch guns,	65 cwt.,	9 feet.
	16 32-prs.	56 "	9¼ "
Quarter Deck and	{ 2 68 "	95 "	10 "
Forecastle	16 32 "	32 "	6¾ "

46 guns. Complement of men, 450.

Note.—All ships of the 4th and 5th rates have, for their boats' armaments, 2 12-pounder heavy howitzers, 10 cwt.; and 2 12-pounder howitzers, light, of 6 cwt., with 1 light 6-pounder brass gun for boats and short practice.

Fifth Rates. "*Pique*" *Class.*

Main Deck	{ 6 8-inch guns,	60 cwt.,	8 ft. 10 in.
	18 32-prs.	56 "	9¼ feet.
Quarter Deck and Forecastle	} 16 32 "	42 "	8 "

40 guns. Complement of men, 350.

Fifth Rates. "*Madagascar*" *Class.*

Main Deck	{ 2 8-inch guns,	60 cwt.,	8 ft. 10 in.
	26 32-prs.	40 "	7½ feet.
Quarter Deck and	{ 4 32 "	45 "	8¼ "
Forecastle	12 32-pr. carronades, 17 cwt.		

44 guns. Complement of men, 320.

Fifth Rates. "*Resistance*" *Class.*

Main Deck	{ 2 8-inch guns,	52 cwt.,	8 feet.
	22 32-prs.	39 "	7½ "
Quarter Deck and	{ 4 32 "	45 "	8¼ "
Forecastle	{ 4 32 "	39 "	7⅜ "
	10 32-pr. carronades, 17 cwt.		

42 guns. Complement of men, 310.

Fifth Rates. "*Castor*" *Class.*

Main Deck	{ 4 8-inch guns,	60 cwt.,	8 ft. 10 in.
	18 32-prs.	56 "	9¼ feet.
Quarter Deck and	{ 2 32 "	50 "	9 "
Forecastle	12 32 "	25 "	6 "

36 guns. Complement of men, 330.

ARMAMENT OF THE BRITISH NAVY.

Fifth Rate. "*Amphion,*" auxiliary Screw (300-*horse power*).

Main Deck . . . { 6 8-inch guns, 65 cwt., 9 feet.
14 32-prs. 56 „ 9½ „
Quarter Deck and { 2 68 „ 95 „ 10 „
Forecastle . . 8 32 „ 25 „ 6 „

30 guns. Complement of men, 330.

Fifth Rates. "*Eurotus*" *Class, S. Guard Ship* (350-*horse power*).

Main Deck . . . 20 42-prs. 66 cwt., 9¼ feet.
Quarter Deck and { 2 56 „ 87 „ 10 „
Forecastle . . 2 10-inch guns, 86 „ 9 ft. 4 in.

24 guns. Complement of men, 320.

Sixth Rates. "*Vestal*" *Class.*

Main Deck . . . { 2 8-inch guns, 52 cwt., 8 feet.
16 32-prs. 40 „ 7¼ „
Quarter Deck and { 2 32 „ 42 „ 8 „
Forecastle . . 6 32 „ 25 „ 6 „

26 guns. Complement of men, 240.

Sixth Rates. "*Trincomalee*" (*Reduced* 46) *Class.*

Main Deck . . . { 10 8-inch guns, 65 cwt., 9 feet.
8 32-prs. 56 „ 9½ „
Quarter Deck and { 2 56 „ 87 „ 10 „
Forecastle . . 4 32 „ 25 „ 6 „

24 guns. Complement of men, 240.

Sixth Rate. "*Amazon.*" *Reduced* 46.

26 32-prs., 50 cwt., 9 feet.
Complement of men, 240.

Sixth Rate. "*Curaçoa.*"

Main Deck . . . { 20 32-prs. 40 cwt., 7½ feet.
2 32 „ 50 „ 9 „
Quarter Deck . . 2 8-inch guns, 52 „ 8 „

24 guns. Complement of men, 230.

Sixth Rate. "*Brilliant.*" *Reduced* 42.

Main Deck . . . { 6 8-inch guns, 52 cwt., 8 feet.
10 32-prs. 50 „ 9 „
Quarter Deck and { 2 56 „ 87 „ 10 „
Forecastle . . 2 32 „ 25 „ 6 „

20 guns. Complement of men, 230.

Sixth Rate. "*Havannah.*" *Reduced* 42.

Main Deck . . { 6 8-inch guns, 52 cwt., 8 feet.
10 32-prs. 50 „ 9 „
Quarter Deck and { 1 56-pr. 87 „ 10 „
Forecastle . . 2 32-prs. 50 „ 9 „

19 guns. Complement of men, 230.

Sixth Rate. "*Dædalus.*" *Reduced* 42.

Main Deck . . { 6 8-inch guns, 52 cwt., 8 feet.
12 32-prs. 50 „ 9 „
Quarter Deck . . 1 56-pr. 87 „ 10 „

19 guns. Complement of men, 230.

Note.—These classes have for their boats' armaments 2 12-pounder brass howitzers, light, 6 cwt., and 1 6-pounder light gun for short practice.

Second-class Sixth Rates. "*Andromache*" *Class.*

Main Deck . . { 2 68-pr. howitzers, 36 cwt.
16 32-prs. 25 „
Quarter Deck and { 2 32 „ 45 „
Forecastle . . 6 32-pr. carronades, 17 cwt.

26 guns. Complement of men, 195.

Second-class Sixth Rates. "*Calypso*" *Class.*

Main Deck . . { 2 8-inch guns, 52 cwt., 8 feet.
14 32-prs. 40 „ 7½ „
Quarter Deck . . 2 32 „ 45 „ 8½

18 guns. Complement of men, 195.

Second-class Sixth Rates. "*Daphne*" *Class.*

Main Deck . . { 2 8-inch guns, 52 cwt., 8 feet.
14 32-prs. 42 „ 8 „
Quarter Deck . . 2 32 „ 50 „ 9 „

18 guns. Complement of men, 175.

Second-class Sixth Rates. "*North Star.*"

Main Deck . . { 2 32-prs., 39 cwt. 7½ feet.
16 32-pr. carronades, 17 cwt.
Quarter Deck . . 4 32-pr. carronades, 17 cwt.

22 guns. Complement of men, 175.

Second-class sixth rates have 2 12-pr. light howitzers for their boats.

Sloops.

"Nimrod" Class { 2 32-prs., 39 cwt., 7¼ feet. Reduced 28.
16 32 „ 25 „ 6 „

18 guns. Complement of men, 145.

APP. F.   ARMAMENT OF THE BRITISH NAVY.   607

"Siren" Class . { 2 32-prs., 39 cwt., 7½ feet.
                  14 32   "   25 "   6   "

16 guns. Complement of men, 130.

"Champion" Class { 2 32-prs., 39 cwt., 7½ feet.
                   12 32   "   25 "   6   "

14 guns. Complement of men, 130.

"Acorn" Class . { 2 32-prs., 39 cwt., 7½ feet.
                  10 32   "   25 "   6   "

12 guns. Complement of men, 130.

"Childers" Class . { 2 32-prs., 32 cwt., 6½ feet.
                     10 32   "   25 "   6   "

12 guns. Complement of men, 130.

"Cygnet" Class . { 2 32-prs., 32 cwt., 6½ feet.
                   6 32   "   25 "   6   "

8 guns. Complement of men, 80.

"Alert" Class . { 2 32-prs., 25 cwt., 6 feet.
                  6 32-pr. carronades, 17 cwt.

8 guns. Complement of men, 80.

"Pantaloon" Class { 2 18-prs., 20 cwt., 6 feet.
                    6 18   "   15 "   5½ "

8 guns. Complement of men, 80.

Brigs.

"Cameleon" Class { 2 32-prs., 32 cwt., 6½ feet.
                   4 18-pr. carronades, 10 cwt.

6 guns. Complement of men, 65.

"Dolphin" Class . { 1 32-pr., 39 cwt., 7½ feet.
                    2 32-prs., 32 "   6½ "

3 guns. Complement of men, 65.

"Griffin" Class . { 1 32-pr., 39 cwt., 7½ feet.
                    2 24-pr. carronades, 13 cwt.

3 guns. Complement of men, 60.

All these classes of vessels have 1 12-pr. howitzer for boats, light, 6 cwt.

(G.)

FLOATING BATTERIES COVERED WITH IRON PLATES.

THE project of covering ships with iron plates, in order to render them proof against artillery (Art. 164, Note), was suggested many years since by Colonel, now General Paixhans, in his work entitled *Nouvelle Force Maritime*, No. 19. The Comité Consultatif de la Marine at that time, having caused the weight of an iron covering, and the capability of ships to bear the load, to be calculated, found that such armour could not be applied to line-of-battle ships of the lowest class, to frigates, nor to smaller vessels. With respect to ships with three decks, the Comité stated in its Report that the great displacement of these would enable them to bear the requisite weight, provided the quantity of artillery on the upper decks were diminished; and observed that the expense was estimated at 600,000 francs (25,000*l*.) for each ship! The inquiry led, however, to no attempt in France to *cuirasse* ships of war, and the project was apparently abandoned as impracticable.

On referring to the note at the foot of page 346, it will be seen that a proposal for constructing floating batteries of iron so thick as to be shot-proof, was entertained by the United States' Government in or before the year 1852, and that the feasibility of the proposition was made the object of an experiment; the result of this being unfavourable, the project fell to the ground. It would seem, however, that the strength of wrought-iron in a solid mass is far greater than that of iron of equal thickness, but formed of laminæ (as was the case in the American experiment), however firmly these may be bolted together. During the months of September and October, 1854, some experiments were carried on at Portsmouth, in order to try the capability of wrought-iron slabs to resist the impact of solid and hollow shot, and the following are the results: the target was a section of a frigate covered with wrought-iron plates, 4½ inches thick, and the projectiles employed were 32-pounder and 68-pounder solid shots, and 8-inch and 10-inch hollow shots. At 400 yards, the 32-pound solid shot, and the 10-inch and 8-inch hollow shot, merely indented the target to the depths respectively of 1½, 2½, and 1 inch; but the 68-pound solid shot, being fired with 16 lbs. of powder, penetrated the plates. These were always split at the bolt-holes, which were about one foot asunder; and in consequence, it was recommended that they should be bored as far apart as possible. The conclusion drawn from the experiments was, that 4½-inch iron plates, applied as a covering to ships, would give protection, during an action, against 8-inch and 10-inch hollow shot, and against 32-pounder solid shot, but very little against solid shot of 68 lbs.

The project of covering ships with plates of iron has recently been adopted in this country: it was, at first, intended to cover with slabs of iron 4½ inches thick, only the bows of steamers, it being supposed that, in the use of their bow pivot-guns, the heads merely of the vessels would be presented to the enemy. The weight of an iron covering 4½ inches thick, applied to the bows of a screw-propelled ship of 1074 tons, the covering extending 36 feet from the head along each bow, and from the deck to 2 feet below the load-water line, is found to be 10 tons, nearly; and this would surely be an enormous burthen, in addition to that which already loads the bows of a steam-vessel.

From the public journals we learn that an armament is now being prepared to operate against the Russians in the Baltic, at the opening of the spring campaign next year, consisting, it is said, of forty floating batteries, whose decks and sides are to be covered with iron plates of such thickness as to render them proof against either shells or solid shot, and they are to be armed with twelve Lancaster guns.[a] They are about 1500 tons burden, flat-bottomed, with round stem and stern, 180 feet extreme length, 56 in width, and 20 in depth, each being propelled by horizontal engines of 200-horse power. They have two decks, the upper being bomb-proof, 8 inches thick, and the lower the fighting deck. The batteries are perfectly encased with upwards of 700 tons of wrought-iron slabs, each slab 4 inches thick, 12 inches broad, and 14 feet in length.

If the figures in the above extract be correct, the enormous weight of iron casing, together with that of the heavy armament will render it scarcely possible, in these vessels, to satisfy the necessary conditions of buoyancy and stability. With such an immense top-weight, a great counteracting weight of *ballast* will be requisite, but for this there is neither sufficient displacement

---

[a] Floating batteries expressly built for such purposes as these should assuredly be armed with the most efficient and unerring battering guns. The 68-pounder gun of 95 cwt. is infinitely preferable to the Lancaster shell-gun, for, though not chambered according to the Lancaster system, it is intended for shell-firing against ships. No shells can, however, be so efficient against stone walls as solid shot of the same diameter, and the least efficient of all are shells armed with percussion fuzes. There is no evidence either experimental or real, certainly none in the trials made during the present war, that can warrant the preference given to these guns and their peculiar shells in the armament of vessels specially designed for battering granite walls.

nor depth of hold; and as the metacentre must be very near the centre of gravity, the equilibrium cannot but be greatly endangered.

These perfectly flat-bottomed floating batteries, resembling the upper section of a large ship, cut by a horizontal plane some feet below the water-line, and having little hold of the water, will be so much affected in any swell, as to be very unfavourable to good gunnery, where the utmost attainable precision is required. By Dynamics, the times of the vibration of floating bodies vary, as the depth of the vertical section below the plane of floatation; and nothing but a very deep false keel can counteract the tendency of these vessels to follow all the undulations of an agitated fluid. But to give them false keels would add to their draught of water, and so vitiate the scheme.

The shape, dimensions, displacement, casing with shot-proof iron plates, and the armament of these floating batteries, form altogether a very difficult affair, which does not appear to have been fully examined before it was adopted. The idea of making the decks shell-proof, by covering them with iron slabs 8 inches thick, is, we find, given up; and oak beams are substituted for the iron; but these would be broken in by shells fired from mortars, and would be penetrated by a plunging fire of solid shot, from batteries having considerable elevation above the sea-level. Nor will iron slabs, $4\frac{1}{2}$ inches thick, be proof against 68-pounder or 84-pounder solid shot, with which it appears the Russians are plentifully provided; and, unless the timbers of these vessels are enormously thick, such heavy shots will not only punch holes in the iron, but may also make great breaches in the wooden sides, by their prodigious battering power.[a] It is not probable that a Russian fortress can be effectually destroyed by such, or any other kind of floating battery. This can only be done after a land-army shall have obtained possession of the place; and if it is meant to attempt the reduction of a Russian fortress with such vessels alone, the author can only refer to what he has already stated upon that subject in Art. 332, 332 c, pp. 347, 353.

---

[a] Supposing the largest battering-ram in use in ancient warfare to have been worked by 500 men, each exerting a force of 70 lbs., the momentum produced by their action, if it were moved at the rate of 1 foot per second, would be represented by 35,000. The momentum of a 68-pounder shot, moving with a velocity of 1500 feet per second, estimated in like manner, would be represented by 102,000.

(H.)

OBSERVATIONS ON THE LANCASTER GUNS.

The author has ever entertained an opinion that Mr. Lancaster's elliptically bored gun is a dangerous piece of ordnance. The reader will find, on referring to Arts. 190 and 335, that, in the earliest trials of that gun, the shells broke to pieces on leaving it; and the author felt confident that if, with the maximum charge and an elevation of 18°, there were propelled from the gun solid shot or shells so strong as not to break, a great risk of the gun bursting would be incurred. This opinion has been verified by the results of the experimental trials recently carried on at Shoebury Ness, of which an account has been received while this sheet is in type, one of Lancaster's 68-pounder guns having burst while firing one of its peculiar shells, though a reduced charge of 12 lbs., instead of 16 lbs. of powder (the maximum charge), was employed. The fragments were thrown to considerable distances; but, happily, no one was injured, the firing party having, from some distrust of the gun's strength (see Art. 335), been placed under cover: had it been otherwise a fatal catastrophe must have ensued, as in the bursting of the guns at Malta and Gibraltar (pp. 291, 292). It was observed that the vent had become much enlarged; but this circumstance, as well as the reduced charge employed, must be considered as having diminished the danger of the gun bursting. This accident, together with those which have since occurred on service, can only be ascribed to the peculiar formation of the bore, which causes the shot, in forcing its way through, to exert a great strain on the gun. In a former experiment a shell stuck in the bore while the gun was being loaded (p. 172); and it is easy to conceive, therefore, that a shell might stick in going out; in which case, if the shell do not break, the gun must inevitably burst.

To the vast force with which the projectile rubs and strikes against the surface of the bore, while following the spiral turn, and thus acquiring the properties of a rifle-shot, may be attributed the frequent breaking of the projectile, even when made of wrought iron. It requires much habitual skill or knack in serving the gun, to introduce and set the shell home, through the turning of the bore, and it is not unlikely that, in some instances, the bursting of the Lancaster guns may be occasioned by the oval ball leaving a space between it and the gunpowder, or by getting fixed in the gun by change of position whilst it is being propelled through the oval bore. Without the utmost care in loading, these guns must be liable to burst at every round, and the firing must be very slow. The strongest gun bursts readily if the ball be obstructed in its progress when near the muzzle, which, no doubt, was the cause of the bursting of one of the Lancaster guns near the muzzle at Sevastopol. With respect to leaden shots fired from the Lancaster elliptically-bored muskets, no such accident can happen with them; but, as has been shown in p. 517, they often *strip*, or pass straight out of the barrel, which must happen when, being heated in the flame, they change their form and do not follow the windings of the spiral bore. Iron shots are incapable of changing their form, and must, therefore, either follow the spiral bore, or cause the gun to burst. The withdrawal of the Lancaster elliptically-bored guns from the "Pelter" gunboat at Portsmouth, and from the Despatch-gunboats "Arrow" and "Beagle" at Sevastopol, and the judicious order to arm all the new gun-boats with the 68-pounder guns of 95 cwt., are necessary consequences of the very unfavourable reports which were made of those guns at Bomarsund, as being deficient in precision, and not to be depended on (p. 376, line 16 from bottom), corroborated, as these reports have been, from very high authority on the spot, of the very bad practice made by the Lancaster guns at 1300 yards at Sevastopol in the land-batteries, and the fact that two of them burst! In firing into the town they are said to have done great damage to the place, when loaded with their own peculiar shells; but few of these having been supplied, the oval guns have been chiefly used in firing round projectiles, shot and shells, grape and canister-shot, all of which would have been more efficiently and appropriately used from 68-pounder guns. So confident were the expectations entertained of the alleged powers and the assumed precision of the Lancaster guns to destroy any works at a distance of 5000 or 6000 yards, that it is actually intended to rebore 68-pounders and 8-inch shell guns into the elliptical spiral form! This transformation must weaken them so much, that the danger of bursting, to which elliptically bored guns are already so liable, will be greatly increased:—it will undoubtedly spoil a capital 8-inch gun, and make a very bad and dangerous elliptical howitzer; because, being chambered, it is incapable of receiving the large charges requisite to produce the long range, in which, together with their alleged superior precision in distant firing, the peculiar merits of the Lancaster guns were supposed to consist; these were, in fact, the sole reasons for which they were introduced into the naval service. Though executed at enormous cost, and equipped with their peculiar shells, they have failed to accomplish on service the special purposes for which they were designed. They cannot, as has been proved, resist the charge (16 lbs.), nor stand the high elevation (18°) necessary to produce the vaunted range of 5600 yards; they are proved to be defective in precision in distant firing, and even at short ranges; and they have been withdrawn from the Despatch and other gun-boats. No other uses that can be made of that particular gun, whether it be to fire spherical shot from its elliptically spiral bore (pp. 363, 364) or, with its own projectiles, to bombard towns, can redeem it from the verdict which men of science in general pronounce:—that they have failed to accomplish the great objects for which expressly they were made. The bombardment of towns can be more effectually accomplished by mortars (p. 361) of equal weight,[a] projecting much heavier shells with equal bursting charges, and producing very nearly equal ranges.

It is no doubt a great desideratum to obtain for the cannon the advantages which the musket has derived from elongated projectiles; but the author believes that this may be obtained by some vastly better, safer, and far cheaper method, than by firing from an elliptically-bored gun elongated shells, manufactured of wrought iron, at a cost of twenty pounds each, and which nevertheless frequently break when fired (Art. 335, p. 357, and Art. 337, p. 363).

[a] The author learns, with great satisfaction, that mortar-ships are at length preparing for service, in accordance with what is stated in Articles 194, 195, and in p. 361, and Arts. 333, 355. Twenty mortar-ships should have been sent with our fleets to the Baltic and Black Seas. It is stated in the public journals (*Morning Chronicle*, Dec. 9th) that these mortar-ships are to be armed with 13-inch mortars slung by the trunnions to iron rods moving upon an iron shaft of 9 inches in diameter, in such manner that by their pendulous weight they may remain undisturbed by the motion of the vessel, and be fired, with certainty, at any elevation, whatever be the pitching or rolling motion of the ship! We old artillerists of the late war remember Congreve's suspended mortar, or howitzer, for it was either, accordingly as it might be placed in the slings; all such contrivances failed, and the author cannot help expressing his entire conviction that such will be the result with respect to these suspended mortars: neither the iron rods, nor the shaft or centering on which they turn, can resist the prodigious shock occasioned by the explosion of a charge of 20 lbs. of powder; and if any such failure do take place, it cannot but be most serious and perhaps fatal to the ship.

The author takes this opportunity of correcting an error of the press in page 188, Art. 206, line 1, in which for the word "cast" *read* "forged." The gun therein described was of wrought iron, made to replace the monster gun which, as stated in Art. 206, burst with such fatal effect on board the U. S. ship "Princeton;" this was made likewise of wrought iron, not forged in the solid and then bored out, but, he believes, by welding bars of wrought iron together, and fortifying them with strong hoops of that material.

(I.)

ON THE NAVAL AND MILITARY OPERATIONS IN THE BLACK SEA.

THE battle of the Alma, and the powerful attacks subsequently made by the Russians in the field, together with the vigorous defence of Sevastopol itself, though in every conflict success has crowned the heroic exertions of the allied armies, have so thinned the ranks of these that their commanders have been compelled to suspend all active operations till reinforcements in troops and materials sufficient to warrant the prosecution of ulterior measures shall be received. The awful effects of the late storm, moreover, by causing the loss of the vessels containing ammunition and stores, of which the troops begin to feel the want, have brought the armies to a very distressing and even perilous condition. The valour, constancy, physical endurance and heroism of the French and British troops cannot be too highly estimated; and it may be permitted the author to enter into an examination concerning the causes of a failure which has so grievously disappointed the sanguine expectations which the public had been led to entertain of rapid and complete success in the Expedition to which this Appendix relates.

The author would not have presumed to make this important matter the subject of a special publication, but having been informed by his bookseller that the work in which he had long previously treated of naval gunnery was out of print, and that it was much in demand, he could not avoid undertaking a new edition. In so doing he felt himself bound to bring down to the present time all that relates to naval operations, whether of fleets alone, or in combination with land forces; and this he has done amidst much severe domestic affliction. Finding, after the most mature deliberation, that he should be compelled, by his convictions, to treat a part of his subject in a manner which would conflict very much with the prevailing opinion that the war would be brought to a speedy and successful termination by our vast and noble fleets alone; convinced that this expectation would be disappointed, and that the country would be involved in a war by land with a colossal military power for which we were little prepared, he could do no less than conclude this work by re-stating his former convictions and his apprehensions, the justness of which he has, during the continuance of the strife, found no occasion to doubt, and which he has distinctly avowed as often as his opinion was asked, throughout the operations of the present year. In proof that these are no *ex post facto* opinions, the author may appeal to the very numerous persons, of every rank and station, to whom he imparted his apprehension, and stated his opinion; this he did with a degree of hardihood from which, in other circumstances, he would have shrunk, but having stated in Art. 332 that attacks of maritime fortresses by a judicious co-operation of military and naval forces are generally successful, he feels bound to show why, with such an army and navy, this great operation should have so far failed. The actual publication of his observations he has however reserved till this moment, when they can produce no discouraging effects upon the operations of the present campaign; and, by showing past errors, they may tend to provide for a happier result in that which is to come.

At the beginning of the year 1854 there remained little hope that the peace of Europe would be preserved, and it was soon afterwards judged necessary to send a British army to the East, in order to co-operate with one from France. By great exertions upwards of 20,000 troops, infantry and cavalry, were shipped and sent off; the guns, military stores, and provisions were to be despatched in proportion as they could be collected. A few field-batteries only, affording, on an average, scarcely one gun for every thousand men, were sent. Gunner-drivers and horses for the train, waggons to carry ammunition, spring-carts for the sick or wounded, sappers and miners with their intrenching tools, and bridge equipments, with all the other indispensable requisites for an army in the field, were scantily supplied, and some were altogether wanting.

Thus, on a small peace establishment, the country was caught in a political storm and involved in a mighty war. There existed some good regiments of infantry and a few over-officered squadrons —they could not be called regiments—of well-appointed cavalry; but all were totally unprovided with the means necessary for enabling them forthwith to take and keep the field. In this state a military force, constituting nearly the whole of our effective strength, was despatched with wonderful promptitude to the contemplated seat of war; but, lacking the establishments which should have given it vitality, it is not surprising that it was not prepared to enter on a campaign till the season propitious for military operations was near its termination.

These deficiencies and disabilities cannot, however, with any justice be charged to the present Government of the country, nor could they be provided at the eleventh hour by any administrative talent on the part of the new Department of War. The evils were too deeply seated to be removed on a sudden; all that zeal, energy, and ability could accomplish has been done by the Government to repair the evils, and supply the deficiencies which resulted from the persistence of Parliament at, and ever since the conclusion of, the late war, in measures dictated by a reckless spirit of economy to abolish or reduce the military establishments of the country, as if war were never again to overtake the nation. All the establishments which are indispensable to enable the army to take the field, of which we now in vain lament the want, had to be restored and reorganised. For this much time must necessarily be lost, and precious opportunities be let slip, ere England can collect and organise her neglected military resources, and be again prepared to buckle on her armour and exert the plenitude of her military strength, whether for defending her own territories or to resist aggressions directed against those of her allies.

The author feels himself justified in making public the above reflections, because they are consonant with what he has written, published, and on some occasions spoken, since the termination of the late war up to the present eventful period. Now that the evil predicted has overtaken us, it is hoped that the country will take warning and become convinced that something more than numerical strength and personal bravery is required to render an army efficient in the field; and that a nation which had so neglected her military establishments as to have permitted her army to fall into such a state of inefficiency for immediate service, could scarcely be counted upon as a military power capable of promptly taking part in a great territorial war, and of immediately sustaining and vindicating the lofty tone which, without reflecting upon our very limited military means, the Government of this country so nobly assumed, and the spirit of the nation so generously and congenially re-echoed, at the breaking out of this great war.

Under these very disadvantageous and inauspicious circumstances, with respect to the small amount of our effective military force, and the late period of the season at which it was so far equipped as to be able to take the field, the allied army, deeply impregnated with the seeds of disease, and, had it even been in an efficient sanitary state, not numerically strong enough, particularly in cavalry, to ensure success, entered on the arduous service in which it is now engaged, the object being to besiege, capture, and destroy the great fortress and naval arsenal of Sevastopol.

There never was a case in which a siege required to be undertaken with greater regard to the relative strength of the besieged and besieging armies, and to the quantity as well as quality of their siege artillery—never one in which a great superiority of the investing army over the forces forming the garrison of the place was so imperative. In estimating the amount of force required to besiege and capture Sevastopol, regard should have been had to the important fact that, in its local character as a military position, that town is a vast fortress situated on both sides of a long harbour resembling a very broad river, and of which the northern side, occupied by the citadel, is elevated above the southern part. The place belongs therefore to the category of a fortress divided into two portions by an unfordable river,[a] in which case the divisions of the investing corps would be prevented from mutually assisting each other. To invest such a place there is required an army twice as strong as would suffice if no such obstruction to intercommunication and mutual support existed. In this case also, the enemy keeping the field with a numerous army of observation, a strong and very extensive line of circumvallation would be necessary.

With respect to the means of defence, with which it is well known that Sevastopol was plentifully provided,—Here is a vast naval arsenal already well fortified, and capable, from the time of being menaced with an attack, of being greatly strengthened in its works and its garrison: it possessed enormous quantities of ordnance and ammunition, which had been accumulated in its magazines; and, exclusive of the artillerymen attached to the ordnance of the place, it had the power of drawing from the fleet in the harbour vast numbers of well-trained naval gunners, all of whom could be rendered available for manning the artillery during the progress of the siege.

No operation in war may be depended upon with so much certainty as the siege of a fortress, provided it be undertaken with sufficient means and be skilfully conducted;[b] but no measure is so disastrous as the undertaking of a siege, as was the case with that of Burgos, at which the author served, in 1812,[c] without the requisite strength in men and materials. The attacking force should be sufficiently numerous to invest the place on every accessible side, so that nothing may be able to get in or out, and it should be equal in amount to about five, and never less than three, times the garrison: there should be, moreover, in the field a covering army, of which a large portion should be cavalry, in order to protect the operations of the siege, and prevent them from being interrupted by an army

---

[a] "Une place partagée par une rivière non guéable exige le double plus de forces pour son investissement qu'une place autour de laquelle des communications faciles permettent à tous les corps qui en forment l'investissement de s'entre-secourir promptement."—Bousmard, *Essai Général de Fortification*, liv. i. chap. ii. p. 65.
[b] Napier, 'History of the Peninsular War,' vol. iv. p. 476.
[c] Ib., vol. v. p. 369.

of observation, which the enemy may bring up while they are being carried on. The allied army in the Crimea found itself manifestly inadequate to the accomplishment of the object in view, and even the victory on the Alma rendered it still less able to compete with the overwhelming power of its opponents. The allied commanders were then compelled to come to a determination on which side, the northern or the southern, Sevastopol should be attacked: on both at once they could not act. The northern side is strongly fortified; the Russians held a formidable position on the Belbec river, and no shelter for the ships could be afforded on that part of the coast; there was no place, in fact, convenient for landing the siege-train, or for establishing a secure point of communication between the army and the fleet, which at that late season could not be expected long to lie at anchor on the open sea. On these accounts it was promptly determined to abandon the precarious base of operations north of Sevastopol; and to turn the enemy's positions on the Belbec and the northern heights, by a flank march to the south, in order to establish in Balaklava Bay a new base, from which the attack might be made on the southern heights.

To invade the Crimea, an integral portion of the Russian empire, and lay siege to Sevastopol at that late period of the season, and, as has been already observed, with an army deeply impregnated with the seeds of disease, was, in the opinion of the author, a desperate and dangerous operation, undertaken, it is said, contrary to the judgment of an eminent engineer, whose opinion should have ruled, but who, when that determination was taken, did all in his power to meet the difficulties of the case by recommending, as a matter of necessity, to abandon the base north of Sevastopol, whose communications with the sea had at that late period of the year become precarious, and to seize upon some bay south of the place, in which a secure base might be established—a change of position which, under the difficulties in which the allied army was placed from insufficiency of force, was apparently the best measure that could be adopted under such circumstances; and this very critical movement was gallantly and successfully accomplished by a somewhat hazardous march of the whole army.

It is, however, much to be regretted that, from want of sufficient force, it should have been necessary to abandon the line of operations by which the place was at first approached, and on which, at the Alma, the army covering Sevastopol in that direction was defeated. The battle was a brilliant deed of arms, most honourable to the allies; but, in consequence of the change of plan, it must be allowed that, except in its moral effects, it was fruitless, and in some important respects disadvantageous. In laying siege

OPERATIONS IN THE BLACK SEA.

to Sevastopol, it may safely be asserted, that the most advantageous point of attack was the northern side; there the ground is most elevated, and the large octagonal work on its summit is its citadel and the key of the place. This taken, the Telegraph and Wasp batteries on the northern heights, Fort Constantine and the forts below, being commanded and attacked in reverse, must have soon fallen; while the town, docks, arsenal, and barracks on the south side of the harbour would be at the mercy of the allies, who, by the fire of their batteries, might have entirely destroyed them all; whereas, by attacking the place from the south, the enemy holding the northern heights, although the works on the crest of the southern heights should be breached and taken, the town, the body of the place, with its docks and arsenals, will not be tenable by the besiegers till the great work on the northern side, and all its defensive dependencies, shall have been taken; and these, no doubt, will be greatly strengthened before the allies are in a condition to direct their attacks against them.

The flank march of the whole army to the south abandoned at once to the enemy a perfectly free communication between the place to be besieged and his army of observation in the field, and left open their line of operation from their base at Perekop; it disclosed the alarming fact that, from want of sufficient force, Sevastopol could not be invested on every side; that the most advantageous point of attack was not to be attacked, but turned; that the enemy's communication with the strongest portion of the town, its citadel, its keep, and the key of the whole position, was to be left open to him; and that, instead of besieging Sevastopol, the allied army was only to attack an intrenched position on the southern heights, supported in its rear by the strongest feature and most formidable works of the place, and open to receive succour or reinforcements to any extent; also that the attack of the place was to be carried on without a covering army, distinct from the besieging force, to protect it from being disturbed in its operations by the enemy in the field, who was thus left in direct and immediate communication with a tête which he might support with all his force. The flank march of the whole army to the south was therefore an error in strategical science, imposed of necessity upon the allied commanders by want of numerical strength to render the attack of Sevastopol safe and successful; and such error can only be justified by the absolute inability of the army to fulfil the conditions on which the siege of a fortress with a large army of observation in the field can be successful.

Had the allied army been strong enough to follow up its success on the Alma by the occupation of Duwankoi and Khuton, or Bakchi Sarai, and to invest the place on the north with a large reserve force which assuredly should have been at hand, by attacking and carrying the small intrenched camp recently established by the Russians to prevent a landing being made good at the mouth of the Belbec river, the state of affairs in the Crimea might at this time have been very different. However formidable the defences of that camp may be, seaward, it might easily have been taken by land, while a part of the allied force moved round and gained possession of Balaklava so as to open its port to the fleet, the latter having on board a sufficient *reserve* to invest the place on that side also. To this it may be said, we sent to the intended seat of war the whole of our effective military force, and could do no more. This unhappily is true; but if, as the author has always thought, and is now undeniable, "our all" was clearly insufficient to do our part to effect the purposes in view, all that can be said is, that so great an operation should not have been undertaken until ample means had been provided, and not on any account at so late a period of the year. But, moreover, if the whole of our very limited means was not sufficient to enable us to provide a contingent adequate, in the stipulated proportions, to form with our allies an army sufficiently strong to enter on a great territorial war in an enemy's country with any fair prospect of success, how was it then—and, till too late—with respect to reinforcements? Where were *our* reserves? No such operation should ever be undertaken without large reserves either at hand or immediately forthcoming. Much it may be feared that those about to be despatched from England, or collected from our stations in the Mediterranean, will not be available during this campaign to do more than fill up the gaps which pestilence and war have made and are still making in the ranks of the allied army. And as to the campaign of 1855! Not the driblets which we are now sending out, and chiefly of newly-raised men, will suffice: 200,000 men at least will be required to retrieve our affairs in the Crimea, and to carry on the war.

The very first principle in strategical science is to keep a retreat open in the event of a failure in the object of an operation; another is not to undertake any military measure without well considering both the unsuccessful and the successful issue. No thought appears, in the present instance, to have been bestowed on either of these maxims: complete and speedy success was deemed certain: that Sevastopol was doomed to fall no one seemed to doubt, and failure in this object was pronounced impossible. The sad disappointment is now attributed to causes which, it is said, could not have been foreseen—the strength of the place, its abundant means of defence, and the determined resistance opposed by the Russian forces: yet all this should have been anticipated; it was, in fact, foretold by the author, whose convictions were founded on what military science and experience have taught him respecting an enterprise attempted in such circumstances. The following extracts from a work published by the author in 1819, will serve to show what, at various times since the invention of gunpowder, have been the consequences of attempting to besiege fortified places with insufficient means:—

"When Beauvais was besieged by Charles the Bold, in 1472, the place not being completely invested, succours were thrown into it, and the king, who discovered his error when it was too late, was obliged to raise the siege." ('Observations on the Motives, Errors, and Tendency of M. Carnot's "Principles of Defence,"' by General Sir H. Douglas, London, 1819, p. 79.) The like error was committed, and the like result followed, when Haarlem was besieged by the Spaniards in 1573. (*Ib.* p. 80.) "When, in 1674, M. Rabenhaupt, with an army of 11,000 men, besieged Grave, in which there was a garrison of 4000 men, the sorties from the place were so frequent, that, according to M. Quincy, in his '*Histoire Militaire de Louis XIV.*,' vol. i. p. 387, it was difficult to pronounce whether M. Rabenhaupt was the assailant or the defender; a circumstance which proves that an error had been committed in undertaking the siege with so small a force." (*Ib.* p. 157 *et seq.*)

The investing force required to attack such a place as Grave, should not be less than 21,000 men. The proportion required for camp and other duties cannot well be less than one-tenth of the whole army, which for 11,000 men, at three reliefs, is 3300 men; the strength of the working parties, at three reliefs, is 3600; there would remain, therefore, only 4100 men as the guard of the trenches, which, at three reliefs, would furnish only 1366 men to oppose the sorties; and these were no doubt made with 3000 men. In this calculation no allowance is made for sickness or casualties; and all the duty is supposed to be performed with three reliefs, which no troops could support except for a very short time and in fine weather.

Numerous other instances of failures arising from this error may be found in military history, nor is it necessary to go far back. In the year 1854 the Russians attacked Silistria without having invested it, and tried to carry the place by assault, but were repulsed with great loss. Omar Pacha succeeded in throwing reinforcements into the fortress, with assurances that he would speedily come to its relief. The Russians made another desperate assault, hoping to take it before it should be relieved; but the Turks, strengthened by the reinforcements thrown in, repulsed the attack, and the Russians were compelled to raise the siege, with a loss of 10,000 men, who had fallen during the forty days that it had lasted.

How the troops now before Sevastopol have endured their labours, and service in the trenches, is a miracle in war. The force for guarding the trenches cannot be regulated by any proportional part of the strength of the garrison, for the place not being invested, that cannot be known; but this we know, that the men on duty in the lines were very nearly half the effective strength of the division that furnished them; and that a very large portion of those who so heroically repulsed the attack of the 5th of November, had just left their night's duty in the trenches.

The force required for guarding the trenches should not be less than three-fourths of the strength of the garrison; and unless this proportion be observed, the operations and works of the siege will be continually exposed to be disturbed or destroyed by the sorties. Frequent sorties from a besieged place are strongly condemned,[a] particularly in the early stages of the attack, when its works are yet distant; because even if partially successful, the loss of one man in a place completely invested, is more serious to the besieged than six or seven would be to the besiegers. But when the garrison is strong, and the besieging army inadequate to the enterprise (which is the case in the attack of Sevastopol, it not being invested), this maxim is reversed: the loss of one man to the allied army was far more serious to it than a much greater loss to the defenders of a position which might be strengthened to any extent commensurate with its force in the field. Under these circumstances the Russians did right to make frequent sorties, and to resort to operations of active defence which they could not have done had the place been invested. In these attacks, though most gallantly repulsed, the allied army has sustained far greater loss than in prosecuting the operations of the siege; and this is a penalty paid in precious blood, for having undertaken a siege with means so inadequate as to invite, and admit of, as we see, those *retours offensifs*, which under usual circumstances are as condemnable as impracticable.

Nothing could justify the attack of Sevastopol at that late period of the year but the certainty of taking it by a coup-de-main; and that this was believed possible, and urged on accordingly, is clear from the general tone of the organs of public opinion, which at the commencement, and throughout the operations of the war, committed the serious error of underrating the force and power of our enemy, and of exaggerating our own. The author knew that the reports of the place having been taken could not be true, and did all in his power to discredit statements which raised the expectations of the people of this country to the highest pitch, thereafter to occasion the most bitter disappointment

---

[a] Vauban, Traité de la Défense des Places, p. 100, Foissac's Edition.

The southern heights may be crowned by our batteries, but lodgments formed on the face of the slope descending towards the town, docks, and arsenal would be so much exposed to the fire of the large octagonal work and of all the batteries which, no doubt, have been established on the opposite side, that the occupation of the place appears to be utterly impracticable without first reducing the works on the northern side; and to effect this will require another siege:—such is the necessary consequence of having attacked the place at the wrong side!

Viewed strategically, the operation of laying siege to Sevastopol commenced inauspiciously: the place is not invested, its communications with the country, with the army in the field, and with its base are free; succours and supplies to any amount can be thrown in, or taken out; the defensive force in the place is in direct communication with the offensive force in the field. The besiegers know not what force they are fighting. The Russian army of observation may one day be increased by large draughts from a very strong garrison; and assaults which, against a garrison greatly reduced in number, and inaccessible to any external support, would be followed by the surrender or capture of the place, will fail in the event of the garrison being strengthened from without; whilst even if the assault of the breaches that may be made by the allies on the outworks of the southern side be successful, this would lead to no such result as would follow when lodgments are made, or breaches opened on the ramparts of the body of a place inaccessible to relief, and from which there is no escape.

Such a place need not and will not capitulate, attacked as it is, however successful that attack may be. The garrison cannot be captured; since, after making the most determined resistance, it may retire to the northern heights, or it may evacuate the place altogether, and unite itself with the army already in the field, after having rendered the town uninhabitable, and destroyed all the warlike stores it contains.

The bombardment of the 17th October satisfied, to some extent, the desire of the commanders, officers, and seamen of the fleets to have an active share in the labours and dangers of the attack of Sevastopol, and to gratify popular clamour against the reserved position in which the Admirals wisely kept their fleets, as at Bomarsund; but that bombardment contributed nothing to the reduction of the place. The co-operation of the fleet could only be useful as a diversion in favour of the land attack, when the army should be prepared to assault the enemy's position at the same time; but under existing circumstances it could produce no such effect, and the severe damage and loss sustained by the ships and their gallant crews was very inadequately compensated by the little injury they inflicted on the enemy's forts and seaward batteries, which not being faced with granite, appeared to be more severely damaged than at Bomarsund. The safety of the whole operation depends very materially on the presence of the fleets, and on their ability to keep the sea, as we shall find hereafter. The ships were greatly short of guns and of hands. The entrance to the harbour was blocked up by the sunken ships, so that the batteries which the fleets engaged could neither be approached sufficiently near, nor turned by forcing an entrance. The effects produced by the ships on the stone forts were far from justifying an opinion that the fleet could have attacked the place with any prospect of reducing it, had not land forces been employed (Art. 332c, p. 352): it shows rather that an attack made by the fleet alone on the seaward batteries, however gallant and successful it might have been in damaging some of the defences, and dismounting some of the guns, would have produced no results commensurate with the losses sustained; the ships would have been vastly more crippled than they were in the attack which actually took place, and there can be no doubt that some ships would have been entirely destroyed and many disabled. The severe effects produced by the Telegraph batteries and the Wasp Fort, by their plunging fire, owing to their elevation, and the very little damage they sustained by the fire of the ships, may well be cited as a practical illustration of what is stated in Art. 327 on the "command" which coast batteries should have over the surface of the sea.

The French ships were drawn up in line against the forts on the south side of the harbour, and partly across its mouth; the British ships were in line opposite the forts on the north side; and the Turkish ships were drawn up between them. The ships were so much *underhanded*, in consequence of no less than 4000 seamen and marines having been landed from the fleet to serve at the siege, that the gallant Admiral would not allow any men to be exposed on the upper deck of the "Britannia" but his staff, signal-men, &c., and walked his poop, dictating signals for the arrangement of the ships in the order of battle. During the action a shell from one of the enemy's batteries exploded close to him. It was a dead calm, and great difficulty was experienced, as well as time lost, in moving the heavy sailing line-of-battle ships, by steamers lashed to their sides. This mode of propulsion was preferred to traction or towing,[1] in order to protect the steamer from the danger of being crippled by the enemy's fire; but in avoiding the danger incidental to towing, other difficulties were incurred, which, together, show that, as stated in Art. 329, no vessels should be employed in attacking land batteries but such as possess steam power inherent in themselves: for it took an hour to turn the "Britannia" into the proper position to advance after her anchor had been weighed!

Notwithstanding the diversion caused by the fire of the besiegers' land batteries, considerable loss, according to Admiral Dundas's Report, was sustained by the British ships; besides the damage to their hulls from shells and red-hot shot, the masts, yards, and rigging were, more or less, cut away.

The damage sustained by the combined fleets was caused chiefly by the Russian shells, which were fitted with time-fuzes, as those were which they used in the affair at Sinope (Art. 302, pp. 294, 296). The "Albion" received several shells close to the water-line; three entered her cockpit, and she was once or twice on fire. The "Retribution" and the "London" had their mainmasts shot away, and being on fire, were obliged to be hauled off. The "Queen" also was forced to withdraw, a red-hot shot having set fire to her. The "Agamemnon" lay near Fort Constantine, and in consequence suffered severely. The French ships appear also to have suffered greatly. The "Ville de Paris," while engaging the Quarantine Battery, received a shell which blew away part of her poop-deck, and killed and wounded a great number of men. These facts, extracted from the general reports of damages, are sufficient to show that the shells fired by the Russians were fitted with time-fuzes. Shells which struck penetrated into the ships, and then exploded; and others which exploded over the ships without striking, could only be time-fuzed shells. Percussion fuzes could not burst their shells without hitting a sufficiently resisting part of the ship to cause explosion; nor could they explode after having penetrated. It is therefore evident that a great error will be committed if we persist in preferring percussion to time-fuzes in shell-firing against ships;[a] against forts they are utterly useless; and the result of this bombardment fully bears out what has been constantly asserted by the author. (See Art. 302, p. 296.)

It is stated that, to the last moment, the Russian guns (upwards of 300) kept up their fire. The effect produced on the masonry works is said to have been very small: the archings of two casemated embrasures were shaken, and these have been since repaired. From Admiral Hamelin's despatch to the French Government, it appears that, after the firing from the French ships had continued about an hour and a half, that of the Russian batteries opposed to them slackened, and the Quarantine Battery was silenced. On the British side, we learn that towards dusk the Russians returned to their guns, and that these batteries re-

---

[a] The author learns from letters written by a naval officer of high rank, and a distinguished artillery officer, who were present in the action on board a line-of-battle-ship, that the observations contained in Art. 328, p. 337, on the severe effects of shell firing with time-fuzed shells upon ships, were remarkably verified in the attack of the Russian forts at Sevastopol.

opened their fire with considerable effect upon the allied fleet. In consequence of the little impression made, apparently, on the forts, when darkness was coming on, the fleets retired to their anchorage.

The loss sustained by the Russians is not yet known; their artillerymen were several times driven from their guns, many of which were dismounted, but as often they resumed their stations: they appear to have served their guns well, and fired bar-shot, rockets, red-hot shot (Art. 300, p. 291), and shells with great precision.

The observations which the author has frequently had occasion to make on the subject of the expedition to the Crimea, are fully borne out by all that has taken place since the siege of Sevastopol commenced. In order to understand the actual state of the belligerent armies, it is necessary to observe that, on the side attacked, the town is fortified with a double *enceinte* of earthworks recently constructed, terminating on the east at the head of the inner harbour, from whence a strong redan line extends northwestward, and ends on the careening bay. This line encloses the docks, and the camp of that part of the Russian army which is in immediate co-operation with the garrison of the town.

The besieging army occupies a development of not less than twelve miles—the French, British, and Turkish troops extending from Streletska Bay on the west, to the river Chernaya on the east, and crossing the Balaklava and Woronzoff roads. The only ground on which the approaches can be carried on, is an open plain about 2200 yards broad, between two ravines, one of which leads to the Quarantine bay, and the other to the inner harbour, which forms the eastern side of Sevastopol. Stretching across this plain, at 1300 yards from the outer line of the ramparts of the place, the first parallel has been formed; the second parallel is scarcely 400 yards in advance of the first.

A chain of batteries extending over a line four miles in length forms a sort of countervallation, which is within 600 yards of he ramparts of the town and the fortifications of the entrenched camp; and several of these batteries are armed with the heaviest ordnance.

The presence of a Russian army of observation in the field has rendered it necessary to form a chain of redoubts and batteries, which constitute a line of circumvallation: these are nearly four miles distant from the works of the town; and immediately beyond them took place the actions of the 25th of October and the 5th of November, in which British gallantry was displayed under very unfavourable circumstances.

The task assigned to the besieging army is precisely that of attacking a strongly intrenched camp in connection with a powerful fortress, which is open to receive supplies to an unlimited amount. The allied army is, in fact, situated between two armies, each apparently as numerous as itself; and it has to contend with the most serious difficulties from the nature of the ground, which is such as scarcely to admit of trenches being dug in it: its works are incessantly exposed to the fire of a powerful artillery in the massive casemated towers of the place, and to the numerous batteries established on the recently constructed lines by which these points of defence are now covered and connected, and harassed continually by the powerful army in its rear; while the brunt of action has to be borne by the French and English troops, who can place little dependence on their Turkish allies.

After two months of open trenches, the besiegers have not even arrived within the distance at which a practicable breach can be made in the works of the place; and, even were such breach effected, they would only be at the point of commencing the most difficult and most murderous part of the attack in advance of the third parallel—the passage of the ditch and the ascent of the ramparts. Nothing less than continuing the approaches to the counterscarp, and laying the whole length of the two lines of rampart in ruins, will allow an assault to be made with any rational hope of success, more particularly if there should be loop-holed walls and stone casemates in the ditches. But, should the rocky nature of the ground prevent the continuance of the approaches by sap, and an assault be attempted, it is plain that an immense loss, as at Badajos, must be sustained: the troops marching over a great extent of open ground, will be opposed in front and flank by the fire from all the works of the place; and should the remains of the weakened and disordered columns arrive at the ditch, they would have to attempt the passage under a deadly fire of musketry and incendiary missiles, as well as of the artillery from the flanking-works of the fortress, all of which, it appears, have been vastly improved, extended, and strengthened since this protracted siege commenced, and especially whilst active operations against the place have been suspended, or prosecuted with little vigour: all this is independent of the resistance which would be made by the troops of the garrison, strengthened as those troops would then be by the army encamped within the lines. Nor does it appear that a successful assault of those outworks would enable the allied armies to take and occupy the town, nor open the port to the ships of the combined fleet, until the commanding position on the northern side shall have been taken likewise. Thus only can the fortress and arsenal of Sevastopol, and all it contains, be captured.

We are taught by a high authority,[a] and the precept is fully

---

[a] Bousmard, *Des Camps Retranchés sous les Places*, chap. vii. p. 224.

illustrated by what is now passing before Sevastopol, that the most effectual mode of retarding the operations of a siege, as well as of rendering them difficult and sanguinary for the besieging army, is to have the place so strongly garrisoned that the defenders may be able to make frequent sorties (*retours offensifs*), and to be, moreover, in immediate connection or to have secure communications with a strongly entrenched camp. This is precisely the state of Sevastopol, which serves as a *grand tête* to the northern position with its citadel and numerous other works; the whole being open towards the rear, by which reinforcements and supplies to any amount may arrive along the great road from Perekop.

Very different is the condition of the Russians at Sevastopol from that of the Austrian army at Ulm, in 1805: that city being in a position which admitted of being surrounded by the French, and far distant from the army which should have supported it, was compelled to surrender. Rather may their circumstances be compared to those of the Austrians at Ölmütz, in 1758, when that city was besieged by the army of Frederick II. On that occasion General Thierheim connected the detached forts about the place by works of earth, so as to convert the city into a strongly retrenched camp, by which the place was enabled to hold out till the king was obliged to retire from it (*Jomini*, ch. x.). The state of the allies before Sevastopol is nearly similar to that of Napoleon I. when, in 1796, he besieged Mantua. That great general, finding himself in danger of being immediately surrounded by the two armies which were advancing to relieve the place, did not hesitate to raise the siege, abandoning even his siege artillery. He threw his whole force on each of the Austrian armies in succession, and, in defeating them, he struck the decisive blow which rendered him master of the north of Italy. (*Jomini*, ch. xxx.)

Active operations against the entrenched position on the southern heights of Sevastopol having been suspended, the safety of the allied army through the winter is become a matter of painful interest. After an unopposed landing, most skilfully and gallantly conducted by Rear-Admiral Sir Edmund Lyons, under the orders of the Vice-Admiral Commanding in Chief, in the manner practised at Aboukir in 1801 (Art. 332, p. 347 *et seq.*), and a series of brilliant exploits in the field, in a few short weeks the allied army, disappointed in its expectations of speedy and complete success, finds itself shut up and besieged in a *cul-de-sac* in the remotest corner of Europe; while large portions of the fleets are to be employed throughout the winter, in a stormy sea and at all risks, in conveying to the beleaguered troops succours of the first necessity, and in which service so many ships have been already lost. Here the whole of the British army, almost to the last man, must remain, depending for every article of subsistence and warlike stores, as well as of shelter from the inclemency of the weather, on supplies sent from France or England.

Whatever may be done to provide for the safety, comfort, and repose of the army throughout the winter, there can be no rest for the fleet. The ships will have to encounter a more formidable enemy than that which menaces the land army, in having at all risks to force their way through a stormy sea, which cannot be navigated with safety at this season, in order to convey to the imprisoned troops the supplies without which they must inevitably perish, or be compelled to surrender to the enemy. As it happened in the blockades during the war with France, so may it happen in this. When the fleets shall be compelled by the weather to get as far from the land as possible, or run for shelter to remote harbours of refuge, opportunities will offer, long before the combined fleets can resume their stations on the coast, for the Russian steam-ships, of which there are many yet unhurt in Sevastopol, to pounce suddenly upon vessels freighted with succour, as they attempt to approach Balaklava Bay. This bay is small, its anchorage is bad, and, from what happened to the ill-fated "Prince," it is evident that it is not easily accessible: thus serious interruptions will take place in the arrival of supplies to the allied army by the only line of communication with their remote bases in France and England. Those persons are seriously mistaken who assert that the command of the sea by the fleets of England and France will always enable the allies to convey reinforcements and supplies of every description to their respective armies in the Crimea with greater promptitude than Russia can send troops there by land: but unless the allied admirals be endowed with power to "ride the whirlwind and direct the storm," the contrary, during the tempestuous months of winter, will be found to be the fact;—of this too ample evidence has been afforded in the fearful wrecks which have lately taken place on the coasts of the Black Sea. It may indeed be feared that reinforcements will reach the enemy in the Crimea, by land, with greater certainty than they can be supplied to the allies by sea from England or France, when snow and frost shall have rendered steppes at present impassable with wheel-carriages, easily and rapidly traversed by sleys and sledges.

Whenever the allied army shall, happily, be well furnished with provisions, stores, and comforts of every description; whenever it shall be strongly reinforced, and re-equipped with all the means necessary to enable it to resume offensive operations—horses and beasts of burthen can scarcely be expected to survive the winter, from want of forage and shelter—those operations must be conducted in a manner very different from that which has ended by placing the army in its present perilous predicament. If it be true, as undoubtedly it is, that the capital error lay in invading the Crimea with so small a force, and in besieging a strongly-fortified place without having previously invested it, a force adequate to the retrieval of those errors should be sent out; but no greater force should be sent to the southern side of Sevastopol than would be sufficient to render the position at present occupied by the allies quite secure: rather it would seem that a force sufficient specially to invest and attack the town on the northern side should be sent out. Eupatoria should be secured: it was useless as a base point when the attack of Sevastopol by the north side was abandoned, but it will be highly advantageous should an attack on that side hereafter take place; and effectual means should be taken to prevent the enemy from communicating with Sevastopol by the line from Perekop. No siege should ever be undertaken in any seat of war till the enemy in the field shall have been defeated, and completely driven back by the covering army of the besiegers, so that the operations of the siege may be carried on undisturbedly. This might have been done by the allies, had the descent on the Crimea been made at an earlier season, with a force larger and better provided with the means of more effectually carrying out the object of the expedition. An army of 70,000 men, of such troops as those of the allies have proved themselves to be, might, as the Duke of Wellington said of his army in Spain, " have gone anywhere and done anything." It would be a great error to land all the force that may be provided for carrying on the war in the Crimea, in 1855, at Balaklava; and strategical combinations very different from those recently made must be formed for the operations of the coming year; but upon this subject, the author, for obvious reasons, declines to enter.

On contemplating the nature of the fortifications about Sevastopol, it may not be out of place to remark, that an experiment is now being tried for the solution of the problem concerning the means of equalizing the powers of attacking and defending fortified places. Since the invention of gunpowder, the superiority has always been in favour of the besiegers; for, whatever be the resources of the garrison in men and materials, they must ultimately be exhausted; but by placing the artillery of the fortress chiefly in casemates of masonry, according to the principles of Montalembert and the latest engineers of Germany, instead of mounting it on ramparts of earth, as in the systems of the French schools, it is supposed by some that the advantage will, as in ancient times, be turned in favour of the defenders. Hitherto no opportunity has existed of bringing this question to the test of experiment.

INDEX TO THE WORDS,

ALPHABETICALLY ARRANGED.

A.	ARTICLE.	PAGE.
Action, naval, between the Chesapeake and the Shannon	106, 399	82, 423
——— Phœbe and Essex	147	116
——— Hornet and Peacock	392	418
——— Avon and Wasp	392	418
——— Frolic and Wasp reviewed	393	418
——— President and Belvidera	398	422
——— Macedonian and United States	459, 472	480, 495
——— Guerrière and Constitution	466	486
——— Java and Constitution	460, 472	481, 496
Actions, naval, will, in future, commence at great distances	264	253
Ajax, disasters on board the	295	286
Algiers, attack on, by Lord Exmouth	330	338
Antoni, Chevalier	48	28
Armament of the British Navy	App. F.	601
——— of the British Army, unsettled state of the	(37) App. A.	530
——— of coast batteries	338	365
Artillery, proposal to arm the men with the Minié carbine, objected to	(43) App. A.	534
Artillery Officers, subjects of instruction for	17	10
Attack, by sea, on fortresses	326 et seq.	335
Atmosphere, density of the, at different heights	77	53
Avon and Wasp, action between the	392	418
Asphixians, projectiles	300	292

B.		
Backstays, table of the lengths of	354 Note.	384
Ballistic pendulum, invention of the	47	28
——— description of the	49, 50	29
——— in France, and in the United States	50	29
——— centre of gravity in the	51	30
——— centre of oscillation in the	52	31
Batteries on the coast, situations for	327	335
Bellone, accident to the	290	283
Billette shell	299 Note b.	288
Black Sea, operations in the	App. I.	614
Blomefield, Sir Thomas, guns constructed by	208	190
Boat brigade, exercise for the	422	441
Bomarsund, small effect produced by the fire from the ships, at	346 Postscript.	375
Bomb-ships preferable to steam-frigates for naval bombardment	195	175
Bombardment more injurious to the inhabitants of a place than to the garrison	193	174
——— the heaviest projectiles should be used in	194	174
Bore, spiral and elliptical, for guns	190, (24) App. A.	170, 516

	Article.	Page.
Bored-up guns, advantages of	197	181
—— defects of, from their great recoil	199	183
—— failure of, on board the Sesostris	202	186
—— unsafe when fired double-shotted	203	186
Boulets asphixians	300	292
Boxer's (Captain) fuze	403	428
Breech-loading guns used in the sixteenth century	227	212
—— rifled, by M. Cavalli	228	213
—— rifled, by M. Wahrendorff	228, 235	213, 218
Breech-loading rifles, objections to	(13) App. A.	510
Breechings, spare, necessary	389	415
Bullet, cylindro-conical, by M. Delvigne	(6) App. A.	505
—— by M. Tamisier	(7) App. A.	505
—— by Mr. Lancaster	(9) App. A.	506
Bullets made by compression	87 Note.	63

C.

	Article.	Page.
Calibre, dimensions of the, are the distinctive designations of mortars and howitzers	345	374
Canons obusiers, different kinds of, in the French navy	220	201
—— double shot fired from	222	207
—— proofs of, in France	223	209
—— experiments with, compared with a British 8-inch shell gun	224	209
—— originally in limited numbers on board of French ships	306	309
Carabine à tige, the Chasseurs d'Orléans armed with the	(18) App. A.	512
—— experiments to determine its merits	(20) App. A.	514
Carnot, vertical fire proposed by	78 Note.	54
Carnot's detached wall	78 Note.	54
Carriage, Gun, by Baron Wahrendorff	236	219
Carriages for the bows and stems of French ships	221	203
Carronades, re-boring of, proposed	121	95
—— chambers of, on the Gomer principle	122	97
—— disadvantages of, compared with long and heavy guns	147	115
—— improper for the armament of large ships	148	116
Cartridge, the initial velocity depends on the diameter of the	154	121
—— for French chambered ordnance	217 Note.	200
—— manner of passing, from the magazine to the deck	426	449
—— inconvenience of greasing	(28) App. A.	521
Case-shot, spherical (Shrapnel's shells), may be used against troops on shore, or ships with crowded decks	400	425
—— should be fired at low elevations	401	425
—— invention of	402	427
Cavalli, Major, rifle gun invented by	179	150
—— gun, description of the	229, 230	213
—— experiments with the	231	215
Cavalry, proposal to train, to act dismounted	(42) App. A.	533
Centre of gravity in a pendulum, and manner of finding it	51	30
—— of oscillation in a pendulum	52	31
Chambered, some objections to naval guns being	218	200
—— ordnance, a designation of 8 and 10-inch shell guns	414	435
Chambers necessary in naval ordnance of large calibre	219, 339	201, 365
—— their inconveniences	123	97
—— forms of, in British shell guns, and in French canons obusiers	217	199, 200
—— reasons for having, in carronades and shell guns	219	201

INDEX TO WORDS. 633

	Article.	Page.
Charge should be such as gives the required range with the least elevation	133	105
Charges, full, circumstances in which, are necessary	96	70
——— full and reduced, for single and double shotting; proportions of	104	80
——— of powder used in shell firing against ships	313	319
Charging, simultaneous	408, 410, 413	431 et seq.
Chasseurs d'Orléans armed with the carabine à tige	(18) App. A.	512
Chesapeake and Shannon, effects of 18-pr. guns and 32-pr. carronades in the action between the	106	82
——— penetrating effects of shot on the	254 and Note.	235
Christian VIII., observations on the mode of firing on board the	367 Note.	394
Circular platforms in the British naval service	221 Note.	206
——— trigger, invented by Lieut. Harris	(36) App. A.	529
Coast batteries, armament of	338	365
——— unchambered ordnance may be used for	340	366
——— should be armed with guns affording the longest range with the least elevation	340	367
——— should be in a commanding position	327	336
Coast guard, exercise for the	422	441
Columbiad, a new gun in the United States' naval service	266	255
Command of a work, an erroneous notion respecting the, corrected	346	374
Commanded, cases in which a work is	346	375
Compression, bullets made by	87 Note.	63
Concentration of fire	366	393
——— objectionable case for the	367	394
Concentric cylinders, gun formed of	211 Note.	194
Concussion fuze (short time), advantages of a	283	277
——— shells defined	257 Note.	241
——— shells not suitable for dismantling	261	246
——— and percussion shells are less destructive than shells with fuzes	302	296
Congreve, Gen. Sir William, guns constructed by	132, 208	104, 190
——— cause of the superiority of a 24-pr. gun by	132	105
——— first introduced rockets in the British service	318	327
Congreve's top sights	362	389
Conical picket (projectile)	(9) App. A.	506
Conoidal shot, by Captain Norton	178	150
——— by Captain Thistle	191	173
——— by Dr. Minesinger	192	173
Copenhagen, attack on	330	341
Cork wad, the use of a	411	434
——— wads, necessity for, in chambered ordnance	339	366
Cotton, Gun	454	476
Cylinder gunpowder	448	473
Cylinder, revolving, the velocity of shot determined by a	58	35
Cylindrical shot, circumstances in which, may be advantageous	100	75
Cylindrical shot gauges recommended	126	100
Cylindro-conical shot by Delvigne	(6) App. A.	505
——— by Captain Norton	178	150
——— by Major Cavalli	179, 233	150, 217
——— by Baron Wahrendorff	179, 233	150, 217
——— defects of, in ricochet firing	234	218
——— by Mr. Lancaster	190	170
Cylindro-conoidal shot, defects of, in ricochet firing	234	218
——— by Captain Minié, with iron cup	(10) App. A.	507
——— disadvantages of, on account of its weight	(16) App. A.	512

	ARTICLE.	PAGE.

D.

	ARTICLE.	PAGE.
Delvigne, M., proposal of, that rifle bullets should be expanded by striking them	(4) App. A.	504
Depôts of instruction, the necessity of	20	11
Despatch gun-boats, dimensions, weights, and armament of	334	356
Deviation of shot, noticed by Dr. Hutton	180	152
Deviations of shot, longitudinal and lateral	84	59
——— experiments to determine the	141	110
——— effects on the, by the position of the shot in the gun	181, 182	152, 153
——— of solid and hollow shot compared	245	225
——— from the rotation of the earth	85	60
——— from the rotation of the shot on its axis	86	60
——— produced by the non-sphericity of shot	87	62
——— caused by the resistance of the air on elongated shot	88	64
——— horizontal and vertical, formulæ for the	89	65
——— most considerable near the extremity of the range	143	113
——— experiments to determine the, at sea	141	110
Dickson, Sir Alexander, letter from, to the author	372 Note.	401
Dispersion of two shots fired at once	163	128
——— of grape shot	167	132
——— of round and grape	168	134
Displacement of ships	249	229
Distances at sea, necessity of knowing the, before coming into action	347	377
——— method of computing	348	378
——— method of computing, by angles taken at the bow and stern	350	379
——— method of computing by angles taken at the top of a mast	351	380
Double shot, irregularities in firing with	102	78
——— experiments with, from canons obusiers	222	207
——— experiments with, from 8-inch guns	226	211
——— and triple shot, experiments with	247	227
——— shotting, unsafe with bored-up guns	202, 203	186
Douglas, Sir Charles, introduced flint locks	370	398
Dundas, Colonel, a 68-pr. constructed by	210	192
——— percussion lock by	373	403

E.

	ARTICLE.	PAGE.
Edinburgh, establishment on board the, for training seamen gunners	36	19
Education in naval gunnery	10	5
Elevation of a gun should be the least possible	133, 200	105, 184
——— expedients used in the United States to obtain it	361	389
Elevating screw used by the French	364	391
Elliptical bore, experiments with guns having an	190	170
——— Lancaster's gun with an	335	357
Elongated shells, by Mr. Lancaster	335	357
Elongated shot, resistance of the air on	88	64
——— the efficacy of the new rifles depends on	(3) App. A.	503
——— should be used with rifle barrels	176	149
——— advantages of	(3) App. A.	503
Enfield rifle, description of the	(31) App. A.	523
Excellent, gunnery school founded on board the	27, 28	14, 15
——— number of seamen gunners trained on board the	32	17
——— school on board the, advantageous in instructing the coast-guard	421	440

INDEX TO WORDS. 635.

	Article.	Page.
Excellent, present establishment of the	34	18
Excentric, manner of rendering shot	186, 188	162, 165
Excentric shot, experiments with, by General Paixhans	181	152
—— advantages of, in increasing the range	182	155
—— tables of experiments with, on board the Excellent	184	158
—— tables of experiments with, at Shoebury Ness	186, 188	161, 166
Excentricity of shot increases its imperfections	189	169
Exercise on both sides of a ship	431	460
—— of shell guns	432, 436	461, 465
—— with the boat brigade	422	441
—— in the French navy	424	446
Explosion on board a ship is principally to be feared	293	285
—— probable cause of, when struck by shot	314	320

F.

	Article.	Page.
Field artillery, new system of	App. B.	536 et seq.
Firing across a valley	140	110
—— at sea, practice of	382	410
—— position in a ship's roll most advantageous for	383	411
—— of shells horizontally, the maximum effect of, with time fuzes	259	244
Flint lock, double, invented by the Author	372	400
Floating batteries cased with iron	App. G.	608
Fortresses, maritime, should be attacked by a naval and military force combined	332	347
—— small effect of shells fired from ships against	326	335
Freeburn's concussion shells, experiments with	313	320
French, naval gunnery of the, deficient near the end of the reign of Napoleon I.	3	2
French navy, abstract of the	App. C.	556 et seq.
Frolic and Wasp, action between the	393	418
Fuzes, Time, frequently fail in horizontal shell firing	256	240
—— divided into classes	277	272
—— may be extinguished on striking a ship or grazing water	260	245
—— of metal superior to those of wood	273	269
—— the lengths of, vary with the distance of the object	276	271
—— manner of examining	311	317
—— machine for driving	257, Note b	242

G.

	Article.	Page.
Gaining twist in a rifle musket	(9) (24) App. A.	507, 517
Gauges, cylindrical, for shot, recommended	126	100
Glatton, action between the, and a French squadron	148	117
Grape-shot may be used in a chase	397	421
Gravity, centre of, in a pendulum	51	30
Great Britain, the naval officers of, were once too confident	4	2
Guerrière and Constitution, action between the	466	486
Gun, a 32-pounder, substituted for the 42-pounder	213	195
Gun-boats necessary in shallow water	196	178
Gun-cotton, experiments with	454	477
—— relative strength of, and gunpowder	455	477
Gunners and bombardiers, necessity of training	37	22
—— master, instruction of, recommended	10	5
Gunnery, school of, formed on board the Excellent	27	14
—— theory of, advantages of cultivating it	40	23

	Article.	Page.
Gunnery, principles of, necessary for naval artillerists	42	24
——— practice, tables of.	App. D.	565 et seq.
Gunpowder explodes by percussion, concussion, and friction.	314	321
——— and gun cotton	440, 454	468, 476
——— characteristics of.	444	470
——— and gun-cotton, manner of testing it.	450 et seq.	474
——— protection of, from damp.	441	469
——— vessels for preserving.	442	469
——— deterioration of	445	471
——— process of drying	447	473
Guns affording the greatest range, with equal calibre, preferred	130	103
——— of wrought-iron proposed	187	163
——— constructed by Congreve, Blomefield, Monk.	208	190
——— constructed by Colonel Dundas	210	193
——— A, B, C, by Mr. Monk	211	193
——— for shells and hollow shot introduced in the British Navy	215	197
——— proportion between the number and weight of, in the armament of ships	249	229
——— proportion of, for shells and solid shot on board of French ships	251	231
——— short and light, formerly preferred	134	106
——— bored-up, advantages and defects of	198, 199	182, 183
——— bored-up, few, in the navies of France and the United States	201	184
——— bored-up, failure of, on board the Sesostris	202	186
——— bored-up, unsafe when used with double shot	203	186
——— perfection of, consists in affording the greatest range with the least elevation	200	183
——— (42-pounders), experiments with, on board the Excellent	214	196
——— necessity of laying, horizontally in close action	355	385
——— of extraordinary size	204, 206	187, 188
——— of extraordinary size for solid shot	205, 207	188, 189
——— of extraordinary size for shells	215	197

H.

Hale, Mr., rocket by, without stick	324	333
——— rocket by, defects of the	325	334
Hay, Col., projectile for the Minié rifle proposed by	(26) App. A.	519
Heavy shot range further than lighter shot	98	72
Hollow shot, advantages of, with respect to the size of fracture	97	70
——— concentric and excentric, experiments on	188	164
——— inferior to solid shot in having greater lateral deviation	239	221
Hollow and solid shot, relative accuracy of practice with	244	225
Homogeneity of shot, advantages in the	187	163
Horizon, balistique	359	387
Horizontal shell-firing, experiments on	256	239
——— case in which the greatest effect of, is produced	259	244
——— great velocity of the shell in, required to injure a ship	258	244
——— difficulties attending	260	245
Hornet and Peacock, action between the	392	418
Howitzers, sea-service, fired on board ship produce severe strains	333	354

I, J.

	Article.	Page.
Jacob, Major, nipple and percussion caps by	380	407
Java and The Constitution, action between the	472	496
Impact, relative, of 8-inch shells and 32-pr. shot fired double	253	233
Incendiary projectiles, disasters experienced by the French from	299	289
—— inhumanity of using	301	293
—— used by the Russians at Sinope	303	296
Increasing spiral	(24) App. A.	517
Initial velocity of shot	61, 63	38, 41
—— French formula for the	65	43
—— maximum of the	66	44
—— formula for, in terms of the range	82	58
Instructors in naval gunnery should be naval men	12	6
Iron plates, effects of shot against	164 Note, 166,170,171 App. G.	129, 131, 135, 608
Iron ships, disadvantages of	173	145
—— unfit for war purposes	172	144
—— may be used as troop-ships	173	145

L.

	Article.	Page.
Lancaster's ordnance, experiments with	190	171
—— pillar breech musket	(9) (19) App. A.	506, 513
—— gun, great range obtained with, at an elevation of 17 degrees	335	357
—— gun, failures of the	App. H.	611
Land batteries, ships are now less able than formerly to contend against	331	343
Le Couteur, Col., experiments by	(32) App. A.	525
Leeward ship, manœuvres of a, in action	467 et seq.	487
Line of metal, pointing by the, not now used in the British service	134	106
Line of sight, practice of firing by the	368, 369	395, 397
Live shells, manner of stowing, on board the Royal Albert	287	280
—— accidents to French ships from having on board	306	310
Loading, comparative times of, with single and double shot	108	84
—— compounded of round and grape shot, effective ranges obtained by	107	83
Locks and tubes for naval ordnance	370 et seq.	397
—— for guns first introduced	370	397
—— percussion, by Col. Dundas	373	403
—— in use among different nations	375 et seq.	405
Lugeol, M., manner of stowing powder by	438	466

M.

	Article.	Page.
Macedonian and United States, action between the	459	480
Magazines, powder, on board a ship	439	466
Magnus, on the resistance of the air to elongated shot	86	61
Manual exercise of naval artillery	421 et seq.	440
—— for the coast guard	422	441
—— for one side of a ship	428	453
Marine artillery, excellence of the, for particular service	21	12
Maritime fortresses, attack of	326	335 et seq.
Master gunners	10, 17	5, 10
Masts, the heights of, in French and American ships nearly equal	349	379

	ARTICLE.	PAGE.
Matelots canonniers, establishment for in the naval arsenals of France	38	22
Maximum range, elevation producing the	90	66
———— equation for the	90 Note.	66
———— velocity, length of charge producing the	66	43
Medea, accident on board the	274	270
Medium guns A, B, C	211	193
Metal fuzes, experiments with	279	273
Minié rifle	(10) App. A.	507
———— French and Belgian	20 App. A.	514
———— objections to the	(24) App. A.	515
———— cause of its fouling	(27) App. A.	520
Momentum of inertia in a pendulum	53	31
Monk, Mr., guns constructed by	208, 209, 211	190, 193
Mortars, the use of, discontinued in the British navy	333	354
Moorsom, Captain, percussion fuze by	284	277
Musket, pillar breech	8 (9 App.A.	506
———— needle prime	(11) A] p. A	508
Musketry, danger of using in the *tops* of ships	(2) App. A.	503
Muskets, rifle, importance of	(7) App. A.	505
———— manner of sighting	35) App. A.	528
———— not sufficiently used in the British navy	1) App. A.	503

N.

	ARTICLE.	PAGE.
Naval cadets, studies of the	30	16
Naval college for officers, established in 1839	28	15
———— subjects studied at the	29	16
———— strength of the	29, 30	16
Naval depôts for instruction recommended	11	5
Naval gunnery, the state of, once unsatisfactory in England	1	1
———— objection to the use of instruments for, obviated	9	4
———— subjects of instruction in, proposed by the author	17	10
———— establishment for, should be under the Admiralty	22	12
———— improvement of, depends on the character of the depôts for instruction	25	14
———— proposed to be an item in the examinations for promotion	11	4
———— cultivated on board the Excellent	31	17
———— advantage of studying the theory	42	24
Naval instructors	12	6
———— officers, the instruction of, recommended	10	5
Navarino, action at	330	341
Navy, necessity of maintaining the high position of the	8	4
———— British, warlike skill in the, formerly deficient	7	4
———— armament of the	App. F.	601
———— of France	App. C.	556
———— of Russia	App. C.	563
Needle-prime musket	(11) App. A	508

O.

	ARTICLE.	PAGE.
Oblong shot, by Captain Thistle and Dr. Minesinger	191, 192	172, 173
———— cylindrical, when advantageous	100	75
Ordnance, Lancaster's, experiments with	190	170
———— large, used by the French against Cadiz and Antwerp	204	188
———— cast for the Pasha of Egypt	205	188
———— cast at Liverpool for the ship Princeton	206	188

INDEX TO WORDS.

	ARTICLE.	PAGE.
Ordnance, ranges obtained with large	207	189
———— of the President and Chesapeake	145	114
Oscillation, centre of	52	31

P.

	ARTICLE.	PAGE.
Paixhans, General, observations of, on the use of incendiary projectiles	305	300
Paixhans guns, different kinds of, in the French navy	220	201
———— restricted to firing hollow shot	223 Note.	209
———— more destructive against large, than against small ships	304	299
Parabolic theory, reasons for noticing the	45	25
———— application of the, when the initial velocities are small	46	27
———— equation for the trajectory on the	45 Note.	26
Pendulum for laying naval ordnance	356 et seq.	386
———— ballistic, invented	47	28
———— ballistic, description of the	49	29
———— ballistic, in France, and in the United States	50	29
Penetrating force, relative, of the canon obusier of 80 and British ordnance	242	224
———— of solid and hollow shot compared	343	370
———— necessity of insuring	105	81
———— with bored-up guns	198	182
———— the importance of, illustrated	253	233
———— importance of, exemplified in the effects produced on board the Chesapeake and Shannon	254 Note.	235
———— of the new muskets, experiments to determine the	(32) App. A.	526
Penetration of shot, formula for the	79	55
———— experiments on the, are discordant	157	122
———— tables of the	App. D.	565 et seq.
Penetration of solid shot into wood, results of experiments on the	158	123
———— of solid shot into water, results of experiments on the	160	125
———— of hollow shot, results of experiments on the	159	124
———— of two shots fired at once	163	128
———— of grape shot into oak	167	132
Percussion shells defined	257 Note.	241
———— not fit for dismantling purposes	261	246
———— may be ignited by a blow from a solid shot	285	278
Percussion fuze, by Captain Moorsom	284	277
———— and concussion shells less destructive than shells with fuzes	302	294
———— locks	373 et seq.	403
———— gun caps, Report of the Naval Committee on, by Major Jacob	381	410
———— muskets and Minié rifles, table of comparative accuracy of	(30) App. A.	522
Pile of shells, experiments to determine the effects arising from a, being struck by shot	281, 282	275
Pillar breech musket (French)	(8) App. A.	506
———— by Lancaster	(9) App. A.	506
Platform, circular, for pivot guns	221 Note.	206
Point blank range, equation for the	81	58
———— the greatest, preferred	130	103
———— table of, for the new rifles*	(29) App. A.	521
Ports, the dimensions of, on board of French ships	365 Note ª.	393
Powder in a shell supposed to become ignited by percussion	280	274
Power of impact should be contemplated	248	228
Practical gunnery, subjects of instruction in, indicated	17	10

* By an oversight the number of this Article has been omitted.

	Article.	Page.
Preponderance, effects of	150	118
———— in Sir W. Congreve's gun	132	104
President and Chesapeake, action between the	106, 399	82, 424
———— ordnance of the	145	114
Probabilities of striking an object increase with the charge	136	107
———— diminish with the increase of distance	137	108
Projectile force, necessity of preserving the	105	81
Projectiles, solid and hollow, when large strike more frequently than when small	262	247
———— cylindro-conical, by Captain Norton	178	150
———— cylindro-conical, by M. Delvigne	(7) App. A.	505
———— cylindro-conoidal, by Cavalli and Wahrendorff	179	150
———— cylindro-conoidal, defects of, in ricochet firing	234	218
———— ellipsoidal	190	171
———— excentric spherical	181 *et seq.*	152
———— incendiary, disasters occasioned by	299	289
———— incendiary, inhumanity in using them	301	293
———— asphixians	300	292
Proof, ordinary, of 8-inch guns in England and of canons obusiers in France	223 Note.	209
Psyche, armament of the	250 Note.	231

Q.

	Article.	Page.
Quarter sights, the use of	360	388

R.

	Article.	Page.
Range, equation for the, in vacuo	45 Note.	26
———— equation for the, in air	80	57
———— equation for the maximum	90 Note.	66
———— cause of a supposed diminution of, when shot is fired over a valley or the sea	140	109
———— increased by the use of excentric shot	182	155
Ranges with long and short guns compared	131	103
———— at sea and land compared	142	111
———— with some very large guns	207	189
———— of English and French guns compared	212	194
———— of solid and hollow shot compared	238, 240	220, 22?
———— from a 68-pounder gun	243	224
———— in the parabolic theory, vary with the squares of the initial velocities	45 Note.	27
———— of a 24-pound shot in air and vacuo compared	92	67
———— effective, of double shot and compound loading	107	83
———— of solid shot exceed those of hollow shot with equal diameters and charges	238, 239, 240	220, 221
———— of heavy shot longer than of lighter shot	98	73
———— point blank, guns affording the longest preferable	130	103
———— great differences in, correspond to small differences in vertical height on an object struck	138	108
———— comparative, of a 56-pr., a 42-pr., and a 10-in. gun	210	192
———— comparative, of an English 32-pr. and a French gun	212	194
———— elevations which give equal, in 32-prs., 8-inch and 10-inch guns	244	225
———— point blank, table of, for the new rifles	(29) App. A.	521
Reaming out guns, originated in 1830	197	181
Recoil, formula for the velocity of	69	48

INDEX TO WORDS. 641

	ARTICLE.	PAGE.
Recoil, effect of, on the range and direction	149	117
——— increases with the elevation of the gun, in some cases	151	119
Recoils, table of	149	118
Red hot shot used by the Russians at Sinope	302	294
——— the charges for, are three-quarters of the full charges	344	372
Reflection of shot from water, results of experiments on the	161	126
Regulation musket, experiments to determine its merits	(19) App. A.	513
——— rifle, experiments to determine its merits	(19) App. A.	513
Resistance of air to the motion of shot	70	48
——— law of the, deduced by the pendulum and whirling machine	74	50
——— expression for the retardation arising from the	59	36
——— manner of finding the, from tables	76	53
Resistance, equation for the solid of least	175	149
——— of the air on the ballistic pendulum	56	35
——— of the air, experiments on the	73	49
Retardative force, arising from the resistance of the air	59, 76	36, 52
Ricochet fire, effective distance for	139	109
——— inferiority of, to direct fire	139	109
——— efficiency of, impaired in long ranges or by the use of elongated shot	(42) App. A.	533
——— from water, effects produced after	162	126
Ricochets, experiments to determine the number of, with the regulation rifle	(33) App. A.	527
Rifle gun, loaded at the breech by Cavalli	228	213
——— loaded at the breech by Wahrendorff	232	216
——— compared with a 32-pr. smooth bore	233	217
Rifle guns, ancient	227	212
Rifle muskets	App. A.	503
——— with tige	(5) App. A.	504
——— with cup	(10) App. A.	507
——— with pillar, by Lancaster	(9) App. A.	506
——— with elliptical bores	(24) App. A.	517
——— objections to, when loaded at the breech	(13) App. A.	510
——— elevations of, to produce different ranges	(29) App. A.	522
——— manner of sighting	(35) App. A.	528
——— needle-prime, or zundnadelgewehr	(12) App. A.	508
——— needle-prime, compared with English rifles	(19) App. A.	513
——— table of the dimensions, &c., of	(23) App. A.	516
——— opinion that they will prevent artillery from keeping the field	(38) App. A.	531
——— objection to arming artillerymen with	(43) App. A.	534
Robins, Mr., maxims of, controverted	129	102
Rocket troop, in what the, consists	321	331
——— stick, use of the	317	326
——— proposed by Mr. Hale	324	333
——— defects of the	325	334
——— height to which a, will ascend, and distance at which the explosion of a, can be seen	317	326
Rockets, nature of, for military purposes	315	324
——— cause of the propulsion of	316	325
——— impediments to the motion of	319	329
——— used at Leipzig, at Flushing, at the passage of the Adour	320	330
——— fired from ships at Tangier in 1844	320	330
——— danger of, on board ship	322	332
Rotation of the earth, deviation produced by the	85	60
——— of shot, deviations produced by the	86	60
——— of shot, effects of, in ricochet firing	183	156
Rotations of solid and hollow shot compared	239	221

2 T

	Article.	Page.
Russian navy, strength of the	267, App. C.	255, 563
———— ships of war draw less water than the English ships	337	362
Rust, necessity of preserving shot from	116	90

S.

	Article.	Page.
School, naval gunnery, on board the Excellent	27	14
———— naval gunnery, on board the Edinburgh	36	19
Seamen, the proficiency of, in warlike practice diminished during the peace	2	2
———— should be engaged for terms of years	13	6
———— importance of having a portion of, trained to warlike practice	5	3
———— gunners, manner of employing	19	11
———— gunners should be induced to renew their engagements	15	7
———— gunners, number of, trained	32	17
———— gunners, inconvenience of disbanding	14	6
Shannon and Chesapeake, action between the	474	500
———— penetrating effects of shot on the	254	235
Shell, effects produced by a, exploding on board a ship	255, 274	238, 270
———— precaution to be used on placing one in a gun	274	269
Shell boxes, dimensions of	272	267
Shell firing, horizontal, experiments on	256	239
———— may be used against near objects, from unchambered guns	339	365
Shell fired horizontally, case in which a, produces the greatest effect	259	244
Shell gun, weight of the, compared with the weight of the canon-obusier	215 Note.	198
———— 8-inch, trial of a, double shotted	226, 246	211, 226
———— 8-inch, compared with a 32-pounder gun	249, 250	230, 231
———— 8-inch, inferior to a 32, 42, or a 56-pounder gun	240	221
———— 8-inch, strength of, to resist the fire of double shot	246	226
———— 8-inch, experiments with a, to determine its strength when solid and hollow shot are fired from it	246	226
———— 10-inch, inferior to a 68-pounder gun, for projecting solid shot	268	258
Shell guns, 8-inch, compared with 32-pounders in respect of weight and number	249, 252	228, 233
———— and solid shot guns, proportion between the numbers of, on board of French ships of war of different classes	251	231
———— proportion of, on board of English ships of different rates	252	233
———— not approved of, for steamers in the United States' service	266	254
———— may be properly designated sea-service howitzers or chambered guns	345	372
Shell rooms in ships of war, situation and dimensions of	271	264
———— the crowns of, are not sufficiently below the water line	296, 297	287
Shells filled with sand or lead	99	73
———— in exploding, act as mines	255	238
———— not suitable for dismantling purposes	261	244
———— with time fuzes unsuccessfully employed by Sir George Collier	257 Note.	243
———— fired horizontally should explode on striking	261	246
———— fired with great velocity less destructive to the material and less to the crew, than such as imbed themselves in the wood	263	249

INDEX TO WORDS. 643

	ARTICLE.	PAGE.
Shells placed on shelves, or triced to beams on the fighting decks	275	270
—— manner of stowing, on board of ships of war . .	286	279
—— number of, on the fighting decks of ships of war .	288	282
—— danger of bringing up, and placing on deck . . .	289	282
—— should be stowed far from the sides of ships . .	298	288
—— 8-inch and 10-inch, costs of	314	323
—— should not be permitted on the fighting decks . .	307	313
—— precautions to be used in filling, on board ship . .	310	317
—— frequently strike an object without exploding . .	312	318
—— experiments with Freeburn's concussion . . .	313	320
—— effects of, when perforating a ship's side	259, 263	244, 249
—— concussion and percussion defined	257, Note.	241
—— effects of when grazing water	260	245
—— manner of filling on board ships	310	317
—— fired from long guns	402	426
—— increased thickness of, recommended	403	428
—— loaded, compared with hollow shot	244	225
Ships of war require armaments of long and powerful guns .	265	253
—— reasons why the lengths of, have been increased .	309	315
Ships, small, preferred to large ones, when armed for shell firing	303	296
—— small, should attack land batteries when in motion .	332 (c)	353
—— great, should attack land batteries when at anchor .	332 (c)	354
Short guns, partiality for, at one time	128	101
Shot, double, irregularities in firing with	102	78
—— double, effective ranges of	107	83
—— triple, limits to the employment of	103	79
—— double and triple, compared	247	227
—— single and double, comparative times of loading with	108	84
—— disposal of on shipboard	127	100
—— cautions for preserving	116	90
—— effects of, when fired against plates of iron . .	164,166-170, App. G.	129-135 608
—— effects of, when fired against masses of cast iron .	165	131
—— effects produced by, on the Lizard	169	134
—— experiments with grape and round, fired together .	168	134
—— formed like the solid of least resistance, advantages of	177	149
—— of wrought iron proposed	187 Note.	163
—— of wrought iron liable to be broken	335	358
—— splintering effects of, vary with the square of the diameter	248	228
—— hollow, advantages of, with respect to fractures .	97	70
—— excentric, advantages of	182	153
—— hollow, excentric and concentric, experiments with .	188	164
—— excentric, deviations of	183	156
—— oblong, by Captain Thistle and Dr. Minesinger . .	191, 192	172, 173
—— solid and hollow, with loaded shells, comparative accuracy of, fired from a 68-pounder gun	243	224
—— splintering effects of, the number of hits being equal	248	228
—— shells and metal fuzes, proportion of, for British ships	270	262 et seq.
—— sphericity and homogeneity of, necessary . . .	188	168
—— heavy, superior range of	98	73
—— reflected from water	161	126
—— elliptical	190	171
—— elongated	{ 176, 177, (3) App. A.	149 504
—— red hot	300	291
—— cylindro-conical	178 et seq.	150
—— cylindro-conoidal	234	218

2 T 2

	ARTICLE.	PAGE.
Shot, cylindro-conoidal, weight of, disadvantageous . . .	(16) App. A.	512
Shrapnel, Major-General, invented spherical case shot . .	402	427
Shrapnel shells, case in which, are useful	400	425
——— cause of the premature explosions of	403	427
Sights, employment of, for rifles	(35) App. A.	528
Simultaneous charging, or loading, recommended	408, 410, 413	431, et seq.
Solid of least resistance, equation for the	175	149
——— advantages of shot formed like the	177	149
Solid shot guns preferable, in fixed batteries, to shell guns .	340	366
——— from long guns, more powerful than hollow shot from shell guns	264	251
Spherical case shot, *see* Shrapnel shells.		
Spherical projectiles, excentric	182, 183	153, 156
Splinters, necessity of obtaining the greatest effects from . .	94, 95	68, 69
Steam navy, British, armament of the, in 1849	269 & App. F.	261 601
——— of France	App. C.	556
Steam tugs, risk in using	329	338
——— vessels, long guns proposed for	265	253
——— warfare, a work on the subject of, proposed . . .	215	197

T.

	ARTICLE.	PAGE.
Tactics, naval, of single actions	457 *et seq.*	479
Tamisier, projectile of, with circular grooves	(7) App. A.	505
Tangent scale	360, 363	388, 390
Terminal velocity of a shot and of a shell	75, 78	51, 53
——— investigated	75	51
Theory of gunnery, advantage of cultivating the	40	23
Tige, by Colonel Thouvenin	(5) App. A.	504
Time of flight, in vacuo	45 Note.	27
——— in air	83	59
Times of loading, with single and double shot, compared .	108	84
Time fuze, composition of a	278	273
——— though capped, may become ignited	282	276
Time fuzes frequently fail in horizontal shell firing . . .	257	241
——— may be advantageously employed in some cases . .	257	242
——— divided into classes	276, 277	271
Toggle and tripping line	364 Note.	392
Top sights, by Sir William Congreve	362	389
Trajectory, equation for the, in vacuo	45 Note.	26
——— equation for the, in air	80	57
——— in air and in vacuo differ in form	91	66
——— of a shot should be nearly horizontal for guns in a coast battery	342	369
——— of a cylindro-conoidal projectile	(40) App. A.	532
Triple shot, limits to the employment of	103	79

V.

	ARTICLE.	PAGE.
Vacuum behind a shot in its flight	70	48
Valmy, French ship of war, accident on board the . . .	322 Note.	332
Velocity of shot in the parabolic theory	45 Note.	27
——— formula for the, in air	82	59
——— formula for determining the, on striking a pendulum	54	32
——— initial value of, greater than that determined by the pendulum	57	35

INDEX TO WORDS. 645

	Article.	Page.
Velocity of shot, law of the decrements of	67	45
——— formula for the initial, and for the velocity at different distances from the gun	60	37
——— formula for the initial, in terms of the charge	61	38
——— terminal, in shot and shells	75 & Note [b]	52
——— formula for the, when discharged from a suspended gun	64	41
——— formula for the, given by French artillerists	65	43
——— maximum, length of charge producing the	66	44
——— empirical formulæ for the	68	46
——— experiments at Washington to determine the	68 Note	46
——— relative, of single shot and two shot fired together	101	77
——— effects of windage on the	110 & Note	84
——— loss of, by windage	111	85
——— not affected by the use of wads	152	119
——— depends partly on the diameter of the cartridge	154	120
Velocity of a gun's recoil	69	47
Velocities of balls of different weights	68 Note.	46

W.

Wad of cork, use of a	411	433
Wads do not affect the velocity of shot	152	119
——— grummet, preferred to those of junk	155	121
——— tight, not to be used in quick firing	153	120
——— for muskets should be of the stiffest material	156	122
Wahrendorff gun	232	216
——— experiments with the	233	217
——— superiority of the, to Cavalli's gun	235	218
Wahrendorff, Baron, a 24-pounder invented by	236	219
——— cylindro-conical projectiles by	179, 233	150, 217
Walcheren, expedition to	332 c	351
Windage, definition of	109	84
——— loss of velocity by, formula for the	111	85
——— of siege and garrison guns	119	93
——— of 8-inch shell guns, anomaly in the	216	199
——— reduced, dangers of, obviated	125	99
——— reduced, advantage of, in brass guns	118	92
——— prejudicial effects of high	112	86
——— of French ordnance	114	88
——— of shells different from that of shot	113	88
——— scale of, proposed	115	89
——— of field guns	117	91
——— amount of, for iron ordnance	124	99
Whirling machine, description of the	72	49
——— experiments with the	73	49
Works, isolated, should be attacked in detail	332 c	351

Z.

Zundnadelgewehr, or needle-prime rifle	(11)(12)(19) App. A.	508, 513

THE END.

PLATE I.

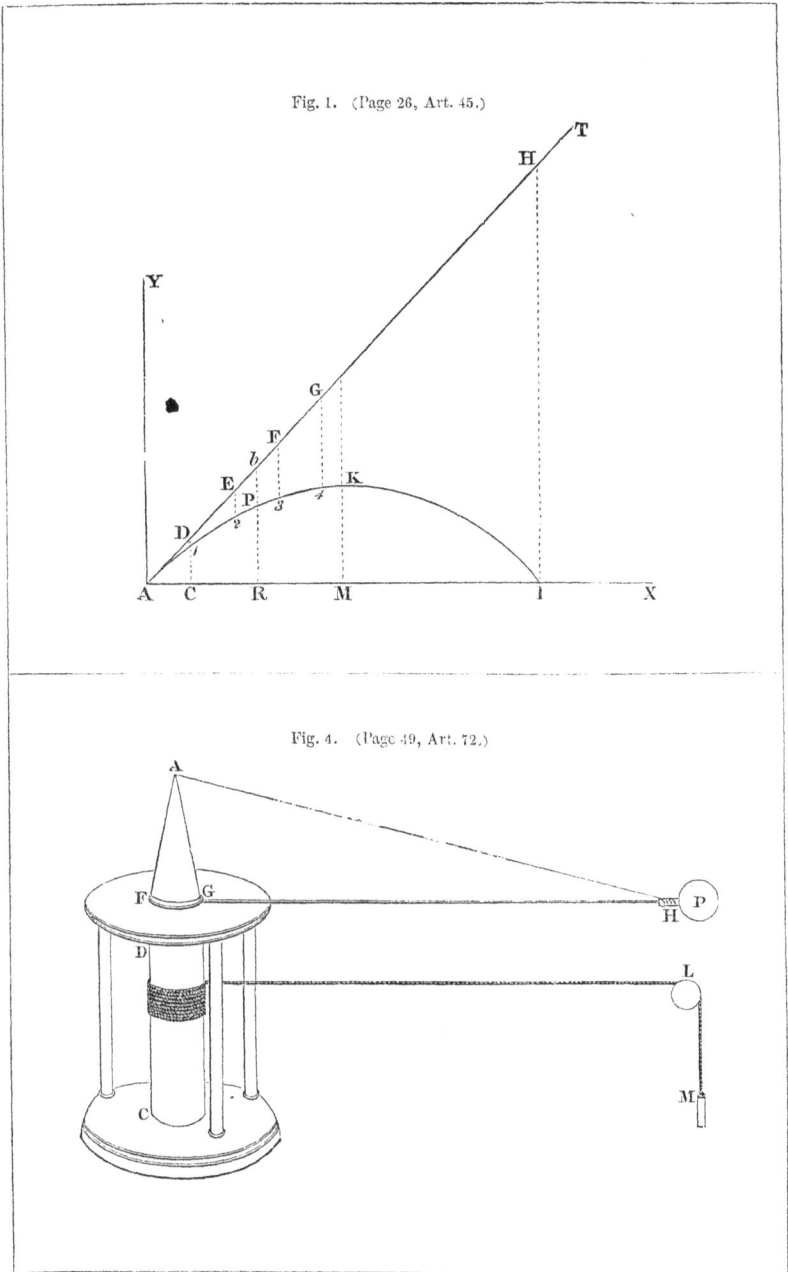

Fig. 1. (Page 26, Art. 45.)

Fig. 4. (Page 49, Art. 72.)

TRANSVERSE SECTION.

BALLISTIC PENDU
(Page 29, Art. 49.)

Fig. 2.

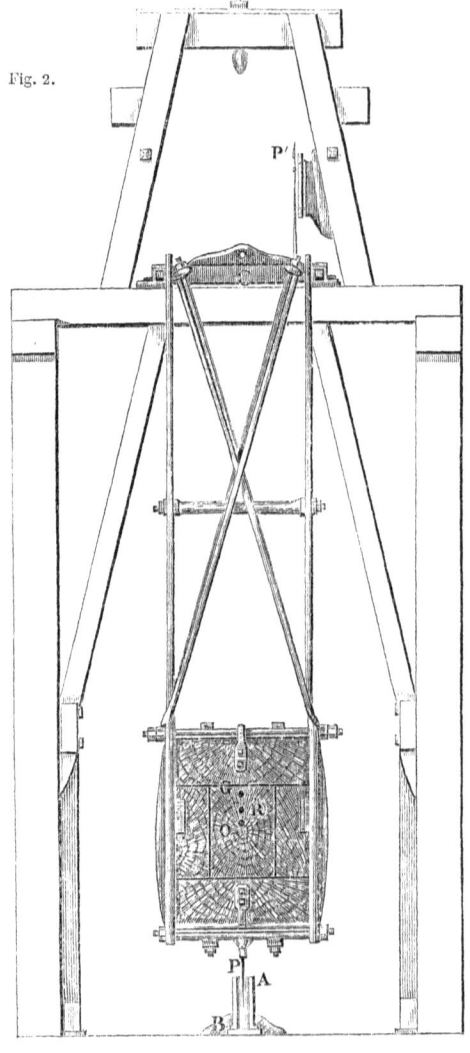

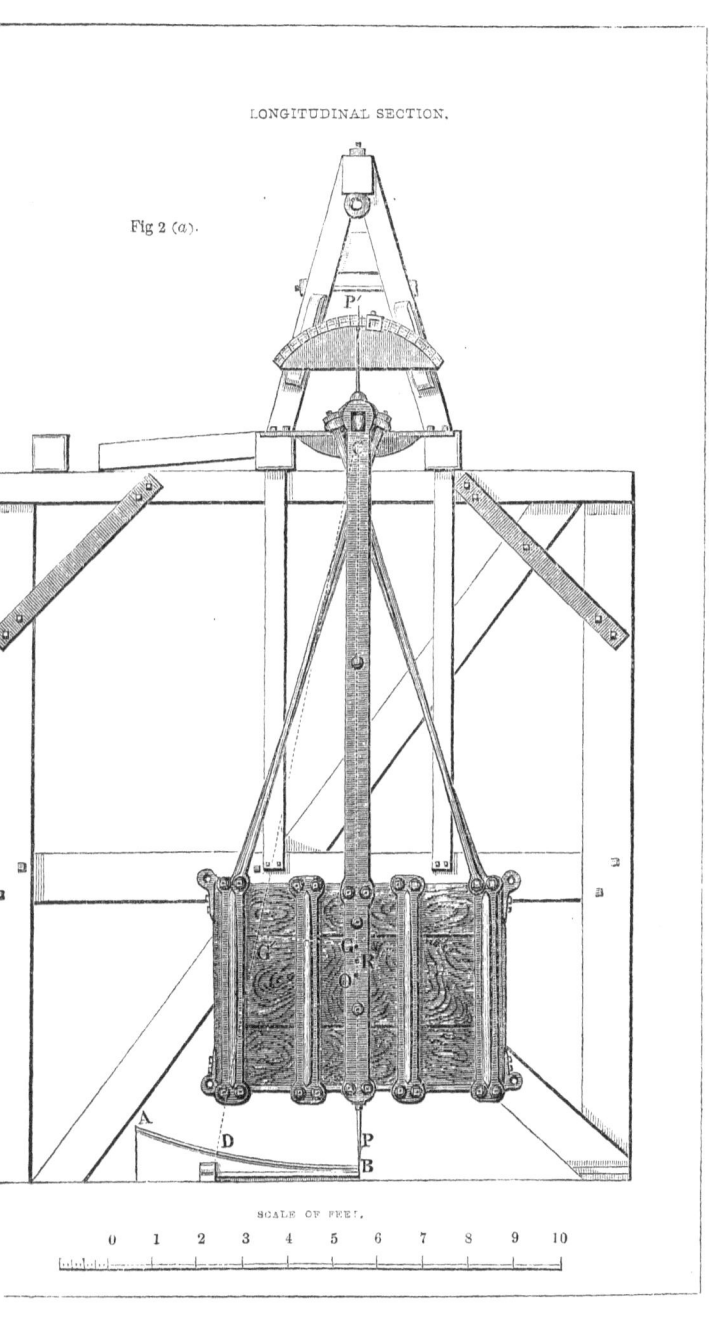

PLATE I.

Fig. 3. (Page 30. Art. 51, Note.)

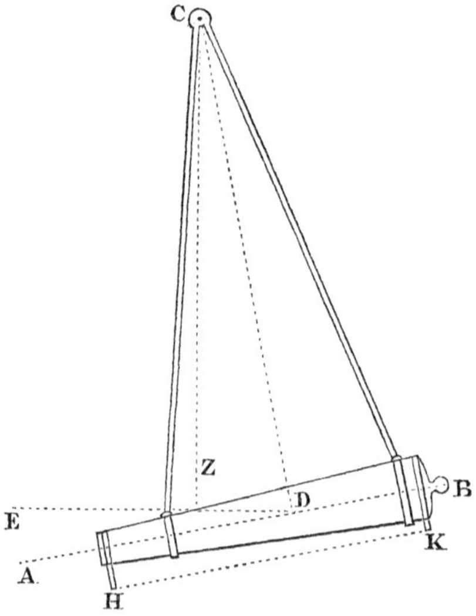

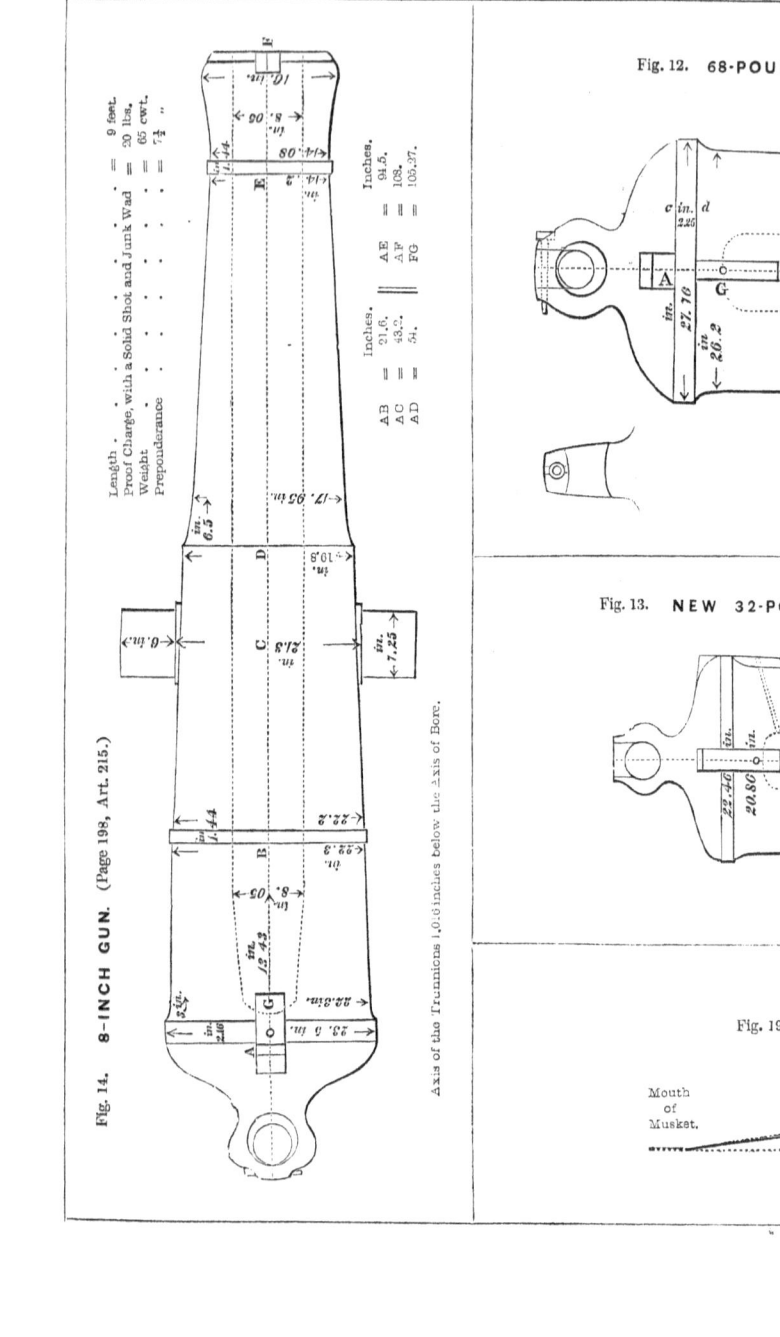

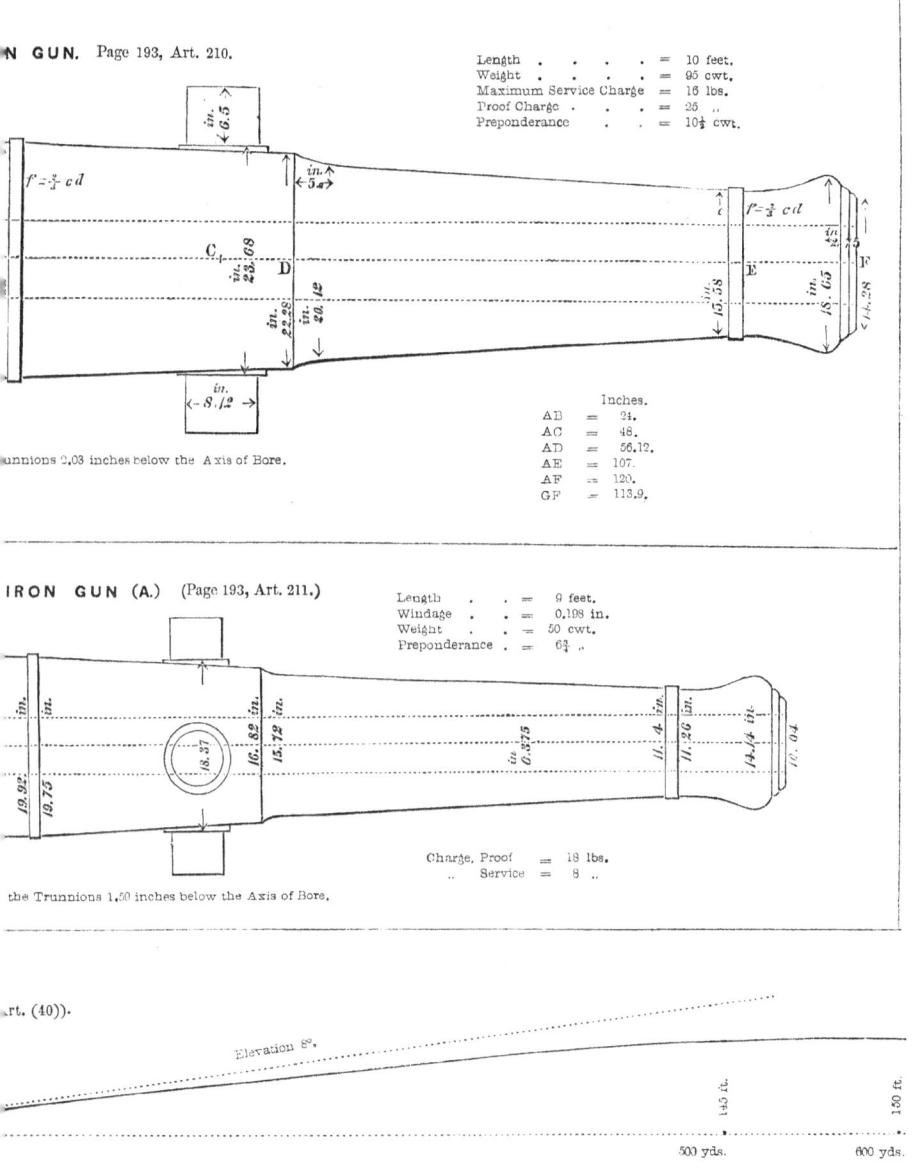

PLATE II.

Fig 15. **THE CAVALLI GUN.** (Page 213, Art. 229.)

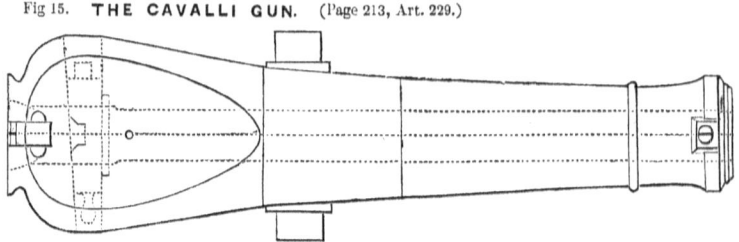

Fig. 16.
THE WAHRENDORFF GUN.
(Page 216, Art. 232.)

Fig. 18.

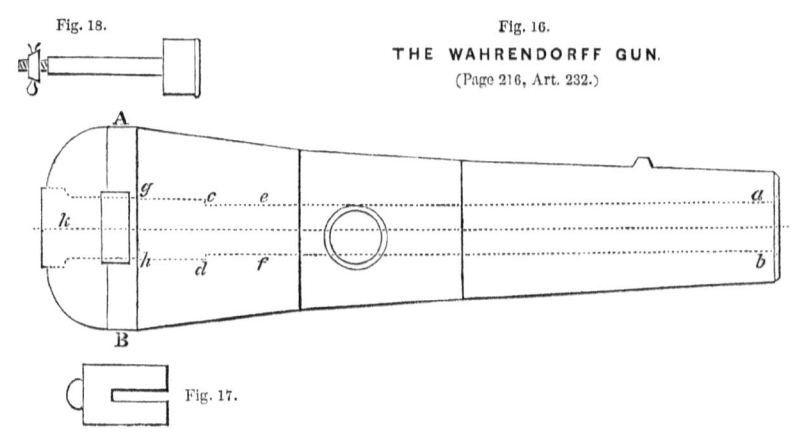

Fig. 17.

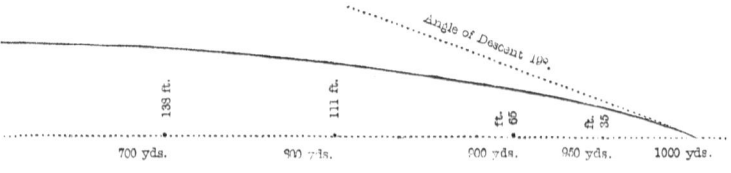

www.ingramcontent.com/pod-product-compliance
Lightning Source LLC
Chambersburg PA
CBHW021712300426
44114CB00009B/112